Springer Geology

The book series Springer Geology comprises a broad portfolio of scientific books, aiming at researchers, students, and everyone interested in geology. The series includes peer-reviewed monographs, edited volumes, textbooks, and conference proceedings. It covers the entire research area of geology including, but not limited to, economic geology, mineral resources, historical geology, quantitative geology, structural geology, geomorphology, paleontology, and sedimentology.

Tatiana Chaplina
Editor

Processes in GeoMedia—Volume IX

Springer

Contents

Ecological Aspects in the Creation and Application of New Oil and Gas Production Technologies 1
V. I. Karev and Yu. F. Kovalenko

Mathematical Modeling of Acoustic Fields in the Black Sea 9
Yu. I. Papkova and S. O. Papkov

Cosmogony of Open Solar System (I): Formation of Giant Planets and Origin of the Moon 31
A. A. Barenbaum

Cosmogony of Open Solar System (II): Model of Secondary Planet Formation, Secondary Accretion Disk of the Sun, Formation of Satellites of Giant Planets, Molchanov's Resonance, Titius–Bode Rule 65
A. A. Barenbaum

Weather Conditions of the Winter–Spring Season in Moscow in Early 2025 109
D. M. Frolov, V. E. Gagarin, A. V. Koshurnikov, and G. A. Rzhanitsyn

Trace Elements of Oil Fields of the Orenburg Region 115
G. A. Ponomareva

Long-Term Variability of the Ice Regime in the Mouth Areas of Tributaries of the Western Coast of the White Sea Against the Backdrop of Climate Change 123
V. N. Baklagin

Tsunamis Are an Urgent World Problem 139
K. V. Pokazeev and T. O. Chaplina

On Lagrangian Description of Standing Surface Gravity Waves 155
V. A. Kalinichenko and A. O. Fomenko

Changes in Total Ozone Content, Ultraviolet Radiation Fluxes, and Cosmic Rays Over Northern Europe 417
A. V. Kholoptsev, R. G. Shubkin, and G. F. Batrakov

Impact of Clay Liner Integrity Breaches on Leachate Migration in Municipal Solid Waste Landfills 435
Ya. N. Parshakova and R. I. Kataev

Experiment on Studying Thermophysical Properties of Snow Cover 451
D. M. Frolov, A. V. Koshurnikov, V. E. Gagarin, and G. A. Rzhanitsyn

Geodesic Control of Technological Object Condition by Laser Scanning Method 457
Dmitriy Neguritsa and Ivan Tereshin

Numerical Experiments on Modeling the Black Sea Circulation in 2016 Using Various Degrees of Polynomial Dependence of Density on Temperature and Salinity 463
Natalia Evstigneeva

Ecological Aspects in the Creation and Application of New Oil and Gas Production Technologies

V. I. Karev and Yu. F. Kovalenko

Abstract The article considers the problems associated with environmental risks in hydraulic fracturing and substantiates the necessity of careful planning of hydraulic fracturing operation before its implementation to minimize these risks. It is shown that in many cases it can be replaced by effective environmentally safe methods of impact on the borehole zone of the reservoir, and the result from their application is often more significant than when using hydraulic fracturing. The technology of directed unloading reservoir developed on the basis of geomechanical approach is given as an example.

Keywords Oil production technologies · Well · Hydraulic fracturing · Ecology · Geomechanical approach · Directional unloading reservoir

1 Introduction

The oil and gas industry is one of the environmentally hazardous industries in the world practice. The harm that hydrocarbon production causes to the environment is made up of a number of factors. These include oil spills during production and transportation, environmental pollution during flaring of associated gas, groundwater pollution during application of enhanced oil and gas recovery methods, and so on. Moreover, the risk of negative impact on the environmental situation during well operations has recently increased manifold.

This is due to the fact that easily recoverable hydrocarbon reserves are being depleted all over the world, which leads to the need to actively develop hard-to-recover oil and gas reserves, two thirds of which are contained in low-permeability

V. I. Karev (✉) · Yu. F. Kovalenko
A.Yu. Ishlinsky Institute of Physics and Mechanics of the Russian Academy of Sciences, Moscow, Russia
e-mail: wikarev1@gmail.com

Institute for Problems in Mechanics of the Russian Academy of Sciences, Moscow, Russia

T. Chaplina (ed.), *Processes in GeoMedia—Volume IX*, Springer Geology,
https://doi.org/10.1007/978-981-96-9302-3_1

reservoirs. Conventional technologies for the development of such fields are of little use and it is required the use of fundamentally new production methods.

2 Hydraulic Fracturing Technology

To date, the main technology that allows to extract oil and gas from such fields is hydraulic fracturing (Yanin 2018; Barkhatov and Yarkeyeva 2017; Prud'homme 2013; Shojaei and Shao 2017). However, environmental problems in the development of fields by hydraulic fracturing are still unresolved. The results of research in this area are ambiguous and require additional research. Thus, the use of hydraulic fracturing has many negative consequences for the environment. The only way to avoid these problems is to use environmentally friendly and safe methods of natural resource extraction.

Hydraulic fracturing is the creation of large-area fractures in the reservoir to increase the filtration surface and increase flow to the well. For this purpose, an overpressure sufficient to fracture the formation rock and create a fracture is established at the bottom of the well with the help of powerful pumps using high-viscosity fracturing fluid. Proppant (polymer beads) or just sand is pumped into the resulting fracture to prevent it from closing after the pressure is removed.

Initially, hydraulic fracturing was applied to vertical wells by creating a single fracture (so-called 1-stage fracturing). In recent years, with the development of horizontal well drilling technologies, multistage hydraulic fracturing (MSHF), which consists in sequential creation of vertical fracturing cracks along the horizontal wellbore (Speight 2016; Medvedev 2019; Kaarov and Lubyshev 2019), Fig. 1, has been used.

Currently, multistage hydraulic fracturing is one of the most widespread methods of oil and gas production, but its use can lead to serious environmental consequences (Bulatov et al. 2018; Borisenko and Smolina 2019; Levina et al. 2018; Tulinov and Andreev 2019). To perform this procedure, a huge amount of water is injected into the reservoir, as well as a "cocktail" of several dozen different chemicals required to increase permeability in the vicinity of the fractures. Most of the chemicals used are toxic and can cause significant groundwater contamination (Fig. 2).

Nevertheless, despite all these disadvantages, hydraulic fracturing technology for fields with low-permeability reservoirs currently has practically no alternative and is widely used.

However, when during drilling or operation of wells contamination and reduction of bottomhole permeability (colmatization) occurs, while the permeability of the reservoir as a whole does not change. The effect of hydraulic fracturing is often minimal, and the costs and environmental damage are high. This is the case, for example, at wells drilled to reservoirs of oil and gas fields with good and average permeability, to formations-objects of underground gas storage, to injection wells.

The reason is that hydraulic fracturing is effective in reservoirs with low permeability, where the effect is achieved by significantly increasing the filtration area

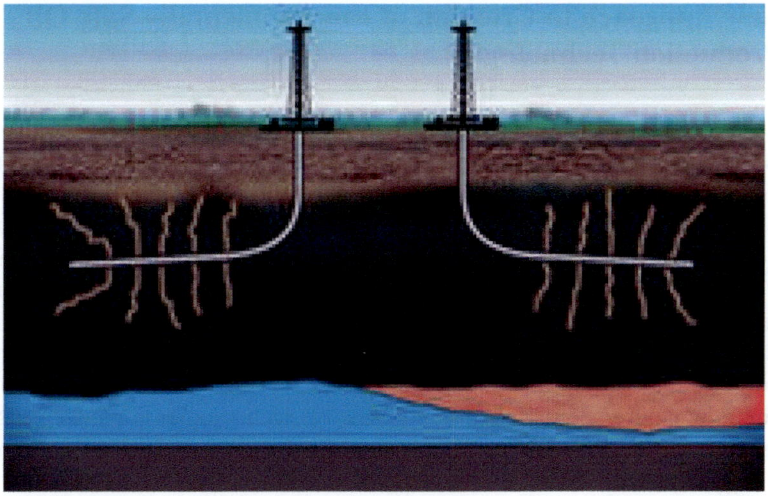

Fig. 1 Schematic diagram of multistage hydraulic fracturing

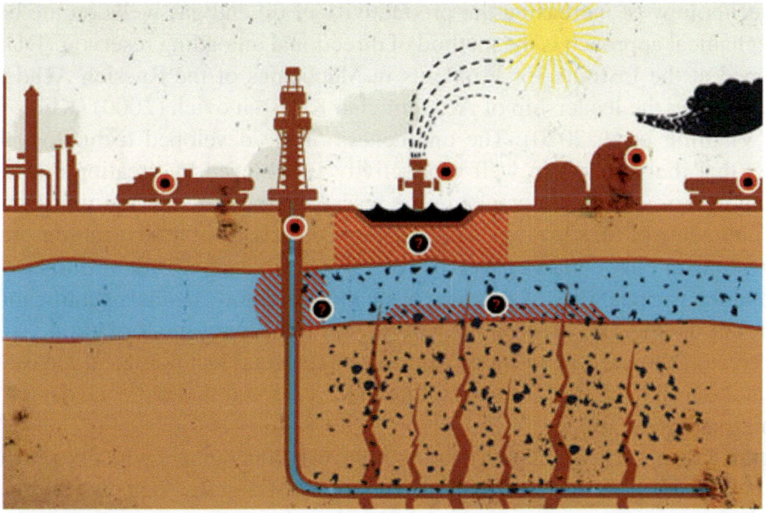

Fig. 2 Scheme of contamination at hydraulic fracturing, breakthrough of reagents into groundwater

through the formation of a large crack. In wells of fields with good to medium permeability, in underground gas storage facilities wells and injection wells, significant permeability reduction occurs in a small zone of the formation around the well, outside of which permeability remains high. Therefore, the task is to restore the permeability of this zone (to eliminate the low-permeability "plug"), which the hydraulic fracture cannot provide.

3 New Approach to Creation of Environmentally Safe Oil Production Technologies

Thus, the problem of creation of new environmentally friendly low-cost methods of well stimulation with long-lasting effect due to restoration of rock permeability in the bottom-hole zone (BHZ) is urgent. And such technologies already exist.

As an example, we can cite the method of wave action on the formation, created at the Mechanical Engineering Research Institute of the Russian Academy of Sciences under the guidance of Academician Kuznetsov et al. (2023). Its approbation was successfully carried out not only in the fields of Russia, but also jointly with the largest Western Companies Shell and British Petroleum in other countries.

However, the most promising direction for solving the above problems and creating new environmentally safe and effective technologies for increasing well productivity seems to be the use of geomechanical approach (Tulinov and Andreev 2019). It is based on the creation of adequate mechanical and mathematical models describing the processes of deformation and destruction of reservoir rocks with low permeability. The most important factor to be taken into account in geomechanical modeling is the influence of stresses arising in rocks on their filtration properties.

A technology of increasing the productivity of oil and gas wells on the basis of geomechanical approach is the method of directional unloading reservoir (DUR) has developed at the Institute for Problems in Mechanics of the Russian Academy of Sciences under the leadership of Academician Khristianovich (2000) (Klimov et al. 2015; Vladimir et al. 2020). The uniqueness of the developed technology lies in the fact that the increase in well productivity is achieved by creating a system of multiple micro- and macro-cracks in the reservoir by means of non-uniform directional unloading of rock from rock pressure and utilization of fluid reservoir pressure. The created system of cracks plays the role of an artificial network of filtration channels, and the permeability of this new system of filtration channels significantly (by an order of magnitude) exceeds the natural permeability of the formation.

To create artificial fracturing in the reservoir, special technological measures are taken at the well to change the stress state in the reservoir and a special design of the well bottom hole is used.

Figure 3 shows schematically a bottomhole reservoir element with two perforation holes before the well was fractured (a) and after (b). Figure 3b shows the resulting multiple fractures acting as an artificial filtration channel system.

Unlike hydraulic fracturing technology, in the process of which a chemically active fracturing fluid and then proppant are injected into the formation under high pressure to form a single fracture, the implementation of the directional fracturing method requires neither of these, because the cracks sytem is achieved not due to additional load on the formation, but due to its uneven unloading from the rock pressure. Therefore, the method of directional unloading reservoir is environmentally safe and does not harm the environment.

In order to adapt the DUR method to the conditions of a particular field, it is necessary to conduct a series of tests of core material from the reservoir of interest on a true

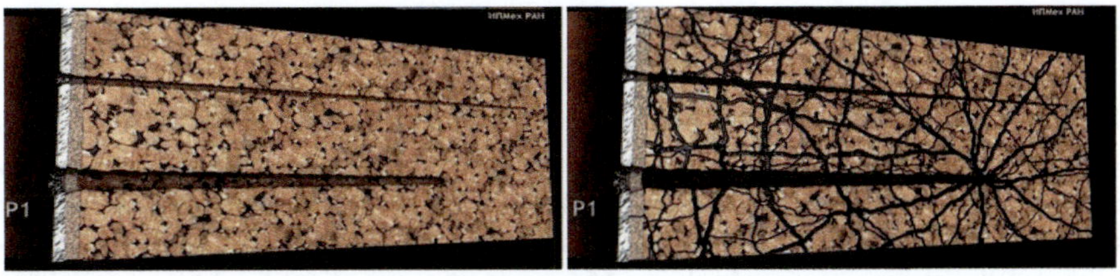

Fig. 3 Formation of artificial fracturing in the bottomhole formation zone as a result of application DUR method at the well

triaxial loading unit, which allows to create in rock specimens real stresses occurring in the reservoir during certain technological operations, to determine the parameters of the mechanical and mathematical model, the dependence of permeability on stresses and to perform the appropriate calculations (Karev and Kovalenko 2013). Such an installation—Triaxial Independent Loading Test System—has been created and successfully used at the Institute for Problems in Mechanics of the Russian Academy of Sciences (Fig. 4).

The DUR method is applicable to any type of field at any depth of formation. To implement the DUR method, standard equipment is required, which is available at almost every field. The DUR method has undergone pilot tests at a number of oil fields in Western Siberia and the Kama region. The positive results obtained can be considered as a proof of its efficiency.

Fig. 4 Triaxial independent loading test system (TILTS)

4 Conclusion

The application of technologies based on geomechanical approach can in many cases significantly reduce environmental risks in the development of oil and gas fields. However, for their effective use it is necessary to conduct preliminary comprehensive studies of deformation and filtration properties of reservoir rocks and their dependence on the stresses acting in them. Moreover, these studies should be carried out in the conditions of true triaxial stress states actually occurring in the BHZ under geomechanical impact, which requires special experimental facilities, such as the Triaxial Independent Loading Test System created at the Institute for Problems in Mechanics of the Russian Academy of Sciences (Vladimir et al. 2020).

The research was carried out within the framework of State Assignment 124,012,500,441–6 (FFGN-2024–0007).

References

Barkhatov EA, Yarkeyeva NR (2017) Efficiency of multi-zone hydraulic fracturing application in horizontal wells. Proc Tomsk Polytech University Eng Georesources 328(10):50–58

Borisenko MY, Smolina IA (2019) Assessment of the impact of hydraulic fracturing on the environment. Ecol Ind Russ 23(2):29–34

Bulatov AA, Tereschenko NA, Ovcharova MM (2018) Ecological consequences of hydraulic fracturing during oil and gas production. Ecol Ind Russ 22(1):22–27

Kaarov ZZ, Lubyshev VS (2019) Experience of application of multistage hydraulic fracturing using TTS technology at oil fields of Western Siberia in conditions of low-permeable reservoirs with complex geological structure. Colloq-J 13–3(37): 135–137

Karev VI, Kovalenko YF (2013) Triaxial loading system as a tool for solving geotechnical problems of oil and gas production. In: True triaxial testing of rocks. CRC Press. Balkema, Leiden, pp 301–310

Khristianovich SA, Kovalenko YF, Kulinich YV, Karev VI (2000) Increase in productivity of oil wells with the help of the method of geoloosening. Oil Gas Eurasia (2):90–94

Klimov DM, Karev VI, Kovalenko YF (2015) Experimental Study of the Influence of a Triaxial Stress State with Unequal Components on Rock Permeability. Mech Sol 50(6):633–640

Kuznetsov YS, Ganiev SR, Sultanov DR, Fedorov VN, Bystrikov SN (2023) Experimental and field studies of the wave technology of bottom-hole cleaning and colmatization of permeable reservoirs. Izvestiya vysshykh uchebnykh savedeniy. Neft i Gas. 6:25–34. https://doi.org/10.31660/0445-0108-2023-6-25-34

Levina EA, Makarenko IA, Bezuglov AV (2018) Environmental risk in the development of gas fields by hydraulic fracturing. Econ Manag Innov 3(18):55–60

Medvedev KS (2019) Review of modern methods of multi-zone hydraulic fracturing at oil fields in Siberia. Bull Sci Educ PART 2. pp 14–19

Prud'homme A (2013) Hydrofracking: what everyone needs to know. Oxford University Press, p 208

Shojaei AK, Shao J (2017) Porous rock fracture mechanics with application to hydraulic fracturing, drilling and structural engineering. LA—1 st ed. Woodland Publishing, p 336

Speight JG (2016) Handbook of hydraulic fracturing. Wiley, Laramie, WY, p 304

Tulinov AL, Andreev IO (2019) Ecological problems of oil and gas fields development by hydraulic fracturing. Sci Tech Bull Inf Technol Mech Opt 19(3, 119):441–447

Karev V, Kovalenko Y, Ustinov K (2020) Directional unloading method is new approach to enhancing oil and gas well productivity. In: Book: geomechanics of oil and gas wells. Ser. Advances in oil and gas exploration and production. Springer Nature Switzerland AG, pp 155–166
Yanin AN (2018) Hydraulic fracturing is a breakthrough technology. Drill Oil (7)

Mathematical Modeling of Acoustic Fields in the Black Sea

Yu. I. Papkova and S. O. Papkov

Abstract The results of in situ measurements of sound speed in the marine environment in the coastal part of the Black Sea are analyzed. Based on the processing of the measurement results, a flat-layered model of waveguide sound propagation is proposed. This model includes a water layer with a non-uniform sound speed profile lying on a layer of liquid sediments (sea silt), which, in turn, lie on an elastic layer of bottom sediments. The lower boundary of the waveguide is assumed to be absolutely rigid, which approximately corresponds to a foundation in the form of rocks. For the first time, a Green's function is constructed for this model, which allows expressing the sound pressure potential as a sum of normal modes. The paper provides examples of numerical modeling.

Keywords Acoustic marine waveguide · Speed of sound · Black Sea · Layered elastic bottom · Normal modes

1 Introduction

Currently, the creation of hydroacoustic systems involves the use of various waveguide models to take into account the factors affecting the propagation of sound in the marine environment. In particular, there are a number of waveguide models that take into account the geoacoustic parameters of the bottom and the sound velocity profile (Brekhovskikh and Godin 1989; Jensen et al. 2011; Papadakis and Keller 1980; Shenderov 1972; Grigoriev et al. 2017; Paul 2018), namely, the Pekeris waveguide (Pekeris 1948), a flat-layered waveguide, etc. Models of these waveguides are usually used as a basis for computer modeling to analyze sound fields and create specialized software. Modeling sound fields in irregular waveguides requires taking into account many aspects: the heterogeneity of the medium, different waveguide geometries, various geoacoustic properties of bottom sediments, etc. (Buckingham 2020; Ballard and

Yu. I. Papkova (✉) · S. O. Papkov
Sevastopol State University, Sevastopol, Russia
e-mail: yulia.papkova@gmail.com

T. Chaplina (ed.), *Processes in GeoMedia—Volume IX*, Springer Geology,
https://doi.org/10.1007/978-981-96-9302-3_2

Sagers 2019; Sertlek and Ainslie 2014), so the relevance of creating flexible mathematical models adapted to specific tasks is becoming increasingly important in connection with the needs. There are a large number of works (Ballard and Sagers 2019; Jensen and Ferla 1990; Chapman et al. 2010; Grili et al. 1998; Vendhan et al. 2010; Lin et al. 2013; Knighly et al. 1987; Tappert and Hardin 1974; Luo and Schmidt 2009) that provide an overview of approaches and methods used to calculate sound fields in irregular waveguides; taking into account specific requirements for the accuracy of solving the problems and the speed of calculations. It should be noted that, despite the existing studies, a comprehensive consideration of the characteristics of the inhomogeneities of the environment and bottom sediments in the Black Sea has not been sufficiently reflected in the literature, so there are prerequisites for further research that can significantly improve the understanding of hydroacoustic processes in this region.

The model of the Black Sea waveguide is proposed to be considered as a special case of a non-uniform acoustic waveguide with sediments and an elastic bottom layer on a rigid foundation, where the seasonal variability in the upper water layer will be taken into account using the refraction term of the speed of sound in seawater. The boundaries of the waveguide are assumed to be plane parallel. Based on the normal mode method, the functions of sound potentials in the water and elastic layers are constructed. The joining of these solutions at the bottom makes it possible to obtain a dispersion equation for determining the complex eigenvalues of the problem in the form of a sixth-order determinant. The solution of the non-uniform problem with a source is obtained based on the Green's function. It is important that for this model, from the system of relations on the boundary of the water layer and the elastic layer, it is possible to accurately derive an analytical expression for the impedance function, which specifies the relationship between the potential in the water layer and its derivative. This, in turn, makes it possible to obtain explicit analytical expressions for all quantities included in the solution and to construct a solution in the form of a sum of normal modes.

The acoustic properties of the marine environment are largely determined by the non-uniform refractive term of the sound velocity profile. For numerical modeling of sound fields in the marine environment (ray method, finite element method, finite difference method, etc.), it is sufficient to know this value in a certain system of nodal points, while for mode analysis, its analytical description is essential. The use of the obtained sound velocity profiles in the marine environment in calculating acoustic characteristics implies, first of all, their smoothing, i.e. approximation by a certain smooth function. Approximation of the sound velocity profile by functions, with the help of which the analytical solution of the Helmholtz equation is constructed, is one of the necessary procedures for solving direct and inverse problems of sound propagation in the marine environment, therefore the article touches upon the issue of constructing approximating functions that allow an analytical solution for a waveguide, in particular the Black Sea one. This fact allows us to obtain an exact analytical solution to the problem of sound propagation for the case of a waveguide with an elastic layer on a rigid foundation that is inhomogeneous in sound velocity, and to conduct a parametric study of the influence of the elastic properties of the bottom layer on the picture of the emerging acoustic fields. Thus, the construction of refined models of the sound velocity profile in geoacoustic marine waveguides, allowing us to obtain an analytical solution, makes it possible to accurately model and predict the behavior of sound in complex irregular waveguides, which is important for many hydroacoustic applications.

2 Fundamentals of the Theoretical Approach

The formation of the sound field in the Black Sea is caused by a combination of salinity, temperature and depth, i.e. the type of thermal structure of seawaters, as well as geoacoustic characteristics of the bottom layer. These factors significantly determine the underwater acoustic waveguide and its characteristics, which are used in various applications, including marine navigation and communication, so it is of interest to consider the specifics of sound propagation in acoustic natural marine waveguides typical of the Black Sea waters. To conduct a detailed analysis of the refractive component of the sound speed of the Black Sea depending on seasonal changes, based on an array of in situ data from the Oceanographic Data Bank obtained in the period 2015–2022, and data from oceanographic stations in the form of drifting measuring buoys ARGO (Array for Real-Time Geostrophic Oceanography). In total, 147434 values of vertical distributions of the sound speed of the Black Sea, stratified by depth, were considered in the analysis, of which 621 measurements were made by ARGO buoys. The location of measurements in the studied area of the Black Sea is shown in Fig. 1. The measurement polygon was constructed with a measurement radius from the coastal zone of about 50 km.

The water layer of the Black Sea consists of the upper, cold intermediate and lower layers. The upper layer of the Black Sea waters is characterized by significant seasonal dynamics, which is caused primarily by seasonal changes in temperature and salinity, hydrological conditions, currents and other factors. The lower water layer is practically independent of the season with increasing depth, starting from 120 m, the sound speed profile changes slightly during the year, thus, at a depth of 120 m, all vertical distributions of the sound speed are combined. This fact is due to the fact that in the lower layer, temperature, pressure, and salinity have less variability. Table 1 presents the data on the sound speed of the Black Sea based on in situ measurements of temperature and salinity in the coastal part for different seasons in different years with increasing depth z of the waveguide. Analyzing the data presented in the table, we can conclude about the seasonal stability of the sound speed field in the lower layer of the waveguide. The increase in the speed of sound with increasing depth, as well as the small intra-annual variability, will allow us to represent it as a linear function of the depth z.

$$c(z) = \begin{cases} 0.0269z + 1466; & 120 \leq z < 300 \\ 0.0174z + 1469; & z \geq 300 \end{cases}.$$

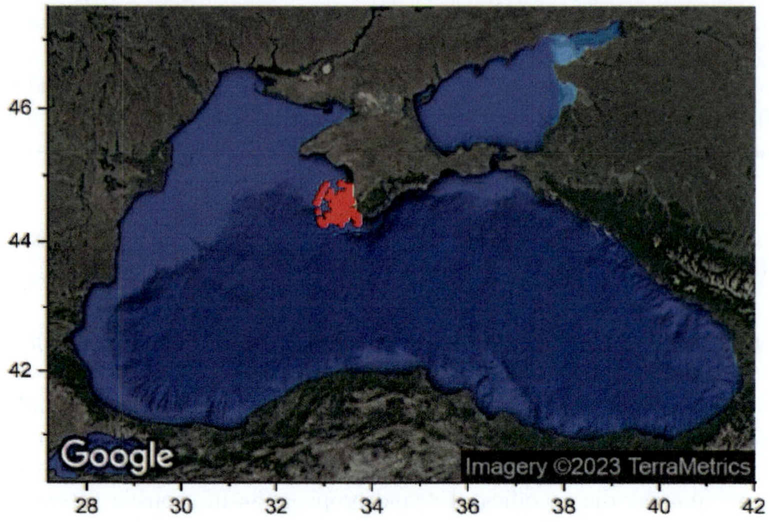

Fig. 1 Location of measurements in the study area of the black sea

It is worth noting that this formula can be used to calculate characteristics in the deep-water part of the Black Sea (Fig. 2). To assess the optimality of the approximation of the speed of sound, the coefficient of determination R^2 was used in the calculations:

$$R^2 = 1 - \frac{\sum_{i=1}^{n} (c_i - c(z_i))^2}{\sum_{i=1}^{n} (c_i - \overline{c})^2},$$

Table 1 Values of sound speed (m/s) in the lower layer of the black sea waveguide

z, м	Date						
	07.07.16	30.04.17	14.12.17	14.06.18	02.09.19	03.01.21	30.06.21
122	1468.203	1467.878	1468.078	1468.154	1468.849	1468.992	1469.077
151	1469.59	1469.542	1469.497	1469.585	1469.988	1470.066	1470.097
203	1471.225	1471.369	1470.968	1471.201	1471.485	1471.561	1471.485
253	1472.647	1472.682	1472.378	1472.555	1472.728	1472.803	1472.756
302	1473.795	1473.754	1473.725	1473.687	1473.885	1473.878	1473.082
402	1475.797	1475.822	1475.754	1475.757	1475.903	1475.895	1475.832
501	1477.633	1477.682	1477.603	1477.629	1477.702	1477.702	1477.688

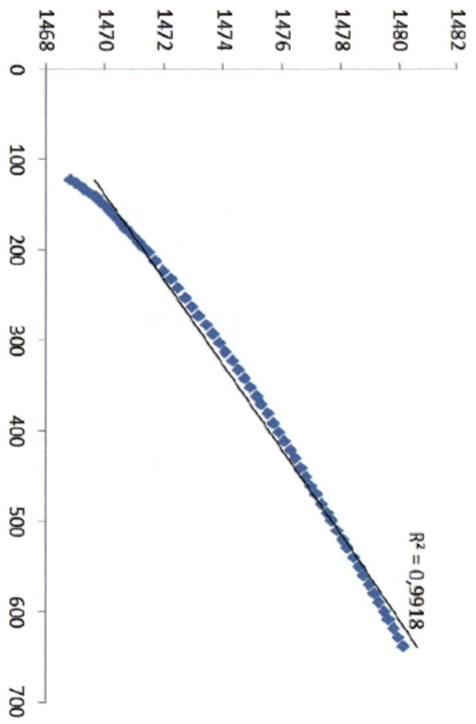

Fig. 2 Linear approximation of the sound speed profile (……… – experimental values, — - approximating line)

where c_i и $c(z_i)$—actual and calculated values of the speed of sound, $\overline{c} = \frac{1}{n}\sum_{i=1}^{n} c_i$. When evaluating models, the value $R^2 = 0.9918$ is interpreted as a match to actual data.

The upper layer of the vertical profile of the sound velocity in the Black Sea in the warm season from May to October has a strong asymmetry. During this period of time, the underwater sound channel in the Black Sea is formed due to the increase in the sound velocity in the near-surface layer and the formation of a minimum at a depth of 50–80 m, which leads to a strong concentration of sound energy near the surface. Table 2 presents the values of the array of summer in situ data on the sound velocity of different years. The nature of the vertical distribution of the sound velocity in the warm season will allow us to conclude that there is an underwater sound channel (USC), to find the depths on the horizons of which the USC axis is located, which in turn creates conditions for the effective propagation of acoustic signals. In particular, from the data in Table 2 it can be seen that in the summer season the USC axis is located at horizons from 55 to 80 m, for example, for in situ data 14.06.2018 y. USC axis $z = 78$ m, 18. 06. 2017 y. $-z = 57$ m, 07.07.2016 y. $-z = 59$ m. From November to April, there is little variability in the vertical distribution of sound speed (Fig. 3), the range of in situ data samples is greatest in November—19.520, the lowest in February—4.229.

Table 2 Vertical values of sound speed (m/s) in summer

z. м	Date							
	06.06.16	18.06.17	14.06.18	12.07.19	02.09.19	27.06.20	24.06.21	28.07.21
3	1516.92	1502.927	1508.278	1511.127	1516.037	1512.472	1503.106	1515.676
7.6	1516.543	1502.761	1504.181	1511.416	1516.119	1510.756	1495.686	1514.508
12.7	1514.976	1479.194	1498.821	1476.487	1516.226	1501.239	1490.758	1514.232
22.6	1493.826	1466.806	1492.237	1471.464	1513.637	1488.736	1480.909	1510.197
32.6	1474.531	1462.770	1477.280	1466.418	1472.580	1472.120	1469.583	1495.515
42.6	1468.049	1460.866	1470.711	1465.019	1467.090	1468.630	1467.555	1483.413
52.6	1465.575	1460.108	1467.509	1464.236	1464.846	1466.719	1466.512	1470.196
62.6	1464.340	1460.235	1465.602	1463.685	1463.950	1466.174	1466.015	1467.822
72.2	1464.737	1461.001	1464.476	1463.887	1464.827	1466.536	1466.076	1467.272
82.1	1465.908	1462.080	1464.354	1464.614	1465.977	1466.346	1466.814	1467.039
83	–	1462.162	1464.503	1464.732	–	–	1466.835	1467.089

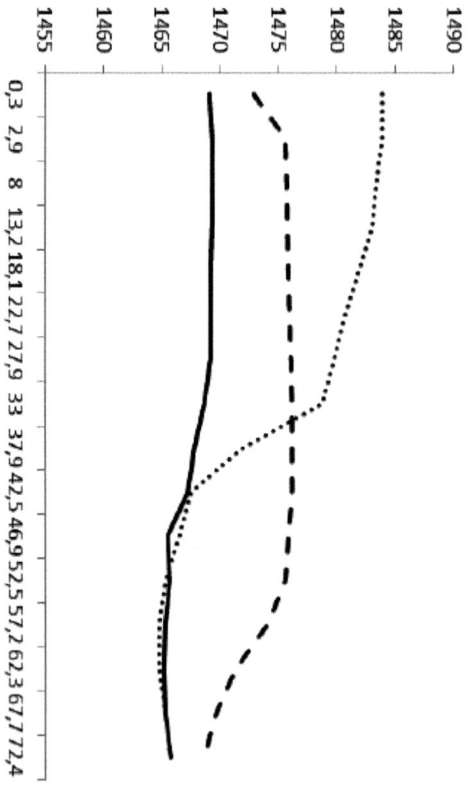

Fig. 3 Vertical profile of the speed of sound of the black sea (….. - November, ___ - February, - - - - April)

Further application in the calculation of a data array of vertical distributions of sound speed in a certain period of time involves the construction of some averaged sound speed profile at the measurement points $\langle c(z_i)\rangle$, and also finding the range of in situ data. Figure 4 shows the vertical profiles of the speed of sound for different years (2016–2018), typical for June, the range of samples is respectively equal to 52.124; 43.091; 44.577. It is possible to construct two types of average profile used for data adjustment, namely the mean and median profiles (Fig. 3). Similarly, it is possible to obtain a set of average profiles for other months.

Averaged sound speed profile $\langle c(z_i)\rangle$ can be used to construct the most probable sound speed profile $c(z_i)$, using in situ values of the speed of sound on the surface $0 < z \leq 40$ according to the formula $c(z_i) = \langle c(z_i)\rangle + \Delta z_i$, where Δz_i—the difference between the experimental and average values at a fixed depth.

An important numerical characteristic is the vertical relative gradient of the speed of sound $\alpha = \frac{|c'(z)|}{\bar{c}}$, which create unique conditions for the transmission of acoustic signals in the Black Sea. Figure 5 shows the gradient values calculated from in situ data in the summer period, since the highest values of the vertical gradient of the speed of sound are observed during this period.

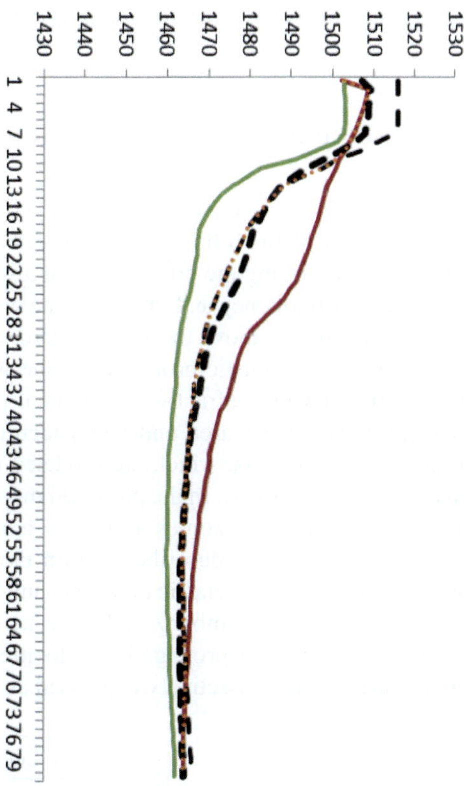

Fig. 4 Vertical profile of sound speed of the black sea in June (- - - average sound speed profile, - - - - median sound speed profile)

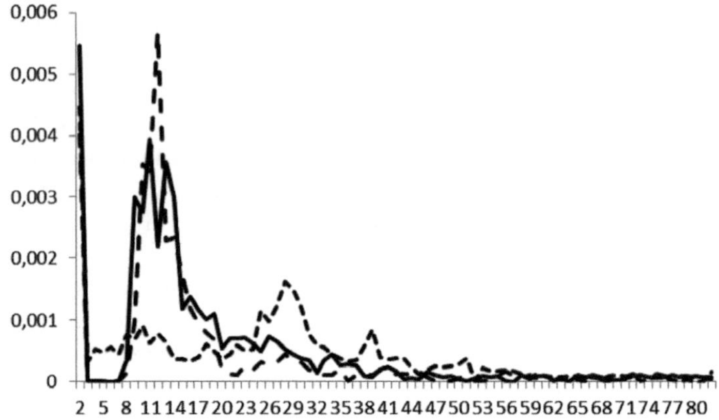

Fig. 5 Values of the vertical gradient of the speed of sound (c^{-1}) depending on the depth

It is worth noting that the anomalously high sound speed gradients in seawater are a feature of the hydrological anomalies of the Black Sea region and are not found in the ocean. Sound speed gradients, both positive and negative, create unique conditions for the transmission of acoustic signals in the Black Sea.

3 Mathematical Model of Sound Propagation

Based on the above-mentioned in situ data and known geological features of the seabed structure, the following flat-layered model of a marine acoustic waveguide is proposed, describing the propagation of a sound signal by a point source at depth z_0 with circular frequency ω. Let's place the origin of the cylindrical coordinate system on the sea surface above the point source, then the coordinates of the point source will be $(0, z_0)$. Acoustic properties of the water layer of depth h_0 are described by the inhomogeneous refractive component of the speed of sound in water $c(z)$ and density ρ. The water area under consideration is characterized by a large layer of bottom silt sediments, which are modeled by a liquid layer with a constant sound speed c_0 and density ρ_0. In the proposed model, the depth of the liquid sediment layer is $h_1 = h - h_0$. Next comes a layer of bottom sediments with pronounced elastic properties, which are described within the framework of the model of an elastic isotropic body by two elastic constants and a density ρ_1. As such constants, one can choose the Poisson number λ and shear modulus G, however, it is more convenient to use the associated propagation velocities of longitudinal and transverse elastic waves, which are respectively expressed as

$$c_1 = \sqrt{\frac{\lambda + 2G}{\rho_1}}, c_2 = \sqrt{\frac{G}{\rho_1}}. \tag{1}$$

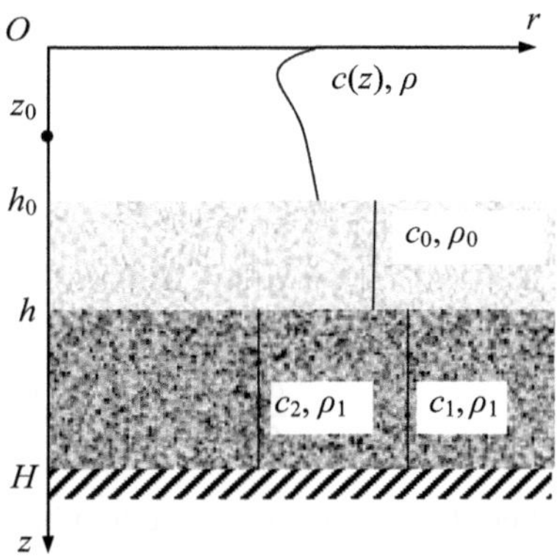

Fig. 6 A model for acoustic marine waveguide

The depth of the elastic sedimentary layer is calculated as $H_1 = H - h$. At the base of the flat-layered waveguide is an absolutely rigid foundation, which can be interpreted as a rocky substrate (Fig. 6).

It is known (Grigoriev et al. 2017; Achenbach 1975; Schneiderwind et al. 2012; Press and Ewing 1950) that the acoustic field in the presented waveguide can be expressed using the potentials of the water layer $\mathbf{\Phi}(r, z)$ $(z \leq h_0)$, potential of the liquid precipitation layer $\mathbf{\Phi}_0(r, z)$ $(h_0 \leq z \leq h)$, and two elastic potentials $\phi(r, z)$ and $\theta(r, z)$ at $h \leq z \leq H$. Then the sound pressure in the water layer is expressed as

$$P = \rho\omega^2\mathbf{\Phi} \tag{2}$$

and in the layer of liquid sediments as

$$P_0 = \rho_0\omega^2\mathbf{\Phi}_0. \tag{3}$$

In this case, the components of the displacement vector, as well as the normal and tangential stresses in the elastic layer are expressed as follows:

$$u_r = \frac{\partial\phi}{\partial r} + \frac{\partial^2\theta}{\partial r\partial z}, \; u_z = \frac{\partial\phi}{\partial z} + \frac{\partial^2\theta}{\partial z^2} + k_2^2\theta \tag{4}$$

$$\sigma_{zz} = 2G\frac{\partial}{\partial z}\left(k_2^2\theta + \frac{\partial^2\theta}{\partial z^2}\right) + 2G\frac{\partial^2\phi}{\partial z^2} - \lambda k_1^2\phi \tag{5}$$

$$\sigma_{rz} = 2G\frac{\partial}{\partial r}\left(\frac{\partial\phi}{\partial z} + \frac{\partial^2\theta}{\partial z^2} + \frac{k_2^2}{2}\theta\right) \tag{6}$$

where wave numbers are introduced for convenience as $k_j = \omega/c_j$ $(j = 1, 2)$.

The potentials must satisfy the Helmholtz equations in each layer respectively.

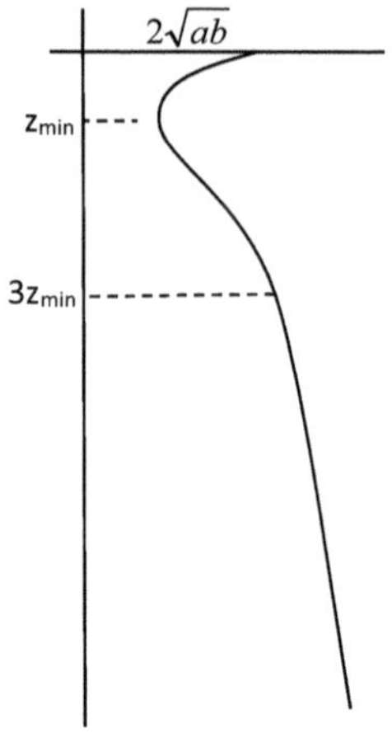

Fig. 7 Graph of approximate function for $c(z)$

In particular, for the dependence of the speed of sound $c(z) = a\sqrt{z+d} + \frac{b}{\sqrt{z+d}}$ (Fig. 7).

the general solution can be represented in terms of Whittaker functions

$$\psi_1(z;\xi) = M_{\kappa,m}\left(2\xi\left(z+\frac{b}{a}+d\right)\right),\ \psi_2(z;\xi) = W_{\kappa,-m}\left(2\xi\left(z+\frac{b}{a}+d\right)\right),$$
$$\kappa = \frac{\omega^2}{2a^2\xi};\ m = \sqrt{\frac{1}{4}+\frac{b\omega^2}{a^3}} \tag{31}$$

which makes it possible to describe the experimental values of sound speed in the water layer quite closely.

Note also that differential Eqs. (20)–(21), due to the fact that the speed of sound in the data layers is constant, allow elementary solutions

$$\Phi_0 = C_3 \sin\mu_0 z + C_4 \cos\mu_0 z \tag{32}$$

$$\tilde{\varphi} = C_5 \sin\mu_1 z + C_6 \cos\mu_1 z \tag{33}$$

$$\tilde{\theta} = C_7 \sin\mu_2 z + C_8 \cos\mu_2 z \tag{34}$$

where $\mu_j = \sqrt{k_j^2 - \xi^2}$, $(j = 0,\ 1,\ 2)$.

Explicit analytical formulas for general solutions of differential Eqs. (19)–(23) make it possible to significantly simplify both the finding of eigenvalues and functions of the corresponding homogeneous boundary value problem and the construction of Green's function in the water layer. Indeed, conditions (27)–(29) allow us to actually exclude elastic potentials from the system $\tilde{\varphi}$ and $\tilde{\theta}$. To do this, we will consider these conditions, substituting into them the general solutions (33)–(34):

$$\left.\begin{array}{l}\mu_1\cos\mu_1 hC_5-\mu_1\sin\mu_1 hC_6+\left(\xi^2-\frac{k_2^2}{2}\right)\sin\mu_2 hC_7+\left(\xi^2-\frac{k_2^2}{2}\right)\cos\mu_2 hC_8=0\\ \sin\mu_1 HC_5+\cos\mu_1 HC_6+\mu_2\cos\mu_2 HC_7-\mu_2\sin\mu_2 HC_8=0\\ \mu_1\cos\mu_1 HC_5-\mu_1\sin\mu_1 HC_6+\xi^2\sin\mu_2 HC_7+\xi^2\cos\mu_2 HC_8=0\end{array}\right\}. \quad (35)$$

It is obvious that relatively unknown $Y_6 = C_6/C_5,\ Y_7 = C_7/C_5,\ Y_8 = C_8/C_5$ this is a system of three linear equations, the solution of which can be expressed explicitly using Cramer's formulas

$$Y_6=\frac{\Delta_6}{\Delta_5},\ Y_7=\frac{\Delta_7}{\Delta_5},\ Y_8=\frac{\Delta_8}{\Delta_5} \quad (36)$$

where the determinant of the system is denoted as

$$\Delta_5=\det\begin{bmatrix}-\mu_1\sin\mu_1 h & \left(\xi^2-\frac{k_2^2}{2}\right)\sin\mu_2 h & \left(\xi^2-\frac{k_2^2}{2}\right)\cos\mu_2 h\\ \cos\mu_1 H & \mu_2\cos\mu_2 H & -\mu_2\sin\mu_2 H\\ -\mu_1\sin\mu_1 H & \xi^2\sin\mu_2 H & \xi^2\cos\mu_2\end{bmatrix}$$

and auxiliary determinants Δ_6, Δ_7, and Δ_8, as is known, are obtained by substituting a column

$$-\begin{bmatrix}\mu_1\cos\mu_1 h\\ \sin\mu_1 H\\ \mu_1\cos\mu_1 H\end{bmatrix}$$

instead of the corresponding column Δ_5.

Then we can obtain analytical expressions:

$$\frac{\tilde{\varphi}}{\tilde{\varphi}'}=\frac{1}{\mu_1}\frac{C_5\sin\mu_1 h+C_6\cos\mu_1 h}{C_5\cos\mu_1 h-C_6\sin\mu_1 h}=\frac{1}{\mu_1}\frac{\Delta_5\sin\mu_1 h+\Delta_6\cos\mu_1 h}{\Delta_5\cos\mu_1 h-\Delta_6\sin\mu_1 h} \quad (37)$$

$$\frac{\tilde{\theta}'}{\tilde{\theta}}=\mu_2\frac{C_7\cos\mu_2 h-C_8\sin\mu_2 h}{C_7\sin\mu_2 h+C_8\cos\mu_2 h}=\mu_2\frac{\Delta_7\cos\mu_2 h-\Delta_8\sin\mu_2 h}{\Delta_7\sin\mu_2 h+\Delta_8\cos\mu_2 h} \quad (38)$$

Next, from conditions (25) and (26) we can find the relation

$$-\frac{\Phi_0}{\Phi_0'}=\frac{4\rho_1}{k_2^4\rho_0}\left(\left(\xi^2-\frac{k_2^2}{2}\right)\frac{\tilde{\phi}}{\tilde{\theta}}+\xi^2\frac{\tilde{\theta}'}{\tilde{\theta}}\right)$$

which taking into account the connection from (27)

$$\tilde{\theta} = -\frac{\tilde{\phi}'}{\xi^2 - \frac{k_2^2}{2}}$$

and can be written as

$$-\frac{\Phi_0}{\Phi_0'} = \frac{4\rho_1}{k_2^4\rho_0}\left(\xi^2\frac{\tilde{\theta}'}{\tilde{\theta}} - \left(\xi^2 - \frac{k_2^2}{2}\right)^2\frac{\tilde{\phi}}{\tilde{\phi}'}\right). \tag{39}$$

Thus, we obtain one impedance condition at the bottom of the liquid sediment layer.

$$\Phi_0(h;\xi) + g_0(\xi)\Phi_0'(h;\xi) = 0 \tag{40}$$

where is the impedance function $g_0(\xi) = \frac{4\rho_1}{k_2^4\rho_0}\left(\xi^2\frac{\tilde{\theta}'}{\tilde{\theta}} - \left(\xi^2 - \frac{k_2^2}{2}\right)^2\frac{\tilde{\varphi}}{\tilde{\varphi}'}\right)$ can be significantly simplified by elementary transformations:

$$g_0(\xi) = \frac{4\rho_1}{k_2^4\rho_0}\times \frac{\xi^2\left(\mu_1^2\mu_2^2 + \left(\xi^2 - \frac{k_2^2}{2}\right)^2\right)\sin\mu_1 H_1 \sin\mu_2 H_1 + 2\xi^2\mu_1\mu_2\left(\xi^2 - \frac{k_2^2}{2}\right) - \mu_1\mu_2\left(\frac{k_2^4}{4} - k_2^2\xi^2 + 2\xi^4\right)\cos\mu_1 H_1\cos\mu_2 H_1}{\mu_1\left(\xi^2\cos\mu_1 H_1 \sin\mu_2 H_1 + \mu_1\mu_2\sin\mu_1 H_1\cos\mu_2 H_1\right)}. \tag{41}$$

Note also that when condition (39) is satisfied, conditions (25)–(29) are satisfied automatically.

The approach described above can also be used to satisfy the remaining three conditions (22)–(24). Indeed, if we take into account that in the sediment layer the potential function is expressed through trigonometric functions (32), then we can obtain the impedance condition at the boundary of the water layer—the liquid sediment layer. Next, (32) is substituted into the continuity conditions (23), (24) and into condition (39):

$$\left.\begin{array}{l}\sin\mu_0 h C_3 + \cos\mu_0 h C_4 + g_0(\xi)\mu_0\cos\mu_0 h C_3 - g_0(\xi)\mu_0\sin\mu_0 h C_4 = 0\\ \rho\Phi(h_0) = \rho_0(\sin\mu_0 h_0 C_3 + \cos\mu_0 h_0 C_4)\\ \Phi'(h_0) = \mu_0(\cos\mu_0 h_0 C_3 - \sin\mu_0 h_0 C_4)\end{array}\right\} \tag{42}$$

where is it easy to find that

$$\Phi(h_0;\xi) + g(\xi)\Phi'(h_0;\xi) = 0 \tag{43}$$

where $g(\xi) = \frac{\rho_0}{\rho\mu_0}\frac{\sin\mu_0 h_1 + g_0(\xi)\mu_0\cos\mu_0 h_1}{\cos\mu_0 h_1 - g_0(\xi)\mu_0\sin\mu_0 h_1}$.

The condition on the water surface (21) can also be satisfied by construction if the eigenfunctions in the water layer are sought in the form

$$\Phi = A(\psi_2(0;\xi)\psi_1(z;\xi) - \psi_1(0;\xi)\psi_2(z;\xi)) \tag{44}$$

where A is an undefined constant.

Therefore, all boundary conditions and field continuity conditions will be satisfied for the waveguide model under consideration if condition (42) is satisfied, where the eigenfunctions in the water layer are chosen in the form (43). Thus, Eq. (42) can be considered as a dispersion equation for finding the eigenvalues $\{\xi_n\}$ of the boundary value problem (19)–(29), while formula (43) gives an expression for the corresponding vertical normal modes in the water layer

$$\Phi_n(z) = \psi_1(z;\xi_n) - \frac{\psi_1(0;\xi_n)}{\psi_2(0;\xi_n)}\psi_2(z;\xi_n). \tag{45}$$

To solve the inhomogeneous differential Eq. (19) under the boundary conditions (21) and (42), we can construct Green's function of the boundary value problem if we introduce the functions into consideration $V(z;\xi)$ and $U(z;\xi)$—solutions (19) that satisfy condition (21) $V(0;\xi) = 0$ and condition (42) $U(h_0;\xi)+g(\xi)U'(h_0;\xi) = 0$. Then Green's function can be written as

$$G(z, z_0;\xi) = \begin{cases} \dfrac{U(z_0;\xi)V(z;\xi)}{W(\xi)}, & z < z_0 \\ \dfrac{U(z;\xi)V(z_0;\xi)}{W(\xi)}, & z > z_0 \end{cases} \tag{46}$$

where $V(z) = \psi_1(z;\xi) - \frac{\psi_1(0;\xi)}{\psi_2(0;\xi)}\psi_2(z;\xi)$; $U(z) = \psi_1(z;\xi) - \frac{\psi_1(h_0;\xi)+g(\xi)\psi_1'(h_0;\xi)}{\psi_2(h_0;\xi)+g(\xi)\psi_2'(h_0;\xi)}\psi_2(z;\xi)$;

$$W(\xi) = \det\begin{pmatrix} U & V \\ U' & V' \end{pmatrix} = \left(\frac{\psi_1(h_0;\xi)+g(\xi)\psi_1'(h_0;\xi)}{\psi_2(h_0;\xi)+g(\xi)\psi_2'(h_0;\xi)} - \frac{\psi_1(0;\xi)}{\psi_2(0;\xi)}\right)\det\begin{pmatrix} \psi_1 & \psi_2 \\ \psi_1' & \psi_2' \end{pmatrix}.$$

From the $W(\xi)$ formula it can be seen that this expression is equal to zero only at the eigenvalues $\{\xi_n\}$ of the boundary value problem (19)–(29). Indeed, given that $\psi_1(z;\xi)$ and $\psi_2(z;\xi)$ represent a fundamental system of solutions (19), one can see that.

$$\det\begin{pmatrix} \psi_1 & \psi_2 \\ \psi_1' & \psi_2' \end{pmatrix} \neq 0$$

On the other hand, it follows from (44) that

$$\frac{\psi_1(h_0;\xi)+g(\xi)\psi_1'(h_0;\xi)}{\psi_2(h_0;\xi)+g(\xi)\psi_2'(h_0;\xi)}-\frac{\psi_1(0;\xi)}{\psi_2(0;\xi)}=\frac{\Phi(h_0;\xi)+g(\xi)\Phi'(h_0;\xi)}{\big(\psi_2(h_0;\xi)+g(\xi)\psi_2'(h_0;\xi)\big)\psi_2(0;\xi)}$$

thus equality to zero $W(\xi)=0$ is equivalent to the dispersion Eq. (42).

It is known that the boundary value problem under consideration has a discrete spectrum, where the eigenvalues $\{\xi_n\}$—simple poles of Green's function $G(z, z_0; \xi)$. Then the potential of the water layer can be written as a contour integral on the plane ξ, where the contour C covers all singular points $G(z, z_0; \xi)$:

$$\mathbf{\Phi}(r,z)=\frac{1}{4\pi}\oint_C G(z,z_0;\xi)H_0^{(1)}(\xi r)\xi d\xi. \tag{47}$$

Calculating this integral as a sum of residues, we obtain an expression for the potential $\mathbf{\Phi}(r, z)$ as a sum of normal modes.

$$\mathbf{\Phi}(r,z)=\frac{i}{4}\sum_{n=1}^{\infty}\frac{\Phi_n(z)\Phi_n(z_0)}{\gamma_n}H_0^{(1)}(\xi_n r) \tag{48}$$

where $\gamma_n=\frac{1}{2\xi_n}\frac{dW}{d\xi}\Big|_{\xi=\xi_n}=\int_0^{h_0}\Phi_n^2(z)dz-\frac{\Phi_n^2(h_0)}{2\xi_n g^2(\xi_n)}\frac{dg}{d\xi}\Big|_{\xi=\xi_n}$.

4 Numerical Results

As an example, let us consider a marine acoustic waveguide with parameters typical for the Crimean coast: a water layer of depth $h_0 = 80$ m with a non-uniform sound velocity profile (see Fig. 6) lies on a layered base. The silt layer is modeled by a layer of liquid sediment with a sound velocity $c_0 = 1600$ m/sec and density $\rho_0 = 1500$ kg/m^3, which corresponds to the average values of silt rocks. Next, we consider an elastic layer with density $\rho_1 = 2200$ kg/m^3, longitudinal velocity of propagation of elastic waves $c_1 = 1800$ m/sec and transverse waves $c_2 = 1000$ m/sec. At the base we assume an absolutely rigid substrate at a depth of 200 m. A study of the spectrum of the boundary value problem was conducted $\{\xi_n\}$ depending on the thickness of the liquid sediment layer h_1. The table shows the values of horizontal wave numbers ξ, calculated using Newton's iterative method, at the frequency $f = 16$ Hz. The waveguide considered is a fairly typical example for the case of shallow water. Note that at a sufficiently high source frequency it contains a large number of undamped modes corresponding to the actual values of ξ. Variation of liquid precipitation layer thickness h_1 and, accordingly, the thickness of the elastic layer of sediments $H_1 = H-h_1-h_0$ for a fixed sedimentary layer thickness H, it shows that there are three propagating modes. Their values are given in Table 3. It can be seen that these values

Table 3 Real eigenvalues $\{\xi_n\}$ depending on the thickness of the liquid sediment layer

h_1	ξ_1	ξ_2	ξ_3
20	0.0538059	0.0441408	0.0256615
40	0.0464514	0.0421137	0.0171688
60	0.0539617	0.0400886	0.0281907
80	0.0573151	0.0535357	0.0370062
100	0.0526027	0.0369873	0.0326691

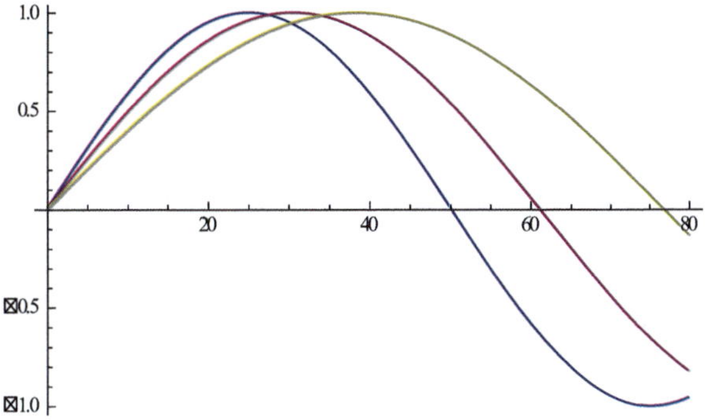

Fig. 8 First, second and third vertical normal modes for considered numerical example

depend nonlinearly on h_1, where there is a pronounced maximum at $h_1 = 80$ m and minimum at $h_1 = 40$ m (Table 3).

Note that, despite significant fluctuations in the first eigenvalues, the corresponding eigenfunctions calculated using formula (44) remain practically unchanged (Fig. 8).

Figure 9 shows the values of horizontal wave numbers ξ for the waveguide under consideration at a frequency of $f = 16$ Hz. Comparison of these results shows that an increase in the thickness of the elastic layer leads to the appearance of a larger number of complex values of ξ; in the extreme case (Fig. 9c), when almost the entire sediment layer is assumed to be liquid, the spectrum of eigenvalues consists of a finite number of real values and an infinite number of almost purely imaginary ones. In the liquid bottom model, this is exactly the distribution of wave numbers: a finite number of real ones corresponding to propagating modes and an infinite number of purely imaginary ones, which correspond to decaying modes. However, it should be noted that even minor variations in the horizontal wave numbers ξ lead to significant differences in the location of the vertical wave numbers μ_j characterizing the soil. In the presented model, they have both a real and an imaginary part, since the attenuation is caused by dissipative losses and energy leakage into the soil. The presence of purely imaginary μ_j is due exclusively to energy leakage into the ground.

As is known, the change in the field amplitude in a waveguide is characterized by the magnitude of the losses during propagation (intensity):

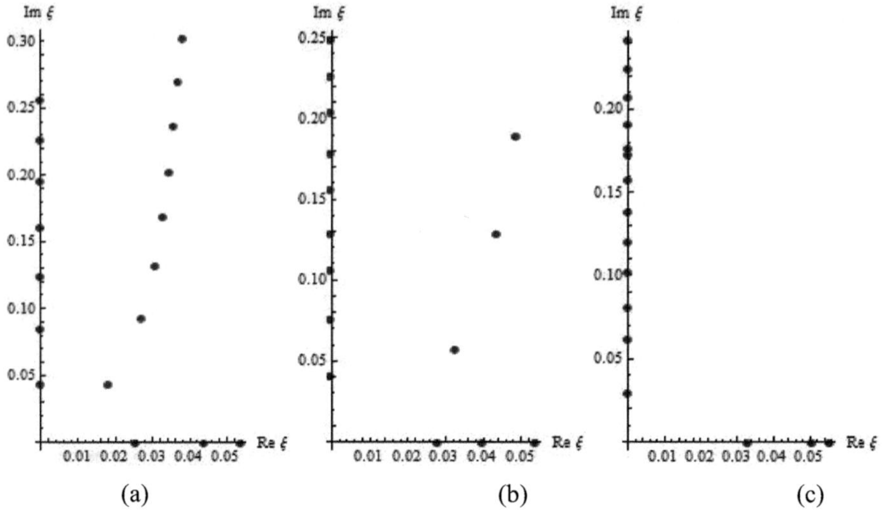

Fig. 9 Values of wave numbers ξ (a–$h_1 = 20$ m; b–$h_1 = 60$ m; c–$h_1 = 115$ m)

$$TL = -20\lg\left|\frac{\Phi}{\tilde{\Phi}}\right|,$$

where $\tilde{\Phi} = \frac{Qe^{ikr}}{4\pi r}$—the amplitude of the velocity potential created by a point source of sound at a distance $r = 1$ m, Q—volumetric vibrational velocity. Figure 10 shows the change TL (r, z) depending on the radial coordinate r at different depths—on the surface of the waveguide $z = 10$ m (normal line), in the middle of the waveguide $z = 40$ m (dashed line) and near the bottom $z = 75$ m (dot-dashed line). Source location $z_0 = 20$ m.

Figure 10 shows that all three graphs have similar fluctuations as r increases. The losses on the surface are minimal and maximal at the bottom of the waveguide.

Figure 11 shows the change in the amplitude of sound pressure $P(r, z)$ depending on r at a fixed depth value $z = z_{min}$. Note that it is at this depth that the axis of the underwater near-surface sound channel is located (see Fig. 7). The depth of immersion of the sound source was varied z_0 (normal line—$z_0 = 10$ m is located near the surface, dashed line—$z_0 = 33$ m is located on the axis of the underwater sound channel and dot-dashed line—$z_0 = 75$ m the source is located at the bottom). It can be seen that when the source is located on the axis of the sound channel, the amplitude of the sound pressure is the greatest, somewhat less when the source is located at the bottom and the least if the source is located at the surface.

For a visual assessment of the influence of the elastic layer on the amount of losses, Fig. 12 is presented, which shows the TL values for two cases: (a) with a silt layer thickness of $h_1 = 20$ m and elastic layer $H_1 = 100$ m, (b) at the thickness of the silt layer $h_1 = 115$ m and elastic layer $H_1 = 5$ m. Although the total precipitation

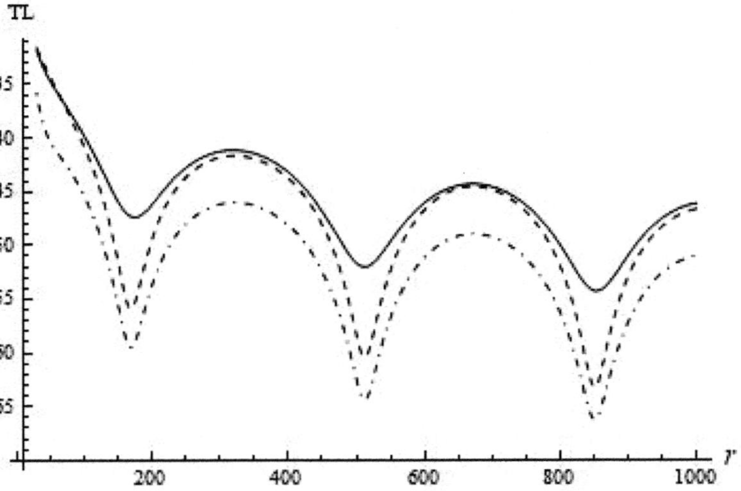

Fig. 10 Change in *TL* (r, z) at distance r for different values of z (10, 40, and 75 m)

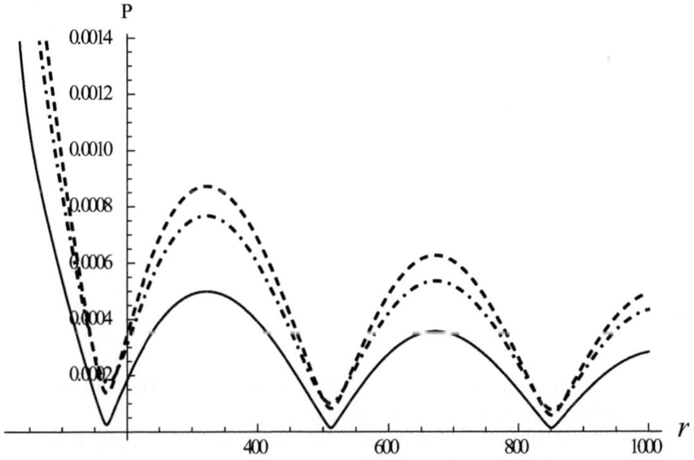

Fig. 11 Change in $P(r, z_{min})$ at distance r for different values of z_0 (10, 33, and 75 m)

thickness remains the same, significant differences in the distribution of sound losses can be seen.

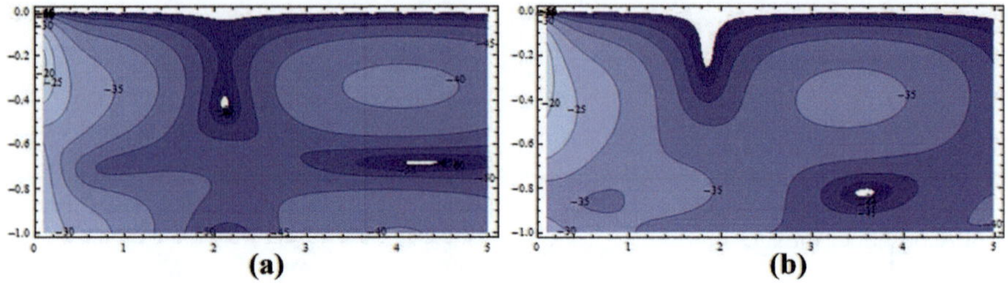

Fig. 12 Change in $TL\ (r, z)$ for different depths of liquid layer

5 Conclusions

Thus, in the work, the analysis of in situ measurements of the speed of sound in the coastal waters of the Black Sea serves as the foundation of theoretical research. To describe the propagation of sound in this water area, it is proposed to use a flat-layered model of an acoustic marine waveguide with a non-uniform profile of the speed of sound. The bottom of the waveguide is modeled as consisting of a layer of liquid sediments and an elastic substrate. An exact analytical solution to this problem is constructed in the form of a sum of normal modes. This paper presents numerical examples that show the influence of the bottom layers of a waveguide on its characteristics.

Acknowledgements This work was supported by the Russian Science Foundation, project No. 24-27-20021.

References

Achenbach JD (1975) Wave propagation in elastic solids. North-Holland Series in Applied Mathematics and Mechanics. Elsevier, Amsterdam

Ballard MS, Sagers JD (2019) Measurements and modeling of acoustic propagation in a scale model canyon. J Acoust Soc Am 146(3):1858–1866

Brekhovskikh LM, Godin OA (1989) Acoustics of layered media, M. Nauka

Buckingham MJ (2020) Wave speed and attenuation profiles in a stratified marine sediment: geo-acoustic modeling of seabed layering using the viscous grain shearing theory. J Acoust Soc Am 148(2):962–974

Chapman C, Hobro J, Robertsson J (2010) Elastic corrections to acoustic finite difference simulations. SEG technical program expanded abstracts 2010. Society of exploration geophysicists, pp 3013–3017

Grigoriev VA, Petnikov VG, Shatravin AV (2017) Sound field in a shallow-water arctic-type waveguide with a bottom containing a gas-saturated sediment layer. Acoust Phys 63:433–448

Grili S, Pedersen T, Stephanisen P (1998) A hybrid boundary element method for shallow acoustic propagation over an irregular bottom. Eng Anal Boundary Elem 21:131–145

Jensen FB, Porter MB, Kuperman WA, Schmidt H (2011) Computational ocean acoustic. Springer-Verlag, N.Y.

Jensen FB, Ferla CM (1990) Numerical solutions of range-dependent benchmark problems in ocean acoustics. J Acoust Soc Am87(4):1499–1510
Knighly GH, Lee D, Mary DF (1987) Higher-order parabolic wave equation. J Acoust Soc Am 2:580–587
Lin YT, McMahon KG, Lynch JF, Siegmann WL (2013) Horizontal ducting of sound by curved nonlinear internal gravity waves in the continental shelf areas. J Acoust Soc Am 133(1):37–49
Luo W, Schmidt H (2009) Three-dimensional propagation and scattering around a conical seamount. J Acoust Soc Am 1:52–65
Papadakis J, Keller J (1980) Waves propagation and underwater acoustics. Mir, Moscow
Paul C (2018) Etter. CRC Press, Underwater Acoustic Modeling and Simulation
Pekeris CL (1948) Theory of propagation of explosive sound in shallow water. Geol Soc Am Mem 27:1–117
Press F, Ewing M (1950) Propagation of explosive sound in a luquid layer over-lying a semi-infinite elastic solid. Geophysics 15:426–446
Schneiderwind DJ, Collis JM, Simpson HJ (2012) Elastic Pekeris waveguide normal mode solution comparisons against laboratory data. J Acoust Soc Am132:EL182
Sertlek HÖ, Ainslie M (2014) A depth dependent formula for shallow water propagation. J Acoust Soc Am136:573–582
Shenderov EL (1972) Wave problems of hydroacoustics, L.: Shipbuilding
Tappert FD, Hardin RH (1974) Computer simulation of long-range ocean acoustic propagation using the parabolic equation method. In: 8th International Congress on Acoustics, Contributed papers. London 5(2):452
Vendhan CP, Diwan GC, Bhattacharyya SK (2010) Finite element modeling of depth and range dependent acoustic propagation in ocean waveguides. J Acoust Soc Am 127(4):3319–3326

In reality, this is not the case. In connection with the discovery of the phenomenon of jet outflow of gas and dust matter from center of spiral galaxies and creation of Galactocentric paradigm (Barenbaum 2010), it turns out that as Sun moves in Galaxy, the Solar System and its planets are quasi-periodically subjected to powerful bombardments by galactic comets (Barenbaum 2023a, b), and the observed comets have nothing to do with the primary gas and dust nebula (Barenbaum 2024). Therefore, the current cosmogony ideas about origin and evolution of Solar System and its objects need to be revised and adjusted accordingly (Barenbaum 2002, 2010).

The article presents a cosmogonic concept of the formation of Solar System, its planets, and their satellites, different from the Kant–Laplace hypothesis, called the "Cosmogony of Open Solar System (COSS)" concept (Barenbaum 2002). The COSS concept, being an important component of Galactocentric paradigm, is intended to complement and develop Kant–Laplace cosmogonic scheme, taking into account the phenomenon of jet outflow of matter in our Galaxy.

The COSS concept is based on two postulates (Barenbaum 1992a, b):

1. The Sun and planets represent an "open" system of cosmic bodies, subject to the powerful influence of galactic comets, which bring large portions of matter and energy into it.
2. The event that occurred 4.567 billion years ago was not the first, but the second main cycle of planet formation in Solar System, caused by death of planet Phaeton, which existed earlier, according to the hypothesis of G. Olbers (Simonenko 1985), in the place of modern asteroid belt.

The first postulate has been convincingly argued in many of the author's articles (see, for example, Barenbaum 2015, 2023a, b, 2024; Barenbaum and Shpekin 2016, 2019, 2021).

Therefore, the main goal of this article is to prove the second of these provisions. This goal, in the author's opinion, is achieved by solving two interrelated problems. The first is to provide new convincing arguments and facts in favor of the former existence of Phaeton based on new factual material using data from meteoritics, cosmochemistry, and geophysics. And the second is to offer concerted explanations of three problematic issues based on the assertion about of Phaeton death: (1) the origin of giant planets, (2) the nature of meteoritic matter, and (3) the cause of emergence and mechanism of the Moon formation. In the modern cosmogony of Solar System, these issues are resolved independently and do not yet have a satisfactory explanation.

2 Phaeton Problem

The idea of Phaeton is historically connected with the problem of the origin of the asteroid belt. Are asteroids fragments of a real planet or so-called planetesimals, that is, "embryos" of a planet that never formed for various reasons?

The attitudes to this issue in modern cosmogony and in COSS concept are opposite. The former stands on the positions of planetesimal hypothesis (Safronov 1969;

Safronov and Ziglina 1990; Dorofeeva and Makalkin 2004; Kuskov et al. 2009; Makalkin et al. 2019), while in COSS concept the existence and destruction of Phaeton is one of the key positions.

The prehistory of the Phaeton hypothesis in modern science is closely connected with the name of I. Kepler (1571–1630). Based on some ancient books, Kepler suggested in 1595 that the number of planets in Solar System exceeded the 5 planets known at that time, i.e., there was another undiscovered planet between Mars and Jupiter. In 1796, at an astronomical conference in Gotha, it was decided to begin searching for "Kepler's planet." The results were not long in coming. In 1801, G. Piazzi (1746–1826) discovered the first "minor planet," the asteroid Ceres, at a distance of 2.8 AU from Sun, and in 1802, G. Olbers (1758–1840) discovered a second asteroid, Pallas, near Ceres. In 1804, in a letter to J. Bode (1747–1826), Olbers, discussing this fact, suggested that these planets were "...only fragments of a previous large planet, blown up by some catastrophe" (Simonenko, 1985). He suggested searching for other fragments of the planet. Further research showed that he was right. In 1804, K. Harding (1765–1834) discovered the third asteroid, Juno, and in 1807, Olbers himself found the fourth asteroid, Vesta.

Over the next century and a half, the Phaeton hypothesis determined the development of scientific thought for many generations of researchers. Based on this hypothesis, many other asteroid bodies were discovered, most of which are concentrated between the orbits of Mars and Jupiter within a single toroidal ring. This hypothesis also played an important role in understanding the nature of "heavenly stones"—meteorites. After the nineteenth century, when the fact of meteorites falling to Earth was officially recognized (Eremeeva and Tsitsin 1989), they began to be considered fragments of a dead planet.

The study of meteorites from this point of view led the French geologist J. Dobré (Rezanov 2004) in 1861 to another fruitful idea, subsequently confirmed, that Earth has an iron core and a system of silicate shells surrounding it.

However, in the early 1960s, the attitude to the hypothesis of the former existence of Phaeton changed dramatically. This idea gave way to the planetesimal hypothesis, which considers asteroids not as fragments of a previously existing planet, but as embryos of a planet that failed to form. The reason for revising the previous ideas was two main circumstances (Proceedings 1951): (1) the absence of an obvious cause for Phaeton death and (2) the difficulty of explaining the chondrite structure of meteorites from the standpoint of the destruction of one planet.

Over the next two decades, the planetesimal hypothesis developed successfully without encountering any difficulties (Safronov 1969). However, in the mid-1970s, a more detailed study of the isotopic composition of meteoritic matter raised a number of very serious questions about this hypothesis. Thus, Sobotovich (1974) established that the isotopic age of one of the groups of iron meteorites exceeds the time of formation of Solar System according to the Kant–Laplace hypothesis by at least 1 billion years. A difference between the "magnesium" and "xenon" ages of meteorites has also been discovered. A number of other cosmochemical problems arise that remain unanswered (Sobotovich and Semenenko 1984, 1985).

Therefore, the planetesimal hypothesis does not currently appear as flawless and correct as it was thought half a century ago. The discovery of a jet outflow in Galaxy has further exposed its shortcomings. In light of this discovery, the arguments against the existence of Phaeton, which previously seemed convincing, not only do not appear so today, but are simply erroneous (Barenbaum 1992a, b; Rezanov 2004). Moreover, it is now possible to substantiate Phaeton hypothesis more broadly and in more detail than half a century ago (Daly 1943; Orlov 1949; Zavaritsky 1951; Fesenkov 1952; Putilin 1953; etc.).

2.1 *Justification of the Phaeton Hypothesis in COSS Concept*

The COSS concept suggests that all planets, including Phaeton, arose simultaneously with Sun in Crux-Scutum galactic arm (Barenbaum 2023a, b). After condensation, Sun moved to an independent elliptical orbit, the line of apses of which slowly rotates in the direction of Sun's orbital motion. In the process of this motion, Sun quasi-periodically crosses spiral arms and jet streams of Galaxy, represented by gas and dust matter, as well as the products of its condensation—dense molecular clouds, comets and stars. Each time at the moments of such intersections Solar system is subjected to bombardments by galactic comets, the intensity of which increases sharply in the star-formation zones of galactic arms in the aphelion section of Solar orbit, coinciding with Galaxy corotation radius.

According to the COSS conclusion, one such event occurred of 4.567 billion years ago, which led to Phaeton destruction. It is assumed that the reason for Phaeton death could have been the passage of Sun through a massive cloud of stars or a close flyby of a jet stream star near Sun. Similar interactions of Solar System with stars occurred later (Barenbaum 2010), but no longer resulted in the destruction of planets.

This interaction disrupted the stability of Phaeton figure, causing its fragmentation to begin. It is known (Savchenko 1975) that in order to destroy a planet in this way, it is necessary to impart a relatively small impulse capable of creating an instantaneous pressure equal to the pressure in its center. In this case, not much energy is transferred to planet, so the overwhelming majority of its fragments remain in original orbit, forming a ring.

The death of Phaeton, with its transformation into a pile of debris that became objects of collisions with galactic comets, turned out to be a completely extraordinary event in Solar System, which initiated a repeated cycle of planet formation.

We will substantiate this conclusion with the following calculation (Barenbaum 1992a). Let us take the density of comet flux in the jet streams of Galaxy of 4.567 billion years ago to be equal to its present value (Barenbaum and Shpekin 2021), let us set the mass of Phaeton to be equal to 2.5 Earth masses and assume that most of its debris, having formed a ring, remained in orbit. Then, given the cross section of collisions of galactic comets with asteroids (Barenbaum 2024), it is easy to calculate that during the first ~ 100–300 million years after Phaeton death, a total number of galactic comets comparable in mass to the Sun could have broken up.

The scattering of such a large amount of new cosmic matter in asteroid belt, along with the growth of Sun's mass, inevitably had to stimulate three more important processes that occurred in different parts of Solar System simultaneously (Barenbaum 1992a):

–fragmentation and intensive mixing of the material of galactic comets and asteroids during their collisions in the ring itself;
–mass ejection from the zone of the ring of cometoasteroids and fragments of deep internal rocks of Phaeton, bombarding the surface of all planets, and
–carrying out from the ring of enormous amount of evaporated gases and dust captured by planets external to the belt.

The composition of meteorites that have fallen to Earth indicates the first process, and a number of patterns that have developed at present in Solar System indicate the widespread development of the last two processes (Fig. 1).

At the initial stage of Phaeton's fragmentation, when the number of its fragments in the belt was significantly greater than today, the activity of these processes was also higher. However, subsequently asteroid belt lost its bodies, and the intensity of these processes has sharply decreased by now.

Among such regularities, we note two. Firstly, the peculiar nature of the change with distance from the asteroid belt of the mass and average density of planets, which allows us to divide all planets into "inner"—Earth-type and "outer"—giant planets.

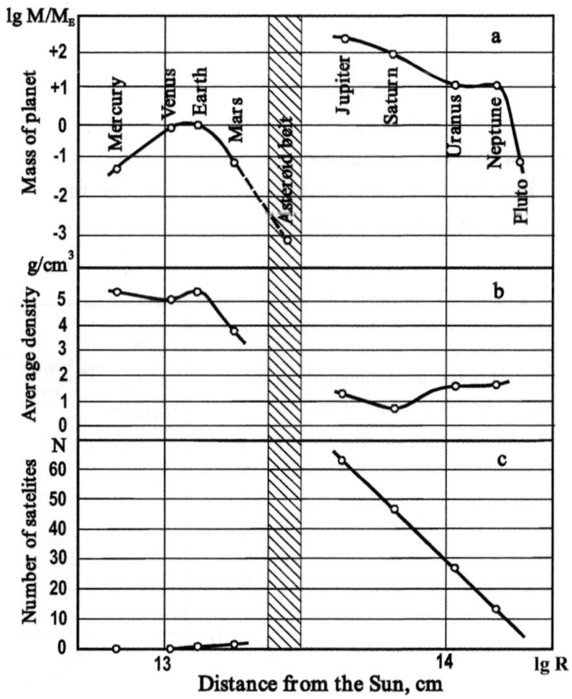

Fig. 1 Dependence on the heliocentric distance: **a** the masses of planets, **b** the average density of their matter, **c** the number of satellites with a diameter of more than 1 km according to data (Astronomy 2007)

And, secondly, the presence of similar families of rings on the outer planets, as well as satellites and comets, the number of members of which decreases with distance from the asteroid belt.

The COSS concept allows that the formation of planets, unlike their satellites and other small bodies of Solar System, occurred, at least, in two stages. In the first, the planets themselves arose. And in the second, after the Phaeton death, arose their satellites, asteroids, and comets.

An illustration of this conclusion can be the presence in Solar System today of a non-stationary “cloud” of comets, which arose during the last stay of Sun in the Orion–Cygnus jet stream (Barenbaum 2023a, b). There is no doubt that the formation of cometoasteroid bodies and meteorites, as well as their ejection from the belt along with gas and dust, occurred in all epochs of Sun’s stay in Galaxy’s jet streams. At the initial stage of Phaeton’s fragmentation, when the number of its fragments in the ring was much greater than today, the activity of these processes was also higher. However, later the asteroid belt lost its bodies, and the intensity of these processes has sharply decreased by now.

2.2 *Two Stages of Planet Formation and Separation*

The initial masses of the planets, as well as the changes in these masses after Phaeton destruction, can be determined with sufficient accuracy for our subsequent calculations using the method shown in Fig. 2.

Here we have presented the dependence of the distributed mass of Solar System substance (Alfvén and Arrhenius 1979) on the heliocentric distance. The distributed mass is defined as $\rho_i(R_i) = M_i/R_i^3$, where M_i and R_i are the mass and distance from Sun of individual planets.

The dividing line that allows us to find the masses of the planets at the first stage of formation is a straight line passing through the points Venus, Earth, Uranus, and

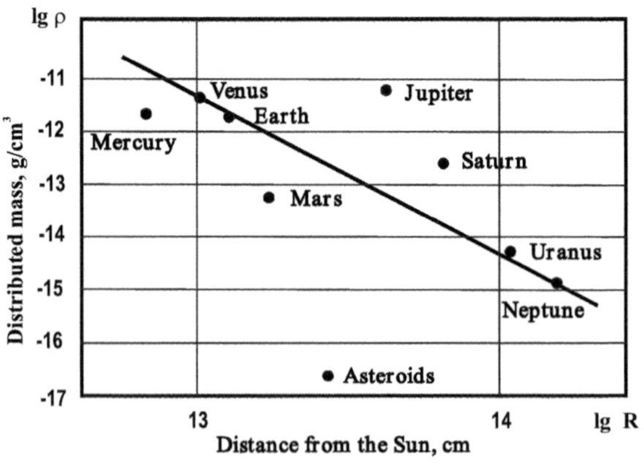

Fig. 2 Distributed mass of solar system matter: the straight line determines values of mass of all planets after the first formation stage

Neptune. This line is drawn in such a way so as to satisfy three requirements at once: (1) the constancy of Venus and Neptune masses $M_i = M_i'$ before and after Phaeton destruction; (2) the power dependence of function $\rho(R)$ at the first stage of planet formation; and (3) the exponential decrease in mass of matter $\Delta M_i = M_i' - M_i$, captured by planets external to the belt, as they move away from it (Fig. 2).

The masses of the planets before Phaeton destruction M_i', calculated on the basis of such an interpretation of data in Fig. 2, as well as the absolute $\Delta M_i = M_i' - M_i$ and relative M_i'/M_i changes in these masses compared to the modern values of M_i are given in Table 1.

Table 1 assumes that initially the masses of all planets except Pluto with distance from the Sun corresponded to the general power dependence inherited by them from the protoplanetary stage. However, after Phaeton death, many planets underwent changes that affected their mass, structure, and average composition of matter. Venus and Neptune, which were at a fairly large distance from Phaeton, practically did not change their masses. These two planets are adjoined by Earth and Uranus, which changed them relatively little.

Unlike these four planets, all the other planets of Solar System changed much more after Phaeton death, which led to a shift in their positions in Fig. 2 from the dividing line. The upward deviation from this line was associated with the capture of new matter by planets—mainly comet gases and asteroid fragments evaporated and ejected from the ring, and the downward shift, on the contrary, was due to the loss of their own mass by planets.

Thus, while the outer planets (Jupiter, Saturn, and Uranus) were gaining new matter and acquiring satellite systems, the inner planets, with the exception of Venus, were losing their own mass. If we do not consider Phaeton, which ceased to exist as a planet, then the planet that underwent the greatest changes was Mars, which is closest to Phaeton. Mars, like Phaeton, was apparently completely destroyed, but was then restored from its own debris with the participation of new cosmic matter,

Table 1 Masses of planets before and after Phaeton death, 10^{27} g

Planet	Contemporary mass M_i	Initial mass M_i'	Change of mass ΔM_i	Mass ratio M_i/M_i'
Mercury	0.333	2.88	−2.56	0.116
Venus	4.87	4.87	0	1
Earth	5.97	6.37	−0.4	0.937
Mars	0.642	8.96	−8.32	0.072
Phaeton	0.0042	14.75	−14.75	$2.85 \cdot 10^{-4}$
Jupiter	1899	24.43	+ 1874.6	77.72
Saturn	568	40.06	+ 528	14.18
Uranus	87.2	70.72	+ 16.5	1.23
Neptune	102	102	0	1
Pluto	0.66	–	–	–

already as a planet with a mass of ~ 7.2% of its original value. Mercury also received severe damage, and all the other planets closer to Sun, if such existed in Solar System before Phaeton destruction, have not survived to this day. The destruction of Phaeton did not pass without a trace for Earth, which lost approximately 6.3% of its mass.

In the COSS concept, all the indicated changes in the masses and composition of the planets after Phaeton's death are in certain quantitative relationships with each other.

Let us analyze these relationships in more detail for the inner and outer planets, including Phaeton itself, subordinating our consideration to three questions that are interconnected in the COSS concept: (1) origin of the giant planets, (2) nature of meteoritic matter, and (3) formation of the Moon.

Note that in modern cosmogony these questions are resolved independently and do not yet have a satisfactory explanation.

3 Origin of the Giant Planets

The COSS concept suggests that as a result of Phaeton destruction, a secondary gas–dust disk of Sun arose on the outer side of modern asteroid belt, from the substance of which the giant planets were formed. At the same time, before the disk appeared, current planets: Jupiter, Saturn, Uranus, and Neptune had masses M_i', indicated in Table 1. Once inside Sun's secondary disk, the first three planets acquired a lot of new cosmic material, which became their shell, and they themselves turned into central cores of modern planets.

As a result of the exponential decrease in disk density with distance from asteroid belt, the closer the planet was to asteroid belt, the greater mass of gas and dust matter it captured. Because of its greater distance from the belt, Neptune captured very little of this matter, Uranus captured more, and Saturn and especially Jupiter captured the most.

To substantiate the idea of a two-stage formation of giant planets (Cameron 1980), Table 2 compares the physical models of internal structure of Jupiter, Saturn, Uranus, and Neptune developed by V. Zharkov and colleagues (Gudkova et al. 1988a, b) and our model (Barenbaum 2002), based on the data in Table 1.

The models of Jupiter, Saturn, Uranus, and Neptune were constructed by V. Zharkov et al. in such a way that they corresponded to the values of masses, radii, periods of axial rotation, as well as the second and fourth gravitational moments of each of the planets known from observations. From the calculations of V. Zharkov et al. it follows that giant planets have at least a two-layer structure. They consist of a central solid core and a surrounding gas–liquid convective shell. Along with gases H_2, He, Ne, etc. (G-component), planets should contain some amount of ices of substances of medium volatility H_2O, CH_4, NH_3, etc. (I-component), as well as heavy non-volatile compounds mainly in the form of oxides of Si, Mg, Fe, unoxidized iron, and Fe–Ni alloy (S-component). According to calculations, depending on the condensation temperature of these substances, the composition and ratio of

Table 2 Internal structure and composition of the giant planets

Main parameters of planets	Jupiter		Saturn		Uranus		Neptune	
	[1]	[3]	[1]	[3]	[2]	[3]	[2]	[3]
Mass of SI-core	5.0	4.1	5.7	6.7	13.72	11.84	16.56	17.07
	18%		17.5%		15%		3%	
Mass of shell	312.8	313.7	89.3	88.4	0.88	2.76	0.51	0
	0.3%		1.0%		–		–	
Component composition of shell (%)	S—3.7 I—16.0 G—80.3		S—1.4 I—24.0 G—74.6		S + I—10 G—90		S + I—10 G—90	

Note The masses of core and shell are expressed in units of Earth mass. References: [1] Gudkova et al. (1988a, b); [2] Gudkova et al. (1988a); [3] Barenbaum (2002). Percentages show deviations of both calculations separately for core and shell. Designations: S, I, G—primary composition of matter (see text)

ice and gas components in the convective shell of planets can change, and the central core will contain different amounts of the L-component.

Physical models lead to the conclusion that the S-substance, which is part of giant planets, along with the central core, should be present in a dispersed form and in a gas–ice shell. When moving from Jupiter to Saturn and further to Uranus and Neptune, the mass of central core and its share in the composition of planets increases, and the content of S-substance in shell decreases. Table 2 also shows the masses of core and shell of Jupiter, Saturn, Uranus, and Neptune, which we found under the assumption that after the Phaeton death, these planets, having a mass of M_i', acquired ΔM of new cosmic matter, which became their shell.

Analysis of the data in Table 2 shows that the two-stage mechanism of giant planets' formation postulated by COSS explains well the features of material composition and internal structure of these planets. Although there are deviations from the "exact" physical models of giant planets, in all cases they are small. For the most massive shell, these deviations are 0.3% for Jupiter, 1% for Saturn, 15% for Uranus, and 3% for Neptune. The largest discrepancies are characteristic of the SI-cores of giant planets, but they do not exceed 18%.

In this regard, we note that the models of Gudkova et al. (1988a, b) are numerical solutions of the inverse problem, which consists in determining main characteristics of internal structure of giant planets based on their external physical parameters. Such solutions, as is known, are not always unambiguous and even more so accurate.

In our opinion, such good agreement between the masses of the shells, obtained on basis of two fundamentally different approaches, cannot be a random coincidence. On the one hand, it indicates a sufficiently high accuracy of the theoretical calculations (Gudkova et al. 1988a, b), and, on the other hand, it can be considered an important argument in favor of the correctness of Cameron's idea (Cameron 1980) about the two-stage formation of giant planets, implemented in the COSS concept. This conclusion is currently the most plausible (Kuskov et al. 2009).

Given the formation method of Jupiter, Saturn, and Uranus, their cores must be planets that formed during the first stage of Solar System evolution. It should be noted that the idea that cores of giant planets may be ordinary terrestrial planets was first put forward by O. Schmidt (1957).

The results of calculations by Zharkov's group, interpreted from standpoint of COSS concept, allow us to note another important circumstance. It consists in the fact that among the substance captured by Jupiter and Saturn, which subsequently became part of their shells, a significant proportion may be rocky fragments of Phaeton. According to the data in Table 2, as the distance from asteroid belt increases, the number of such fragments decreases from 3.7% in Jupiter's shell to 1.4% in Saturn's shell.

4 Nature of Meteoritic Matter

The most convincing evidence of the former existence of Phaeton is contained in meteorites (Barenbaum 1992a, b). In the history of the formation of meteoritic matter, at least three stages can be distinguished: (1) the stage of chemical differentiation in the depths of Phaeton, which lasted more than ~ 0.5–1.5 billion years, (2) the time of intensive destruction of Phaeton ~ 100–300 million years, and 3) the stage of asteroid evolution, which has not ended to this day.

There are good reasons to believe that the collisions of galactic comets with Phaeton's debris could not completely neutralize the influence of primary differentiation of its deep rocks on the mineral and chemical composition of the matter of asteroids and meteorites. Therefore, meteorites falling to Earth carry information not only about the physical and chemical changes in Phaeton's rocks that followed its destruction, but also about the processes that participated in formation of composition of this matter even before Phaeton destruction.

The issues of the formation of asteroid and meteorite matter before and after the Phaeton death are beyond the scope of this article. Therefore, they are touched upon below only to the extent necessary for understanding the conditions of formation and evolution of our own planet, as well as the processes that may be occurring on it at the present time.

The main goal of our consideration is to identify the connection between shell structure of modern Earth and meteorite matter, if the latter is considered as matter from internal zones of Phaeton, which changed in a certain way during collisions with galactic comets and interrupted own evolution at a comparatively early stage.

4.1 Meteorite Classes and Earth Structure

Table 3 compares the probability of meteorites of the main classes and groups falling on Earth according to data (Sears and Dodd 1988) with the volumes of the internal shells of our planet according to the chondrite model Hart et al. (1977).

This model takes into account the known values of the mass and rotational moment of Earth, the frequency spectrum of its natural oscillations, as well as the propagation speeds of longitudinal and transverse seismic waves at different depths, right down to the center of planet. These parameters are currently measured for Earth with such accuracy that they impose strict restrictions (Zharkov 1983) on the possible chemical and mineral composition of rocks, as well as the internal structure of our planet body.

The model of R. Hart et al. is based on the assumption that Earth is entirely composed of carbonaceous chondrite matter, i.e., more than 90% consists of five chemical elements: oxygen (32%), iron (25%), silicon (14.5%), magnesium (12.5%), and sulfur (8%).

It is known (Brown and Mussett 1984) that if these elements are given the opportunity to form compounds under conditions of high temperatures and pressures, then, taking into account their ratio in the mixture, chemical affinity and electro-negativity properties, they should form three successive layers in the planet body. The upper layer is lithophilic with a predominance of Mg and Fe silicates, the middle layer is chalcophilic, represented by Fe–sulfide, and the inner layer is siderophilic from metallic iron.

As a result, the outer zone of Earth will have a predominantly silicate composition, followed by a layer rich in sulfides, and below it, excess iron accumulates, forming the

Table 3 Probability of meteorite falls in comparison with volumes of Earth main internal shells

Number of group	Class of meteorites	Percent of falls	Shell of Earth (lower boundary, km)	Volume of shell, %
0	Carbonaceous chondrites (C)	4.3	–	–
I	Achondrites	7.2	Earth's crust (33)	1.55
II	Amphoterites (LL)	8.0	Upper mantke (420)	16.96
III	Olivine-hypersthene (L)	38.8	Transition zone (670)	9.84
IV	Olivine-bronzite (H)	33.0	The same (1050)	13.39
V	Enstatite (E)	1.6	Lower mantle (2885)	41.87
VI	Siderolites	1.1	External nucleus (5155)	15.69
VII	Siderites	5.1	Internal nucleus (6371)	0.7

central core. All other unaccounted metallic (M) and non-metallic chemical elements are distributed among these layers in accordance with their electrochemical properties. During the entire active stage of Earth's development, equilibrium reactions of the following type must occur in its body:

$$\begin{aligned} &M - \text{sulfide} + Fe - \text{silicate} \Leftrightarrow M - \text{silicate} + Fe - \text{sulfide};\\ &M + Fe - \text{sulfide} \Leftrightarrow M - \text{sulfide} + Fe \end{aligned} \quad (1)$$

These reactions gradually sort the specified categories of elements, distributing them into corresponding zones. According to modern models reflecting the presented ideas (Zharkov 1983), our Earth is covered with a comparatively cold and rigid crust to depths of 30–40 km on continents and 5–10 km under oceans, which isostatically "floats" on a layer of partially molten rocks of olivine-pyroxene composition. At a depth of 420 ± 20 km, at a pressure of P = 12 ± 1 GPa and a temperature of T = 1600 ± 50°C, a phase transition of olivines into a β-spinel structure occurs, accompanied by an increase in rock density by 7%. At the boundary of 560–700 km, where P = 20 GPa and T = 2200°C, further compaction of the silicate mineral lattice occurs, associated with their transition to post-spinel modifications. And even lower, at a depth of 1500 km at P = 40 GPa, all silicates acquire a pyrovskite structure, which remains stable right up to the boundary of Earth's core.

At high temperatures and pressures, Fe and Mg replace each other in framework of rocks. Therefore, according to reaction (1) and due to the peculiarities of phase diagrams of FeMg-silicates, we should expect their regular depletion of iron with depth and the release of the latter, as a less refractory and more mobile and heavy component, into a separate phase migrating to center of planet. In this case, iron is capable of dissolving and entraining all available sulfur (Brown and Mussett 1984).

At a depth of 2885 km, where pressure is ~ 140 GPa and temperature reaches 1800–3900°C, the boundary of metallic core is located. At this mantle-core boundary, density of rocks increases sharply from 5.5 to 9.9 g/cm^3. The substance of outer core, represented by the FeS + Fe system, is in a molten state, and below the boundary of 5155 km it becomes solid. The density of inner core, equal to 12.6–13.0 g/cm^3, cannot be provided by Fe alone and requires the presence of 10–20% Ni, which forms an alloy with iron.

The wide range of possible temperatures allowed for Earth's core is due to the difficulties in accounting for the content and distribution of sulfur in outer zone of core. This is due to the fact that the FeS + Fe system forms a eutectic mixture with a low melting point. This mixture is the first to enter the melt. The sulfur content in the eutectic at the core-mantle boundary is estimated to be ~ 17.5% and decreases to ~ 15% toward boundary of outer and inner cores (Usselman 1975).

Such ideas about the structure of the Earth's interior, according to D. Brown and A. Mussett, correspond well to empirical data, requiring clarification only for the intervals of chemical and mineralogical changes in rocks. This is, first of all, Earth's crust and the first 1000 km of mantle, the lower mantle region at a depth of 2780–2885 km above boundary with core, as well as the zone between outer and inner cores at a depth of 4590–5155 km.

It is known that for lower mantle this model systematically underestimates the density of rocks by 3–5%. As Press (1968) showed, this shortcoming of the model can be eliminated if we abandon the assumption of chemical constancy of mantle and admit that the ratio of iron to magnesium changes with depth from 0.25 in the upper part of mantle to 0.6 in its lower part. According to experimental data (Zharkov 1983), such changes should be expected, first of all, at the depths of phase transitions in mantle silicates.

4.2 *Phaeton and Earth*

The above ideas about the structure of Earth and patterns of change in its rocks with depth are confirmed by the analysis of meteorites, if we consider their main classes as substance of different internal shells of Phaeton, although to some extent changed by the processes that followed its destruction. To make this conclusion obvious, we will pay attention to the fact that all classes of meteorites in Table 3 were arranged from top to bottom in accordance with the rules of Prior (1920), i.e., in order of increasing total iron content and decreasing Fe content in silicate part of rocks.

It is clearly visible that each of the main classes of meteorites, with the exception of carbonaceous ones, has its own terrestrial analogue in chemical composition: the heterogeneous group of magmatic strongly brecciated achondrites corresponds to Earth's crust; stony chondrites with very low (LL), low (L) and high (H) Fe content, as well as enstatite (E) meteorites—respectively, rocks of upper, middle and lower mantle, and stony-iron and iron meteorites—material of outer and inner core. Undifferentiated carbonaceous chondrites have no terrestrial analogue and are therefore assigned to zero place. Meteorites of this class are represented by gas-dust weakly altered low-temperature cometary condensate (Barenbaum 1990). They were formed under different conditions and later than meteorites of other types.

A clear similarity can also be traced in structure of cores of Earth and Phaeton. Evidence of presence of liquid and solid phases in metallic core of Phaeton is presence among iron meteorites of two their types, differing in iron composition and age (Sobotovich 1974). Meteorites of the first group FeI with the most primitive lead, represented by octahedrites (6–14% Ni), have an age close to the solidification time of most chondrites 4.6–4.3 billion years. Meteorites of the second group FeII contain radiogenic lead, are characterized by an increased range of changes in contents of Ni = 4–26% and Pb = 10^{-8}–10^{-7}, and are 0.3–1.9 billion years older than FeI meteorites (Sobotovich 1974). According to (Sobotovich and Semenenko 1984, 1985), the presence of radiogenic Pb in FeII meteorites cannot be attributed to either the decay of uranium, the concentration of which is many orders of magnitude less than required or to their contamination with terrestrial lead when it reaches Earth.

Interpretation of these facts in accordance with the scheme of Earth formation (Vinogradov 1974) allows us to assume that formation of Phaeton core began no later than 6 billion years ago (Barenbaum 2002). The process of core growth occurred under adiabatic conditions at temperatures and pressures insufficient for the complete melting of its material. The initial basis of solid core was probably FeII iron, strongly and unevenly enriched in many siderophilic elements: Ni, Co, P, etc. Unlike the FeII material, FeI iron entered the core later as a product of processing and differentiation of the original substance, which had undergone melting and separation into metallic and silicate phases. Simultaneously with the increase in volume of liquid core of Phaeton, all five other mantle shells were also forming.

A comparative analysis of the data (Table 3) also allows us to formulate a number of general considerations regarding the nature of the evolution of interiors of terrestrial planets after accretion. Assuming the frequency of meteorite falls on Earth to be proportional to volumes of corresponding internal shells of Phaeton and comparing these data with the modern structure of Earth according to geophysical data, we can conclude that differentiation of planetary interiors continues actively after the formation of planets, and proceeds in the direction of a decrease in size and volume of the external and an increase in internal shells of planets.

This process, according to chemical transformations (1), is accompanied by melting of Fe and S from silicate minerals, which migrate to the center of planet. The migration of both elements through the thickness of rocks is facilitated by their combination into the structure of troilite, which enters the core through separate, probably permeable zones.

We find signs of high activity of migration processes in the Phaeton depths and, above all, in its lower mantle and outer core, directly in the composition of meteoritic matter. In our opinion, the division of enstatite meteorites into EH and EL types (Sobotovich and Semenenko 1984), which differ significantly in the content of Fe, S, P and other elements, speaks of intensive migration of substance in Phaeton lower mantle. And the specific variety of iron meteorites—sorotites, represented by a mixture of FeNi alloy of octahedrite structure with troilite, testifies in favor of the same processes in core of Phaeton. Both the content of troilite in sorotites (25–47%) and the content of Fe and S in high-iron EH meteorites are entirely consistent with composition of the eutectic melt in outer core of Earth (Brown and Mussett 1984).

According to modern concepts, the process of differentiation of matter continues today in body of our planet. Judging by geophysical data, it is mainly localized in boundary layer (2780–2875 km) at the base of mantle and in the zone between inner and outer core of Earth. In these areas, a certain decrease in seismic wave velocities has been detected, which is interpreted (Bolt 1985) in favor of physical heterogeneity of the environment. Such heterogeneity is a sign of intensive migration of matter melts through boundaries of these zones. Migration of matter in deep interior can occur until the mass of Earth it inner core becomes ~ 10% of all mass of our planet. In this case, the central core will consist entirely of iron, and include all the available nickel and a pure eutectic mixture will remain in the outer core (Brown and Mussett 1984).

Considering Phaeton and Earth as planets that are similar to each other in many ways, one cannot help but note their obvious differences. The difference between our planet and Phaeton is manifested, first of all, in: (1) a lower iron content in its mantle silicates, (2) less variability of the chemical composition of matter at boundaries of phase transitions, (3) the predominance of volume of inner mantle shell over the outer shells, and, finally, (4) the larger size and mass of its metallic core. All these differences are quite explainable by a longer period of differentiation of Earth's matter compared to Phaeton, which interrupted its evolution at a relatively early stage.

Thus, the undoubted analogy of the internal structure of both planets and, as a consequence, their close thermobaric states allow us to conclude that even in depths of Phaeton, the substance of meteorites underwent a series of profound physical and chemical transformations. As a result, firstly, two of its main types emerged, represented by silicates and an iron-nickel alloy. And, secondly, silicates turned out to be differentiated by composition into four main chemical groups: LL, L, H and E, and metal into three: FeI, FeII and FeS (Dodd 1986). It is also obvious that before Phaeton destruction in its depths this material could have been heated to several thousand degrees and was under high pressure.

Let us now consider the question on meteorite matter formation. Let us recall that it was precisely the lack of a satisfactory answer to this question half a century ago that became one of the reasons for rejecting the hypothesis of Phaeton existence.

4.3 Formation of Meteoritic Matter

According to the COSS concept, after the Phaeton death, collisions of its fragments with galactic comets and with each other in the belt began to play an important role. These collisions destroyed the colliding bodies and caused mechanical and chemical mixing of their matter.

In the period 4.6–4.4 billion years ago, when mass of Phaeton fragments in belt was ~ 3500 times greater than its present value (Table 1), and the fragments themselves were heated to the state of melting, such collisions were accompanied by ejection of huge quantities of evaporated cometary and melted asteroid matter into interplanetary space. In cosmic conditions, drops and splashes of melt with dust and gas particles stuck to them quickly cooled and fell on asteroids as solid, crystallized formations. These processes led to formation of a layer of strongly altered rocks of a specific chondrite structure on the surface of large, not yet cooled fragments of Phaeton. Fragments of these rocks, affected to a greater or lesser extent by the processes of chondroformation, metamorphism and melting, make up the bulk of meteorites falling to Earth today.

A more detailed classification of meteorites, based on the assumption that they were formed by mixing material from various deep rocks of Phaeton (Table 3) and cometary matter, is given in Table 4.

In order to simplify the classification scheme, only two components of meteoritic matter are taken into account—main material and predominant admixture. Although the actual processes were much more complex, Table 3 quite satisfactorily reflects the known genetic relationships of meteorites (Dodd 1986) of different classes and

4.4 Isotopic Ages of Meteorites

The duration of some important periods of formation of meteoritic matter can be determined using cosmochronology methods. These methods are based on measuring isotopic shifts in the composition of a number of chemical elements caused by decay of radioactive nuclides: ^{44}Pu, ^{238}U, ^{235}U, ^{232}Th, ^{187}Re, ^{129}I, ^{87}Rb, ^{40}K, ^{26}Al, etc.

All of the listed isotopes are formed in reactions of thermonuclear synthesis of chemical elements during stellar explosions (Nuclear Astrophysics 1986). In particular, it has been established that the now-extinct nuclides ^{26}Al, ^{129}I, and ^{244}Pu owe their presence in meteoritic matter to inclusions of solid particles of interstellar dust that have not been processed by solar plasma (Sobotovich and Semenenko 1984).

However, estimates of meteorites formation time based on the ^{26}Mg isotope, formed as a result of the decay of ^{26}Al, and the ^{129}Xe isotope, formed during the decay of ^{129}I and the fission of ^{244}Pu, do not coincide. The "magnesium" age of meteorites is $t_{Mg} \sim 10^6$ years, and the "xenon" age is $t_{Xe} = (0.3-8) \times 10^8$ years with a maximum in distribution of ~ 100 million years (Sobotovich and Semenenko 1985).

Within the framework of existing cosmogonic concept, these authors interpret t_{Mg} as the time that has passed between the end of nucleosynthesis and the formation of condensates and the value of t_{Xe} as the time interval between the last nucleosynthesis and formation of meteorites. Other researchers do not make such distinctions, attributing the age of t_{Xe} to a nearby supernova explosion. It is believed that as a result of this explosion, new portions of radioactive matter could have entered the protosolar nebula.

To explain the nature and age difference of t_{Mg} and t_{Xe} in the COSS concept there is no need to resort to such artificial assumptions as random supernova explosions near the Sun. According to Galactocentric paradigm ideas, during evolution of Galaxy, a continuous explosive destruction of stars occurs in its center. The resulting gas–dust particles, enriched in nuclides ^{26}Al, ^{129}I, ^{244}Pu, etc., accumulate in Galaxy nuclear disk. Here, the gas–dust particles spend some time and are then carried out of disk in the course of a jet outflow. In particular, as part of galactic comets, they reach Solar System limits, where, during collisions of comets with asteroids, they mix with asteroid material and participate in the formation of meteoritic matter.

Therefore, the "magnesium" age of meteorites, as we believe, determines the characteristic time of formation of dust particles during stellar explosions. Its low value indicates that dust condensation during the death of stars occurs quite quickly—in a time shorter than or at least comparable to the half-life of ^{26}Al. A higher "xenon" age determines the time from the moment of explosion of individual stars in Galaxy core to the moment of mixing of their gas–dust products with asteroids matter. In other words, this is the time of accumulation of dust particles in Galaxy nuclear disk and their subsequent delivery to Sun by comets or jet streams.

The age value t_{Xe} in our model has strict lower and upper limits. The first is $R_{\odot}/V_S \approx 30$ million years, where $R_{\odot} = 8.35$ kpc is the average radius of solar orbit; $V_S = 300$ km/s is velocity of galactic comets, and the second should not exceed lifetime of stars in Galaxy core ~ 7×10^8 years (Barenbaum 2010). Both of these values are well confirmed by observations. The real ages of meteorites $t_{Xe} = (0.3-8) \times 10^8$ years with a maximum in the distribution at $t_{Xe} \sim 10^8$ years indicate that dust particles spend a significant part of their lives in central region of Galaxy, even before being included in cometary ices.

It is important to emphasize that, unlike carbonaceous meteorites, most ordinary chondrites do not show an excess of radiogenic isotope ^{129}Xe (Sobotovich and Semenenko 1985). This means that their substance was formed or experienced the last significant heating (~500°C) after decay of ^{129}I and ^{244}Pu, which are part of its composition. Thus, the "magnesium" and "xenon" ages of meteorites carry information about the time of formation of dust particles in Galaxy and are not associated with Phaeton destruction. The time of Phaeton's death can be determined by the content of long-lived radioactive isotopes in meteorites with half-lives of ~ 10^9-10^{11} years. Analysis of such data leads to the conclusion that fragmentation of this planet and, accordingly, the formation time of meteorite matter were not a one-time event, but a cyclically repeating process that intensified during the epochs of Sun's presence in jet streams and spiral arms of Galaxy.

The most active phase of Phaeton fragmentation took place 4.6–4.4 billion years ago. The indicated age range mainly covers (Sobotovich and Semenenko 1984) the times of solidification of meteorites of all classes and groups. The exceptions are iron meteorites of the FeII type, which are more ancient, as well as ordinary meteorites that experienced complete or partial melting later than the main cycle of meteorite formation.

4.5 On Initial Stage of Earth Formation

In concluding the analysis of composition and properties of meteoritic matter, we will dwell on some important details of the formation and subsequent internal evolution of Earth, resulting from a joint consideration of meteoritic and geophysical data.

Such an analysis allows us to assume the following:

1. The formation of Earth and Phaeton from protoplanetary matter began with the separation and growth of their solid metallic core (Vinogradov 1975), the basis of which was iron FeII, strongly and unevenly enriched in many siderophile elements (Ni, Co, P, etc.). Iron FeI entered the core later as a product of significant processing and differentiation of matter accumulated by planets, after its melting and separation into metallic and silicate phases. Simultaneously with increase in volume of the core, the formation of all five mantle shells of planets probably occurred.
2. Having identified FeII meteorites with solid metallic core of Phaeton, which did not undergo complete melting during Phaeton death, the lower estimate of time formation of Sun and planets can be taken as equal to age of FeII meteorites. Since the upper limit of the FeII age according to measurements of Sobotovich (1974)

completely corresponds to the time Sun spent in the Crux-Scutum galactic arm, the most probable age of Earth and Solar System as a whole should be considered 6.567 billion years.

3. The stratification of mantle rocks into individual silicate shells does not necessarily occur only at the post-accretion stage of planetary evolution as a result of phase transitions in iron-magnesium silicates. Most likely, this process begins at the stage of planet formation during heterogeneous accretion (Vinogradov 1974; Kuskov and Khitarov 1982) due to a change in chemical composition of accumulating matter, primarily due to a decrease in the iron content in it.
4. Transformations of planetary interior actively continue throughout the life of planets. Moreover, they proceed in direction of decreasing volume of outer zones and increasing the inner ones. At the same time, at the boundaries of phase transitions, there is a decrease in the contrast of rocks of neighboring silicate shells in chemical composition. These processes are accompanied by melting of Fe and S from silicate minerals, which, apparently, in the form of a eutectic mixture migrate along separate permeable channels to the center of planet.
5. Considering Earth as a close analogue of perished Phaeton, we will once again emphasize its modern differences from Phaeton. These are, first of all, the lower iron content of its mantle silicates; lesser variability of chemical and mineral composition of rocks at phase boundaries; the predominance of the inner mantle shells of Earth over the outer zones; and, finally, the large size and mass of its metallic core.

Taking into account point 4, all these features of the internal structure of Earth are quite explainable by the longer time of its development compared to Phaeton, which interrupted its evolution 4.6 billion years earlier.

Now let us turn to the question of the Moon origin and see what new can be added to what has been said about Earth's evolution by analyzing the properties and composition of the substance of this close satellite of our planet.

5 Origin of the Moon

There are four main groups of hypotheses regarding the Moon origin (Ruskol 1991): (1) Moon was captured by Earth after its formation; (2) Moon formed simultaneously with Earth; (3) Moon separated from Earth at an early stage of its evolution; and (4) Moon separated from Earth as a result of a large impact event.

Presently, the last hypothesis is preferred. According to one of its most developed versions (Hartman and Davis 1975; Cameron and Ward 1976), Moon was formed from terrestrial matter as a result of a glancing blow to our planet by a cosmic body with a mass of ~ 10% of the mass of Earth, i.e., the size of Mars.

This point of view, however, is disputed by Ringwood (1988, 1989). Guided primarily by geochemical considerations, he believes that formation of Moon was caused by the fall of not one large body, but many smaller bodies with a mass of ~

10^{-3}–10^{-2} Earth masses. Later, others came to the same conclusion (Vityazev and Pechernikova 1996).

Let us recall that COSS concept links the origin of Moon with bombardment of Earth by Phaeton's fragments during the era of its most active fragmentation. It is assumed that as a result of falls of fairly large Phaeton's fragments, the surface layer of rocks represented by the substance of its crust and upper mantle was "torn off" from Earth, part of which, having ended up in near-Earth orbit, accumulated on Moon.

5.1 Thickness of Rock Layer Lost by Earth

In support of A. Ringwood's hypothesis, we will consider the thickness of rock layer lost by Earth, as well as its density, total mass, and composition of the substance. We will find the thickness of this layer in two different ways (Barenbaum 1992a).

The first way: Having accepted, on the basis of meteorite data, that our planet existed for at least 0.5–1.5 billion years before Phaeton's destruction, we will proceed from the fact that 4.6 billion years ago Earth had mantle shells similar to modern ones (Table 3), although with a higher iron content in silicates.

Since our planet lost ~ 6.3% of its substance as a result of the falls of Phaeton's debris (Table 1), only the surface layer of its rocks, represented by crust and upper mantle, could have been "lost." Therefore, this material (plus some of material from Phaeton and galactic comets) should today determine the chemical composition and average density of Moon. If now all this missing substance is mentally "returned" back and evenly distributed over Earth surface, taking its density to be equal to the average density of Moon's rocks, then we can roughly estimate the original dimensions of our planet and its density. Taking the density of material lost by Earth to be equal to the current density of Moon's rocks $\rho_M = 3.35$ g/cm^3 (Allen 1977), we get that after such an "addition" the radius of our planet should increased by

$$\Delta r_E = \frac{\Delta M}{4\pi \rho_M r_E^2} = 235\text{ km}, \tag{2}$$

where $r_E = 6371$ km is the current radius of Earth, $\Delta M = 4 \times 10^{26}$ g is the mass of rocks (Table 1) lost by Earth due to fall of Phaeton's debris.

Thus, before Phaeton's destruction, Earth with a mass of $M_E{}' = 6.37 \times 10^{27}$ g would have had an average radius of 6606 km and an average density of $\rho_E{}' = 5.2$ g/cm^3.

The second method: There is an approach to studying the internal structure of Venus based on the assumption that its interior is in all respects similar to that of Earth (Hubbard 1987). Using this approach, the known density section of Earth is recalculated for Venus by integrating the equations of hydrostatic equilibrium. It is

Knowing the value of the index γ, we have

$$W_M = \frac{1}{2}\int_{V_1}^{V_2} V^2 \frac{dN}{dV} dV = 3.3 \times 10^{30}\text{J}, \tag{8}$$

$$W_E = \frac{1}{2}\int_{0}^{V_1} V^2 \frac{dN}{dV} dV = 5.4 \times 10^{30}\text{J}. \tag{9}$$

Estimates (8) and (9) show that the total energy W of cosmic bodies bombarding the Earth should be ~ 10^{32} J.

In this case, not only very large fragments of Phaethon could have been the energy carriers, but also its smaller fragments. These fragments could have acquired highly eccentric orbits (e $\approx$ 1) and then, after many thousands and even millions of years, like modern long-period comets, return to Solar System. Their speed of meeting Earth $V = 25\sqrt{(1+e)/(1-e)}$ could have been and thus reach many tens of kilometers per second.

Comparing the value of W_M with the heat content of Moon in a state of complete melting 1.6×10^{29} J (Galkin 1978), we find that this energy is sufficient not only to form a satellite of lunar mass near Earth, but also to significantly heat its surface due to accumulation processes. Considering that during impacts approximately the first few percent of the energy of impactors is spent on melting rocks, as well as the possibility of Moon formation in just a hundred years (Melosh 1994), we come to the conclusion that energy of W_M and W_E was sufficient to melt the upper layer of rocks of both planets to a great depth. On Earth, this zone apparently reached $\approx$100 km, and on Moon, it probably amounted to $\approx$300 km (Latham et al. 1975).

5.3 Angular Momentum of Earth–Moon System

An important consequence of the bombardment of Earth by cosmic bodies could be the transfer of additional rotational moment to it (Barenbaum 1992a, b). Following (MacDonald 1963), we will assume that for planets that have not experienced strong external influences at the post-accretion stage, the axial rotational moment L_i and mass M_i are related by the dependence (Fig. 3):

$$L_i \propto M_i^{1.8}. \tag{10}$$

Using this dependence, we will estimate angular momentum L_E' and period of axial rotation T_E' of Earth before the separation of matter from it. Taking value of M_E' from Table 1, we find that angular momentum and period of Earth rotation could initially be $L_E' = 1.6 \times 10^{41}$ g cm^2/s and $T_E' = 10$ h. Comparing value of L_E' with

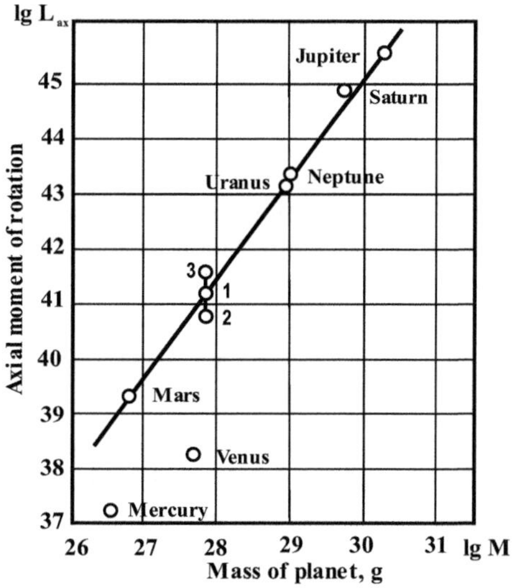

Fig. 3 The relationship between axial moment of rotation of planets and their mass. The line corresponds to formula (10). Designations: 1—current rotation moment of Earth; 2—moment rotation of Earth before Moon formation; 3—current rotation moment of the Earth–Moon system Point "1" is the assumed value of angular momentum of Earth before formation of Moon

current angular momentum of the Earth–Moon system $L_{E+M} = 3.5 \times 10^{41}$ g cm^2/s (Allen 1977), we conclude that this system today has an excess angular momentum $\varDelta L = L_{E+M} - L_E{}' = 1.9 \times 10^{41}$ g cm^2/s (Fig. 3).

As a possible option, let us assume that the moment $\varDelta L$ was transferred to Earth by Phaeton fragments that bombarded it 4.6 billion years ago, and since then total moment of Earth–Moon system L_{E+M} has not changed in any way. In this case, it should also be assumed that moment $\varDelta L$ was close to Earth's rotation moment $L_E{}'$ not only in magnitude, but also in direction. If the vectors $\varDelta L$ and $L_E{}'$ were anticollinear, then the direction of rotation of Earth could well change to the opposite, as with Venus. And if they were perpendicular, then the tilt of Earth's axis would change sharply and could become, for example, like Uranus.

Taking into account the comments made, we will estimate the total mass (m) of bodies bombarding Earth and the probable velocity V of their impact on Earth's surface. In the case of a "tangential" fall of cosmic bodies into the equatorial region of our planet, we will have

$$\Delta L = mVr_{E'} = 1.9 \times 10^{41}\text{g cm}^2/\text{s}, \tag{11}$$

$$W = \frac{mV^2}{2} \approx 10^{32}\text{J}. \tag{12}$$

From the system of these equations, we find $V \approx 70$ km/s and $m \approx 5 \times 10^{25}$ g. Thus, the number of Phaeton fragments bombarding Earth required for Moon formation and, consequently, the mass of largest of them is equal to $\approx 5 \times 10^{25}$ g. This value is 12.5% of the mass of matter torn away from our planet or less than 1% of Earth's mass, which agrees well with A. Ringwood's assumption.

5.5 Internal Structure and Evolution of Moon

According to the mega-impact hypothesis, it is believed (Latham et al. 1975) that Moon accumulated in near-Earth orbit from blocks of terrestrial rocks as a generally cold and weakly cemented body, the surface of which was melted by accretion processes to a depth of up to 300 km.

So the entire subsequent geological history of Moon was connected mainly with cooling of this molten layer and differentiation of its substance. The process of solidification of Moon's surface was often disrupted, especially in the first billion years, by fall of large cosmic bodies that crushed and mixed the solidifying layer, causing processes of local magmatism and metamorphism.

The study of rock samples returned from Moon generally confirms these ideas. At the same time, chemical analysis of the rocks shows that composition of Moon's lavas changed regularly over time. If in the early period (4.6–3.9 billion years) mainly the anorthosite crust melted, then ~ 3.6 billion years ago titanium-rich ($TiO_2 > 8\%$) lavas began to erupt onto Moon's surface—products of rock melting at depths of $\approx$100 km. And later (3.4–3.1 billion years ago), erupting melts mainly rise from depths of 200–400 km and are characterized by a low Ti content. At the turn of 2.6 billion years, a new impulse of basaltic magmatism occurred with the effusion of medium-titanium lavas. It is believed that at this time, the melts less affected by differentiation rose upward from even greater depths (Hayes and Walker 1975).

According to (Ruskol 1991), the surface layer of Moon's rocks has now completely cooled. The thickness of lunar crust varies from several kilometers to 50 km. Moreover, on Moon side facing Earth, the crust is thinner, and on the far side of Moon, it is thicker. The crustal material changes from anorthosite at the top to more gabbroic at the bottom. Below 300 km lie mantle rocks of olivine-pyroxene composition (80–85% olivine). With depth, they show an increase in Fe content and iron-magnesium ratio. The 300 km boundary separates the once molten outer zone of Moon from its cold central part. Below this boundary, according to (Latham et al. 1975), lie rocks of the original composition, slightly changed by differentiation.

The interval of 200–500 km is characterized by low electrical resistance of ~ 10^4 Ohm·m. It is assumed (Urey and MacDonald 1973) that the high electrical conductivity of these rocks is due to their increased iron content. Note that a similar interval of increased electrical conductivity exists at depths of 40–180 km on Earth. The high viscosity ($\geq 10^{23}$ Pa·s) and quality factor ($Q > 3000$) of the rocks of the outer shells of Moon indicate that the substance of its lithosphere is at solidus temperatures. The first signs of rock melting expressed in strong absorption of transverse and weakening of longitudinal seismic waves, as well as a sharp decrease in viscosity and quality factor of rocks, appear at depths of ~ 800 km on Moon. By analogy with Earth, this region is called the asthenosphere.

The boundary between the lithosphere and asthenosphere, which occupies a depth range of 800–1000 km on Moon, is very blurred and is characterized by large gradients of physical characteristics. Most of the centers of tidal (deep-focus) moonquakes are confined to this boundary, which gravitate toward several seismicity belts of a complex deep profile. The presence of such belts is explained by the initial heterogeneity of blocks of matter accumulated on Moon (Galkin 1978). Seismic data indicate the presence of a metallic core on Moon with a diameter of 330–360 km, surrounded by a partially molten shell with a diameter of ≈480 km. This shell, in our opinion, may consist of troilite melt, which is unlikely to form a continuous zone on Moon, but most likely fills individual cavities.

5.6 About the Ancient and Modern Earth's Crust

Analysis of the conditions of Moon formation from standpoint of COSS concept allows us to better understand the origin and properties of Earth's crust, as well as the features of its modern structure.

Calculations using formulas (2) and (3) lead to conclusion that before Phaeton destruction, Earth was covered with a thicker layer of anorthosite crust than is observed today. Given the similar sizes of Earth and Venus, the average thickness of Earth's crust initially probably reached about 80 km, as is the case with modern Venus. Consequently, the surface layer of material torn from Earth, 245 ± 10 km thick, should have consisted of ~ 1/3 crustal rocks and ~ 2/3 upper mantle rocks.

The latter conclusion also follows from seismic sounding data of Moon. If we assume that the substance that formed Moon was initially well mixed, then the composition of rocks of its previously melted 300-km layer should correspond to the average composition of rocks (excluding volatiles) torn from Earth. Since after cooling and differentiation, the surface layer of Moon to depths of ~ 100 km is represented by crustal matter, which by volume makes up ~ 1/3 of all previously completely melted rocks, then Earth's matter, obviously, should have been lost in the same proportion.

As a result of Phaeton's death, most of the ancient Earth's crust was lost, and what remained melted and mixed with mantle matter (Barenbaum 2023a). After cooling, this matter again experienced separation into crustal and mantle parts. However, due to the absence of ~ 2/3 of primary anorthosite substance, modern Earth's crust is no longer able to "cover" entire surface of globe.

The deficit of Earth's crustal substance, which arose on Earth in a single process with Moon formation, is today manifested in the presence of our planet's "continental" and "oceanic" hemispheres, the existence of lithospheric plates moving along its surface, and is also reflected in the structure and nature of evolution of Earth's tectonosphere (Khain 2003).

An idea of the ancient Earth's crust is to some extent given by a comparison of compositions of modern crust of Earth and Moon (Ringwood 1982). However, it should be borne in mind that continental crust existing on Earth today, which has an average thickness of ≈ 40 km, is approximately two times thinner than the original and contains less iron and sulfur. In addition, on Earth and Moon we are dealing with the products of differentiation of once molten rocks on surface of both celestial bodies. But if Moon has now cooled to a depth of ~ 800 km, then surface layer of

Earth's rocks, identified with lithosphere, is still mobile and is in a state of convective mixing and ongoing differentiation. The main role in maintaining the lithosphere in this state is played by bombardments of Earth by galactic comets. Comet falls split and move the plates, and also cause strong uneven heating of asthenospheric rocks, along which the plates and lithospheric shell as a whole move (Barenbaum 2023a).

In connection with the problem of Earth's formation and evolution, one cannot ignore the attempts that became more frequent in the 1980s and 1990s to connect the insufficient amount of crustal matter and the spreading of its blocks in rift zones with increase in size and mass of our planet (Carey 1991; Milanovsky 1984; Kropotkin and Efremov 1992; Larin 2005). Supporters of this hypothesis assume that Earth size has changed many times since its formation, having increased by tens of percent, or even several times, by now. The adherents of this idea are not embarrassed by the fact that they do not know of an acceptable mechanism for changing Earth size that does not contradict the known laws of physics.

Calculations using formulas (2) and (3) of the thickness of rock layer lost by our planet give the same result only under the assumption that Earth, Venus, and Moon have not changed their mass, size, and density of matter in any noticeable way in the last 4.6 billion years. A change in at least one of these parameters for Earth requires a simultaneous change of the same parameters for Venus and Moon by the same amount. It is absolutely impossible to explain the synchronous identical "swelling" or "pulsations" in sizes of such different celestial bodies as Earth, Venus, and Moon. This circumstance reveals, in our opinion, the complete inconsistency of the idea that tries to explain smaller amount of ancient anorthosite crust on Earth, compared to Moon and Venus, as well as the observed movement of continental plates by a change in physical size and mass of our planet.

Criticism of the "expanding" Earth hypothesis from other positions is also contained in many other works (see Sorokhtin and Ushakov 1991).

5.7 Main Results and Conclusions

The article revives the hypothesis of Phaeton existence, of which for half a century, it would seem, has been abandoned forever in the Solar System cosmogony. Compelling arguments are given in favor of this hypothesis and its development is proposed within the framework of the concept of "Cosmogony of Open Solar System (COSS)" (Barenbaum 2002, 2010).

The COSS concept eliminates the fundamental drawback of the Kant–Laplace hypothesis, which does not take into account the influence of processes in Galaxy on formation and evolution of Solar System and its planets. From the standpoint of COSS concept, agreed solutions are proposed for a number of problems that have not yet received a satisfactory explanation in Kant–Laplace cosmogony.

The main results and conclusions of article are as follows:

- There were at least two main cycles of planet formation in Solar System. The first, the main cycle, is associated with the formation of Sun and planets from protosolar substance, and the second, which began after death of planet Phaeton, led to a change in mass and matter composition of most planets. So the age of Sun and planets is at least 2 billion years older than is commonly believed today.
- Before the Phaeton fragmentation 4.567 billion years ago, the masses of all planets in Solar System were subject to a common power-law dependence acquired by them at protoplanetary stage. The division of planets into "inner"—Earth-type and "outer"—giant planets occurred at the second stage of planet formation. During this process, the inner planets mainly lost their own substance, and the outer ones increased their mass and acquired satellite systems.
- The meteorite substance bears clear signs of its formation in the depths of Phaeton. The properties of meteorites allow us to assert that at the time of its destruction Phaeton was a planet of terrestrial type already stratified into a system of mantle silicate shells and already having a solid iron–nickel core.
- This conclusion fully applies to Earth. It has been shown that the Moon did not form as a result of a collision with Earth of a large celestial object (the size of Mars), but, according to the hypothesis of A. Ringwood (1998), based on geochemical data, as a result of smaller bodies falling to Earth. Following COSS concept, these bodies were large fragments of Phaeton, which most intensively bombarded Earth and other planets during the first hundred million years after beginning of Phaeton's fragmentation 4.567 billion years ago.

In conclusion, we also note that this article is a translation into English of the first part of the chapter devoted to solving problematic issues of Solar System cosmogony from standpoint of the Galactocentric paradigm (see Barenbaum 2002). The second part of this chapter is presented in the next article of this collection (Barenbaum 2025).

References

Alfven H, Arrhenius G (1979) Evolution of the solar system. Mir, Moscow, p 512

Allen KW (1977) Astrophysical quantities. Mir, Moscow, p 446

Astronomy (2007) The 21st century. Surdin VG Fryazino (ed) "Vek 2" 608

Avsyuk YuN (1996) Tidal forces and tidal processes. United Institute of Physics of Earth of Russian Academy of Sciences, Moscow, p 188

Barenbaum AA (2010) Galactocentric paradigm in geology and astronomy. LIBROKOM, Moscow, p 544

Barenbaum AA, Shpekin MI (2016) To the development of the mechanism of interaction of galactic comets with the terrestrial planets. J Phys Conf Ser 774:012096. https://doi.org/10.1088/1742-6596/774/1/012096

Barenbaum AA, Shpekin MI (2019) Origin and formation mechanism of craters, seas and mascons on the Moon. J Phys Conf Ser 1147:012057. https://doi.org/10.1088/1742-6596/1147/1/012057

Barenbaum AA, Shpekin MI (2021) Estimation of the flux density of galactic comets in the Orion-Cygnus branch based on number of shield volcanoes on Venus, craters on Mars and marine

basins on the Moon. J Phys Conf Ser 1787:012021. https://doi.org/10.1088/1742-6596/1787/1/012021
Barenbaum AA (1990) On the nature of C-asteroids and carbonaceous chondrites. Cometary Circ Kyiv. 419:9–10
Barenbaum AA (1992) Origin of asteroids and meteorites (new cosmological concept). Conceptual foundations of geology. Notes of the St. Petersburg Mining Institute 134:9–27
Barenbaum AA (1992a) The solar system in the epoch of the phaeton explosion. Origin of the Moon. Conceptual Foundations of Geology. Notes of St. Petersburg Mining Institute 134:95–106
Barenbaum AA (2002) Galaxy, solar system, earth. subordinate processes and evolution. GEOS, Moscow, p 393
Barenbaum AA (2015) On the mechanism of energy dissipation of galactic comets bombarding terrestrial planets. J Phys Conf Ser 653:012073. https://doi.org/10.1088/1742-6596/653/1/012073
Barenbaum A (2023) Solution of problem questions of astronomy and geology using the optimized galactic model. In: Physical and mathematical modeling of earth and environment processes—2022. https://doi.org/10.1007/978-3-031-25962-3
Barenbaum A (2023a) On the issue of "True polar wander" phenomenon and its alternative physical interpretation based on galactic model. In: Processes in GeoMedia—vol. 7. Chapter 5 (Springer) https://doi.org/10.1007/978-981-99-6575-5_5
Barenbaum A (2024) Origin of solar system comets. In: Chaplina T (ed) Processes in GeoMedia—vol. 8. Springer Geology, pp 167–198
Barenbaum AA (2025) Cosmogony of open solar system (II): model of secondary planet formation, secondary accretion disk of the Sun, formation of satellites of giant planets, Molchanov's resonance, Titius-Bode rule
Barnes C, Clayton D, Schramm D (eds) (1986) Nuclear Astrophysics. Mir, Moscow, p 519
Binder FB (1982) The Moon: its figure and orbital evolution. Geophys Res Letter 9(1):33–36
Bolt B (1985) In depths of the earth. Mir, Moscow, p 189
Brown GC, Mussett AE (1984) The inaccessible earth. Mir, Moscow, p 263
Cameron AGW, Ward WR (1976) Origin of the Moon. Lunar Planet Sci Pt 7:120–122
Cameron A (1980) Formation of the outer planets and satellites. In: Satellites of the Planets. Mir, Moscow, pp 506–516
Carey W (1991) In search of patterns of development of the earth and the universe: history of dogmas in the earth sciences. Mir, Moscow, p 447
Daly RA (1943) Meteorites and the earth-model. Bull Geol Soc Amer 54:401–456
Dickey JH, Bender PL, Faller JE et al (1994) Lunar laser ranging: a continuing legacy of the Apollo program. Science 265:482–490
Dodd RT (1986) Meteorites. Mir, Petrology and Geochemistry. Moscow, p 384
Dorofeeva VA, Makalkin AB (2004) Evolution of the early solar system. In: Cosmochemical and physical aspects. Editorial URSS, Moscow. p 264
Drozdova NYu, Kiselev VM (1995) Evolution of the earth-moon system according to empirical data and the MacDonald-Goldreich theory. Astron J 72(2):767–772
Eremeeva AI, Tsitsin FA (1989) History of astronomy: textbook. Moscow State University, Moscow, p 349
Fesenkov VG (1952) Nature and possible origin of meteorites, zodiacal light and asteroids. Quest Cosmogony 1:92–130
Galkin IN (1978) Geophysics of the moon. Nauka, Moscow, p 176
Goettel KA, Shields JA, Decker DA (1981) Density constraints on the composition of Venus. Proc Lunar Sci Conf 12B:1507–1516
Goldreich P (1975) History of the lunar orbit. In: Tides and resonances in the Solar system. Mir, Moscow, pp 97-128
Gudkova TV, Zharkov VN, Leontiev VV (1988) Models of uranus and neptune with partially mixed envelopes. Astron Vestnik 22(1):23–40

Gudkova TV, Zharkov VN, Leontiev VV (1988a) Models of jupiter and saturn with a two-layer molecular envelope. Astron Vestnik 22(3):252–262
Hart RS, Anderson DL, Kanamori H (1977) The effect of attenuation on gross earth models. Geophys Res 82:1647–1654
Hartman WK, Davis DR (1975) Satellite-sized planetesimals and lunar origin. Icarus 24:504–515
Hayes JF, Walker D (1975) Igneous lunar rocks. In: Cosmochemistry of the Moon and planets. Nauka, Moscow, pp 274–282
Hubbard W (1987) Internal structure of the planets. Mir, Moscow, p 326
Kerr RA (1989) Making the moon, remaking earth. Science 243(4897):1433–1435
Khain VE (2003) The main problems of modern geology. Nauchnyy Mir, Moscow, p 348
Kropotkin PN, Efremov VN (1992) Changes in the Earth's radius in the geological pas. Geotectonics 4:3–14
Kuskov OL, Khitarov NI (1982) Thermodynamics and geochemistry of the Earth's core and mantle. Nauka, Moscow, p 278
Kuskov OL, Dorofeeva VA, Kronrod VA, Makalkin AB (2009) Systems of jupiter and saturn: formation, composition and internal structure of large satellites. Editorial URSS, Moscow, p 576
Larin VN (2005) Our Earth. Origin, composition, structure and development of the initially hydride Earth. PH "Agar", Moscow, p 242
Latham G, Nakamura I, Dorman J et al (1975) Results of the passive seismic experiment under the Apollo program. In: Cosmochemistry of the moon and planet. Nauka, Moscow, pp 299–311.
MacDonald GJF (1963) The internal constitutions of the inner planets and the moon. Space Sci Rev 2:473–557
MacDonald GJ (1975) Tidal friction. Tides and resonances in the Solar system. Mir, Moscow, pp 9–96
Makalkin AB, Ziglina IN, Artyushkova ME (2019) Topical problems in the theory of planet formation: formation of planetesimals. Phys Earth (1):104–121
Mason B (1965) Meteorites. Mir, Moscow, p 306
Melosh G (1994) Formation of impact craters. Geological process. Mir, Moscow, p 336
Milanovsky EE (ed) (1984) The problem of expansion and pulsation of the earth. Nauka, Moscow, p 195
Orlov SV (1949) Asteroids and meteorites. Meteoritics 5:3–113
Press F (1968) Earth models obtained by Monte Carlo inversion. J Geophys Res 73:5223–5234
Prior GT (1920) The classification of meteorites. Mineral Mag 19:51–63
Proceedings of the first conference on cosmogony (1951) USSR Academy of Sciences, Moscow p 372
Putilin II (1953) Minor Planets. Gostekhteoretizdat, Moscow, p 412
Rezanov IA (2004) History of the exploded planet. Nauka, Moscow, p 184
Ringwood AE (1982) Origin of the earth and the moon. Nedra, Moscow, p 292
Ringwood AE (1988) Lunar origin: single giant impact or multiple large impacts? Abstr. Paper 19th Lunar Planet Sci Conf Houston Pt 3:982–983
Ringwood AE (1989) The earth-moon connection. Z Naturforsh A 44(10):891–923
Runcorn SK (1988) Ancient magnetic field of the Moon. In the world of science (2):18–27
Ruskol EL (1991) Models of the structure and origin of the moon. Astron Vestnik 25(4):408–421
Safronov VS (1969) Evolution of the protoplanetary cloud and the formation of the earth and planets. Nauka, Moscow, p 244
Safronov VS, Ziglina IN (1990) Origin of the asteroid belt. In: Marov MY (ed) Solar system. Nauka, Moscow
Savchenko KN (1975) Kant's cosmogony and the problem of the origin of small bodies of Solar system. In: Problems of the origin of bodies of solar system. Nauka, Moscow-Leningrad, pp 8–243
Schmidt O.Yu. (1957) Four lectures on the theory of the origin of the earth. USSR Academy of Sciences, Moscow, p 140

where M_O is mass of central star and G is gravitational constant.

If the motion is not completely ordered, then value of L will be smaller. Moreover, during formation process the planet should shift toward star. In formulas (1) and (2), this circumstance can be taken into account to some extent, formally moving to new values of R_O and $\tilde{r}$.

We will find solutions (1) and (2) under the condition $\partial\rho / \partial R = 0$ in a small neighborhood of the point R_O, equal to $|R–R_O| \leq \tilde{r}$, i.e., at $\eta = 0$ for the entire accretion region of planet.

The last requirement corresponds to the case when the forming planet introduces a strong perturbation into distribution $\rho(R, t)$, in which the behavior of matter in planet's sphere of influence $\tilde{r}$ depends weakly on the function ρ outside this region. This situation can occur if the planet "scoops out" matter from its sphere of influence faster than the filling of this sphere with matter from an external source.

Integrating (1), we find

$$M = \frac{4}{3}\pi\rho(R_0)\tilde{r}^3. \tag{3}$$

Solution (2) is more cumbersome. Having expanded it in a series by parameter $\tilde{r}/R_O < < 1$, for the first two terms of the expansion we have

$$L = \frac{4}{3}\pi\rho(R_0)\tilde{r}^3\sqrt{GM_OR_0}\left\{1 + \alpha\left(\frac{\tilde{r}}{R_0}\right)^2 + \ldots..\right\}, \tag{4}$$

where α is numerical coefficient.

The expression before the curly brackets (4) is the orbital moment of the emerging planet:

$$L_{orb} = \sqrt{R_0GM_OM}. \tag{5}$$

The presence of a factor other than unity in the curly brackets of formula (4) indicates the occurrence of an uncompensated moment of axial rotation L_{ax} in the forming planet. In accordance with (4), the relationship between orbital and axial moments of planet has the form:

$$L_{ox} = \alpha\left(\frac{\tilde{r}}{R_0}\right)^2 L_{orb}. \tag{6}$$

Note that formula (6) is obtained not only in the case of $\eta = 0$, but also for an arbitrary value of this indicator. In this case, the value and sign of coefficient α depend on choice of value η.

The latter circumstance, as follows from rigorous theoretical calculations (Giuli 1968; Fernandez et al. 1976; Kiladze 1986), is due to the fact that axial moment of rotation of the forming planet turns out to be very critical to the features of accepted accretion mechanism, and depending on its details, the planet acquires rotation either in the forward or in the reverse direction.

Therefore, in order not to introduce poorly substantiated assumptions into the model, we will determine the value of α based on the known (Alfven and Arrhenius 1979; Gorkavy and Friedman 1994) values of orbital and axial moments of rotation of real planets of Solar System.

However, first we should define an expression for the value $\tilde{r}$. The parameter $\tilde{r}$ in our model characterizes the radius of planet's sphere of action, at which its gravitational attraction becomes greater than the force of attraction of star. When calculating the radius $\tilde{r}$ of Solar System planets, the following formula is most often used:

$$\tilde{r} = R_O \left(\frac{M}{M_O} \right)^{2/5}. \tag{7}$$

We will require that our model satisfies the most general assumptions regarding the nature of the change in the planets' masses over time and the distribution function of substance feeding them. Therefore, we will assume that relation (6) is satisfied for all stages of planet formation, including the present one.

Substituting expression (7) instead of $\tilde{r}$ into formula (6), we have

$$\alpha = \frac{L_{ax}}{L_{orb}} \left(\frac{M}{M_O} \right)^{-4/5}. \tag{8}$$

Thus, the coefficient α in expansion (4) can be directly established from obser vational data on the mass, orbital, and axial moments of rotation of planets and the mass of central star.

2.2 Axial Moment of Planets' Rotation

Figure 1 shows the values of parameter α calculated for Solar System planets using formula (8).

We see that for most planets the parameter α decreases with their heliocentric distance, obeying the general pattern:

$$\alpha = Z R_0^{-0.5}, \tag{9}$$

where $Z = 2.6 \times 10^9$ km.

From the structure of expression (8), however, it follows that in its physical meaning quantity α can only be a numerical coefficient. Therefore, the obtained functional dependence on R_O must be attributed not to coefficient $\tilde{r}$, but to another parameter. Only the quantity can act as such a parameter. Therefore, we can write

Thus, the developed phenomenological model allows us to make at least three important conclusions:

1. The formation of planets in stars accretion disk occurs simultaneously with an increase in mass of star itself.
2. The main reason for turbulence in accretion disk of stars and the long lifetime of disk itself, necessary for the formation of planets in it, is, apparently, the rotational instability of stars.
3. The thickness of gas–dust disks of stars at the stage of planet formation is $Z = 2.6 \times 10^9$ km and, apparently, weakly depends on mass of star itself, i.e., on type of central object.

We will now apply our model to the study of the formation of planets in the "secondary" accretion disk of Sun.

3 Secondary Accretion Disk of Sun

The identical nature of the $\alpha(R)$ dependence for all planets in Fig. 1 gives reason to believe that the growth processes of giant planets at the second stage of Solar System evolution after Phaeton death differed little from the same processes at the first stage of Solar System formation. There is no doubt that the second cycle of planet formation in Solar System, like the first, passed through formation a gas–dust accretion disk of Sun with effective thickness Z.

At the same time, there are clear differences in the conditions of first and second cycles of planet formation. These differences are clearly visible in Fig. 4 in the double logarithmic coordinate system $\{L_{orb}, M/M_O\}$. The planets of group I: Venus, Earth, Uranus, and Neptune (see Fig. 1 in (Barenbaum 2025)), which did not change their mass or changed it relatively little, form their own trend, which is sharply different from the trend of planets of group II, which includes Mercury, Mars, Jupiter, and Saturn.

The most important differences between the primary and secondary accretion disks of Sun were related to the sizes of disks and distribution of matter within them. If the distribution of matter in primary disk was power law and it covered the entire region of localization of planets, then the secondary gas–dust disk was formed on outer side of asteroid belt, and with distance from Sun the density of matter in it decreased exponentially.

As for the inner region of Solar System, here the secondary disk was characterized by "negative," if I may say so; values of integral density of matter, which led not to an increase, but to a decrease in masses of previously formed planets (see Table 1 in (Barenbaum 2025)).

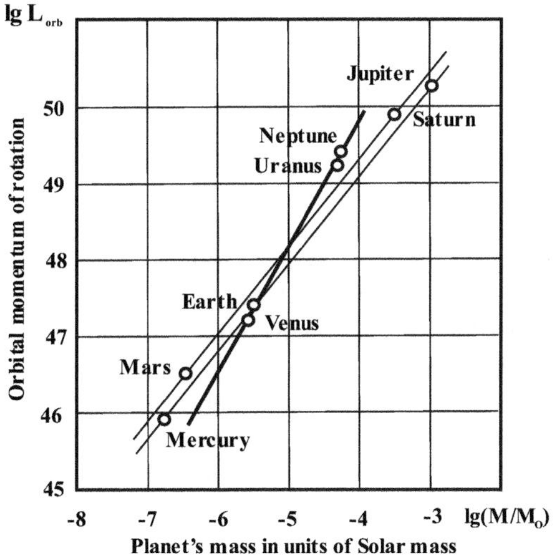

Fig. 4 Position of the planets in the coordinate system $\{L_{orb}, M/M_O\}$

3.1 Parameters of Secondary Accretion Disk of Sun

Changes in masses of Jupiter, Saturn, and Uranus during second stage of Solar System formation (see Fig. 1 and Table 1in (Barenbaum 2025)) give reason to believe that integral density of matter in secondary accretion disk can be approximated with sufficient accuracy by an exponential function:

$$\rho(R) = \rho_0 \exp\left\{-\frac{1}{L}(R - R_\alpha)\right\}, \tag{16}$$

where constant $L = 3.4 \times 10^{13}$ cm, and value of integral density of substance ρ_0 is defined as

$$\rho_0 = \begin{cases} 0\,\text{at}\,R \leq R_a = 4 \times 10^{13}\text{cm} \\ 3.6 \times 10^{-5}\text{g/cm}^3\,\text{at}\,R > R_a \end{cases}. \tag{17}$$

Formulas (16) and (17) somewhat conventionally reflect our conclusion that secondary accretion disk of Sun was located on outer side of asteroid belt, and there was no disk in the inner region of Solar System. We associate such a distribution of matter in secondary disk with accumulation in this region of Solar System of huge amounts of gas and dust, which came here from asteroid belt as a result of collisions of galactic comets with Phaeton fragments.

It is also assumed that during the active phase of Phaeton destruction, as in our days (Barenbaum 2002, 2024), asteroid belt served as a barrier that prevented dust particles that diffused to Sun from entering central region of Solar System. Therefore, the central radius of modern asteroid belt R_α is taken as inner boundary of secondary

Table 2 Average masses of giant planets at second stage of planet formation

Characteristics of planets	Jupiter	Saturn	Uranus	Neptune
M_1, g	2.44×10^{28}	4.01×10^{28}	7.07×10^{28}	1.03×10^{29}
M_2, g	1.90×10^{30}	5.68×10^{29}	8.72×10^{28}	1.03×10^{29}
$\Delta M = M_2 - M_1$, g	1.87×10^{30}	5.28×10^{29}	1.65×10^{28}	0
M_2/M_1	77.87	14.16	1.23	1.00
$\tilde{M} = \beta M_2$	3.74×10^{29}	1.75×10^{29}	7.86×10^{28}	1.03×10^{29}
β	0.197	0.308	0.9	1

masses of planets are also indicated there (see Table 1 in (Barenbaum 2025)). The last line shows the value of coefficient β, which characterizes magnitude of effective (average) mass of planets at second stage of formation in units of their present-day mass.

We see that, despite the large differences in initial and final masses of Jupiter, Saturn, and Uranus, their average masses differ less and are approximately the same as mass of Neptune, which has remained virtually unchanged.

3.3 Redistribution of Rotational Moment

The capture by Sun of large masses of new cosmic matter, which formed secondary disk where giant planets arose, could not help but change the rotational moment of entire Solar System. Therefore, the distribution of rotational moment observed in Solar System today, when the planets (mainly Jupiter and Saturn) account for 98% of total rotational moment and only 2% belongs to Sun, should be associated exclusively with the second stage of planet formation.

Let us pay attention to the following circumstance. Having compared masses of Jupiter, Saturn, and Uranus with mass of secondary disk of Sun (18), in accordance with the data (Table 1 in (Barenbaum 2025)), we can conclude that ~ 0.17% of disk matter should have gone into formation of shells of these planets. If all the disk matter revolved around Sun in the forward direction and strictly in ecliptic plane, then the angular momentum associated with disk would have been:

$$L_d = 2\pi Z \int_{R_\alpha}^{\infty} \sqrt{M_O G}\rho(R) R^{3/2} dR = 1.8 \times 10^{53} g \times cm^2/s. \tag{24}$$

Taking into account the mass of matter acquired by Jupiter, Saturn, and Uranus after Phaeton destruction, we find that together with new substance these planets should have inherited 1.7×10^{-3} part of disk rotation moment, which reaches 3.07×10^{50} g·cm²/s.

Table 3 Some data on regular satellites of giant planets (Astronomy 2007)

№	Satellite name	$m \times 10^{26}$, g	D, km	ρ, g/cm^3	$R \times 10^3$, km	T_{orb}, day	i, deg	ε	T_{ax}
Jupiter									
1	Jo	893	3643	3.53	422	1.769	0.04	0.0041	S
2	Europe	480	3122	3.04	671	3.551	0.47	0.0094	S
3	Ganymede	1482	5262	1.94	1070	7.155	0.17	0.0011	S
4	Callisto	1076	4821	1.83	1883	16.69	0.19	0.0074	S
Saturn									
1	Mimas	0.38	397	1.2	185.6	0.942	1.57	0.0206	S
2	Enceladus	1.04	500	1.6	238.1	1.370	0.01	0.0001	S
3	Tethys	6.18	1060	1.0	294.7	1.888	0.17	0.0001	S
4	Diona	11.0	36 × 30	1.5	377.4	2.737	0.00	0.0002	S
5	Rhea	23.2	1530	1.2	527.1	4.518	0.33	0.0009	S
6	Titan	1346.5	5150	1.9	1121.9	15.95	1.63	0.0288	S?
7	Hyperion	0.11	360 × 226	1.1	1464.1	21.28	0.57	0.0175	C
8	Iapetus	19.5	1440	1.3	3560.8	79.33	7.57	0.0284	S
Uranus									
1	Miranda	0.66	472	1.2	129.9	1.413	4.34	0.0013	S
2	Ariel	13.5	1158	1.7	190.9	2.520	0.04	0.0012	S
3	Umbriel	11.7	1170	1.4	266.0	4.144	0.13	0.0039	S
4	Titania	35.2	1578	1.7	436.3	8.706	0.08	0.0011	S
5	Oberon	30.1	1573	1.6	583.5	13.46	0.07	0.0014	S
Neptune									
1	Triton	214	2707	2.1	354.8	-5.877	156.8	0.0000	S
2	Nereid	0.3	340	1.0	5513.4	360.14	7.2	0.7512	S

Designations: *m*—satellite mass, *D*—its diameter, ρ—matter density, *R*—orbit great semi-major axis, T_{orb}—orbital period, *i*—orbit inclination to Laplace plane (epoch 1999.0), ε—orbit eccentricity, T_{ax}—axial rotation period: *S*—synchronicity of daily and orbital periods, *C*—chaotic nature of rotation.

The latter value agrees quite well with total rotation moment of all three planets, equal to 2.63×10^{50}g·cm^2/s (Allen 1977). The closeness of calculated and actual rotation moments of planets cannot be considered accidental. From the standpoint of our model, it means that gas shells of giant planets should have formed from matter with a well-ordered motion around Sun.

However, such a large rotation moment of secondary disk as (24) cannot be introduced into Solar System only due to galactic comets. The jet stream comets that scattered on Phaeton debris were characterized on average by zero angular momentum relative to Sun. Therefore, disk formed from their matter as a whole should have had a low intrinsic angular momentum. Roughly speaking, half of disk's matter should have rotated around Sun in one direction, and second half in the other.

Obviously, only Sun could provide secondary disk with the necessary rotational moment. At that, Sun had to give away its own rotation moment, forcing substance in disk to rotation in desired direction.

The physical mechanism for such a transfer of rotational moment is well known—it is the entrainment of cosmic plasma particles by magnetic field of rotating central star (Alfven 1942; Hoyle 1960; Alfven and Arenius 1979; Shklovsky 1984; and others). However, if the entire moment of secondary disk (more precisely 0.5 L_d) is attributed only to Sun, then rotation velocity of Sun itself before disk formation would exceed all permissible limits.

The logical way out of the situation is as follows. Along with existence of secondary disk, it should be assumed that this disk mainly consisted of ionized gas and dust matter, which was in a state of partial or complete corotation, i.e., almost completely carried away by magnetic field of rotating Sun. As a result, gas–dust particles of disk after recombination acquired an ordered motion around Sun in forward direction. Most of them then fell onto Sun, giving it their rotational momentum. Therefore, the expenditure of Sun's rotational momentum to impart rotation to disk in desired direction was partially compensated by fall of disk's matter onto Sun.

Such a mechanism for transferring the Sun's rotational moment to the disk and vice versa could be effective only if two conditions were met: (1) secondary disk had a predominantly gas–dust composition, allowing for possibility of ionization of its matter and (2) overwhelming majority of disk's matter fell onto Sun, which led to an increase in its mass.

The first of these requirements is well confirmed by composition of giant planets' shells (Table 2 in (Barenbaum 2025)). The second conclusion will be substantiated below.

3.4 Sun's Mass Before and After Phaeton Death

Since only a small part of mass of secondary disk went to form giant planets and their satellites, the rest of disk's matter presumably fell mainly onto Sun, transferring to it not only its moment rotation, but also mass. Therefore, at second stage of planet formation, along with growth of planets, the mass of Sun itself increased. Due to large mass of substance in secondary disk (18), the increase in Sun mass must have been quite significant.

Let us calculate the increase in Sun's mass in two different ways.

The first method: This method is based on the analysis of planet's position in coordinate system of Fig. 4. It was previously noted that planets of group I (Venus, Earth, Uranus, and Neptune), which did not change or slightly changed their mass, in Fig. 4 fit quite well on one straight line. Whereas planets of group II (Mercury, Mars, Jupiter, and Saturn), in contrast to them, do not have a common trend.

This circumstance can be explained by non-simultaneous formation of planets of both groups under conditions of growth of Sun mass at second stage of Solar System evolution. If we assume that Mercury and Jupiter began to form even before Sun mass changed greatly, and Mars and Saturn later, when Sun mass approached its modern value, then between eras of formation of the first and second pairs of planets, Sun mass could well have increased by 1.5 ± 0.1 times.

From here, as a lower estimate of Sun mass at the first stage of its formation, we will take mass of Sun $M'_O \approx 0.67\ M_\odot$, which is 67% of its current value.

The second method: According to the law of conservation of angular momentum, the product of Sun's mass by radius of planetary orbit $M_O R_0$ should not depend on change in Sun mass. Therefore, the density of substance of secondary disk (23) and its mass (18), assuming a constant disk thickness Z, should depend on Sun's mass as

$$M_d(M_O)\mu M_O^{3.45} \tag{25}$$

In the previous calculations, Sun's mass was assumed to be constant and equal to its current value $M_\odot = 1.99 \times 10^{33}$ g. Therefore, the values of M_d and L_d of secondary disk turned out to be somewhat overestimated. Now we will clarify these estimates.

Assuming for certainty that all matter of secondary disk was spent on increasing mass of Sun, we will have

$$M'_O + M_d(M'_O) = M_\odot. \tag{26}$$

Solving this equation taking into account dependence (25), we find

$$M'_O = 0.7M_\odot \textit{ and } M_d = 0.3M_\odot. \tag{27}$$

The obtained estimates of M'_O and M_d are consistent with Sun's mass before Phaeton destruction, found by the first method, as well as with mass of secondary disk, calculated using formula (18) for Sun's mass M'_O after the first stage its formation. Thus, both calculation methods lead to the conclusion that after Phaeton death, as a result of acquiring new gas and dust matter, the Sun was able to increase its mass by approximately half.

In connection with obtained conclusion, as well as the conclusion about constant "feeding" of Sun with new cosmic substance, one cannot ignore by silence and fail to mention the problem of solar neutrinos in the "standard" Sun model.

It is known that luminosity of Sun and flux of solar neutrinos registered by a chlorine detector are determined by proton–proton reaction in center of Sun. However, according to experiments of Davis (1968), the neutrino flux from Sun is lower than it follows from theoretical models, at least by two to three times (Bakal 1993). This discrepancy has fundamental nature and, according to J. Bakal, cannot be caused by errors in either calculation or experiment.

The discrepancy between the "standard" Sun model and measurements can be explained by: (1) transformation of one type of neutrino into another (Mikheev and

Therefore, before analyzing the distribution of satellites on graphs in Fig. 6, let us discuss the question of how satellites are formed around giant planets.

4.3 Large and Small Satellites of Giant Planets

It is known that along with regular large satellites with a diameter of thousands of kilometers, many small bodies tens to hundreds of kilometers or less in size revolve around giant planets today. The properties of these bodies, such as their non-spherical shape, their frequent retrograde motion around planet, their large orbital eccentricity, and orbits themselves, which do not necessarily lie in equatorial plane of planet, indicate that the small satellites are mostly asteroid bodies, captured in one way or another by gravity of planets (Cruikshank et al. 1985).

Theoretical analysis shows (Goldreich 1975) that after an asteroid is captured by a planet, it tends to move it into the plane of its rotation and give satellite's orbit a circular shape. Therefore, the very fact of existence of small satellites around giant planets suggests that these satellites could have been captured by planets relatively recently, and have not yet had time to move into the equatorial plane of planet that captured them and acquire a circular orbit.

The COSS concept states that each time the Sun enters the galactic comet stream, many secondary cometoasteroids are ejected from the asteroid belt into interplanetary space, which can be captured by gravitational pull of one or another planet giant and become part of its cometary family (Barenbaum 2002, 2024). Considering that this process has been going on for 4.6 billion years, these cometoasteroid bodies could have been captured by outer planets both immediately after Phaeton death, when asteroid belt was surrounded by dense accretion disk of Sun, and later, when density of matter in this disk dropped sharply. Therefore, the main differences between large and small satellites can only be associated with different times of their appearance at planets.

Large (regular) satellites are very old objects that could have arisen immediately after Phaeton death, when a powerful secondary disk of Sun existed beyond outer edge of asteroid ring. In contrast, small satellites are younger bodies that were captured later. By time they were captured, density of matter in secondary disk had decreased so much that satellite theses were never able to acquire a sufficiently thick "coat" of gas and dust matter. Therefore, in Fig. 5, they are all located at $m/M_O < 10^{-7}$, i.e., outside accretion disk region of planet that captured them.

The graphs in Fig. 6 show that formation of regular satellites was also far from simultaneous. Satellites that emerged first were able to accumulate a larger amount of gas–dust matter and therefore have, on average, larger sizes and masses. Satellites that emerged later did not experience such a significant growth; they are all characterized by smaller values of mass and orbital momentum. Due to increase in planet's mass over time, the later satellites in Fig. 6 are located above than satellites that arose first.

Let us point out one more important circumstance connected with planet growth. Due to the conservation of angular momentum, distance of satellites from it decreases proportionally. In accordance (Barenbaum 1992) for Uranus is 23%. However, for Saturn and, especially, for Jupiter, the changes in distances of satellites from planet could be 14.18 and 77.7 times, respectively.

Therefore, it should be assumed that of all regular satellites we observe today only those either arose "late," when the mass of giant planets approached their present-day mass, or were captured, albeit "early," but at a great distance from planet, i.e.:

$$R(t) > R_R \frac{M_2}{M(t)}, \tag{35}$$

where M_2 and $M(t)$ are, respectively, present mass of planet and its mass at the time of satellite's formation; R_R is the Roche limit, which determines minimum distance of satellite from planet, starting from which it is destroyed by tidal forces of planet.

The value of Roche limit can be calculated by formula $R_R = 2.46 r_{pl} (\rho_{sat}/\rho_{pl})^{1/3}$, where r_{pl} is planet radius, ρ_{pl} and ρ_{sat} are density of substance of planet and satellite.

Taking into account that regular satellites nearest to Jupiter, Saturn, and Uranus are today moved off their planets at distances prevailing only a little Roche limit, one should suppose that nearer and greater satellites of these planets (surely if such were earlier) have already ceased their existence. The satellites, as they grew, came so close to planet that they entered its Roche lobe and were destroyed, forming rings. Many such rings, apparently formed during this process, exist today around all the giant planets (Gorkaviy and Friedman 1994).

4.4 Mutual Growth of Satellites and Planet

Consider a small satellite forming in the field of influence of greater planet under conditions of their simultaneous growth from a matter of circumplanetary disk (Fig. 7). Let at some moment of time t_O the satellite location in log–log coordinate system $\{L_{orb}, m/M_O\}$ is given by a point "O." Let us explore different options for mutual change of masses of both bodies on coordinate plane of this figure.

Let us assume that only the mass of central planet's M_O grows, and satellite mass does not change at all. Then, with the growth of planet's mass, the satellite will be forced to move on graph in Fig. 7a to the left parallel to the abscissa axis. Where at some point in time t' ($t' > t_O$) it will occupy position "A." It is obvious that orbital momentum of satellite is preserved, but satellite itself, due to the momentum conservation condition $L_{orb} = M_O R = M_O' R'$, where R and R' are initial and final distances of satellite from planet, will reduce its distance to planet by the same amount of times.

Let the mass of central planet remain unchanged, and only the mass of satellite increase. Let us consider two extremes, but very important cases: (1) the satellite moves around planet in the same direction as disk matter and (2) the rotation of satellite and rotation of disk matter occur in opposite directions. Depending on this, two different situations will take place.

(3), and Callisto (4), as well as Uranus's satellites: Oberon (5), Umbriel (3), and Miranda (1) appeared later. At the time of appearance of Callisto and Miranda, the mass of their central planets could have increased significantly, exceeding the initial value of M1 for Jupiter by ~ 1.5 times and for Uranus by ~ 1.1 times.

As for the regular satellites of Saturn, their formation, as we believe, was delayed in time even more. According to constructions of Fig. 8, of all satellites of Saturn, the massive Titan (6) was apparently the first to form, then, probably, the triad of satellites Rhea (5)— Dione (4) —Tethys (3) appeared, as well as the pair Enceladus (2) — Mimas (1), and only then Iapetus (8)— Hyperion (7). The last of regular satellites of giant planets, as we believe, was satellite Nereid (2), captured by Neptune.

Thus, our analysis shows that satellites that appeared first tend to be larger in size, have greater mass, and have higher average densities. The satellites that appeared later are clearly inferior to the first ones in all these parameters.

4.6 *Rate of Mutual Growth of Planets and Satellites*

The arrangement of points in Fig. 8b is compatible with assumption that the formation of regular satellites of giant planets occurred during several active phases, shifted in time from one another. Each of the phases was marked by appearance of a new group of large satellites.

Table 4 shows the relative growth rates of giant planets and their satellites, calculated using formulas (20) and (21) under the assumption that satellites arose in pairs. For comparison, the table also shows an estimate of relative rate of mass gain by planets and Sun at the initial stage of Solar System formation.

Without dwelling on a detailed analysis of data in Table 4, we will briefly outline some of conclusions that follow from our consideration:

1. The formation of satellites of outer planets began immediately with the beginning of Phaeton destruction. This process apparently had several active phases. According to the data we have, at least two such phases can be distinguished in Jupiter system, three in Uranus system, and four in Saturn system. The numbers of phases for Jupiter and Uranus are clearly underestimated. The total number of satellite formation cycles was probably more than four.
2. Most of the large satellites of giant planets probably formed in groups, most often in pairs. In our opinion, such pairs are: Io (1)—Ganymede (3) and Europe (2)—Callisto (4) in Jupiter system; Ariel (2)—Titania (4) and Umbriel (3)—Oberon (5) in Uranus system and Mimas (1)—Enceladus (2) and Hyperion (7)—Iapetus (8) in Saturn system.
3. In each pair, the satellite closest to planet has smaller dimensions and mass, but usually has a higher average density of matter. Satellites that appeared first, all other things being equal, are larger and denser than those that appeared later.
4. The masses and sizes of regular satellites, as well as the sequence of their appearance near different planets, indicate that the "embryos" of satellites could have

Table 4 Relative growth rates of satellites and planets

Satellite formation cycles	Growth parameters		
	μ	ν	V_{sat}/V_{pl}
Jupiter			
I. Io (1)–Ganymede (3)	2.3599	1.9096	1.554
II. Europe (2)–Callisto (4)	2.4037	1.6506	2.174
Uranus			
I. Ariel (2)–Titania (4)	2.4579	1.4810	4.354
II. Umbriel (3)–Oberon (5)	2.4541	1.4899	4.301
III. Miranda (1)	–	–	–
Saturn			
I. Titan (6)	–	–	8.002
II. Tethys (3)–Dione (4)–Rhea (5)	2.7158	1.2147	
III. Mimas (1)–Enceladus (2)	2.5646	1.3179	5.863
IV. Hyperion (7)*–Iapetus (8)	2.6749	1.2354	7.423
Solar system (Stage I)	2.4131	1.6130	3.731

* Hyperion shows signs of significant mass loss after accretion. Therefore, estimates for Cycle IV in Saturn system are approximate

been not only fragments of Phaeton captured by outer planets, but also matter of these planets themselves.

5. It can be assumed that as a result of falls of Phaeton's fragments on outer planets, they, like Earth and Mercury, also lost surface layer of their substance. This substance, which ended up in near-planetary orbits, could have been the center of formation of large regular satellites.
6. The formation of large satellites of giant planets according to the "lunar" scenario (see article I) or their formation as a result of "planetesimal" growth in secondary accretion disk of Sun—these are, in our opinion, the two main ways of satellite formation. Moreover, for satellites that arose early, the first mechanism played a dominant role, and for satellites born later—the second. A formation method similar to the Moon should be allowed, first of all, for Triton—satellite of Neptune, satellites of Jupiter Io (1) and Europe (2), and also to some extent for Titan—satellite of Saturn.
7. The rate of increase of satellites' mass relative to rate of planet growth is higher, the smaller the value of latter. This rate is low in Jupiter system and higher in Uranus and Saturn systems. We associate the high rate of satellites' growth in Saturn system with their later time of appearance around this planet.
8. Comparison of the growth rate of satellites in Jupiter, Uranus, and Saturn systems with the growth rate of planets in Sun accretion disk at the first stage of planet formation gives grounds to believe that Solar System planets began to form when the protosun mass reached 0.1–0.3 of present value. Therefore, Solar System

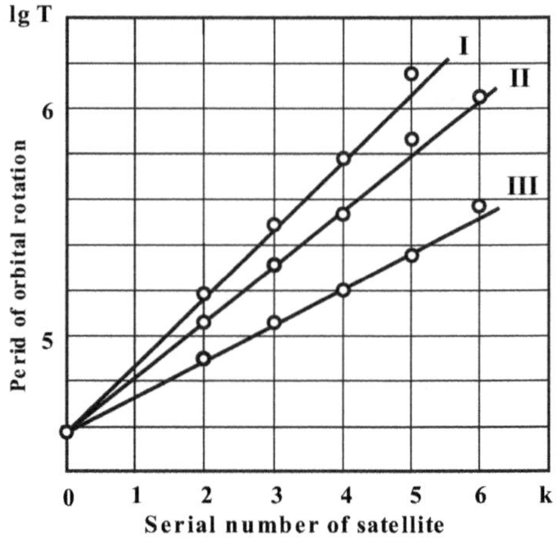

Fig. 9 Dependence of orbital periods on ordinal number of regular satellites in systems of Jupiter (I), Uranus (II), and Saturn (III)

5.2 *Actual Data*

Table 5 shows the ratios of orbital periods of individual satellites according to data (Gorkavy and Friedman 1994), written as ratio of two small integers *k* and *m* under condition max{k, m} < 9, adopted by Molchanov (1966).

Since condition (42) is not strictly fulfilled in the case of parametric resonance, the selection of numbers k and m can, in principle, be ambiguous. Let us also take into account that the width of instability zones of such resonances is a priori unknown. Therefore, Table 5 shows several possible combinations of numbers *k* and *m*, which are arranged in order of increasing deviation of value *k*/*m* from actual ratio of periods.

The measure of this deviation is value of ε_{ij} (indicated in brackets), calculated as

$$\varepsilon_{i,j} = 100\left|\frac{T_i}{T_j} - \frac{k}{m}\right| \Big/ \frac{T_i}{T_j}, \tag{45}$$

where i and j are the serial numbers of satellites ($i \neq j$).

According to data in Table 5, the motion of satellites of all planets is organized in such a way that their orbital periods with an error of several percent—the width of instability zone—are related as some integers. The set of combinations of these numbers for each satellite system is limited and is determined by the ratio of periods of neighboring bodies, which on average weekly depends on serial number of satellite.

Let us designate the "main" resonant ratio as

$$\Gamma = < \mathrm{T_{i+1}} / \mathrm{T_i} > \tag{46}$$

where the angle brackets indicate the averaging procedure.

Based on formula (46) for the scale parameter λ we will have

Table 5 Commensurabilities in periods of satellites of giant planets

k	Name of planets and satellites	Rotation period, days	Satellite number						
			0	1	2	3	4	5	6
Jupiter system									
0	Jupiter	0.4097		4:9 (4.1) 1:2 (7.9)	2:9 (4.1) 1:4 (7.9)	1:9 (3.7)			
1	–	0.8845	9:4 (4.2) 2:1 (7.4)		1:2 (0.0)	1:4 (0.4)	1:8 (1.1) 1:9 (10.1)		
2	Io	1.769	9:2 (4.1) 4:1 (7.4)	2:1 (0.0)		1:2 (0.4)	1:4 (1.1)	1:9 (4.8)	
3	Europe	3.551	9:1 (3.8)	4:1 (0.4)	2:1 (0.4)		1:2 (0.8)	2:9 (4.4) 1:5 (6.0)	
4	Ganymede	7.155		8:1 (1.1) 9:1 (11.3)	4:1 (1.1)	2:1 (0.7)		3:7 (0.04) 4:9 (3.7)	
5	Callisto	16.689			9:1 (4.6)	9:2 (4.3)	7:3 (0.04) 9:4 (3.5)		
Uranus system									
0	Uranus	0.450		5:9 (0.5) 4:7 (2.3)	1:3 (4.4)	1:6 (6.7)	1:9 (2.1)		
1	–	0.806	9:5 (0.5) 7:4 (2.2)		4:7 (0.0)	1:3 (4.3)	1:5 (2.6)		
2	Miranda	1.410	3:1 (4.3)	7:4 (0.0)		4:7 (2.1)	1:3 (2.2)	1:6 (2.5)	1:9 (5.7)
3	Ariel	2.521	6:1 (7.1) 5:1 (10.7)	3:1 (4.1)	7:4 (2.1)		4:7 (6.3)	2:7 (1.3)	1:5 (6.4)
4	Umbriel	4.135	9:1 (2.1)	5:1 (2.6)	3:1 (2.3)	5:3 (1.6) 7:4 (6.7)		1:2 (4.9)	2:7 (7.1) 1:3 (8.4)

(continued)

Table 5 (continued)

k	Name of planets and satellites	Rotation period, days	Satellite number						
			0	1	2	3	4	5	6
5	Titania	8.675			6:1 (2.5)	7:2 (1.2) 3:1 (12.8)	2:1 (4.7)		2:3 (3.0) 4:7 (11.7)
6	Oberon	13.407			9:1 (5.3)	5:1 (6.0)	3:1 (4.3) 7:2 (7.7)	3:2 (2.9)	
Saturn system									
0	Saturn	0.426		5:8 (1.3)	4:9 (1.7)	1:3 (7.2)	2:9 (1.5)	1:6 (7.1)	
1	–	0.673	8:5 (1.3)		5:7 (0.0)	1:2 (1.8)	3:8 (5.2) 1:3 (6.5)	1:4 (1.7)	1:7 (4.1)
2	Mimas	0.942	9:4 (1.8)	7:5 (0.0)		2:3 (3.0) 5:7 (3.9)	1:2 (0.2)	1:3 (3.2)	1:5 (4.1)
3	Enceladus	1.370	3:1 (6.7)	2:1 (1.7)	3:2 (3.4) 7:5 (3.7)		5:7 (1.6) 3:4 (3.4)	1:2 (0.1)	2:7 (5.8) 1:3 (9.9)
4	Tethys	1.888	9:2 (1.5)	8:3 (4.9) 3:1 (6.9)	2:1 (0.1)	7:5 (1.6) 4:3 (3.2)		2:3 (3.4) 5:7 (3.6)	3:7 (2.6) 2:5 (4.3)
5	Diona	2.737		7:1 (4.3)	3:1 (3.3)	2:1 (0.1)	7:5 (3.4) 3:2 (3.5)		3:5 (1.0)
6	Rhea	4.518			5:1 (4.2)	7:2 (6.1) 3:1 (9.0)	7:3 (2.5) 5:2 (4.5)	5:3 (1.0)	

Note Along with orbital periods of regular satellites, the table also indicates period of axial rotation of central planet (k = 0) and period of orbital motion of a fictitious satellite (k = 1), which is in strict resonance with orbital period of satellite k = 2

$$\lambda = \ln \Gamma \tag{47}$$

It seems obvious that with strict fulfillment of resonance all commensurabilities of system must obviously satisfy the following immutable rule:

(i, j)	0	1	2	…	i	…	k
0	Γ^0	Γ^{-1}	Γ^{-2}	…	Γ^{-1}	…	Γ^{-k}
1	Γ^1	Γ^0	Γ^{-1}	…	Γ^{1-i}	…	Γ^{1-k}
…	…	…	…	…	…	…	…
i	Γ^j	Γ^{j-1}	Γ^{j-2}	…	Γ^{j-i}	…	Γ^{j-k}
…	…	…	…	…	…	…	…
k	Γ^k	Γ^{k-1}	Γ^{k-2}	Γ^{k-i}	Γ^0	…	…

(48)

In the idealization (48), each matrix element of this table must represent the number Γ, raised to an integer power, the value of which is determined by place of given element in table.

Obviously, in this ideal case, the exponential law of planetary distances (43) and (44) will be fulfilled exactly, and for all satellites at once.

The real state of satellite systems, as follows from data in Fig. 9, is close to this case. Therefore, using rule (48), from the actual ratio of orbital periods of individual satellites, we can find the estimates of number Γ that interest us:

$$\Gamma_{ij} = \left(\frac{T_i}{T_j}\right)^{i-j}, \; i \neq j. \tag{49}$$

Partial estimates of the main resonance ratio, as well as its average estimates obtained with different degrees of averaging of number Γ_{ij} values, are given in Table 6.

5.3 Discussion of Results

The data in Table 6 show that not only is the resonance number $< \Gamma >$ itself close to the ratio of some small numbers, but the square $< \Gamma^2 >$ and cube $< \Gamma^3 >$ of this number correspond to this remarkable feature to an even greater extent:

$$\begin{aligned} \Gamma_S : \Gamma_U : \Gamma_J &= 7/5 : 7/4 : 2; \\ \Gamma_S^2 : \Gamma_U^2 : \Gamma_J^2 &= \;\; 2 \;\; : \;\; 3 \;\; : 4; \\ \Gamma_S^3 : \Gamma_U^3 : \Gamma_J^3 &= \;\; 3 \;\; : \;\; 5 \;\; : 9, \end{aligned} \tag{50}$$

where Γ_S, Γ_U, and Γ_J are the resonance relations in systems of Saturn, Uranus, and Jupiter, estimated in Table 6.

Table 6 Calculation of the main resonance ratio

k	Object	0	1	2	3	4	5	6	Γ_k*
Jupiter system									
0	Jupiter		2.1589	2.0779	2.0541	2.0443	2.0989		2.0868
1	–	2.1589		2.0541	2.0037	2.0074	2.0842		2.0508
2	Io	2.0779	2.000		2.0073	2.0111	2.1130		2.0419
3	Europe	2.0541	2.0037	2.0073		2.0149	2.1679		2.0496
4	Ganymede	2.0443	2.0074	2.0111	2.0149		2.3325		2.0820
5	Callisto	2.0989	2.0842	2.1130	2.1679	2.3325			2.1593
									2.0784**
Uranus system									
0	Uranus		1.7904	1.7701	1.7760	1.7411	1.8087	1.7807	1.7743
1	–	1.7904		1.7500	1.7689	1.7249	1.8114	1.7548	1.7667
2	Miranda	1.7701	1.7500		1.7879	1.7125	1.8324	1.7560	1.7681
3	Ariel	1.7760	1.7689	1.7879		1.6402	1.8550	1.7455	1.7622
4	Umbriel	1.7411	1.7249	1.7125	1.6402		2.0799	1.8006	1.7862
5	Titania	1.8087	1.8114	1.8324	1.8550	2.0799		1.5455	1.8252
6	Oberon	1.7807	1.7548	1.7560	1.7455	1.8006	1.5455		1.7251
									1.7725**
Saturn system									
0	Saturn		1.5799	1.4872	1.4762	1.4514	1.4507	1.4823	1.4879
1	–	1.5799		1.4000	1.4269	1.4109	1.4201	1.4635	1.4502
2	Mimas	1.4872	1.4000		1.4543	1.4165	1.4269	1.4799	1.4441
3	Enceladus	1.4762	1.4269	1.4543		1.3796	1.4135	1.4885	1.4398
4	Tethys	1.4514	1.4109	1.4165	1.3796		1.4481	1.5461	1.4421
5	Diona	1.4507	1.4201	1.4269	1.4135	1.4481		1.6507	1.4683
6	Rhea	1.4823	1.4635	1.4799	1.4885	1.5461	1.6507		1.5185
									1.4644**

* The column shows the results of averaging number Γ_{kj} by index j
** The values obtained by averaging number Γ_{kj} by both indices are indicated

The system of commensurabilities (50) actually generalizes the results of Dermott (1968), who was the first to draw attention to existence of second of these relations.

From relations (50), it is evident that the law of geometric progression (43) and (44) is a consequence of parametric resonance of the entire satellite system as a whole, and not just the interaction of neighboring bodies. Therefore, the scale parameter λ reflects not any relations (47), but only those that correspond to the rule of integerness (50).

Another important property of real satellite systems is that relations (50) are valid for them only in the form of commensurabilities, i.e., they are fulfilled approximately. For none of them can one select such values of the number $< \Gamma >$, at which relations (50) would be strict. This means that in satellite systems of giant planets, the ideal collective resonance (48) is fundamentally unachievable. Therefore, the periods of satellites must experience some oscillations around the number $< \Gamma >$ as the equilibrium position of entire satellite system.

The study of such systems with a quasi-stable equilibrium state shows (Arnold 1990) that their resonance has character of strange attractor with very complex, in most cases, non-periodic oscillations. The details of oscillations are extremely sensitive to small changes in initial conditions, while average characteristics of regime do not depend on them. One of the first to confirm this character of motion of Solar System planets by numerical calculations was Hills (1970).

This general position of the theory is illustrated in Fig. 10 by dependences of average estimates of the Γ_i value on ordinal number of satellite.

We see that for all satellite systems the stable state of equilibrium is divided into 2 – 3 closely located levels. The difference between one level and another is on average 2%, which is obviously less than the width of resonance excitation zone. According to the data in Table 6 and graphs in Fig. 10, the resonance excitation zone is about 6%. The satellites of central part of system are most closely adjacent to the lines of stable equilibrium. While its extreme members are removed from these levels. This character of resonance curves definitely indicates the presence of dissipative forces in system.

Of particular importance in Fig. 10 is the fact that for all three satellite systems the partial estimates of the number $< \Gamma >$ lie somewhat above the lines of stable resonance. This fact gives grounds to conclude that in all cases we are dealing with a self-oscillating system, fed with energy from the outside—primarily, due

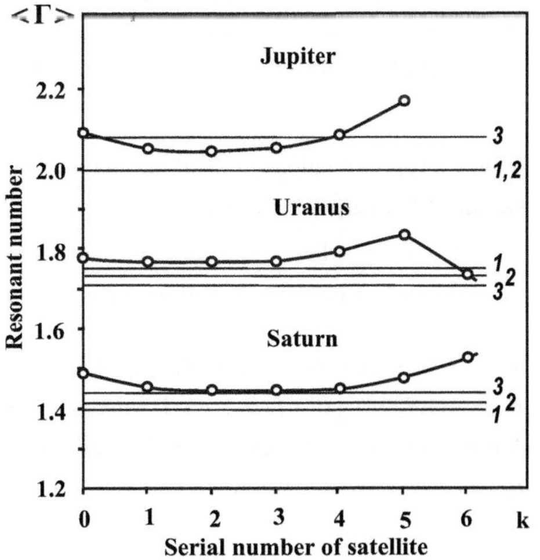

Fig. 10 Dependence of the estimate of magnitude of resonance ratio Γ on serial satellite number k: the horizontal lines indicate positions of resonance levels in accordance with formula (50): 1—level $< \Gamma >$, 2—level $(<\Gamma^2>)^{1/2}$, and 3—level $(<\Gamma^3>)^{1/3}$

to the reserve of energy of axial rotation of central planet. Thus, our analysis shows that the basis for quantization of orbits of satellites of giant planets is essentially phenomenon of collective parametric resonance, theoretically established for weakly coupled mechanical systems by A. Molchanov.

At the same time, it also turns out that the families of satellites of Jupiter, Saturn, and Uranus exhibit all the necessary qualities that allow them to be classified as self-organizing systems (Nicolis, Prigogine 1979; Haken 1980; Gorkavy 1990).

6 Titius–Bode Rule

Due to the two-stage formation of Solar System, the dependence of distances of planets from Sun turns out to be more complex than in systems of satellites of giant planets (Table 7).

The first approximation of this dependence was proposed in 1766 by J. Titius (1729–1796) and widely promoted by J. Bode (1747–1826). Later, the Titius–Bode formula was repeatedly modernized. The most accurate analytical expressions, applicable not only to planets but also to satellites of giant planets, were proposed by Blagg (1913) and Richardson (1945).

Figure 11 shows the dependences of orbital periods of planets on the ordinal number of planet from Sun. The dependences are constructed according to the same rules as the graphs for satellites of giant planets (Fig. 9). The planet closest to center, Mercury, as in families of satellites of giant planets, is assigned the number k = 2.

Since the rotation of Sun slows down over time, at the point k = 0 two different periods of Sun axial rotation are shown: the modern $T = 2.18 \times 10^6$ s, marked with a circle, and ≈50 times smaller, shown as a triangle, corresponding to the beginning of the second cycle of planet formation.

We see that, compared to the graphs for satellites of giant planets (Fig. 9), the similar dependence for planets in Fig. 11 is characterized by at least two differences. In our opinion, both differences are due to the two-stage formation of Solar System.

Table 7 Approximations of distribution function of planetary orbital radii

Orbital radii distribution function	Author
$R_n = 4 + 3 \cdot 2^n$, where $n = -\infty, 0, 1, 2$, etc. with Earth's orbit radius $R_1 = 10$	Nieto (1976)
$R_n = \alpha + b2^n$, where $n = -\infty, 0, 1, 2$, etc.; $\alpha = 0.387$ and $b = 0.293$ with $R_1 =$ 1	Wurm (1787)
$R_n = \alpha + bc^n$, где $n = -\infty, 0, 1, 2$, etc.; α, b, and c are constants	Gilbert (1802)
$R_n = A(1.7275)^n[B - f(\alpha + n\beta)]$, where $n = -\infty, 0, 1, 2$, etc.; A and B are constants, f—periodic function with 2π period, α and β are constant angles	Blagg (1913)
$R_n = (1.728)^n \rho_n(\theta_n)$, where $\theta_n = n(4\pi/13) \equiv n\theta_1$, ρ_n—function with 2π period	Richardson (1945)

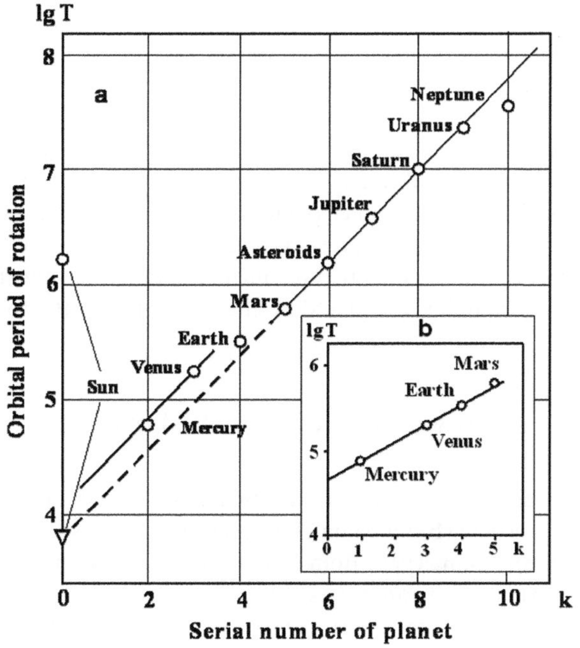

Fig. 11 Dependence of orbital periods of planets on their ordinal number: a—for all planets; b—for the system of inner planets; for Sun it is assumed that k = 0; the Sun's rotation period at the beginning of second cycle of planet formation is shown by a triangle

The first difference is the existence of two lines of collective resonance: one for the "inner" planets (Mercury, Venus, Earth, and Mars), and the second for the "outer" planets, which include giant planets and destroyed Phaeton, represented today by asteroid belt. Another difference is that, due to the transfer by Sun its axial moment rotation to giant planets, the collective resonance for the inner planets (Fig. 11b) is determined not by the current period of Sun rotation, but by its period at the beginning of second cycle of planet formation.

Thus, we come to the conclusion that Phaeton death led to the two different "zones" of parametric oscillations in Solar System. The outer zone, in which, as A. Molchanov rightly noted, Jupiter plays the main role. And the inner one, caused by Sun axial rotation. At result, the main resonance relationships in systems of the inner and outer planets turn out to be different.

For the outer zone, the first three resonance ratios, similar to (50), can be represented as

$$\Gamma : \Gamma^2 : \Gamma^3 = 5/2 : 6 : 15 \quad (51)$$

For the inner zone, where Sun has a decisive influence on planets, the set of resonance numbers is different

$$\Gamma : \Gamma^2 : \Gamma^3 = (3/2,\ 5/3) : 5/2 : 4. \quad (52)$$

The value of number Γ in formula (52), given in brackets, can be represented on equal grounds by two different ratios of small integers 3/2 and 5/3. Due to the competition of these ratios, this state of Sun planetary system cannot be stable. Perhaps this circumstance explains the absence of once more planet between Mercury and Venus, as well as the anomalously large eccentricity of Mercury's orbit.

From formulas (51) and (52), it follows that the resonant number 5/2 is common for the inner and outer planets. Therefore, the Titius–Bode planetary distance rule and its analytical expressions (see Table 7) reflect with varying degrees of accuracy the regime of collective parametric oscillations existing in Solar System, which is more complex than that described by S. Dermott's formula (43).

7 Main Results and Conclusions

In addition to the issues of the two-stage formation of Solar System, considered in the first article (Barenbaum 2025), this article considers two phenomenological models (Barenbaum 2010). Both models allow us to study the formation of planets in accretion disk of forming stars in the conditions of "open" systems, where the application of strict physical theories and methods of celestial mechanics is significantly limited. The models are designed to study processes in Solar System at the second stage of its formation after Phaeton destruction.

The first model allows us to study the general conditions for the formation of an accretion disk in young stars, in which the growth of star and its planets occurs at the stage of planet formation. The model is used to quantitatively estimate the change in mass of Sun and giant planets in the secondary accretion disk of Sun, as well as in the analysis of the sequence of formation of regular satellites in families of giant planets in phase of most active Phaeton's fragmentation.

The second model considers the evolution of already formed Solar System and is intended to study the mechanism of quantization of the orbits of planets and their satellites, leading to regularities such as the Titius–Bode rule.

The main results obtained using both models are.

- The integral parameters of the primary (protosolar) and secondary accretion disks of Sun were established. The disks differed in the distribution of average density of matter as they moved away from Sun, obeying a power law in the first case and an exponential law in the second. The secondary disk, where giant planets and their regular satellites were formed, was smaller than the primary one and was located beyond the outer edge of asteroid belt. The effective thickness of both disks was 2.6×10^9 km.
- At the time of Phaeton's destruction (4.567 billion years ago), the Sun's mass was ≈70% of its present mass. Therefore, the planets arose in primary disk at a greater distance from Sun than they are located today. This conclusion is even more applicable to satellites of giant planets. The formation of giant planets, like Sun itself, is a complex two-stage process.

- Regular satellites of giant planets were most intensively formed in the first ~ 10^8 years after beginning of Phaeton's destruction, and they were formed in groups. Several active phases can be traced in their formation, caused by cyclic bombardments of Solar System by galactic comets.
- The "embryos" in formation of regular satellites were both cometoasteroid bodies leaving asteroid belt and the substance of giant planets themselves, ejected when large fragments of Phaeton fell on them. An example of the second process is the mechanism of the Moon formation from Earth's substance (Barenbaum 2025).
- Using the phenomenon of collective parametric resonance, it has been shown that quantization of the orbital periods of Solar System planets (Molchanov 1966) and satellites of giant planets (Dermott 1968) is possible only in the form of commensurabilities, i.e., it is carried out approximately. Quantization occurs in mechanically weakly coupled (according to V. Molchanov) "open" self-oscillating systems, fed by the energy of an external periodic source. Such a source in the systems of giant planet satellites is the reserve of energy of the axial rotation of central planet.
- The phenomenon of collective parametric resonance can also explain the Titius–Bode rule of planetary distances. However, due to the two-stage formation of Solar System, it has two different zones of parametric oscillations: "external" and "internal." In the external zone, which includes giant planets, the source of oscillation energy is primarily orbital motion of Jupiter. And in the internal zone, which includes terrestrial planets, the axial rotation of Sun. As a result, the Titius–Bode rule for planets is described by more complex expressions than Dermott's formula for satellites of giant planets.

References

Alfven H (1943) On the cosmogony of the solar system II. Stockholm's Observ Ann 14(5):3–32

Alfven H, Arrhenius G (1979) Evolution of the solar system. Mir, Moscow, p 512

Alfven H (1942) Remarks on rotation of a magnetized sphere with application to solar rotation. Arkiv for Mat Astr och Fysic 28A(6)

Allen CW (1973) Astrophysical quantities. Third edition. University of London. The Athlone Press.

Arnold VI (1990) Theory of catastrophes. Nauka, Moscow, p 128

Artemyev AV, Radzievskiy VV (1965) On the origin of axial rotation of planets. Astron J 42(1):124–128

Artemyev AV (1963) On the origin of the axial rotation of the Earth. Uchenye Zapiski Yaroslav. ped.in-ta. Astronomy. Yaroslavl 56:9–86

Astronomy (2007) The 21st century. In: Surdin VG, Fryazino (eds) PH Vek 2, p 608

Backus GE (1969) Critique of "The resonant structure of solar system" by A.M. Molchanov. Icarus 11:88–92

Bakal J (1993) Neutrino astrophysics. Mir, Moscow, p 623

Barenbaum AA (1992) Origin of asteroids and meteorites (new cosmological concept). In: Conceptual foundations of geology. Notes of the St. Petersburg Mining Institute 134:9–27

Barenbaum AA (1992a) The solar system in the epoch of the phaeton explosion. Origin of the moon. In: Conceptual foundations of geology. Notes of St. Petersburg Mining Institute 134:95–106

Barenbaum AA (2010) Galactocentric paradigm in geology and astronomy. LIBROKOM, Moscow, p 544
Barenbaum AA (1998) Effective thickness of gas-dust disks of stars at the stage of planet formation. In: Proceedings of the IV congress of astronomical society: proceedings of GAISh. Moscow State University, Moscow, pp 118–125
Barenbaum AA (1998a) Multifrequency parametric resonance in systems of satellites of giant planets. Ibid, pp 292–299
Barenbaum AA (2002) Galaxy, solar system, earth. In: Subordinate processes and evolution. GEOS, Moscow, p 393
Barenbaum A (2024) Origin of solar system comets. In: Chaplina T (ed) Processes in GeoMedia—volume 8. Springer Geology, pp 167–198
Barenbaum A (2025) Cosmogony of open solar system(I): formation of the giant planets and origin of the Moon. Processes in GeoMedia—Volume IX (in print)
Blagg MA (1913) On a suggested substitute for Bode's Law. Month Not Roy Astron Soc 73:414–422
Cruikshank DP, Degewij J, Zellner BG (1985) Outer satellites of Jupiter. Satellites of Jupiter. Mir, Moscow 1:147–164
Davis R, Harmer DS, Hoffman KC (1968) Search for neutrinos from the sun. Phys Res Lett 20:1205–1209
Dermott SF (1968) On the origin of commensurabilities in the solar system. II. The orbital period relation. Month Not Roy Astron Soc 141:363–376
Dorofeeva VA, Makalkin AB (2004) Evolution of the early solar system. In: Cosmochemical and Physical Aspects. Editorial URSS, Moscow, p 264
Fernandez JA, Horjales E, Zamalvide C (1976) Critical remarks to some theories on the origin of planetary rotation. Publication Department Astronomy Fishery University Montevideo (45):1–13
Fesenkov VG (1951) The criterion of tidal stability and its application in cosmogony. Astron J 28(6):492–517
Gibson E (1977) The quiet sun. Mir, Moscow, p 408
Gilbert LW (1802) Ist der Ophion (ein Planet jenseits der Uranusbahn), ein noch unbekannter Weltkörper? Ann der Phys 11:482–485
Giuli RT (1968) Gravitational accretion of small masses, attracted from large distances as a mechanism for planetary rotation. Icarus 9(1):168–190
Goldreich P (1975) History of the lunar orbit. In: Tides and resonances in the solar system. Mir, Moscow, pp 97–128
Gorkavy NN, Friedman AM (1994) Physics of planetary rings: celestial mechanics of continuum. Nauka, Moscow, p 348
Gorkavy NN (1990) Self-organization of the solar System. Scientific information of Astronomy Council of the USSR Academy of Sciences. Zinatne, Riga, (68):66–71
Gurevich LE, Lebedinsky AI (1950) Law of planetary distances and rotation of planets. Doklady AN SSSR 74(6):1061–1064
Ter Haar D (1950) Further studies on the origin of the Solar system. Astrophys J 111(1):179–190
Haken H (1980) Synergetics. Mir, Moscow, p 405
Harris AW (1977) An analytical theory of planetary rotation rates. Icarus 31(1)168–174
Henon M (1969) A comment of "the resonant structure of solar system" by A.M. Molchanov. Icarus 11:93–94
Hills IG (1970) Dynamic relaxation of planetary system and Bode's Law. Nature 225:840–842
Horedt GP (1979) Cosmogony of Solar system. Moon Planet 21(1):63–121
Hoyle F (1946) On the condensation of the planets. Mont Not Roy Astron Soc 106(5):406–422
Hoyle F (1960) On the origin of the solar nebula. Quart J Roy Astron Soc 1:28
Jeffreys H (1929) Collision and origin of rotation in the solar system. Mont Not Roy Astron Soc 29(7):636–641
Kiladze RI (1986) Modern rotation of planets as a result of the development of near-planetary swarms of small particles. Publishing House Metzniereba, Tbilisi, p 247

Kiladze RI (1965) On the question of the rotation of planets around their own axis. Bull Abastumani Astrophys Obs (32):231–234
Lyttleton RA (1972) On the formation of planets from a solar nebula. Mont Not Roy Astron Soc 158(4):463–483
MacDonald GJF (1963) The internal constitutions of the inner planets and the moon. Space Sci Rev 2:473–557
McClurg RL (1967) On an empirical relationship between diameter and mass of the Sun and the diameters masses and periods of axial rotation of the planets. Mont Not Roy Astron Soc South Afr 21(5):61–62
Migulin VE, Medvedev VI, Mustel ER, Parygin VN (1988) Fundamentals of the theory of oscillations. Nauka, Moscow, p 392
Mikheev SP, Smirnov AY (1987) Resonance oscillations of neutrinos in matter Uspekhi fizicheskikh nauk 153:3–58
Mitra V (1970) Axial rotation of the planets. Astron Asrophys 4(2):263–267
Molchanov AM (1966) Resonances in multifrequency oscillations. Doklady AN SSSR 168(2):284–287
Molchanov AM (1968) The resonant structure of Solar system. The law of planetary distances. Icarus 8:203–215
Molchanov AM (1969a) Resonances in complex system. A reply to critiques. Icarus 11:95–103
Molchanov AM (1969a) The reality of resonances in the Solar system. Icarus 11:104–110
Nieto MM (1976) Titius-bode law. Mir, Moscow, p 190
Nikolis G, Prigogine I (1979) Self-organization in nonequilibrium systems. Mir, Moscow, p 512
Physics of Space (1986) Little Encyclopedia. In: Sunyaev RA (ed) 2nd ed. Moscow: PH: Sov. Entsiklopediya. 783 p
Radzievskiy VV (1952) Origin of the moon in light of cosmogonic theory of O. Yu. Schmidt. Bull All-Russ Astron-Geod Soc (11):3–14
Rein NF (1934) On condensations of dust nebulae. 2. On rotation of condensations. Astron J 11(4):330–345
Richardson DE (1945) Distances of planets from the sun and of satellites from their primaries in the satellite systems of jupiter, saturn and uranus. Pop Astron 53:14–26
Safronov VS (1969) Evolution of the protoplanetary cloud and the formation of the Earth and planets. Nauka, Moscow, p 244
Schatsman E (1947) Surun probleme de mecanique rationnelle en relation avec la formation des plkanetes. Bull Astron 14:69–74
Schmidt OYu (1950) The emergence of planets and their satellites. Izvestiya AN SSSR Ser Phys 14(1):29–45
Shklovsky IS (1984) Stars: their birth, life and death. Nauka, Moscow, p 384
Shmidt OYu (1957) Four lectures on the theory of the Earth origin. PH AN USSR, Moscow (in Russian)
Staler SU (1991) Early stages of star evolution. Zh. V mire nauki, No.9, p 16–24 (in Russian).
Vityazev AV, Pechernikova GV, Safronov VS (1990) The terrestrial planets. Nauka, Origin and Early Evolution. Moscow, p 295
Vityazev AV, Pechernikova GV (1981) Solution of the problem of planet rotation within the framework of statistical accumulation theory. Astron J 58(4):869–878
Voitkevich GV (1979) Chemical evolution of the solar system. Nauka, Moscow, p 176
Williams IP, Bhattacharjes SK. (1979) The distribution of the spin angular momentum of the planets. Asrophys Sp Sci 66(2):319–324
Wurm WJF (1787) Verschidene astronomische Bemerkungen und eine Abhand-lung uber mogliche Planeten und Kometen unsers Sonnensystems. In: Decker GJ, Hofbuch-drucker K (eds) Astronomisches Jahrbuch (fur das Jahr 1790). Berlin 15:162–173

Weather Conditions of the Winter–Spring Season in Moscow in Early 2025

D. M. Frolov, V. E. Gagarin, A. V. Koshurnikov, and G. A. Rzhanitsyn

Abstract In 2024, Russia recorded its second warmest year since 1891, second only to 2020. The main consequences of climate change for Russia are melting glaciers and increased frequency of hazardous weather events, which makes the study of winter weather conditions relevant. The 2024/2025 winter season in Moscow was characterised by heterogeneity: snow cover was established in late November and existed until January, reaching a height of 20 cm, but then collapsed. January 2025 was the second January in history with a positive average temperature (+0.1 °C), more than 6 °C above normal, with 16 days of abnormally high temperatures. February 2025 was slightly warmer than normal with less precipitation. The average monthly temperature was −4.8 °C and precipitation was 6 mm. Snow cover was erratic during the month, reaching its maximum height only in the middle of the month. A special system with thermal sensors was used for temperature measurements, and the data were transferred for analysis and graphing in MATLAB, allowing us to observe changes in the thermal state of the air and snow.

Keywords Snow · Thermal monitoring · Glaciers · Winter · Ski slopes

According to the Russian Hydrometeorological Centre, 2024 was the second warmest year in Russia's meteorological record since 1891. 2020 was even warmer. The most harmful consequences of climate change for Russia are considered to be glacier melting and especially the increase in the frequency of hazardous weather events. Therefore, the study of winter weather conditions is necessary and important.

The winter season 2024/2025 in Moscow was characterised by heterogeneous weather conditions. Thus, the snow cover was established in late November–early December and existed until mid-January, reaching a maximum height of 20 cm. Then the snow cover broke up. The snow cover in Moscow was restored for a short time only in the middle of February.

D. M. Frolov (✉) · V. E. Gagarin · A. V. Koshurnikov · G. A. Rzhanitsyn
M.V. Lomonosov Moscow State University, Moscow, Russia
e-mail: denisfrolovm@mail.ru

T. Chaplina (ed.), *Processes in GeoMedia—Volume IX*, Springer Geology,
https://doi.org/10.1007/978-981-96-9302-3_5

More extraordinary in Moscow was probably only the winter of 2019/2020, when during the entire cold period in the city there was no stable snow cover.

Thus, in 2025, for the second time in the history of continuous instrumental meteorological observations conducted since the end of the nineteenth century, the average January temperature in Moscow turned out to be positive and amounted to 0.1 °C. This was more than 6 °C above the norm. Earlier, January 2020 went down in history with the same value.

In January 2025, there were 16 days with positive average temperature, of which in 11 cases even the minimum temperature remained above zero.

In January 2025, cloudy weather prevailed, precipitation was less than usual—74% of the norm. The sky was mostly covered with layered cumulus and layered clouds characteristic of warm and humid air masses.

February 2025 by temperature regime was slightly warmer than usual with the amount of precipitation much less than a multiyear average. According to observations at the EANE (Exhibition of Achievements of National Economy) weather station, the average monthly air temperature was −4.8 °C, which is 1.1 °C above the norm. February was also the coldest month of winter 2024/25.

On 1–8, 22 and 28 February, the anomaly of mean daily temperature was positive (3–10 °C 1–8, 1 °C 22 and 28). On 15–18, 20, 23–25 February it was negative (2–5 °C). On the other days, the mean daily temperature corresponded to the norm. On 1 February, the daily absolute maximum air temperature was exceeded. The new record value is 4.8 °C. This is the highest air temperature recorded in the past month. The night of 24 February was coldest (-14.2 °C).

Precipitation for the month was 6 mm (14% of normal). There were only 11 days with precipitation $\geq$ 0.1 mm.

On 1–14 February, stable snow cover was absent. The snow cover, which appeared in the following days, reached its maximum height by the morning of 20 February (5 cm, compared to the normal 28 cm). By the end of the month, its height decreased to 2 cm and it became not solid again.

To study the temperature in the surface air layer, we used a measuring system consisting of a measuring rod with temperature sensors, which are located at a distance of 5 cm from each other. The measuring system also consisted of a portable battery logger and a charger (Fig. 1).

The logger was equipped with an appropriate short cable for connecting the stylus. The logger had a solid polycarbonate (rectangular parallelepiped) housing. The cover was removable and sealed. Battery 1S1P Boston Swing (3.6 V 5300 mAh, Li-On). Permissible discharge temperature (operating temperature) −40.... + 70.

The electronics unit (located inside the case) is completely sealed.

The data memory is realised on a chip (flash memory), the volume is not less than 1 Mb and allows to register up to 16,384 measurements.

The logger measuring rod was placed vertically on the study site. Measurements were made every 4 h. The measurement data were then transferred to a computer and graph-diagram of changes in the thermal state of surface air and possibly snow were plotted in the MATLAB software environment (Fig. 2).

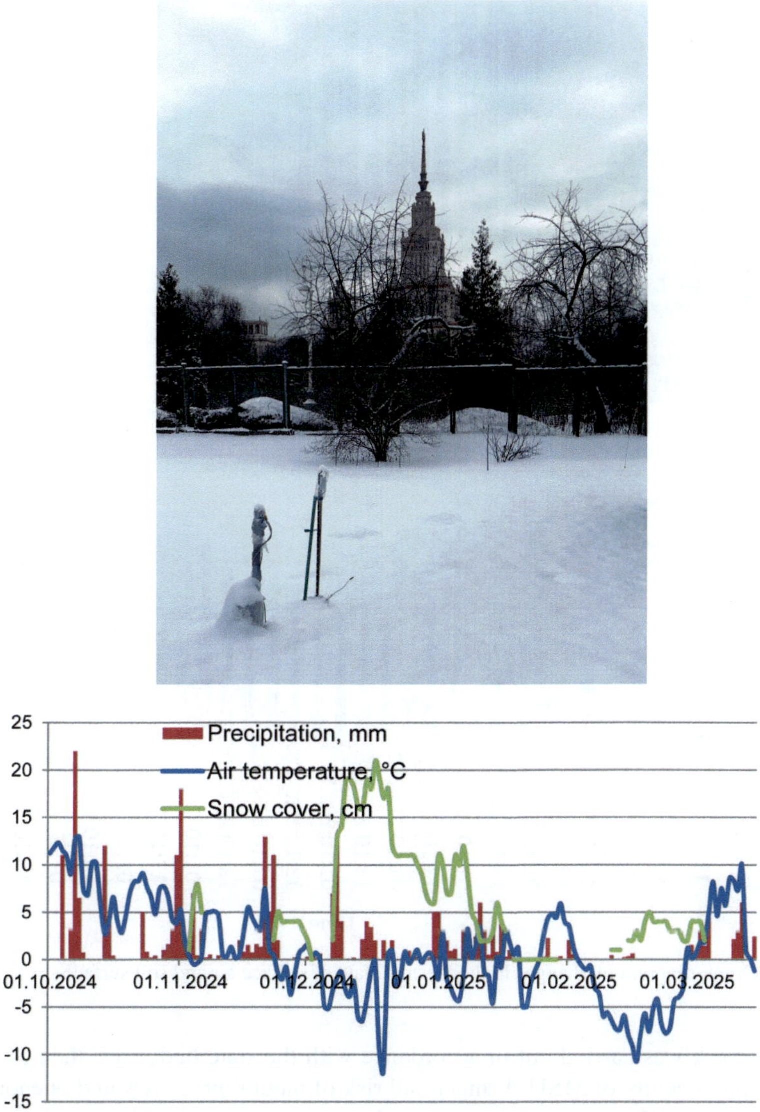

Fig. 1 Measuring system at the MSU meteorological site: measuring rod with temperature sensors spaced 5 cm apart and a portable battery logger

In Fig. 2, the graph shows, in particular, both periods of predominant cyclonic weather, when the difference between day and night temperatures is not so great and when the temperature difference at a height of 1 m and on the ground is not so great, and periods of predominant anticyclonic weather, when these differences are great.

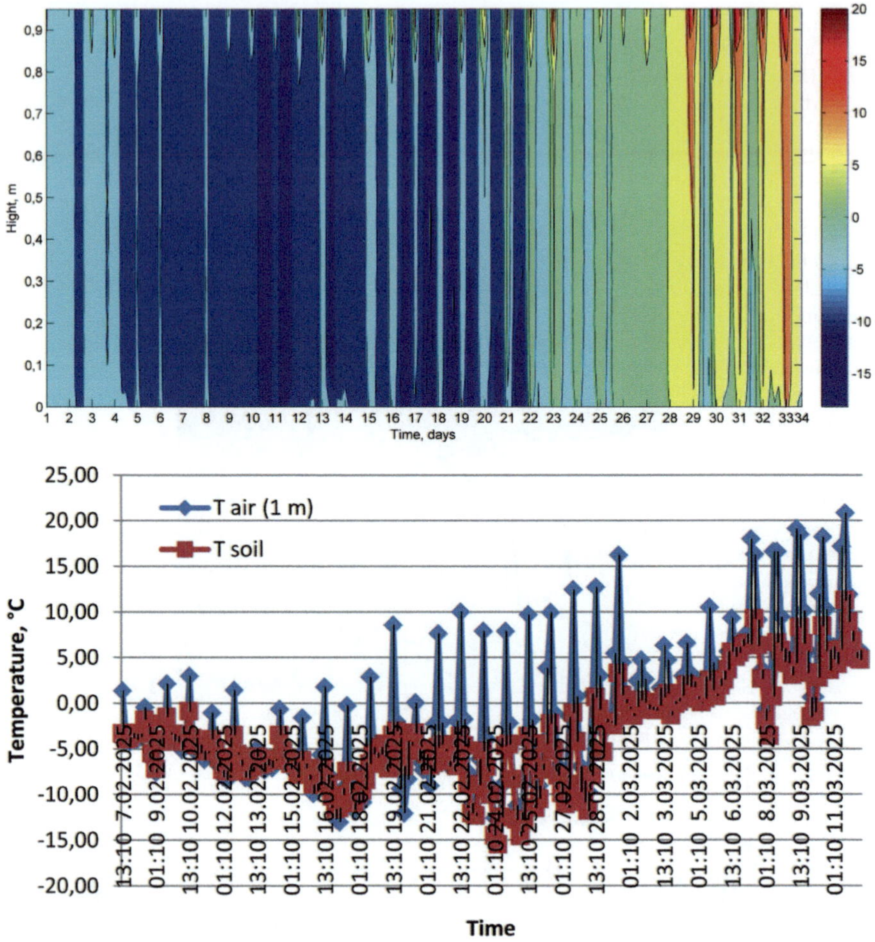

Fig. 2 Graph-diagram of changes in the thermal state of surface air and soil surface

The work was carried out in accordance with the state budgetary themes of the Geography Faculty of MSU 'Danger and risk of natural processes and phenomena' (121,051,300,175–4) and 'Evolution of the cryosphere under climate change and anthropogenic impact' (121,051,100,164–0).

References

Abele G (1990) Snow roads and runways. US Army Cold Reg Res Eng Lab 90(3)

Frolov DM, Komarov AY, Seliverstov YG et al (2019) Study of spatial and temporal heterogeneity of the snow column at the MO MSU site in winter 2018/2019. In: Ecological and climatic characteristics of the atmosphere of Moscow in 2018 according to data from the Lomonosov Moscow State University Meteorological Observatory. MAKS Press LLC, Moscow, pp 225–230

Frolov DM, Seliverstov YG, Sokratov SA et al (2023) Cryological studies at the meteorological site of MSU in winter 2022/23. Arctic and Antarctic (1):1–13. https://doi.org/10.7256/2453-8922.2023.1.40448

Frolov DM, Seliverstov YG, Koshurnikov AV et al (2023) Investigation of evolution and thermal state of the snow cover thickness at the observation site of the lomonosov msu in winter of 2022/2023. In: Processes in GeoMedia—vol. 7. Springer Geology. Springer Singapore, Singapore, pp 135–138. https://doi.org/10.1007/978-981-99-6575-5_12

https://climate.copernicus.eu/global-climate-highlights-2024#:~:text=2024%20had%20a%20global%20average,exceed%201.5%20above%20that%20level

https://meteoinfo.ru/novosti/99-pogoda-v-mire/20778-2024-god-v-rossii-vtoroj-samyj-teplyj

https://wmo.int/media/news/wmo-confirms-2024-warmest-year-record-about-155degc-above-pre-industrial-level

https://www.un-glaciers.org/

Komarov AY, Seliverstov YG, Sokratov SA, Grebennikov PB (2016) Study of spatial and temporal inhomogeneity of the snow column using a high-resolution penetrometer snowmicropen on the territory of the meteorological observatory of MSU. In: Shilovtsev OA (ed) Ecological and climatic characteristics of the atmosphere in 2015 according to the data of the Lomonosov Moscow State University Meteorological Observatory. MAKS Press, Moscow, pp 201–210

Komarov AY, Seliverstov YG, Sokratov SA, Grebennikov PB (2017) Study of spatial and temporal heterogeneity of the snow column at the site of the MSU Meteorological Observatory in winter 2016/2017. In: Nezval EI, Soshinskaya IV (eds) Ecological and climatic characteristics of the atmosphere in 2016 according to the data of the Lomonosov Moscow State University Meteorological Observatory. MAKS Press LLC, Moscow, pp 190–202

Komarov AY, Seliverstov YG, Sokratov SA, Grebennikov PB (2017) Study of spatial and temporal heterogeneity of snow strata using snowmicropen penetrometer. In: III International symposium physics, chemistry and mechanics of snow: collection of papers, Part I, Yuzhno-Sakhalinsk, 2–6 October 2017. Sakhalin branch of FGBUN FGBUN Far Eastern Geological Institute FEB RAS Yuzhno-Sakhalinsk, pp 64–68

Komarov AY, Seliverstov YG, Sokratov SA et al (2018) Study of spatial and temporal heterogeneity of the snow column at the MSU Meteorological Observatory site in winter 2017/2018. In: Lokoschenko MA (ed) Ecological and climatic characteristics of the atmosphere of Moscow in 2017 according to the Lomonosov Moscow State University Meteorological Observatory. MAKS Press LLC, Moscow, pp 199–203

Komarov AY, Seliverstov YG, Grebennikov PB, Sokratov SA (2018) Spatial and temporal heterogeneity of the snow column according to the snowmicropen penetrometer data. Ice Snow 58(4):473–485. https://doi.org/10.15356/2076-6734-2018-4-473-485

Richter GD (1945) Snow cover, its formation and properties Institute of Geography of the USSR, p 121

Sokratov SA, Seliverstov YG, Komarov AY et al (2018) Simultaneous use of different techniques in assessment of spatial-temporal variability of the characteristics of snow cover. In: Fischer J-T, Adams M, Dobesberger P, Fromm R, Gobiet A, Granig M, Mitterer C, Nairz P, Tollinger C, Walcher M (eds) Proceedings of the international snow science workshop ISSW 2018, Innsbruck, Austria, 7–12 October 2018. Innsbruck ISSW, pp 373–376

Sokratov SA, Komarov AY, Vasilchuk YK et al (2023) Spatial and temporal heterogeneity of δ18o values and the structure of the snow column on the territory of the MSU Meteoobservatory. Ice Snow 63(4):569–582. https://doi.org/10.31857/S2076673423040154

Voytkovsky KF (1999) Fundamentals of glaciology. Nauka, Moscow, p 255

Wang Z, Zhang S, Zhang Z, Xie C, Zhou K, Yuan M (2025) Study on the hydrothermal effect in ice materials under the polar glacier runway dynamic load. Case Stud Constr Mater 22:e04214. https://doi.org/10.1016/j.cscm.2025.e04214.

Weather Schedule website. https://rp5.ru/. (Date of application 22 Feb 2025)
Weather and Climate Website. http://www.pogodaiklimat.ru/. (Date of application 22 Feb 2025)

Trace Elements of Oil Fields of the Orenburg Region

G. A. Ponomareva

Abstract The work is conditioned by the need to study the concentrations of micro-components of hydrocarbon raw materials, which makes it possible to consider oil and its derivatives as valuable organomineral raw materials, as well as to solve a number of fundamental problems such as the origin of metals in oil, oil itself, and a number of others. To date, more than 70 different elements have been found in oil, including metals, the concentrations of which change very widely: from traces to hundreds of grams per ton of oil, which became possible as a result of the development of analytical methods. The increased content of trace elements is usually characteristic of highly viscous heavy oils and natural bitumen. Thus, significant amounts of vanadium, titanium, and a number of other metals are contained in oil, which may allow expanding their mineral resource base in the near future. Vanadium and nickel will probably play the most prominent role, as technologies for extracting them from oil already exist. On the other hand, the presence of these metals in oil, petroleum products, natural bitumen, asphaltites, and other naphthoid products greatly complicates the refining processes. It should also be noted that many forms of metals contained in oil, oil waste, and petroleum products, primarily vanadium, pose a serious environmental hazard. Metal compounds are toxic, they can affect the respiratory, digestive, circulatory and nervous systems, as well as cause inflammatory and allergic skin diseases. In this regard, the concentrations and distribution of metals in the oil fields of the Orenburg region of the Volga-Ural oil and gas province were studied. As a result of the conducted research, additional data was obtained on the content of a number of metals in the oils of the Buzuluk depression deposits—copper, lead, chromium, titanium, molybdenum, zirconium, vanadium, nickel, cobalt, gold, silver, palladium, and platinum. It has been established that the oil from the Orenburg deposits is enriched with a number of metals, such as titanium, vanadium, and zirconium. The age levels differ in the nature of the accumulation and distribution of the studied metals. Comparatively high concentrations of metals in oil are found in the deposits of the lower and middle carboniferous.

G. A. Ponomareva (✉)
Orenburg State University, Orenburg, Russia
e-mail: galy.ponomareva@mail.ru

T. Chaplina (ed.), *Processes in GeoMedia—Volume IX*, Springer Geology,
https://doi.org/10.1007/978-981-96-9302-3_6

Keywords Metals · Oil · Distribution patterns · Environmental impacts · Extraction

1 Introduction

More than 70 different elements have been found in oil, the concentrations of which vary widely: from traces to hundreds of grams per ton of oil, which has become possible as a result of the development of analytical methods. An increased content of trace elements is usually characteristic of highly viscous heavy oils and natural bitumen. The study of natural concentrations of metals in oil has both theoretical and applied significance. Thus, Orenburg oil contains significant amounts of vanadium, titanium, and a number of other metals (Ponomareva 2012, 2024), which may allow for the expansion of their mineral resource base in the near future. Apparently, the most noticeable role will belong to vanadium and nickel (Punanova 2014), since technologies for their extraction from oil already exist.

The presence of these metals in oil, petroleum products, natural bitumen, asphaltites, and other products of the naphthoid series greatly complicates the processes of oil refining. The presence of metals in hydrocarbon raw materials has a negative effect on purification operations due to the fact that they pose a danger to the catalysts used in many processes of oil and petroleum product refining (catalytic cracking, hydrogenation, hydrodesulfurization, etc.). Thus, vanadium found in petroleum fuel catalyzes the oxidation of sulfur oxide (IV) SO_2 to sulfur oxide (VI) SO_3, which ultimately leads to corrosion, causes great damage to oil refining equipment, installations, reduces the service life of various engines and boiler installations, and also leads to pollution of the natural environment. It should also be noted that many forms of metals contained in oil, oil waste, and petroleum products, primarily vanadium, pose a serious environmental hazard. Metal compounds are toxic, they can affect the respiratory, digestive, circulatory, and nervous systems, and also cause inflammatory and allergic skin diseases. The problems of environmental consequences during the development of oil and gas fields are obvious (Punanova and Vinigradova 2016).

Precious metals (gold, platinum, palladium, silver) are highly liquid raw materials, and it is very important to study their content in oils. Platinum and palladium are unique catalysts in many oil refining processes, and the economic feasibility of studying their concentrations is obvious. In addition to the fact that various types of fuels, lubricating oils, paraffin, fuel oil, tar, and many other chemicals are produced from oil, oil can be an unconventional source of obtaining various metals. It is known that sulfur is extracted from oil, as well as copper, vanadium, nickel, mercury, and other metals (Marakushev 2007). In the USA, two-thirds of vanadium production is associated with its extraction from oil (Marakushev 2007; Lazarenkov 2001). In a number of oil fields in the Orenburg region, it contains up to 150 g/t of vanadium, up to 100 g/t of titanium and slightly less nickel and other metals, and also contains platinum, palladium, gold, and silver (Ponomareva 2012, 2024, etc.). From a theoretical point of view, metals in oil are also of interest in connection with the study of the metal content of carbon-containing formations in the Eastern Orenburg region and the solution of questions about the genesis and nature of oil itself (Ivanov 2018; Sverjensky et al. 2014; Talovina 2024). Particular attention is paid to vanadium, nickel, copper, lead, titanium, gold, palladium, platinum, and other elements (Shpirt and Punanova 2017).

2 Materials and Methods of Research

The purpose of the work was to determine the metals content in oil, assess their content, distribution patterns, and prospects for their subsequent extraction for the purpose of integrated use of organomineral raw materials and to reduce the negative environmental impact on the territory of oil and gas regions. In this regard, their content and distribution of metals in oil from the Buzuluk Depression fields (South Buzuluk and Mukhano-Erokhovsky oil and gas regions) in the Orenburg region were studied. It should be noted that the largest hydrocarbon deposits in the region are concentrated here.

The analysis of metal distribution features is based on data on oil fields in the Orenburg part of the Volga-Ural oil and gas province.

In regional terms, they are confined to the Buzuluk Depression (first-order structure along the surface of the crystalline basement and sedimentary cover) (Panteleev 1997). We have studied oil samples for metal content from the following layers: Bobrovskoye field—2 layers, Pokrovskoye—3, Tananykskoye field—1, Dolgovskoye field—1, Pronkinskoye—1, Sakharovskoye—1, Garshinskoye—3, Rostashinskoye—4 (Ponomareva 2012; Panteleev 1997).

3 Results and Their Discussion

The average contents of a number of metals in oil from the Orenburg Region fields are presented in Table 1. Atomic absorption spectrometry (AAS) was used to determine gold, silver, palladium, platinum, cobalt, and nickel in oil (Lumex MGA-915 spectrometer with an electrothermal atomizer). A 20 g sample of oil was treated with acids and then ashed in a muffle furnace. Since metals are determined from solutions on the spectrometer, metals were transferred from the ash residue to mineral solutions using acid etching. At low contents of noble metals, concentration was used, which also made it possible to reduce the influence of matrix elements. The measurement technique MVI M 01-29-2006 was used to determine metals. To determine precious metals, the QCA method No. 430-X was used, using our own patented method of oxidation of the carbon base (Patent of the Russian Federation No. 2409810). The contents of other metals were determined by emission spectral semi-quantitative analysis (ESQA). Specially prepared dry oil residue was analyzed on an STE-1 spectrograph. Verification of the ESQA results for nickel was carried out using the atomic absorption method and showed good reproducibility of the analysis results (Ponomareva 2012, 2024). The average values of metals concentrations in 15 samples from Devonian and Carboniferous deposits were (in mg/t): Pt—29; Pd—9.5; Au—22; Ag—31; Co—12, and (in g/t): Ni—13 (AAS), 4 (ESQA); Cu—2; Pb—2; Cr—3; V—43; Ti—76; Mo—3; Zr—18.

According to the obtained data, the oil is enriched with titanium, vanadium, zirconium (76, 43, and 18 g/t, respectively). The content of copper, lead, molybdenum, and chromium is about 2–3 g/t. The concentration of cobalt in oil is the lowest of the listed metals. Chromium was found in only seven oil samples. Oil, as a result, is conventionally divided into geochemical types (nickel, vanadium, etc.), while the oils of the Volga-Ural oil and gas province belong to the vanadium type (Punanova 2014, 2017; Punanova and Vinigradova 2016; Babenko 2023, etc.), which is confirmed by the obtained data. The content of noble metals in oil is about n * 10 mg/t, which exceeds their average concentrations in sedimentary rocks (in black shales and coals) – n * 0.1 – n * 1.0 mg/t (Lazarenkov 2001). The nature of metal distribution in oils of Devonian and Carboniferous reservoirs of the Orenburg region deposits is shown in Figs. 1 and 2. The diagrams clearly show the association of elements Ti + V + Zr + Ni in the form of their maximum. In Carboniferous deposits, increased contents of, as a rule, all metals are observed compared to the Devonian (the exceptions are Cu, Mo). The presence of Cr has been established in almost all Carboniferous reservoirs, whereas it is absent in Devonian deposits. The association of noble metals in Carboniferous deposits (Pt + Ag + Au) differs from that of the Devonian (Pt + Au) (Fig. 1).

The diagram shows the nature of metal distribution in oil from carbonate and terrigenous deposits of the Carboniferous age, which is similar in many ways (Fig. 2a), but in terrigenous reservoirs there is an accumulation of Ag, Ni, V, and Cr, and in carbonate ones Au and Pd.

In the terrigenous Devonian, the content of most metals in oil is higher than in the Devonian carbonate facies. The exceptions are Ni and Cu (Fig. 2b).

Study of the geochemistry of metals in the lithological and stratigraphic sections of the fields allows us to identify the following. In the Rostashinskoye oil and gas condensate field, their highest contents (Au, Pd, Pt, Ni, Cu, Ti) are confined to the D3

Table 1 Metal content in oil from Orenburg region fields, g/t

Objects of research	Metal content, g/t								
	Cu	Pb	Ni	Cr	Co мг/т	V	Ti	Mo	Zr
Месторождения Бузулукской впадины	1–4* 2	3–4 2	3.7–79.9 13	3–15 3	1–63 12	5–150 43	50–100 76	1.5–5 3	10–30 18

* The numerator indicates the range of metal contents, g/t, the denominator indicates the average value

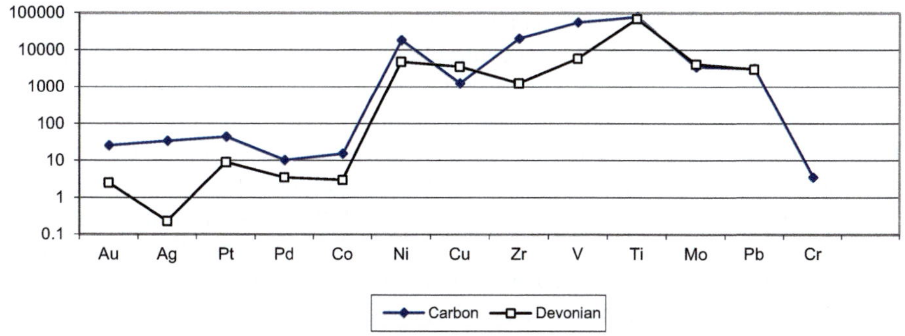

Fig. 1 Metals in oil fields of the Orenburg region (ordinate axis—average metal contents for Carbon and Devonian (mg/t), logarithmic scale)

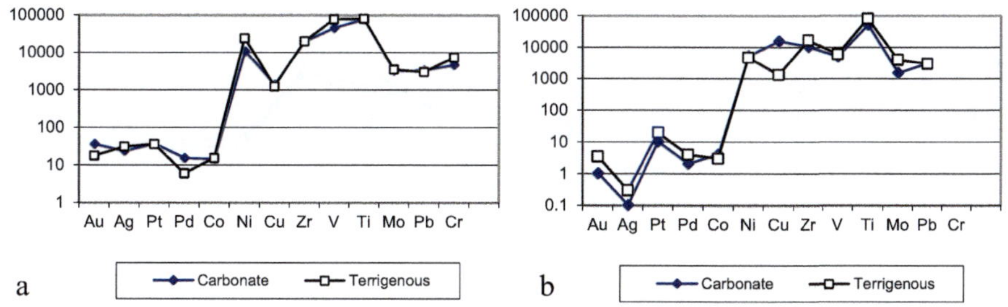

a – Carbon, b – Devonian

Fig. 2 Metals in oil fields of the Orenburg region (ordinate axis—average metal contents by fields in carbonate and terrigenous reservoirs of the Carbon and Devonian (mg/t), logarithmic scale)

layer (terrigenous deposits) of the Upper Devonian, and higher up the section (terrigenous deposits of the Lower Carboniferous (B2 layer)) they, as a rule, decrease. The nickel content practically does not change from the D5 layer (carbonate deposits) to the B2 layer, although an unclear maximum occurs in the B2 layer. At the Pokrovskoe field, the maximum concentration of metals is in the A4 layer (carbonate rocks of the middle Carboniferous), compared to the A3 and B2 layers of the lower and middle Carboniferous (terrigenous deposits). The oil of the A4 layer is also characterized by a higher nickel content.

The oil of the Tananykskoye field of the B2 layer is characterized by increased contents of almost all metals: nickel, chromium, vanadium, titanium, molybdenum, copper, zirconium, and noble metals—platinum and gold.

Thus, the accumulation of these metals in the studied deposits occurred in the Upper Devonian and the Lower and Middle Carboniferous.

The similar nature of the distribution of heavy metals and their close contents and physicochemical properties of oil in the O2 and O3 layers (Ponomareva 2012; Panteleev 1997) (lower Carboniferous carbonate deposits) of the Bobrovskoye field may indicate similar conditions of their formation (Fig. 3). When analyzing the metal contents in the reservoirs of Devonian and Carboniferous sedimentary formations, a correlation is revealed in the concentrations of heavy metals in the D4 layers (terrigenous deposits) of the Sakharovskoye and Rostashinskoye fields (Fig. 4), in the B2 layers (terrigenous deposits) of the Tananykskoye and Pokrovskoye fields (Fig. 5). The B2 layer of the Rostashinskoye field correlates with the B2 layers of the Pokrovskoye and Tananykskoye fields in almost all metals.

It should be noted that when burning oil and oil products, the concentration of heavy metals contained in the feedstock occurs. The ash content of oils is from hundredths of a percent, thus, the ash contains an order of magnitude or more metals compared to the feedstock. At the same time, obtaining metals from oil and oil products does not require opening up deposits, blasting operations, removing ore from quarries, crushing, enrichment of ore, and other technological processes. Therefore, at present, it seems economically feasible to search for technologies for obtaining metals from oil and products.

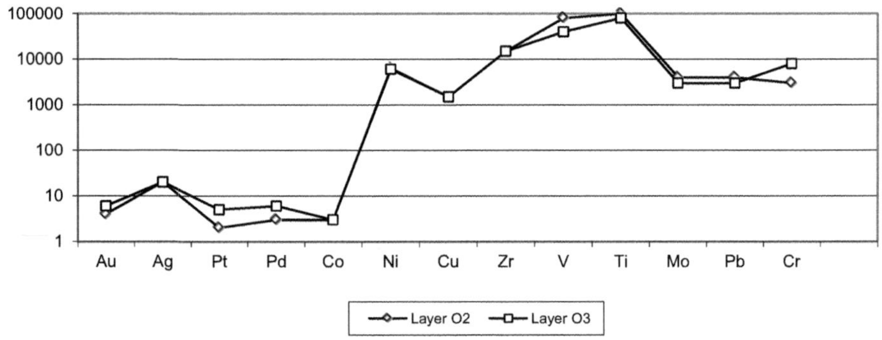

Fig. 3 Metal concentrations (in mg/t) in the O2 and O3 layers of the Bobrovsky deposit of the Buzuluk depression (ordinate axis—logarithmic scale)

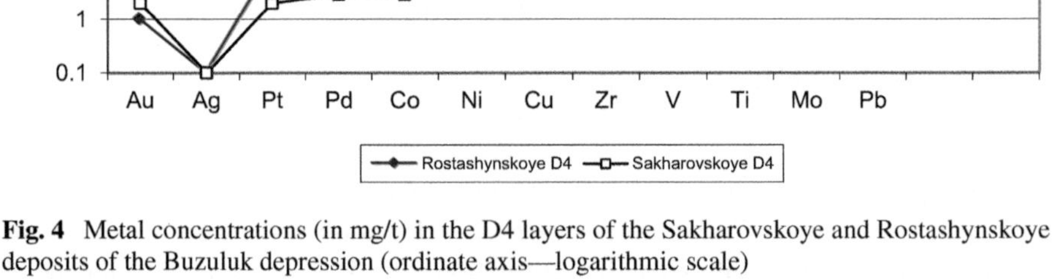

Fig. 4 Metal concentrations (in mg/t) in the D4 layers of the Sakharovskoye and Rostashynskoye deposits of the Buzuluk depression (ordinate axis—logarithmic scale)

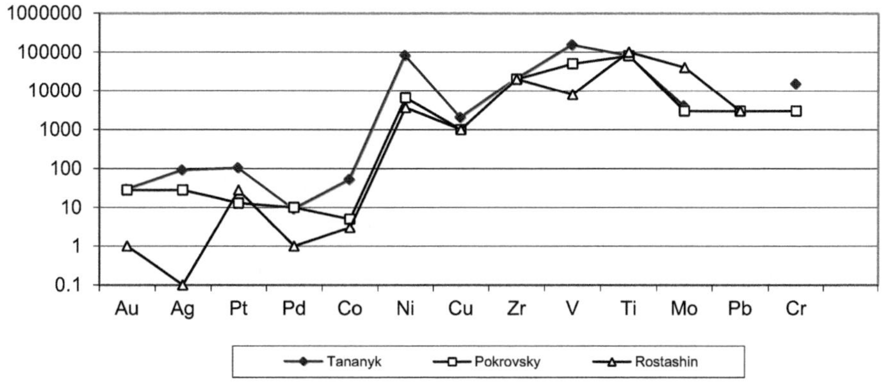

Fig. 5 Metal concentrations (in mg/t) in the B2 formations of the Tananyk, Rostashin and Pokrovsky fields of the Buzuluk depression (ordinate axis—logarithmic scale)

About 70% of vanadium in the world is extracted from sludges obtained during the processing of high-viscosity oils and bitumen sands. Some countries (Canada, Japan) obtain vanadium entirely from heavy oils. Therefore, the increased concentrations of metals in the Orenburg oil fields are also of interest as associated components. Work is also underway in our country to find technologies for extracting metals from oil and petroleum products (Nuranbaeva et al. 2013).

4 Conclusion

As a result of the conducted studies, additional data were obtained on the content of a number of metals (copper, lead, chromium, titanium, molybdenum, zirconium, vanadium, nickel, cobalt—biogenic according to Bowen (1966), as well as gold, silver, palladium, and platinum) in oils from the Buzuluk Depression fields of the Orenburg part of the Volga-Ural oil and gas province. Analysis of the metal content in these fields allows us to draw conclusions about the distribution of the studied elements:

- Oils from the Orenburg fields are enriched with a number of metals, such as titanium, vanadium, and zirconium.
- The highest metal contents are found in the Tananykskoye, Pokrovskoe, Pronkinskoye, and Bobrovskoye oil and gas fields.
- Levels of different ages differ in the nature of accumulation and distribution of the studied metals. Relatively high concentrations of metals in oil were found in the lower and middle carboniferous deposits. The obtained results allow us to consider oil and oil products of the Orenburg platform region as valuable organomineral raw materials.

References

Babenko IA (2023) Informative value of platinum group elements as geochemical indicators of oil nature. In: Babenko IA, Talovina IV, Kanimbue LS (eds) Young scientist: collection of articles of the international scientific and practical conference, Penza. Science and Education, Penza (Gulyaev IP, Yu G), pp 27–30

Bowen HJM (1966) Trace elements in biochemistry. Academic Press, 241 p

Ivanov KS (2018) On the possible maximum depth of oil deposits. In: Ivanov KS (ed) Bulletin of UGMU 4(52):41–49. https://doi.org/10.21440/2307-2091-2018-4-41-49

Lazarenkov VG (2001) Geochemistry of platinum group metals. In: Lazarenkov VG, Talovina IV (eds). St. Petersburg, Galart 266, p 45

Marakushev AA (2007) Parageneses of ore metals of hydrocarbon specificity. In: Marakushev AA, Pane-yakh NA, Rusinov VL, Zotov IA (eds) Bulletin of higher educational institutions. Geology and exploration, no 6, pp 33–40

Nuranbaeva BM, Akhmedzhanov TK, Ismagilova LT (2013) Method and process flow charts for extracting vanadium and other metals from oil and oil products during their preparation. In: Modern science-intensive technologies, no 4, pp 49–52

Panteleev AS (1997) Geological structure and oil and gas potential of the Orenburg region. In: Panteleev AS et al (eds). Orenburg book publishing house, Orenburg, 272 p

Ponomareva GA (2012) Microelement composition of oil from the Orenburg fields. In: Ponomareva GA, Pankratyev PV, Khalzov AA (eds) Bulletin of the Orenburg State University, no 1. OSU, Orenburg, pp 125–131

Ponomareva GA (2024) Features distribution of platinoids in oil and gas fields of the Orenburg region. In: Proceedings of the 9th international conference on physical and mathematical modelling of earth and environmental processes, Springer Proceedings in Earth and Environmental Sciences, pp 63–68. https://doi.org/10.1007/978-3-031-54589-4

Punanova SA (2014) Supergene transformed naphthides peculiarities of trace-elements composition. Geochem Int 52(1):64–75

Punanova SA (2017) Trace elements of naphthides in the process of hydrocarbon ontogenesis in connection with oil and gas potential. In: Punanova SA (ed). IPNG RAS, Moscow, 259 p

Punanova SA, Vinigradova TL (2016) Geochemical features of mature hydrocarbon systems and indicators of their recognitions. Geochem Int 54(9):852–859

Shpirt MY, Punanova SA (2017) Oil and shale as a source of industrial production of potentially valuable elements. Oil Ind (4):58–62

Sverjensky DA, Stagno V (2014) Fang Huang. Important role for organic carbon in subduction-zone fluids in the deep carbon cycle. Nat Geosci 7:909–913. https://doi.org/10.1038/ngeo2291

Talovina IV (2024) Elements of the platinum group as geochemical indicators in the study of oil polygenesis. In: Talovina IV, Ilalova RK, Babenko IA (eds) Notes of the Mining Institute, vol 269, pp 833–847

Long-Term Variability of the Ice Regime in the Mouth Areas of Tributaries of the Western Coast of the White Sea Against the Backdrop of Climate Change

V. N. Baklagin

Abstract The paper presents the results of statistical analysis of data on the main elements of the ice regime of the rivers of the western coast of the White Sea (Republic of Karelia and Arkhangelsk Region) for the period 1956–2020. General patterns of long-term variability of ice regime formation processes along with climate changes that occurred during the period under review are revealed.

Keywords Rivers · Ice regime · Characteristics dates · Freeze-up · Period with ice phenomena · Climate change · Cold season · White Sea

1 Introduction

The winter regime of rivers is an integral phase of the hydrological regime of watercourses and is of particular interest to various sectors of the national economy. Information on the processes of formation and destruction of ice cover on watercourses, their forecasting, serve as the basis for such activities as shipping, fishing, organization of ice crossings, etc., and are also used in hydroforecasting to prevent and avoid emergency situations that cause damage to the environment, life support facilities and the population. Such phenomena can occur due to dangerous hydrological phenomena that occur when the water level in the river rises as a result of ice jams and early ice formation (Banshchikova and Sumachev 2021).

Since errors in forecasting the spatio-temporal characteristics of ice phenomena on rivers can lead to negative consequences for industry and the national economy (damage from floods, dam breaks, destruction of hydraulic structures, etc.), the study of the timing and phases of the ice regime of rivers is currently a very important topical scientific and practical task. The main factors influencing the formation of the ice regime of rivers are climatic conditions (temperature, wind speed and direction, etc.),

---

V. N. Baklagin (✉)
Northern Water Problems Institute Karelian Research Centre Russian Academy of Sciences, Petrozavodsk, Republic Karelia, Russia
e-mail: slava.baklagin@mail.ru

T. Chaplina (ed.), *Processes in GeoMedia—Volume IX*, Springer Geology,
https://doi.org/10.1007/978-981-96-9302-3_7

as well as the water content of the river, the height of the groundwater level (groundwater supply), flow rates and other natural and climatic factors, which, in general, depends on regional characteristics and the geographical location of watercourses (Surface water resources of the USSR 1972; Agafonova et al. 2016). However, to the greatest extent, the processes of ice formation on rivers depend on the prevailing meteorological conditions (temperature regime) preceding the onset of the development of ice processes, and the formed hydrological regime (Debolsky et al. 2008).

The study of ice phenomena on the rivers of the Arctic zone of Russia by Russian hydrologists and scientists was started in the twentieth century. The research results based on the collected long-term data are summarized in the form of statistical characteristics and standardized maps of the ice regime of the rivers of the USSR, and are also presented in a number of fundamental works (Ginzburg 1973; Donchenko 1987). A number of studies are devoted to the study of the influence of changes in climatic conditions on the formation of the ice regime of the rivers of the Arctic North of Russia. Thus, based on the results of the assessment of the ice regime in the estuary areas of the rivers of the Russian Western Arctic (the Severnaya Dvina, Mezen, Pechora, Ob rivers), an increase in the difference between the earliest and latest dates of the onset of autumn ice phenomena was established for the periods of 1930–1960 and 1961–2000, for example, for the Severnaya Dvina River the increase is 14 days (Debolsky et al. 2008). Studies of hydrometeorological conditions in the river basin. The data of the Malaya Severnaya Dvina River showed a decrease in the maximum ice thickness and a reduction in the duration of ice complete on the rivers of the basin, as well as a decrease in the maximum snow reserves and a reduction in the duration of snow cover in this area for 1988–2017 compared to previous decades (1958–1987) (Georgievsky et al. 2021). According to the results of a comparative analysis of the data from water gauges located in different parts along the Pechora River, for the periods 1950–1980 and 1981–2018, against the background of climate change, a shift in the average dates of ice appearance by 3–8 days towards later dates was revealed (Sumachev and Banshchikova 2021). In the period 1981–2018, due to a shift in the opening dates towards earlier dates, a decrease in the duration of ice complete by 6 days was noted compared to the period 1950–1980 on the rivers of the Kola Peninsula (Banshchikova and Sumachev 2021).

Modern research works, including the study of ice phenomena on the rivers of the Arctic zone of Russia, are highly relevant and necessary from the point of view of developing approaches and methodology for predicting and preventing damage from hazardous hydrological phenomena associated with ice formation (Agafonova et al. 2016) in the context of global climate change (warming) and anthropogenic impact (Agafonova 2019; Dzhamalov et al. 2015; Kalinin 2020).

The issue of the ice regime of the tributaries of the western coast of the White Sea located within the territory of the Republic of Karelia and the impact of ongoing climate change on it has not been sufficiently studied—there are not many studies conducted in this part of the White Sea basin. An example of such studies is the work (Frolova et al. 2018), where the authors found that the average duration of the period with ice phenomena for the rivers of the Republic of Karelia (the White Sea basin) is on average 180 days. The work also presents the results of a comparative analysis of the average timing of the onset of characteristic ice phenomena on the rivers of this territory for the period 1960–1990 and in modern conditions (1991–2014). It was found that in recent years (1991–2014), the timing of ice appearance and establishment of freeze-up has shifted by 5–8 days towards later dates relative to the previous period, and ice breaking up and clearing has shifted to earlier dates by 3–7 days, while the overall duration of the period with ice phenomena has decreased by 11–12 days relative to the period 1960–1990.

In general, the analysis of a number of existing studies has shown a noticeable impact of climate changes occurring in the last few years on the processes of ice cover formation in the rivers of the European North of Russia, in connection with which clarifications and revision of methods in hydroforecasting are required, which are largely based on outdated data. Therefore, the purpose of this work is a comprehensive study of the long-term variability of the main elements of the ice regime in the estuary sections of rivers—tributaries of the western coast of the White Sea (Republic of Karelia and Arkhangelsk Region) taking into account the influence of the temperature background in this region for the period 1956–2020.

2 Materials and Methods

It should be noted that, along with air temperature, a significant factor influencing the processes of ice regime formation in rivers is their water content. However, in this work, special attention is paid to the increase in the air temperature background that has been occurring in recent decades for water bodies in the Arctic zone (global warming) and its impact on the processes of ice formation in rivers, in connection with which the factor under study is air temperature. The initial information in the work was based on data on average daily air temperature for the period 1955–2020 according to the marine hydrometeorological coastal stations (HS) Gridino, Kem and Onega of the state observation network of Roshydromet, located on the western coast of the White Sea within the territory of the Republic of Karelia and the Arkhangelsk Region (Fig. 1). These data were obtained from the portal of the Russian Institute of Hydrometeorological Information-World Data Center (RIHI-WDC), which are in the public domain (http://meteo.ru/contacts). Since the formation of ice cover on rivers in the territory under consideration usually occurs from the end of October to the first 10 days of May, the study was subject to a comparison of the main characteristics of ice and temperature regimes during the cold season—from October to May.

Information on the main characteristics of the ice regime (characteristic dates of ice phenomena, their duration) of the tributaries of the western coast of the White Sea for the period 1956–2020 is presented by data from nine hydrological posts (HP), territorially located in the Republic of Karelia and the Arkhangelsk region at the mouths of the Gridina, Kuzema, Pongoma, Kem, Shuya, Nizhniy Vyg, Suma, Nyukhcha and Maloshuyka Rivers (Fig. 1). To form continuous series of the main characteristics of the ice regime of the above-mentioned rivers and their further

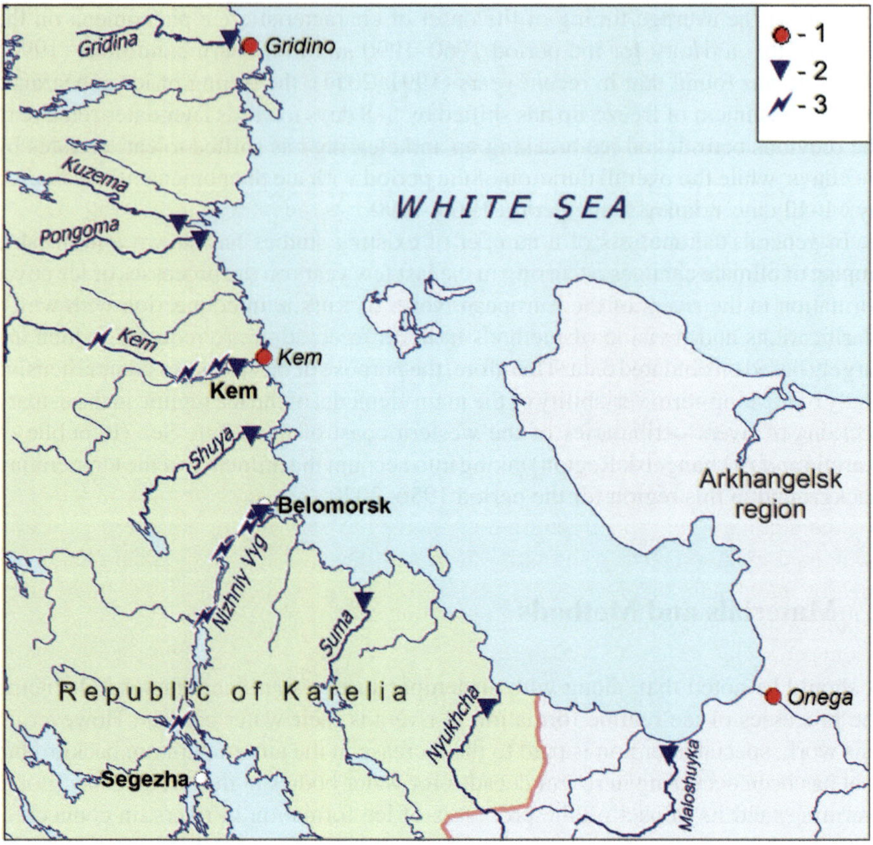

Fig. 1 Location of meteorological stations and hydrological posts on the western coast of the White Sea (Republic of Karelia and Arkhangelsk region): 1—weather station, 2—hydrological station, 3—hydroelectric power station

study, the materials of the State Water Cadastre of the Russian Federation for 1956–2007 available in the library of the Karelian Scientific Center (KarRC RAS) and the Unified Digital Database of the Scientific and Technical Library of the State Hydrological Institute (State Hydrological Institute) (https://www.hydrology.ru/edinyi-fond-dannyx) were used. Data for 2008–2020 were obtained from the web service 'Automated Information System of State Monitoring of Water Bodies', which is in the public domain (https://gmvo.skniivh.ru/).

In the work, the dates of occurrence of typical ice phenomena are considered as the main characteristics of the ice regime of rivers: the date of appearance of autumn ice phenomena (the date of the starting ice freezing, ice drift, slush drift, freeze-up, etc. is adopted); the date of the freeze-up (the date of the first long term, at least 20-day freeze-up is adopted); the date of the starting of spring ice phenomena (start ice breaking)—the first day of ice drift, in its absence—gullies, edges, leads, water on ice, ice movement; the date of the end of ice phenomena (the last day with ice phenomena is adopted); and the duration of ice complete and the period with ice phenomena (the difference between the dates of their starting and end).

The influence of the temperature regime on the processes of ice cover formation for each studied river was assessed by determining the sums of accumulation of negative (positive) air temperatures from the date of transition of the average daily air temperature through 0 °C towards negative (positive) values until the date of occurrence of the sought characteristic of the ice regime. The stable transition of air temperature through 0 °C to negative values in autumn was taken as the first day of the period when the sum of negative average daily temperature values of any of the subsequent periods exceeded the sum of positive temperatures and vice versa for spring, where the stable transition of average daily air temperature through 0 °C to positive values was determined.

Since the meteorological and hydrological observation points are located at different distances from each other (Fig. 1), the temperature regime indicators (average daily air temperature values) affecting the ice formation conditions on each of the studied rivers were determined by interpolation using the inverse distance weighted method (IDW). The share of the influence of meteorological data on the nearby and distant meteorological station (MS) on the calculation of the desired air temperature value is expressed as a weighting coefficient ω. The weighting coefficients $\omega_{MS_i-HP_j}$ for each i MS are calculated inversely proportional to the geometric distance from the j HP under consideration, raised to the power p ($p = 2$ was used in the work):

$$\omega_{\mathrm{MS}_i-\mathrm{HP}_j} = \frac{1}{r^p_{\mathrm{MS}_i-\mathrm{HP}_j}},$$

where $r_{\mathrm{MS}_i-\mathrm{HP}_j}$ is the geometric distance between the i MS and the j HP.

Thus, the average daily air temperature values T in the j HP are calculated using the formula:

$$T_{\mathrm{HP}_j} = \frac{\sum_{i=1}^{N}\left[\omega_{\mathrm{MS}_i-\mathrm{HP}_j} \cdot T_{\mathrm{MS}_i}\right]}{\sum_{i=1}^{N} \omega_{\mathrm{MS}_i-\mathrm{HP}_j}},$$

where T_{MS_i} is the average daily air temperature in the i MS and N is the number of MS. The significance of the calculated characteristics of the linear regression was assessed using the Fisher p-value criterion. The obtained *p-value* was compared with the accepted significance level in this study (0.05), on the basis of which a decision was made on the significance of the trend. The calculation of the values of the confidence intervals Δ was performed based on the assumption of the normal distribution of regression errors.

For comparative analysis of heterogeneous time series of data, the method of normalization of difference-integral curves of modular coefficients was used, which are calculated for the entire series according to the formula:

$$k_i = \frac{x_i}{\overline{x}},$$

where x_i is the value of the hydrometeorological characteristic in the i year and $\overline{x}$ is the average value of the hydrometeorological characteristic over a long-term period. The construction of the difference-integral curve is performed by successively summing the deviations of the modular coefficients from the middle $(k - 1)$:

$$f(t) = \sum_{i=1}^{n} (k_i - 1),$$

where n is the number of values of the time series.

Thus, the difference-integral curve represents the cumulative sum of deviations of the modular coefficients from the average long-term value of the series $\overline{x}$ at the end of each i year.

3 Results and Discussions

As a result of processing data series on the estuary sections of the Gridina, Kuzema, Pongoma, Shuya, Suma, Nyukhcha and Maloshuyka Rivers, averaged characteristics of the ice regime of the tributary rivers of the western coast of the White Sea (Republic of Karelia and Arkhangelsk Region) were obtained for the period 1956–2020 (Table 1, Fig. 2).

Generalized calculations for the territory under consideration are given without taking into account data on ice phenomena in the estuaries of the Nizhniy Vyg and Kem Rivers, since these rivers are regulated by hydraulic structures located on them (hydroelectric power station cascades), and due to their operation, the appearance of autumn ice phenomena and the establishment of freeze-up, as a rule, occurs much later (up to one and a half months) relative to watercourses with a hydrological regime formed under natural conditions (Table 1). For example, in the estuary of the Nizhniy Vyg River near the city of Belomorsk in recent years (2013, 2016–2020), ice freeze-up was not observed.

The starting of ice freezing on the rivers of the western coast of the White Sea (Republic of Karelia and Arkhangelsk region) occurs mainly from the third 10-day period of October to the third 10-day period of November. During early cold snaps in some years (1961, 1974), the start of ice freezing in question period is observed at the end of the first—starting of the second 10-day period of October, in warm winters (1968, 1997, 2006)—in the last days of November. At first, ice cover forms on the watercourses of the northern part of the coast, up to a week later—on more southern rivers (rivers Nyukhcha, Maloshuika) (Table 1). This is explained primarily by the fact that the ice regime of the rivers of the territory in question is formed in the conditions of the transition zone between the Western European maritime and

Table 1 Average statistical indicators of the main characteristics of the ice regime on the tributary rivers of the western coast of the White Sea (Republic of Karelia and Arkhangelsk Region) for the period 1956–2020

Hydrological post	Average dates				Average duration, days	
	The start of ice freezing	The freeze-up	The start of ice breaking	The breakup	Ice complete	Period with ice phenomena
Gridina River–Gridino	31.10	16.11	22.04	03.05	158	185
Kuzema River–St. Kuzema	29.10	15.11	24.04	05.05	161	188
Pongoma River–Pongoma	29.10	14.11	25.04	04.05	163	188
Kem River–Kem	13.11	01.01	20.04	02.05	109	170
Shuya River–Shueretskoe	04.11	09.11	22.04	01.05	164	179
Nizhny Vyg River–Belomorsk	12.11	01.01	18.04	06.05	108	175
Suma River–Sumsky Posad	05.11	19.11	24.04	01.05	156	178
Nyukhcha River–Nyukhcha	05.11	01.12	18.04	30.04	138	176
Maloshuika River–St. Maloshuika	01.11	13.11	18.04	01.05	157	182
Western coast of the White Sea (Republic of Karelia, Arkhangelsk region)	02.11	17.11	22.04	02.05	156	182

Eurasian continental climate. Marine air masses coming from the Atlantic Ocean have a significant impact on the formation of ice cover, its stability and duration.

The sum of negative average daily air temperatures required for the occurrence of primary ice phenomena is 4–6 °C (with a stable transition through 0 °C). The uniformity of ice freezing on the rivers of the territory under consideration varies from 1 (1964) to 37 days (1978), with an average of 12 days. The uniformity of ice freezing (freeze-up) is understood as the difference in days between the first and last dates of ice freezing (freeze-up) throughout the territory under consideration (Surface water resources of the USSR 1972). The duration of the transition period from the onset of primary ice phenomena to the formation of stable freeze-up is on average 15 days, the shortest being 3 days (1965, 1988, 2017), the longest being 46 days (1983). The rivers of Karelia are characterized by the absence of autumn ice drift. In the territory under consideration, ice freeze-up occurs first (first–second 10 days of November) on the reach, calmer rivers (rivers Kuzema, Pongoma, Shuya, Suma), later—ice forms on the riffles and rapids, regulated by cascades of hydroelectric power stations (rivers Nizhny Vyg, Kem)—in January, February. This is explained by the fact that

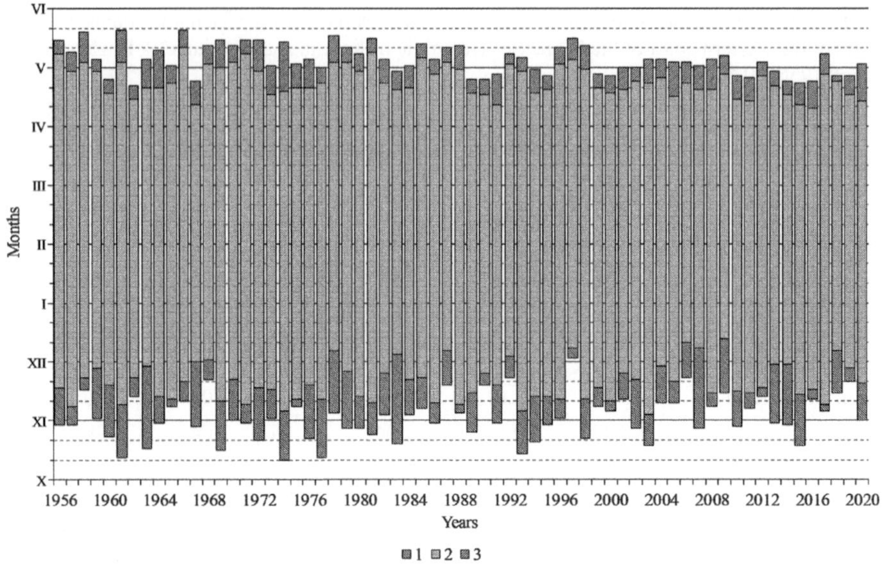

Fig. 2 Dates of phases of the ice regime of rivers on the western coast of the White Sea (Republic of Karelia, Arkhangelsk region): ice freezing (1), ice complete (2), ice breakup (3) for 1956–2020

ice formation is largely determined not only by weather conditions, but also by geomorphological ones. Sometimes closely located sections of one river freeze at different times, which depends on the morphometry, features of the longitudinal profile and other characteristics of the riverbed. In case of early and stable cold snap, ice freeze-up on rivers is formed in the first 10 days of November (1957, 1961, 1969, 1971, 1974, 1981, 1993, 2000, 2017), in warm winters (1978, 1983, 1987, 1992, 1997, 2006, 2007, 2009, 2018)—in December. Early terms (October 21) of freeze-up formation differ from the average (November 17) by 27 days, late terms (December 12)—by 25 days. The sum of average daily negative temperatures required for ice freeze-up is 32–45 °C. The uniformity of ice freeze-up varies from 3 (1960) to 83 days (2007), the average is 27 days. The average duration of ice freeze-up fluctuates from 131 to 192 days (156 days).

The start of ice breaking usually is in the second half of April, with the average date of starting of ice breaking being April 22 (Table 1). The earliest dates for the start of ice breaking differ from the average by 14 days, and the latest by 16 days. The sum of the positive average daily air temperatures required for starting of ice breaking is 12–19 °C. The average uniformity of ice breaking is 11 days, with prolonged intermittent springs the duration of ice breaking is 31 days (2000), and with uniform springs it is 2 days (1968, 1979). Spring ice drift on the rivers of Karelia is poorly developed, and breaking does not occur simultaneously along the entire length of the river. First, areas near the sources of lakes and rapids areas are freed from ice, then pools and lake-like expansions. In recent years, the ice cover on the Gridina, Kem, Suma and Nizhny Vyg Rivers occurred without ice drift.

Ice breakup on rivers (end of ice phenomena), as a rule, occurs at the beginning of May. The earliest ice breakup dates differ from the average by 13 days, the latest by 15 days. The sum of positive average daily air temperatures necessary for ice breakup is about 38–47 °C. The average uniformity of ice breakup is 11 days, with prolonged intermittent springs the duration of ice breakup is up to 26 days (1958), in friendly springs—2 days (1977).

As a result of statistical analysis of time series of average cold season air temperatures ($\overline{T}_{X-V}$) on the western coast of the White Sea and the duration of periods with ice phenomena (D_i) on the rivers under consideration, difference-integral curves were obtained (Fig. 3). Analysis of difference-integral curves (Fig. 3) showed that since the end of the 80s of the twentieth century the temperature background on the western coast of the White Sea has increased significantly, and the duration of the period with ice phenomena has decreased, which indicates a violation of the stationarity and homogeneity of the observation series (both air temperatures during the cold season and the duration of the period with ice phenomena) with a turning point in 1988. However, to apply the results of the work to general hydrometeorological studies, the division of the time series was performed by 1990 to characterize the statistics relative to the WMO period 1991–2020.

Statistical analysis of long-term fluctuations in the average cold season air temperature ($\overline{T}_{X-V}$) (Fig. 4) showed that the temperature background in recent years (1991–2020) is significantly higher (by 1.4 °C) than in the previous series of years (1956–1990) (Table 2). The results of calculating the linear trends in the average cold season air temperature ($\overline{T}_{X-V}$) showed that for 1956–2020 and 1991–2020 there are statistically significant positive linear trends (α), the values of which are 0.042 ± 0.018 °C/year and 0.079 ± 0.050 °C/year, respectively. No statistically significant linear trend (p-value > 0.05) was found for 1956–1990 (Table 2). In general, according to the results of the regression analysis, over the period under consideration (1956–2020), the average air temperature for the cold season ($\overline{T}_{X-V}$) increased by 2.7 °C (Fig. 4).

It was also revealed that the greatest increase in air temperature in the cold season over the past several years (1991–2020) compared to the previous period (1956–1990) was noted in December–April (1.5–2.5 °C), with no significant changes in the temperature background in the autumn months and in May (Table 3).

Regression analysis of the time series of the duration of the period with ice phenomena for 1956–2020 showed the presence of a statistically significant negative trend (α), which is − 3.3 days/10 years (Table 4). In general, over the period 1956–2020, according to the results of the regression analysis, the duration of the period with ice phenomena decreased by 21 days (Fig. 5).

In addition, it is noted that in the last number of years (1991–2020), the average value of this characteristic on the rivers of the territory under consideration has decreased compared to the period 1956–1990 to almost 2 weeks (by 11 days) (Table 5).

The analysis of long-term variability of ice regime characteristics showed that in the last number of years (1991–2020) at the mouth areas of the rivers of the western coast of the White Sea, there has been a shift in the average dates of formation of stable ice phenomena by a week (6 days) towards later dates compared to the previous

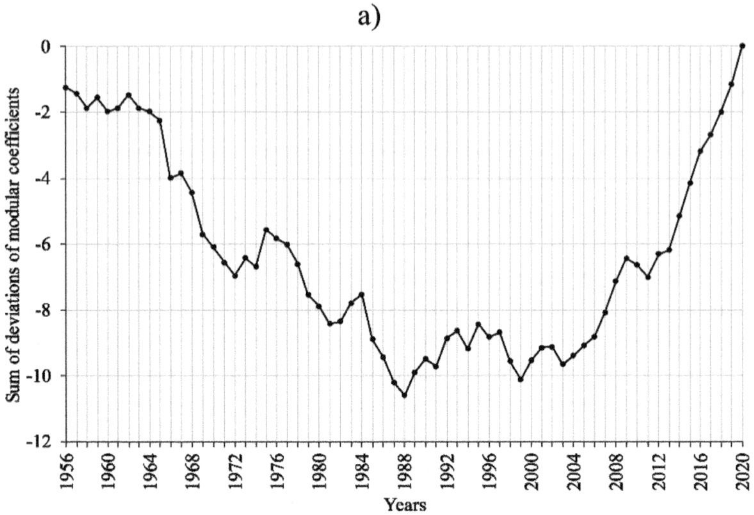

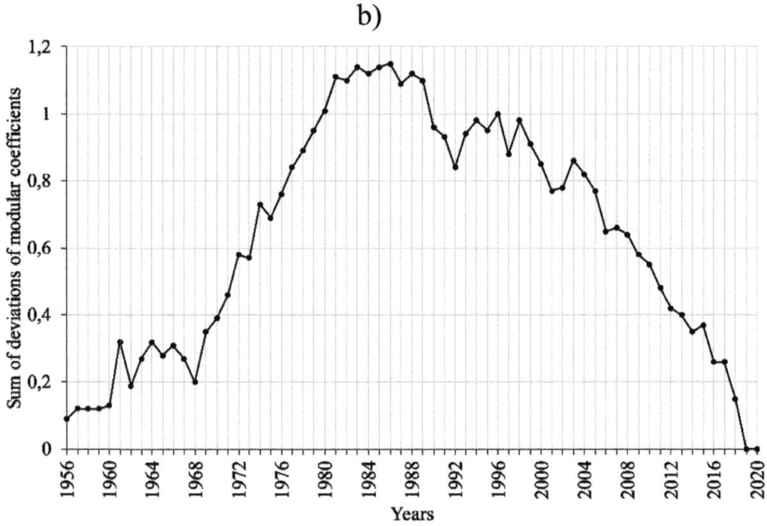

Fig. 3 Difference-integral curves of average air temperature for the cold season (**a**) and the duration of ice phenomena on rivers (**b**) on the western coast of the White Sea (Republic of Karelia, Arkhangelsk region) for 1956–2020

period (1956–1990); the early date has not undergone any significant changes, while the late date of ice formation has shifted by 11 days towards the winter months. The average and early dates of ice formation remained the same, while the late date has shifted by 6 days towards the winter months. The average dates of ice breaking and ice breakup have shifted by up to a week (5 days) towards early dates, the early dates of ice breaking and ice breakup have not changed, and the late dates have shifted by 4–6 days towards the winter months (Table 5). It should be noted that

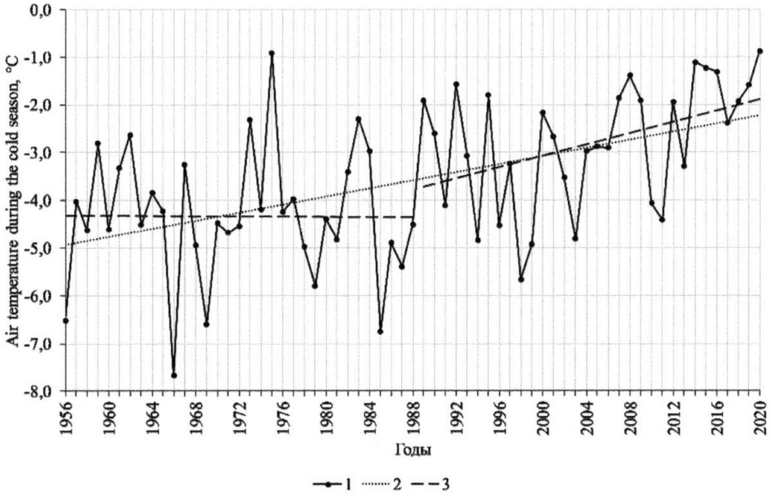

Fig. 4 Long-term fluctuations of average air temperature during the cold season on the western coast of the White Sea (1) and its linear trends for the time periods 1956–2020 (2), 1956–1990 and 1991–2020 (3)

Table 2 Parameters of the regression analysis of the average air temperature ($\overline{T}_{X-V}$) for the cold season on the western coast of the White Sea (Republic of Karelia and Arkhangelsk Region) for different time periods

Period	$\overline{T}_{X-V}$, °C	$\alpha \pm \Delta$, °C/year	R^2	p-value	Significance
1956–1990	–4.2	–	0.02	0.457	–
1991–2020	−2.8	0.079 ± 0.050	0.27	0.003	+
1956–2020	−3.6	0.042 ± 0.018	0.27	0.000	+

Note α—trend value; Δ—confidence interval; R^2—determination coefficient; p-value—Fisher criterion estimate

Table 3 Change in average monthly air temperatures of the cold season on the western coast of the White Sea (Republic of Karelia and Arkhangelsk Region) for the period 1991–2020 in relation to the previous period 1956–1990

Period	Months								Cold season
	10	11	12	1	2	3	4	5	
1956–1990	2.3	−3.4	−8.7	−11.9	−10.9	−6.2	−0.8	5.4	−4.3
1991–2020	3.0	−2.8	−6.3	−9.4	−9.2	−5.0	0.5	6.2	−2.9
ΔT, °C	0.7	0.6	2.5	2.5	1.6	1.2	1.2	0.8	1.4

Table 4 Parameters of the regression analysis of the main elements of the ice regime on the rivers of the western coast of the White Sea (Republic of Karelia and Arkhangelsk Region) for different time periods

Main elements of the ice regime		Period	Average value	$\alpha \pm \Delta$, days/year	R^2	p-value	Significance
Dates	The start of ice freezing	1956–1990	30.10	–	0.01	0.535	–
		1991–2020	05.11	–	0.02	0.443	–
		1956–2020	02.11	0.160 ± 0.147	0.07	0.033	+
	The freeze-up	1956–1990	16.11	–	0.00	0.927	–
		1991–2020	18.11	–	0.11	0.080	–
		1956–2020	17.11	–	0.02	0.235	–
	The start of ice breaking	1956–1990	24.04	–	0.01	0.615	–
		1991–2020	19.04	−0.300 ± 0.272	0.15	0.032	+
		1956–2020	22.04	− 0.166 ± 0.095	0.16	0.001	+
	The breakup	1956–1990	04.05	–	0.03	0.311	–
		1991–2020	29.04	−0.244 ± 0.237	0.14	0.044	+
		1956–2020	02.05	−0.176 ± 0.090	0.19	0.000	+
Duration, days	Ice complete	1956–1990	160	–	0.00	0.733	–
		1991–2020	153	−0.765 ± 0.518	0.25	0.005	+
		1956–2020	157	−0.265 ± 0.184	0.12	0.006	+
	Ice phenomena period	1956–1990	187	–	0.04	0.272	–
		1991–2020	176	–	0.10	0.095	–
		1956–2020	182	−0.330 ± 0.162	0.21	0.000	+

Note α—trend value; Δ—confidence interval; R^2—determination coefficient; p-value—Fisher criterion estimate

the shift in the average dates of the onset of autumn and spring ice phenomena as well as the end of the period with ice phenomena on the rivers of the western coast of the White Sea (Republic of Karelia, Arkhangelsk Region) in comparison with the periods of 1956–1990 and 1991–2020 is in good agreement (the difference is 1–3 days) with the results of the work (Frolova et al. 2018), where similar time intervals were considered (1960–1990 and 1991–2014). The average duration of ice complete and period with ice phenomena on the rivers of the western coast of the White Sea (Republic of Karelia, Arkhangelsk Region) decreased by 8 and 11 days, respectively, the longest duration of the period with ice phenomena by 16 days. Moreover, the above-mentioned work (Frolova et al. 2018) also notes a reduction in the duration of the period with ice phenomena by 12–13 days, but a greater reduction in ice formation (by 5–6 days) was recorded compared to the results obtained in this work.

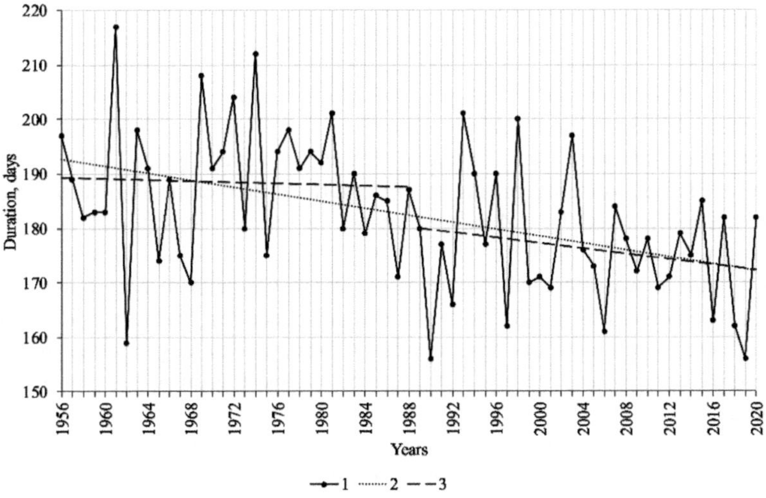

Fig. 5 Chronological graph of the duration of the period with ice phenomena on the rivers of the western coast of the White Sea (1) and its linear trends for the time periods 1956–2020 (2), 1956–1990 and 1991–2020 (3)

Table 5 Statistical characteristics of the main elements of the ice regime of the rivers of the western coast of the White Sea (Republic of Karelia and Arkhangelsk Region) for different time periods

Statistical characteristics of the main elements of the ice regime			Period			
			1956–1990	1991–2020	Shift, days	1950–2020
Dates	The start of ice freezing	Average	30.10	05.11	6	02.11
		The earliest	11.10	14.10	3	11.10
		The latest	21.11	02.12	11	02.12
	The freeze-up	Average	16.11	18.11	2	17.11
		The earliest	22.10	21.10	−1	21.10
		The latest	06.12	12.12	6	12.12
	The start of ice breaking	Average	24.04	19.04	−5	22.04
		The earliest	09.04	08.04	−1	08.04
		The latest	08.05	02.05	−6	08.05
	The breakup	Average	04.05	29.04	−5	02.05
		The earliest	19.04	20.04	1	19.04
		The latest	17.05	13.05	−4	17.05
Duration, days	Ice complete	Average	160	152	−8	156
		Minimum	131	132	1	131
		Maximum	192	181	−11	192
	Ice phenomena period	Average	187	176	−11	182
		Minimum	156	156	0	156
		Maximum	217	201	−16	217

4 Conclusion

The results of the analysis of long-term fluctuations in the air temperature regime during the cold season on the western coast of the White Sea (Republic of Karelia and Arkhangelsk Region) showed that over the past several years (1991–2020), the average air temperature according to observations at meteorological stations has increased significantly (by 1.4 °C) compared to the previous period (1956–1990) and has a tendency to increase by 0.42 °C/10 years. This fact corresponds to the trends of global warming, which are noted by researchers in regional studies for reservoirs and rivers of the Arctic zone in recent decades. Such significant climate changes have not left without consequences of the course of ice processes in the rivers of the study area, where there is a reduction in the duration of the period with ice phenomena by − 3.3 days/10 years, which corresponds to a decrease in this period by almost 3 weeks (21 days) over the considered series of years (1956–2020). This fact is caused equally by both the shift in the dates of the starting of ice freezing towards later dates (by 1.6 days/10 years) and the shift in the dates of the end of ice phenomena (ice breakup) towards earlier dates (by − 1.7 days/10 years). The results obtained in the work can be used to clarify the generalized reference data of Roshydromet, as well as in the development of methods used in hydroforecasting, taking into account the climate changes that have occurred in recent decades.

Acknowledgements The work was carried out within the framework of the theme of the state assignment № AAAA-A18-118032290034-5.

References

Agafonova SA (2019) Studies of the ice regime of rivers in the Arctic zone of Russia in the 20th and 21st centuries. Arktika i Antarktika. Arctic Antarctic 1:81–94 (In Russian). https://doi.org/10.7256/2453-8922.2019.1.29530

Agafonova SA, Frolova NL, Vasilenko AN, Shirocova VA (2016) Ice regime and dangerous hydrological phenomena on rivers of the arctic zone of European Russia (Bulletin of Moscow University. Episode 5). Geography 6:41–49 (In Russian)

Automated information system for state monitoring of water bodies. https://gmvo.skniivh.ru/. Accessed 16 Mar 2024 (in Russian)

Banshchikova LS, Sumachev AE (2021) The air temperature variability during the cold period on the Kola Peninsula as modification factor of the ice regime. Arctic: Ecol Econ 11(3):397–405 (In Russian). https://doi.org/10.25283/2223-4594-2021-3-397-405

Debolsky VK, Debolskaya EI, Kotlyakov AV, Derbenev MV, Zamyatina EHV (2008) Modern spatiotemporal variability of ice phenomena in the estuarine areas of rivers in the Russian Western Arctic. News of the Russian Academy of Sciences. Geogr Ser 2:36–42 (In Russian)

Donchenko RV (1987) Ice regime of rivers in the USSR. Hydrometeoizdat, Leningrad, 246 p (In Russian)

Dzhamalov RG, Frolova NL, Kireeva MB, Rets EP, Safronova TI, Bugrov AA, Telegina AA, Telegina EA (2015) Modern groundwater and surface water resources represent parts of Russia: formation, definition, use. GEOS, Moscow, 315 p (In Russian)

Frolova NL, Magritsky DV, Kireeva MB, Agafonova SA, Povalishnikova ES (2018) Runoff and ice phenomena dynamics on the rivers of Russian arctic due to anthropogenic and climate changes. Geogr Quest 145:233–251 (In Russian)

Georgievsky MV, Goroshkova NI, Khomiakova VA, Georgievsky DV, Plenkina AK (2021) Impact of climate change in autumn-winter period on hydrological regime of the rivers in the Small Northern Dvina river basin. J Hydrometeorol Ecol 64:466–479 (In Russian). https://doi.org/10.33933/2713-3001-2021-64-466-479

Ginzburg BM (1973) Probabilistic characteristics of the timing of freezing and breaking up of rivers. Proc Hydrometeorol Center USSR 40:21–40 (In Russian)

Kalinin VG (2020) On the selection of a representative calculated period of observation in the study of river ice regime. Geogr Bull 1(52):120–129 (In Russian). https://doi.org/10.17072/2079-7877-2020-1-120-129

Russian Institute of Hydrometeorological Information—World Data Center. http://meteo.ru/contacts. Accessed 16 Mar 2025

State Hydrological Institute. https://www.hydrology.ru/edinyi-fond-dannyx. Accessed 16 Mar 2024

Sumachev AE, Banshchikova LS (2021) Ice regime of the Pechora River in modern climatic conditions and principles of forecasting the highest water level during the period of spring ice drift. Adv Curr Nat Sci 10:75–80 (In Russian). https://doi.org/10.17513/use.37701

Surface water resources of the USSR (1972) V. 2. Pt. 1. Leningrad: Hydrometeoizdat: 527 p (In Russian)

Tsunamis Are an Urgent World Problem

K. V. Pokazeev and T. O. Chaplina

Abstract Tsunamis belong to the group of the most terrible natural disasters. The catastrophic tsunami events at the beginning of the third millennium demonstrated the need to improve existing tsunami warning systems and to conduct a wide range of scientific research on tsunamis. The recent development of warning systems should significantly improve their effectiveness and make living in coastal regions safer.

Keywords Gravity waves · Tsunamis · Earthquakes · Tsunami warning systems

1 Introduction

Tsunamis are commonly understood as gravity waves arising in the sea due to large-scale, short-lived disturbances (underwater earthquakes, underwater volcanic eruptions, underwater landslides, meteorites, rock fragments falling into the water, explosions in the water, abrupt changes in meteorological conditions, etc.).

The characteristic duration of tsunami waves is 10–100 min; the length is 10–1000 km, the speed of propagation is 10–200 m/s, and the height when rolling on the shore can reach tens of metres. These waves are very long, and the 'shallow water' theory is applicable to them to a first approximation.

The increase in tsunami amplitude near shore is primarily due to changes in the bottom topography. Due to the decrease in wave propagation velocity, the wave is compressed. A wave that is small in amplitude in the open sea increases to catastrophic heights near the shore: often the wave reaches 20–30 m in height and penetrates several kilometres onto land (Ferrer and González-de-Vallejo 2024).

The suddenness of a tsunami causes many casualties, and the powerful wave action results in significant destruction and great economic losses. Tsunami ranks

K. V. Pokazeev (✉)
M.V. Lomonosov Moscow State University, Moscow, Russia
e-mail: pokazeevk@bk.ru

T. O. Chaplina
A.Yu. Ishlinsky Institute of Physics and Mechanics of the Russian Academy of Sciences, Moscow, Russia

T. Chaplina (ed.), *Processes in GeoMedia—Volume IX*, Springer Geology,
https://doi.org/10.1007/978-981-96-9302-3_8

fifth after floods, typhoons, earthquakes, and drought in terms of the number of deaths per year from natural disasters on Earth. The distribution of tsunamis by regions is characterized by strong spatial heterogeneity, with the bulk of tsunamis—about 75% occurring in the seas of the Pacific Ocean, 9% in the Atlantic Ocean, 12% in the Mediterranean Sea, and only 3% in the Indian Ocean.

The 12-point Papadopoulos-Imamura tsunami intensity scale is now widely used. By reflecting the main results of tsunami impact on the population, coastal infrastructure and nature, and the damage caused, it allows a detailed classification of events.

The more intense the tsunami, the less frequently they occur. Intensity 4 tsunamis occur once every 10 years (in the Pacific), intensity 3 once every 3 years, intensity 2 once every 2 years, intensity 1 once a year and intensity 0 four times a year (Gusyakov 2014).

To get an idea of tsunamis, here are the characteristics of the major tsunamis that have occurred over the last 120 years (Table 1).

The 2004 disaster is the largest known tsunami disaster. A strong magnitude 9.3 earthquake in the Indian Ocean generated a giant wave, with heights of up to 36 m

Table 1 A list of the largest tsunamis that have occurred in the last 120 years

Date	Area	Cause of tsunami	Height, m	Number of casualties
27.08.1883	Indonesia	Volcanic eruption	45	36000
15.06.1896	Japan	Earthquake	30	30000
20.02.1902	Guatemala	Earthquake	10	200
31.01.1900	Ecuador	Earthquake	5	1000
10.11.1922	Chile	Earthquake	9	200
2.03.1933	Japan	Earthquake	28	3000
04.11. 1952	USSR	Earthquake	19	2300
10.07.1958	Alaska	Collapse	525	2
22.05.1960	Chile	Earthquake	25	1300
28.03.1964	Alaska	Earthquake	30	122
17.08.1976	Philippines	Earthquake	9	8000
12.12.1976	Colombia	Earthquake	5	259
16.10.1979	France	Landslide	3	6
12.12. 1992	Indonesia	Earthquake	26	1500
12.07.1993	Japan	Earthquake	32	300
04.10.1994	Kurils	Earthquake	10	
14.10.1994	Philippines	Earthquake	7	70
17.07.1998	Papua	Earthquake	10	2100
	New Guinea			
26.12.2004	Sumatra	Earthquake	36	300000
11.03.2011	Japan	Earthquake	7	20000

near the source off the coast of Sumatra, and significant wave heights have been recorded in many parts of the Indian, Pacific and Atlantic Oceans. The epicentre of the earthquake was located at a depth of 20 km about 200 km west of the northern tip of the Indonesian island of Sumatra (http://www.ngdc.noaa.gov/hazard/tsu_db.shtml).

The length of the centre was about 1200 km. The vertical displacement of the Earth's crust at the epicentre was 8–10 m. Only three major underwater earthquakes—the Kamchatka earthquake in 1952, the Aleutian earthquake in 1957 and the Chilean earthquake in 1960—had such huge sources in the last century. However, the 2004 tsunami had the maximum spreading area. After the end of the movement, the nuclear tracking service in Vienna recorded about 700 aftershocks (smaller earthquakes). The number of casualties exceeded 300,000. Material damage is enormous. The tsunami disrupted the ecosystem of atolls and coral reefs, and mangrove forests were damaged. It will take 20–50 years to restore the underwater world.

The main causes of tsunamis are earthquakes, volcanic island explosions and submarine volcanic eruptions, landslides and rockslides. For example, for the Pacific Ocean, earthquakes account for 79% of tsunamis, landslides for 6%, volcanic origin for 5%, and meteorological causes for 3%. The causes of the remaining tsunamis have not been identified.

Let us briefly consider the mentioned causes of tsunami generation separately.

1.1 Earthquakes

The most important task in forecasting tsunamis of seismic origin is to establish the signs of earthquake tsunamigenicity. It is generally assumed that if the magnitude of an earthquake exceeds a certain threshold value Mp and the origin is located below the seafloor in a certain depth range, the earthquake will be tsunamigenic.

Empirical formulae have been proposed for Japan to relate the magnitude of tsunamigenic earthquakes and the source depth H (in kilometres) (Trukhin et al. 2005):

$$M = 5{,}6 + 0{,}01H$$

Similar relationships linking the magnitude of tsunamigenic earthquakes and source depth have been proposed for other regions.

No more than 0.1 of the energy released by an earthquake is converted into tsunami energy.

The tsunami energy is estimated from data on tsunami wave propagation; from data on vertical displacement in the tsunami nucleation region and from empirical relationships between tsunami energy and tsunami magnitude. To calculate tsunamis, it is necessary to know the parameters of the residual deformation of the seafloor. Authors (Dotsenko 1998) proposed the following relations for determining the average radius of the tsunami source R_0 [km] and the amplitude of the bottom displacement in the source zone ξ_0 [m] depending on the magnitude:

$$\lg R_0 = (0{,}50 \pm 0{,}07)M - (2{,}1 \pm 0{,}6)$$

$$\lg \xi_0 = (0{,}8 \pm 0{,}1)M - (5{,}6 \pm 1{,}0).$$

These relations allow us to estimate the characteristic size of the initial disturbance on the water surface. Tsunami sources are more often elliptical in shape. In the 26.12.2004 tsunami, the maximum residual displacements in the uplift zone were 7.2 m and in the subsidence zone 3.4 m.

As a result of analyses of numerous field data, the following properties of the origin of tsunamigenic earthquakes have been established. The energy propagates mainly along the normal to the main axis of the origin. The degree of directionality depends on the elongation of the source. Large tsunami sources are usually highly elongated. Their axes are oriented parallel to the nearest coast, depression or island arc, so the main source of energy is directed seaward. The ratio of wave amplitude along the fault to wave amplitude in the direction perpendicular to the fault is approximately 1/15. Measurements of individual tsunamis confirm the above estimates. For example, the 1964 Alaska earthquake, the tsunami waves from which were recorded at several seismic stations in the Pacific Ocean, which made it possible to construct a fairly detailed tsunami orientation diagram.

1.2 Volcanic Island Explosions and Submarine Volcanic Eruptions

There are about 900 active volcanoes on earth, two-third of them are located on the shores and islands of the Pacific Ocean. About 200 underwater volcanoes are known. The eruptions of these volcanoes often lead to tsunamis. The main causes leading to tsunami formation during volcanic eruptions are the following: collapse of the caldera, earthquakes of volcanic origin, landslides, and the release of a huge volume of various substances into the water. The intensity of the generated tsunami is proportional to the erupted volume and inversely proportional to the duration of the eruption and depends on the location of the volcano, primarily the depth of the ocean near the island.

In calculations of tsunamis caused by volcanic eruptions, the initial elevation of the water surface is given by formulas of various kinds. However, the main characteristics of the initial disturbance or equivalent source—the source radius R or its area, the amplitude of the initial level displacement A—are defined through the explosion energy.

Similar models are used in the calculation of waves caused by artificial explosions or underwater volcanic eruptions.

1.3 Landslides and Explosions

The most impressive example of tsunami wave formation during a landslide took place on 10.07.1958. Lituya Bay (Alaska) is about 11 km long, 3 km wide and about 300 m deep. The avalanche of 300 million m^3 from the steep slopes of Mount Fairweather into Lituya Bay created a 524 m high splash on the opposite side of the bay, and a 60 m high wave rolled along the bay.

In 1930, tsunamis up to 15 m high were generated by the fall of rock fragments from a height of 200 m on the island of Madeira. In Norway in 1934, tsunamis of 37 m high were generated by the fall of a rock weighing 3 million tonnes from a height of 500 m.

Submarine landslides, manifested as density or turbidity flows, are observed in areas of intense sediment accumulation, a process that can take hundreds of years. Annual sediment accumulation in some canyons reaches 10^{10} m^3. The energy of large density flows reaches a huge value up to 10^{18} J. Under certain regimes (velocity, volume or power of the flow) the movement of a landslide induces a long wave. Underwater landslides are triggered by a variety of causes. Offshore engineering activities can trigger landslides and consequently tsunamis. Landslides, mudflows are often observed on the mainland slope of the ocean, and the role of indicators of the formation and passage of landslides or mudflows is played by ruptures of cables and pipelines.

1.4 Determination of Tsunami Parameters

Three stages are usually distinguished in tsunami development: 1—wave formation and propagation near the origin; 2—wave propagation in the deep open ocean and 3—transformation, reflection and destruction of waves on the shelf, their run-up on the shore, resonance phenomena in bays and on the shelf. The investigation of these stages of tsunami evolution varies greatly. The second stage is the most studied, while the first and third stages are less well studied. This is due to the complexity of the processes occurring during tsunami generation and tsunami propagation in shallow waters and the insufficiency of data required to obtain the necessary information.

The linear theory is usually used to describe tsunami propagation in the open ocean. This approach is reasonable because the nonlinearity parameter, equal to $\frac{A}{H}$, and the frequency dispersion parameter, equal to $\frac{H^2}{\lambda^2}$, where A, λ—amplitude and wavelength, H—depth are much less than unity. The estimation of the Ursell parameter $Ur = \frac{A\lambda^2}{H^3}$, equal to the ratio of the nonlinearity parameter to the dispersion parameter, shows that the dispersion effect ($Ur \ll 1$) prevails in the propagation of tsunamis in the deep open ocean, and the nonlinearity can be neglected. In the third stage, when the waves move on the shelf and run into the shore, the effect of nonlinearity ($Ur \gg 1$) dominates.

To solve the hydrodynamic problem of calculating tsunami waves, it is necessary to set initial conditions: displacement and velocity fields in the source. These data can be obtained by direct measurement of initial disturbances in the source, tsunamis in the deep ocean, or through the characteristics of tsunami occurrence in the coastal zone. In principle, only remote space methods can provide a record of initial disturbances. The first tsunami records under the direction of S.L. Solovyov were made off the South Kuril Islands (Zhak and Solov'ev 1971). Later, tsunami waves were recorded in the open ocean.

Then Japanese (JAMSTEC project) and American scientists (DART project) started to record tsunamis in the open ocean. In recent years, Japan has developed sophisticated systems—DONET1, DONET2 and S-Net observatories—to reliably record tsunamis in the centre of the tsunami source (Kaneda 2014). Recording a wave in the open ocean increases the reliability of forecasts, provides significant lead time and reduces the likelihood of false forecasts.

There is a third method—the principle possibility of determining the parameters in the source by solving the inverse problem—on the basis of a few tsunami manifestations on shore to determine its parameters in the source. The idea of this method is based on the law of conservation of wave energy, since it is assumed that the energy in the source can be estimated based on the calculation of energy in the wave zone. The implementation of this method is difficult, first of all, because of the small number of tsunami records in the coastal zone and their low accuracy.

In order to predict the occurrence of tsunamis in the coastal zone and other engineering applications, it is necessary to know the change in height, period and direction of the wave front due to refraction. Refraction diagrams, which show the position of wave crests (fronts) at different distances at the same time, or the position of the same wave crest at different times, serve this purpose. Rays (orthogonal to the position of the fronts) are drawn on the same map. By assuming that the energy flux between the two orthogonals is conserved, the change in wave height can be estimated. The intersection of the rays results in an unbounded increase in wave height. The power carried by harmonic waves along a wave tube of width b is equal to:

$$C_{g0} b_0 h_0^2 g/8 = C_g \mathrm{bh}^2 \mathrm{g}/8,$$

where g—acceleration of gravity, C_{g0} и C_g— group velocity of the wave at the beginning and end of the wave tube, b_0 and b—initial and final widths of the wave tube and h_0 and h—initial and final heights of the wave.

A refraction diagram for a tsunami is constructed from the initial position of the front, (in the particular case of two rays and the distance between them) and a depth map. The result is the wave rays, the refraction coefficient, which shows the change in the width of the wave front section, the wave height and the wave run-up time.

Special maps, called tsunami wave run-up maps, are constructed for run-up times. These maps are the basic maps used in the simplest tsunami warning schemes. There are two types of run-up maps. The first type shows the run-up time isolines (isochrones) from the tsunami source to the shore, while the second type shows the run-up time isolines from a specific point on the shore to potential tsunami sources.

The calculation of run-up time is based on ray theory:

$$t = \int_{s} ds/c_g,$$

where s is the curvilinear path of tsunami propagation and c_g is the group velocity.

1.5 Laboratory Tsunami Modelling

Hydrodynamic modelling is widely used in wave run-up studies. Let us obtain a similarity criterion for modelling long-wave run-up. Considering that in the case of run-up the greatest role is played by nonlinearity, therefore dispersion and dissipation in the shallow water equations can be neglected:

$$\begin{aligned} &\frac{\partial u}{\partial t} + u\frac{\partial u}{\partial x} + g\frac{\partial \zeta}{\partial x} = 0 \\ &\frac{\partial \zeta}{\partial x} + \frac{\partial}{\partial x}[(H(x) + \zeta)u] = 0. \end{aligned}$$

If the x-axis is directed towards the shore, then $H(x) = \alpha x$, where α—the slope angle.

Let us introduce dimensionless variables:

$$\mathrm{t}^1 = \varpi \mathrm{t}, \mathrm{x}^1 = \alpha \mathrm{x}/\mathrm{H}, \zeta^1 = \zeta/\mathrm{H}, \mathrm{u}^1 = \mathrm{u}\alpha/\varpi \mathrm{h}.$$

Here h—the characteristic height of the wave and ϖ—its frequency. In dimensionless variables (omitting the index), the shallow water equations take the form:

$$\begin{aligned} &\frac{\partial u}{\partial t} + u\frac{\partial u}{\partial x} + \frac{1}{Br}\frac{\partial \zeta}{\partial x} = 0 \\ &\frac{\partial \zeta}{\partial x} + \frac{\partial}{\partial x}[(-x + \zeta)u] = 0 \\ &Br = h\varpi^2/g\alpha^2. \end{aligned}$$

Thus, there is one dimensionless parameter Br, which defines the modelling condition: the equality of the parameter Br under in situ and laboratory conditions.

The parameter Br is found to determine not only the similarity condition but also the type of wave run-up. There are the following main types of run-up: splashing swell, diving swell, collapsing swell and surging swell. The splashing swell is observed at low slope gradients and wave heights. The diving swell is observed at higher slopes than the previous type of swell, in which case the top of the wave overtakes the bottom of the wave and the wave slopes downwards. In the case of a collapsing swell, the wave collapses near the foot of the wave, this type of run-up is observed at even greater slopes than the previous run-up types. On steeper slopes, the wave rolls up without collapsing in the form of a surging swell. The variation of the Br parameter within certain limits determines one or another type of run-up.

Tsunamis are more characterized by run-up without collapse because they are long and do not have time to topple over (Pelinovsky 1996). The value of Br = 1 is critical: at Br < 1 the wave comes ashore without collapse, Br > 1 the wave rolls ashore with collapse. In real conditions, dissipation and dispersion can slow down the collapse process, which will increase the critical value of Br.

As already mentioned, laboratory modelling of wave run-up has become widespread. The results of detailed experimental studies of the transformation and rolling of solitary waves on flat slopes are described in Dotsenko (1998). The experiments were carried out in a glass tray 25 m long, 0.6 m wide, 0.5 m high at water depths of 16, 19, 22, 25 cm (Fig. 1). The initial wave height in different experiments was set from 3 to 8 cm. The wave profile had the form of a soliton. Registration of the waveform and the speed of its progress through the channel was carried out by string wavegraphs. In the area of transformation on the slope profile filming was carried out. To determine the velocity and length of the run-up, a plan film survey was carried out. In experiments to study the transformation of the wave slope varied in the following limits: 5 < ctg a < 15. Studies were carried out for smooth and rough slopes, and several types of roughness were used. At large slope angles, wave transformation occurred without collapse, while wave collapse was observed at small slope angles. At the moment of collapse, the wave height reached its maximum value.

Studies of soliton-type wave run-up have also been carried out for smooth and rough slopes with different roughness characteristics and different initial values of the nonlinearity parameter. It has been shown that with the growth of nonlinearity the value of roll-up increases. It is established that the dependence of the roll-up value on ctg α has a maximum: as the slope angle increases, there is a transition from the roll-up with collapse to the roll-up without collapse; the maximum value of the roll-up is reached at the transition from one type of roll-up to another. Rolling velocities also reach their maximum values at this point.

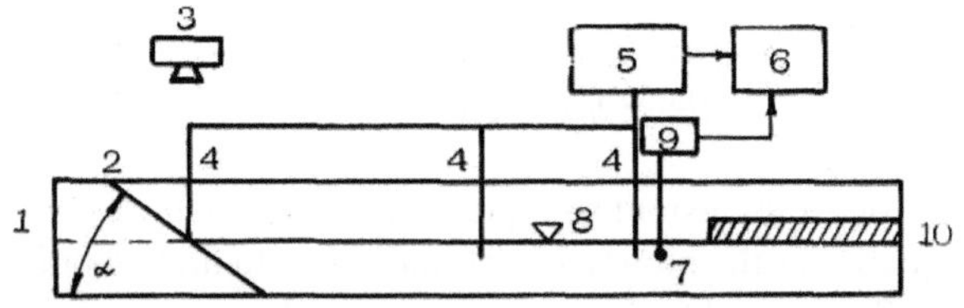

Fig. 1 Schematic diagram of the experimental setup for the study of wave transformation and run-up: 1—chute, 2—slope with changing angle of inclination, 3—film camera, 4—wavegraph sensors, 5—wavegraph, 6—recorder, 7—speed sensor, 8—water level, 9—power supply, 10—wavemaker

1.6 Tsunami Warning Systems

The specificity of tsunami impact factors makes it impossible to create a coastal infrastructure completely invulnerable to the impact of tsunami waves. This necessitates the creation of a system for early warning and evacuation of the population. Since false tsunami warnings lead to significant costs and erroneous warnings to catastrophic consequences, there is a need to develop improved methods of wave registration and pre-calculation at different stages of wave evolution.

Tsunami Warning Systems (TWS) began to appear in the mid-twentieth century. The first system was established in Honolulu in 1948. Then an international warning system was established on its basis. Later, regional tsunami warning systems began to appear. The first of them, the Japanese one, has been in operation since 1952. The Soviet warning service has been in operation since 1958. In 1967, the Alaska Centre in Palmera was established. Since 1971 and 1977 there are two regional centres in Hawaii. A warning system has been in operation in French Polynesia since 1985. A warning system has been in operation in Chile since 1987. The warning systems were created at different times, but they have much in common—similar systems of seismic and tidal data collection, operational information processing systems and decision-making procedure. Modernization of almost all TWSs started after the tsunamis of 2004 and 2011.

Improvements in TWS and tsunami calculation methods are important for tsunami zoning, which is growing in importance as coastal zones become more populated and infrastructures develop. Tsunami zoning encompasses a wide range of issues:

- seismic zoning of oceans and seas (delineation of seismic active zones, determination of earthquake recurrence, etc.);
- determination of hydrodynamic characteristics of tsunami sources (water surface displacement field and velocities of water particles);
- calculation of tsunami wave motion in the ocean with real bathymetry;
- assessment of tsunami intensity manifestation on shore.

1.7 International Tsunami Warning System

Pacific Tsunami Warning System

The first tsunami warning service was established in the USA in 1948, following the 1946 tsunami that hit Hawaii hard. The organization of the TWS was entrusted to the US National Weather Service, which assigned its observatory in Honolulu, Hawaii.

Since August 1948, this observatory has been operating as the Pacific Tsunami Warning Centre (PTWC). The centre had its first test on 8.09.1948 after the earthquakes off the island of Tonga. In 1950, there were several successful forecasts. After the devastating Chilean earthquake of 1960 (which generated tsunamis along many parts of the Pacific coast), the need for a TWS was felt in several Pacific countries. The TWS became international: it uses seismic and tidal data from different countries and in return transmits tsunami information free of charge. In 1966, under the auspices of the Intergovernmental Oceanographic Commission of UNESCO, an international Group on Tsunami Warning Systems in the Pacific was established, which includes almost all countries of the region, including the Russian Federation.

The main objectives of PTWC are as follows:

- Assessment of the tsunami hazard of earthquakes in the basin.
- Observation of generated tsunamis.
- Forecasting the time of run-up and, if possible, wave heights.
- Issuing timely warnings.

In 1995, the system included 31 seismic and 33 tide stations.

The system works as follows. A network of seismic stations continuously monitors the seismic situation. When earthquakes occur, seismic data are received via communication channels to PTWC (time of arrival of longitudinal and transverse waves, their periods and amplitudes). When data are received from 5–7 stations, the epicentre position and magnitude of the earthquake are determined. If it exceeds the threshold value (different in different regions), the centre sends a request to the nearest tidal stations. If at least one tide station reports a tsunami, a tsunami warning is issued. According to the rules of the centre, the time between the warning and the earthquake must be less than 60 min.

The principle of the centre's work is oriented on operative observation of strong trans-Pacific tsunamis, which exist for many hours. This arrangement minimizes the probability of false alarms. However, the population of coastal areas remains practically unprotected by local tsunamis, the arrival time of which can be 15–20 min or less. Regional TWSs are being established to address this problem.

The PTWC issues the following communications:

1. Tsunami Warning—issued after tsunami generation is confirmed by tide gauge observations.
2. Regional Tsunami Alert—issued only when seismic information is used and provides warning of the possibility of a tsunami.
3. Earthquake Research Bulletin—issued to provide information on an earthquake and transmitted hourly until cancellation or until it becomes a tsunami warning.
4. Earthquake information bulletin.
5. Tsunami Threat Message—given monthly to test the operation of the PTWC system and to maintain staff skills.

TOWS was automated in 1970:

- automation of message preparation and transmission,
- automation of seismic data processing,
- telemetric communication with remote stations was introduced.

The digital channels of the GOES satellite system, which covers almost the entire Pacific Ocean, are used for communication between PTWC and seismic and tide gauge stations.

The capabilities of PTWC have been significantly expanded by the American DART project, under which autonomous deep-sea tsunami recording stations are being installed. The system includes a bottom platform with high accuracy level recorders. From the bottom platform, the information is transmitted to the anchored buoy and further on via satellite communication channel. DART stations are installed at depths of up to 6000 m in areas where tsunamis may be generated, and the network of DART stations is constantly being modernized and expanded.

1.8 Japan's Tsunami Warning System

The Japanese tsunami warning system has been in operation since 1952 and is constantly being modernized and expanded. A distinctive feature of the Japanese tsunami warning system is the use of a very dense network of seismic stations. Although Japan has a very dense network of seismic stations (more than 500). The Japanese Tsunami Warning Service does not use their data at the regional and national level. The reason for this is that in the Japanese islands, tide gauges cannot provide the necessary advance warning. The tide gauge data are only used to cancel a tsunami alert. Japan is characterized by the problem of recording tsunamis from nearby sources. For example, the tsunami in the Sea of Japan in May 1983: the time of signalling was 13 min, and the first waves approached the shore in 9–10 min. Several dozens of people died. A similar tragedy was repeated in 2011. In Japan, there are zones in which a tsunami wave can arrive in a period of less than 10 min. To solve the problem of safety in these zones, local warning systems are created, which operate independently of regional and national centres, where the possibility of a tsunami is estimated based on seismic data at one single or several points.

Japan's warning system was last upgraded in 2013 to take into account the analysis of the aftermath of the Tohoku tsunami (2011). The 2011 disaster was the largest tsunami in Japanese history, caused by one of the strongest earthquakes ever recorded in the world. One of the reasons for the high death toll was an underestimation of the magnitude of the earthquake, which led to an underestimation of tsunami wave heights. The conclusions were drawn, and the warning system is now operating under revised rules, although the principle of operation remains the same.

Japan has a national tsunami warning system, which was modernized in 2013 in response to the Tohoku tsunami impact analysis and is now the most advanced system. The Japan Meteorological Agency (JMA) has carried out unique work to simulate about 100,000 tsunami variants, each based on hypothetical earthquakes of varying intensity, with epicentres at different depths at 4,000 locations along crustal faults off the coast of Japan. The database was compiled from calculations of expected wave heights and tsunami arrival times following a tremor at each of these epicentres. After the actual tremor is recorded, the database searches for the closest coordinates, depth, and intensity values, making it possible to send a warning as early as 3–4 min after the first tremors.

As soon as an earthquake occurs, the Japan Meteorological Agency (JMA) estimates the probability of a tsunami based on seismic data, with more than 1,000 seismic stations included in the system. If a tsunami is expected, the JMA issues a tsunami warning for each region; there are 66 such regions in Japan. Tsunami information such as estimated time of arrival and wave heights are calculated and adjusted based on numerous measurement systems. Wave height estimates are quite accurate because they are made using all possible monitoring systems, including data from the DART, DONET and S-Net systems, which provide tsunami wave records from bottom sensors. The information from these stations is used in numerical tsunami calculation schemes, which significantly increases the reliability of forecasts of run-up magnitudes and tsunami arrival times.

The Japanese warning system is based on the division of warnings into three types: tsunami advisory, tsunami warning, and major tsunami warning. The division into types is based on the estimated wave heights—from 20 cm to 1 m, from 1 to 3 m and from 3 m to over 10 m, respectively. Even with a minimal tsunami hazard assessment, people are instructed to leave the water and coastal areas until the tsunami warning is cancelled. The centuries-old experience and mentality of the Japanese people play a major role in ensuring the effectiveness of warning systems.

1.9 Alaska Tsunami Warning Centre

The Alaska Tsunami Warning Centre was established in 1967 in Palmer as a regional tsunami warning centre for Alaska and the Aleutian Islands. In 1982, the west coast of the United States and Canada was added to its area of responsibility. Since submarine earthquake sources in the Aleutian Islands and Alaska area are located close to shore (10–20 min tsunami approach time), a telemetric system of observations and real-time transmission of seismic and tide gauge signals to the Palmer Operations Centre was established.

The Alaska Centre observation network was one of the largest and most advanced telemetry geophysical networks in the world before DONET and S-Net were established. It includes 140 seismic and 8 tide stations. These stations belong to 4 agencies, but this does not prevent them from working together. At present, the system works in an automated mode. The earthquake parameters are determined within 5–7 min after the earthquake is registered.

For earthquakes in the Aleutian-Alaska zone, the warning centre issues a tsunami alert for magnitude greater than 6.75 without waiting for tsunami confirmation from tide stations because the earthquake sources are very close. For earthquakes in the Commander Islands, the west coast of the USA and Canada with magnitudes greater than 7.5, a warning is issued and an alert is issued only if a tsunami is confirmed by tide gauge stations. This differentiated approach reduces the number of false alarms.

1.10 Indian Tsunami Warning Centre

Following the tragic events of 2004, the Indian National Centre for Ocean Information Services (INCOIS) was established in India with responsibility for tsunami warnings (https://incois.gov.in). The centre receives data from more than 35 level sensors, dozens of buoys from bottom pressure sensors, and a network of seismographs (more than 400) from both Indian and other countries in the Indian Ocean basin. The Centre uses information from the Oceansat-2 satellite system. The combination of national measurement systems and modern international systems has resulted in an advanced tsunami warning system.

Let us note some common features of tsunami warning systems. The seismic method was previously used as the basis for all TWSs that existed in the Pacific. The hydrophysical method was not widely used. Geophysical data were used in TWSs to confirm the occurrence of tsunamis or to cancel alarms. In the twenty-first century, the situation has changed radically, DART-type stations have become widespread, showing high efficiency in tsunami wave registration, and DONET- and S-Net-type systems built in Japan allow to register tsunami waves at all stages from the moment of generation on a large part of coastal territory. This is complemented by advanced communications systems, including satellite communications and high-speed methods for analysing the geophysical information received. The incorporation of geophysical information from these advanced systems into tsunami warning systems has increased the reliability of forecasts.

Much of the progress in the development of tsunami warning systems was accelerated by the 2004 tsunami. The tragic events of 26 December 2004 demonstrated the urgent need for a global tsunami detection and warning system using satellite remote sensing capabilities. The modernization of TWS is now underway everywhere.

Lomonosov Moscow State University has created an automatic system for assessing earthquake tsunami hazard—'Tsunami Observer' (http://ocean.phys.msu.ru/projects/tsunami-observe). Since 2018, the system has been monitoring earthquakes in progress and assessing their tsunami hazard. The system uses earthquake data from two sources: USGS (US Geological Survey) and GFZ (German Earth Research Centre). When an earthquake with a magnitude greater than 6 is reported, the coseismic vector field of bottom and/or land deformations is calculated, and the initial elevation of the water surface in the tsunami source is calculated. Then the tsunami intensity is estimated using the Solovyov-Imamura scale; if the tsunami energy exceeds the threshold value, numerical modelling of the tsunami is carried out, resulting in the spatial distribution of the wave amplitude, the direction of wave energy propagation, and an estimate of the maximum tsunami height.

Since the Tsunami Observer system uses USGS and GFZ data, which provide seismic data with a significant delay from the time of origin, it cannot be used for operational forecasts. In addition, the system cannot record tsunamis generated by volcanic eruptions, landslides and meteotsunamis because these types of tsunamis are not accompanied by earthquakes. The system can be used, first of all, to develop methods of tsunami calculation on the basis of received primary information about events and subsequent measurements of the results of wave manifestation, which is of undoubted value.

The analysis of 'Tsunami Observer' system operation showed satisfactory agreement of forecast data with observations, but seismic data provided by USGS and GFZ are insufficient for building detailed models of the tsunamigenic earthquake origin (Kolesov and Nosov 2019).

2 Conclusion

The tsunami problem is still relevant. For some regions, tsunamis pose a serious threat to the sustainable development programme. The catastrophic tsunamis of 2004 and 2011 showed the necessity to improve the existing tsunami warning systems worldwide. The use of measurements of tsunami parameters at DART, DONET and S-Net stations has significantly expanded the capabilities of tsunami warning systems and increased the reliability of forecasts. It is reasonable to use the experience and principles of tsunami warning systems for the development of monitoring and warning systems for other natural disasters, and for the improvement of methods for zoning of regions prone to natural disasters.

The research was carried out within the framework of State Assignment 124012500441-6 (FFGN-2024-0007).

References

Dotsenko SF (1998) Estimation of tsunami hazard level of the Black Sea. Vestnik of Moscow University. Ser. 3. Physics, Astronomy, № 4, pp 19–23

Ferrer M, González-de-Vallejo LI (2024) Global historical megatsunamis catalog (GHMCat). GeoHazards 5:971–1017. https://doi.org/10.3390/geohazards5030048

Global Historical Tsunami Database/NGDC/WDS (National Geophysical Data Center, NOAA/WorldData Service). https://doi.org/10.7289/V5PN93H7. http://www.ngdc.noaa.gov/hazard/tsu_db.shtml

Gusyakov VK (2014) The strongest tsunamis of the world ocean and the problem of safety of sea coasts. Bull Russian Acad Sci. Phys Atmosp Ocean 90(5):496–507

http://ocean.phys.msu.ru/projects/tsunami-observe

https://incois.gov.in

Kaneda Y (2014) DONET: a real-time monitoring system for megathrust earthquakes and tsunamis around Southwestern Japan. Oceanography 27(2):103–103. https://doi.org/10.5670/oceanog.2014.45

Kolesov SV, Nosov MA (2019) Operation experience of the automatic system of tsunami hazard assessment for earthquake 'Tsunami Observer'. VMU. Series 3. Physics. Astronomy, № 6, pp 97–106

Konovalov AV, Nagornykh TV, Safonov DA (2014) Modern investigations of earthquake focal mechanisms of Sakhalin Island. Sakhalin. Dalnauka, Vladivostok, 252 p

Pelinovsky EN (1996) Hydrodynamics of tsunami waves/IPF RAS. Nizhny Novgorod, 276 p

Trukhin VI, Pokazeev KV, Kunitsyn VE (2005) General and environmental geophysics. Fizmatlit Publishing House. ISBN:5-9221-0541-8, p 576

Zhak VM, Solov'ev SL (1971) Remote registration of weak waves of tsunami type on the shelf of the Kuril Islands. DAN USSR, vol 198, no 4, pp 816–817

nonlinear problems of hydromechanics, asymptotic methods are used (Sekerzh-Zenkovich 1951), which allow to obtain with sufficient accuracy an approximate solution of the problem in elementary functions. This method of solution allows investigating nonlinear effects more deeply due to the possibility of analytical representation of approximate solutions.

The purpose of this work is to obtain asymptotic approximations of surface nonlinear standing wave profiles in Lagrangian variables and compare them with experimental data for different liquid depths, which determine the limits of applicability of the theoretical model.

An important part of any study is the experimental verification of the results derived from the mathematical model. While the influence of nonlinear effects on the formation and breaking of Faraday waves as a function of liquid depth has been investigated in Kalinichenko et al. (1995), Kalinichenko and Sekerzh-Zen'kovich (2007), Kalinichenko (2024), the present work utilizes experimental data obtained in the study as well as wave profiles (Kalinichenko and Sekerzh-Zen'kovich 2007).

2 Derivation of Wave Profiles in Lagrangian Variables

We consider free-standing two-dimensional waves on a liquid surface, and the problem is formulated in Lagrangian variables (Sekerzh-Zenkovich 1951; Nesterov 1969). Consider a rectangular vessel of length L with liquid filling it up to the height h. Let us introduce a rectangular coordinate system Oxy rigidly connected with the vessel, with Ox directed along the unperturbed water surface and Oy against the direction of free-fall acceleration. We take the pressure on the surface of the liquid equal to p_0 (Fig. 1).

Let us write down the equations of motion of the ideal fluid and the continuity equations in Lagrangian form, where (a, b, t) are Lagrangian variables. We obtain a system of three equations:

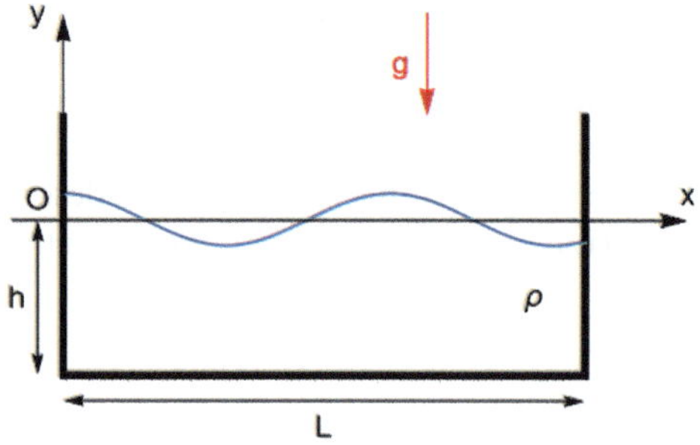

Fig. 1 Model of a vessel with liquid

$$\begin{cases} \frac{\partial^2 x}{\partial t^2}\frac{\partial x}{\partial a} + \frac{\partial^2 y}{\partial t^2}\frac{\partial y}{\partial a} = \frac{\partial \Pi}{\partial a} \\ \frac{\partial^2 x}{\partial t^2}\frac{\partial x}{\partial b} + \frac{\partial^2 y}{\partial t^2}\frac{\partial y}{\partial b} = \frac{\partial \Pi}{\partial b} \\ \frac{\partial x}{\partial a}\frac{\partial y}{\partial b} - \frac{\partial x}{\partial b}\frac{\partial y}{\partial a} = 1 \end{cases} \quad (1)$$

$$\Pi = -\frac{p - p_0}{\rho} - gy$$

In these equations, the following notations are accepted: $x = x(a, b, t), y = y(a, b, t)$—coordinates of an individual particle, $p = p(a, b, t)$—hydrodynamic pressure of liquid, g—free-fall acceleration, ρ—constant density of liquid, and $\Pi(a, b, t)$—auxiliary pressure function. Variables a, b are chosen in such a way that at $t = 0$ the equalities are fulfilled:

$$x(a, b, 0) = x_0 = a$$
$$y(a, b, 0) = y_0 = b$$

Let's require that the particles of liquid, which were on the surface at the moment $t = 0$ were on it and any other moment of time, and also that the particles on the walls of the vessel remained on them all the time. As a result, we obtain the boundary conditions:

$$\Pi(a, 0, t) = -gy(a, 0, t)$$
$$y(a, -h, t) = -h$$
$$x(0, b, t) = 0$$
$$x(L, b, t) = L$$

Assuming that the solution of the system is a standing wave, additional conditions are imposed on the solutions: the functions $x(a, b, t), y(a, b, t), \Pi(a, b, t)$ must be periodic in t with period $= 2\pi/\omega$, and periodic in a, with period $\lambda = 2\pi/\varkappa$, where ω is the oscillation frequency and $\varkappa$ is the wave number. We also assume that the motion is potential (there is a velocity potential).

We consider a standing wave of finite but small amplitude, so we introduce a small parameter ε and make a transition to new functions $(\xi, \eta, \tilde{p})$ by the formulas:

$$\begin{cases} x(a, b, t, \varepsilon) = a + \varepsilon\xi(a, b, t, \varepsilon) \\ y(a, b, t, \varepsilon) = b + \varepsilon\eta(a, b, t, \varepsilon) \\ \Pi(a, b, t, \varepsilon) = \varepsilon, \tilde{p}(a, b, t, \varepsilon) \end{cases}$$

New $\xi = \xi(a, b, t, \varepsilon), \eta = \eta(a, b, t, \varepsilon), \tilde{p} = \tilde{p}(a, b, t, \varepsilon)$ are the new unknown functions, finding which is the solution to the problem. Let us substitute the new functions into the system (1) and into the conditions constraining our solutions, then we obtain

$$\begin{cases} \frac{\partial^2 \xi}{\partial t^2} + \varepsilon \left[\frac{\partial^2 \xi}{\partial t^2} \frac{\partial \xi}{\partial a} + \frac{\partial^2 \eta}{\partial t^2} \frac{\partial \eta}{\partial a} \right] = \frac{\partial \tilde{p}}{\partial a} \\ \frac{\partial^2 \eta}{\partial t^2} + \varepsilon \left[\frac{\partial^2 \xi}{\partial t^2} \frac{\partial \xi}{\partial b} + \frac{\partial^2 \eta}{\partial t^2} \frac{\partial \eta}{\partial b} \right] = \frac{\partial \tilde{p}}{\partial b} \\ \frac{\partial \xi}{\partial a} + \frac{\partial \eta}{\partial b} = -\varepsilon \left[\frac{\partial \xi}{\partial a} \frac{\partial \eta}{\partial b} + \frac{\partial \xi}{\partial b} \frac{\partial \eta}{\partial a} \right] \end{cases} \quad (2)$$

and boundary conditions:

$$\tilde{p}(a, 0, t) = -gy(a, 0, t)$$
$$\eta(a, -h, t) = 0$$
$$\xi(0, b, t) = 0$$
$$\xi(L, b, t) = 0$$

Obviously, the functions $\xi(a, b, t, \varepsilon), \eta(a, b, t, \varepsilon), \tilde{p}\ (a, b, t, \varepsilon)$ satisfy the same periodicity conditions. We require that when the solutions of our system coincide with the results of linear theory.

Now, suppose that our new functions $\xi(a, b, t, \varepsilon), \eta(a, b, t, \varepsilon), \tilde{p}(a, b, t, \varepsilon)$ also have a power series expansion by ε. Let us write out the first two terms of the expansion for each function:

$$\begin{cases} \xi = \xi_0 + \varepsilon \xi_1 \\ \eta = \eta_0 + \varepsilon \eta_1 \\ \tilde{p} = \tilde{p}_0 + \varepsilon \tilde{p}_1 \end{cases}$$

We substitute these expansions into system (2), group the summands by powers of ε (substitutions $\psi = \omega t, \alpha = \varkappa a, \beta = \varkappa b$) and obtain the following system of equations:

(1) at the free member:

$$\begin{cases} \frac{\partial^2 \xi_0}{\partial \psi^2} = \frac{\partial \tilde{p}_0}{\partial \alpha} \\ \frac{\partial^2 \eta_0}{\partial \psi^2} = \frac{\partial \tilde{p}_0}{\partial \beta} \\ \frac{\partial \xi_0}{\partial \alpha} + \frac{\partial \eta_0}{\partial \beta} = 0 \end{cases} \quad (3)$$

(2) at ε:

$$\begin{cases} \frac{\partial^2 \xi_1}{\partial \psi^2} + \frac{\partial^2 \xi_0}{\partial \psi^2} \frac{\partial \xi_0}{\partial \alpha} + \frac{\partial^2 \eta_0}{\partial \psi^2} \frac{\partial \eta_0}{\partial \alpha} = \frac{\partial \tilde{p}_0}{\partial \alpha} + \frac{\partial \tilde{p}_1}{\partial \alpha} \\ \frac{\partial^2 \eta_{0_1}}{\partial \psi^2} + \frac{\partial^2 \xi_0}{\partial \psi^2} \frac{\partial \xi_0}{\partial \beta} + \frac{\partial^2 \eta_0}{\partial \psi^2} \frac{\partial \eta_0}{\partial \beta} = \frac{\partial \tilde{p}_0}{\partial \beta} + \frac{\partial \tilde{p}_1}{\partial \beta} \\ \frac{\partial \xi_1}{\partial \alpha} + \frac{\partial \eta_1}{\partial \beta} = -\left[\frac{\partial \xi_0}{\partial \alpha} \frac{\partial \eta_0}{\partial \beta} - \frac{\partial \xi_0}{\partial \beta} \frac{\partial \eta_0}{\partial \alpha} \right] \end{cases} \quad (4)$$

The boundary conditions for systems (3), (4) take the form:

$$\begin{aligned}
\tilde{p}_{0,1}(\alpha, 0, t) &= -g\eta_{0,1}(\alpha, 0, t)\\
\eta_{0,1}(, -\varkappa h, t) &= 0\\
\xi_{0,1}(0, \beta, t) &= 0\\
\xi_{0,1}(\varkappa L, \beta, t) &= 0
\end{aligned}$$

The solution of the first system (3) is the known linear approximation of standing waves (further all terms of the expansion of the function $\tilde{p}$ will not be written out, since they are not so important for the study):

$$\begin{cases}
\xi_0 = -A\frac{ch(\beta+\varkappa h)}{sh(\varkappa h)}\sin\alpha\cos\psi\\
\eta_0 = A\cos\alpha\cos\psi
\end{cases}$$

The solution to the second system:

$$\begin{cases}
\xi_1 = \frac{A^2}{8sh^2\varkappa h}\sin 2\alpha(1+\cos 2\psi) - \frac{3}{16}\frac{A^2 ch2(\beta+\varkappa h)}{sh^4\varkappa h}\sin 2\alpha\cos 2\psi + \frac{A^2}{4}\frac{ch2(\beta+\varkappa h)}{sh^2 2\varkappa h}\sin 2\alpha\\
\eta_1 = \frac{A^2}{8sh^2\varkappa h}sh2(\beta+\varkappa h)(1+\cos 2\psi) + \frac{3}{16}\frac{A^2}{sh^4\varkappa h}sh2(\beta+\varkappa h)\sin 2\alpha\cos 2\psi - \frac{A^2}{4sh2\varkappa h}\cos 2\alpha
\end{cases}$$

We substitute the obtained expansion terms into the expressions for x and y and reduce them to a dimensional form to write the equation of the wave profile in Lagrangian variables:

$$\begin{aligned}
x(a, 0, t) = {} & a - \varepsilon A\frac{ch\varkappa h}{sh\varkappa h}\sin\varkappa a\cos\psi + \varkappa\frac{(\varepsilon A)^2}{8sh^2\varkappa h}\sin 2a(1+\cos 2\psi)\varkappa\\
& - \frac{3}{16}\frac{(\varepsilon A)^2 ch2\varkappa h}{sh^4\varkappa h}\sin 2\varkappa a\cos 2\psi + \varkappa\frac{(\varepsilon A)^2 ch2\varkappa h}{4sh^2 2\varkappa h}\sin 2\varkappa a
\end{aligned} \tag{5}$$

$$\begin{aligned}
y(a, 0, t) = {} & \varepsilon A\cos\varkappa a\cos\psi + \varkappa\frac{(\varepsilon A)^2 sh2\varkappa h}{8sh^2\varkappa h}(1+\cos 2\psi)\\
& + \varkappa\frac{3}{16}\frac{(\varepsilon A)^2 sh2\varkappa h}{sh^4\varkappa h}\cos 2\varkappa a\cos 2\psi - \varkappa\frac{(\varepsilon A)^2}{4sh2\varkappa h}\cos 2\varkappa a
\end{aligned} \tag{6}$$

3 Transition to Eulerian Variables

Using asymptotic expansion, we find the approximate equation of the wave profile, in Eulerian variables, of the dependence $y(x, t)$. We have eliminated the variable b from the equations by equating it to zero $b = 0$. Now we need to eliminate the variable a. It is convenient to do this if, from Eq. (5), we find a as a function of x, ε in the form of a series with terms:

$$x, \left(\frac{\partial a}{\partial \varepsilon}\right)_{\varepsilon=0}, \ldots, \frac{1}{n!}\left(\frac{\partial^n a}{\partial \varepsilon^n}\right)_{\varepsilon=0},$$

We also look for the explicit profile in the form of the series:

$$y(x, \varepsilon, t) = y(x, 0, t) + \left(\tfrac{\partial y}{\partial \varepsilon}\right)_0 \varepsilon + \tfrac{1}{2!}\left(\tfrac{\partial^2 y}{\partial \varepsilon^2}\right)_0 \varepsilon^2 + \ldots \tag{7}$$

Given that $y(x, 0, t) = 0$, finding all derivatives of ε and substituting them into (7), we obtain the formula for the wave profile using the substitution $\zeta = \varepsilon A$:

$$\begin{aligned} y(x, t, \zeta) = {} & \zeta \cos(\varkappa x)\cos\psi - \varkappa\zeta^2 \frac{ch\varkappa h}{4sh\varkappa h}(1 - \cos 2\varkappa x)(1 + \cos 2\psi) \\ & + \varkappa\zeta^2 \frac{sh2h}{8sh^2 h}(1 + \cos 2\psi) + \varkappa\zeta^2 \frac{3}{16}\frac{sh2\varkappa h}{sh^2\varkappa h}\cos 2\varkappa x \cos 2\psi \\ & - \varkappa\zeta^2 \frac{1}{4sh2\varkappa h}\cos 2\varkappa x \end{aligned} \tag{8}$$

4 Checking the Integral Relations

Let's check the obtained formulas for fulfillment of integral relations:

$$\int_{-\frac{\pi}{\varkappa}}^{\frac{\pi}{\varkappa}} y(x, t, \zeta)dx = 0 \text{ – in Eulerian variables} \tag{9}$$

$$\int_0^L y(a, 0, t)\tfrac{\partial x(a,0,t)}{\partial a} da = 0 \text{ – in Lagrange variables} \tag{10}$$

All calculations were performed in Wolfram Mathematica 13.1 and are presented in the Tables 1 and 2.

As follows from the tables, for the explicit formula, the value of the integral always remains zero, in contrast to the implicit formula. This is due to the fact that the calculation of the second integral discarded all terms of order higher than ε^2,

Table 1 Explicit formula for the second wave mode $h = 20$ cm, $H = 7.2$ cm, $\varkappa = 0.125$ 1/cm, $\omega = 5.939$ 1/s

t, sec	Value of the integral (9)
0	0
2	0
4	0
6	0
8	0

Table 2 Implicit formula for the second wave mode $h = 20$ cm, $H = 7.2$ cm, $\varkappa = 0.125$ 1/cm, $\omega = 5.939$ 1/s

t, sec	Value of the integral (10)
0	0.000019
2	0.000118
4	0.002715
6	0.001514
8	0.000115

which were obtained by multiplication of the two functions. It should be noted that even with this assumption, the accuracy of the solution in Lagrangian variables is of order up to the third decimal place.

5 Analysis of the Behavior of Wave Nodes

Now let's find the expression for the wave nodes (first wave mode). As we know, the standing wave nodes satisfy equation:

$$y(x, t, \zeta) = 0$$

We will look for the solution of this equation in the form:

$$x = x_0 + \zeta x_1$$

Substituting this relation into the explicit expression for the standing wave profile and decomposing the trigonometric functions into a Taylor series up to the first two terms, we obtain (in unitless form):

$$\begin{aligned} \zeta \cos\psi(\cos x_0 - \zeta x_1 \sin x_0) &- \zeta^2 \frac{ch\varkappa h}{4sh\varkappa h}(1 - \cos 2x_0 + 2\zeta x_1 \sin 2x_0)(1 + \cos 2\psi) \\ &+ \zeta^2 \frac{sh2\varkappa h}{8sh^2\varkappa h}(1 + \cos 2\psi) + \zeta^2 \frac{3}{16}\frac{sh2\varkappa h}{sh^2\varkappa h}(\cos 2x_0 - 2\zeta x_1 \sin 2x_0)\cos 2\psi t) = 0 \\ &- \zeta^2 \frac{1}{4sh2\varkappa h}(\cos 2x_0 - 2\zeta x_1 \sin 2x_0) = 0 \end{aligned}$$

We compare the terms at the same powers of ζ (discarding all powers above the second) and get:

$$x_0 = \frac{\pi}{2}$$

$$x_1 = \frac{sh2\varkappa h}{4sh^2\varkappa h}\cos\psi$$

The final expression for the nodal points for the first mode in unitless form:

$$x(\zeta, t) = \frac{\pi}{2} + \zeta \frac{sh^2 \varkappa h}{4sh^2 \varkappa h} \cos \psi$$

This formula shows that already the first terms of the expansion are time dependent. This obviously proves that the equation for the nodes has no time-independent solutions, i.e., our profile has no fixed nodes. It can also be seen that the nodal points do not move horizontally, but along the initial curved profile of the wave.

One more result follows from the obtained from the obtained formula: the maximum free surface displacements (wave crests) will be at the points $x = 0$ and $x = \pi$ (i.e., at $x = 0$ and $x = L$ in unit quantities). They will be reached at $t = k\pi/\omega$, where k is any positive integer number.

6 Comparison with Experimental Data

The experiments were carried out on the stand "Studies of Dynamics and Structure of Oscillating Currents" of IPMech RAS according to the method (Kalinichenko et al. 1995; Kalinichenko and Sekerzh-Zen'kovich 2007; Kalinichenko 2024). The standing gravitational waves excited at the main parametric resonance are close to the free waves on the liquid surface described in Sect. 1. During the experiments, the liquid depth $h = 5$–20 cm was varied for the second and first wave modes ($n = 1, 2$).

In Figs. 2, 3, 4, 5 and 7, column I shows photographs of the wave profiles corresponding to time instants $t = 0$, $T/4$, and $T/2$ (T is the wave period). Column II each shows the calculated profiles using Lagrangian variables (solid curve) and Eulerian variables (dashed curve), with some figures showing digitized experimental profiles (square symbols).

The discrepancy between the Lagrangian and Eulerian description of the wave profile increases with decreasing liquid depth h—Figs. 3, 6, and 7. The profiles constructed in Lagrangian variables correspond more accurately to the experimental data. It follows from Fig. 4 that for the first mode in the case $h = 5$ cm, $H = 4$ cm, the presented solution in both Lagrangian and Eulerian variables does not describe the experimental data.

As follows from Figs. 2 and 5, the calculated profiles in Lagrangian and Eulerian variables practically coincide with each other and describe well enough the experimental data for the first and second wave modes on the liquid surface with depths of 20 and 15 cm, respectively.

The results of comparison of experimental and calculated wave profiles depend on both the liquid depth h, wavelength λ (mode number n), and wave height H, i.e., on the relative liquid depth h/λ and wave steepness H/λ. According to Bridges (1986), the limits of applicability of the theoretical model taking into account the joint influence of the relative depth of the liquid and wave steepness can be estimated by the value of the Ursell number (Stokes parameter), defined as follows:

$$Ur = (H/l)/(h/l)^3$$

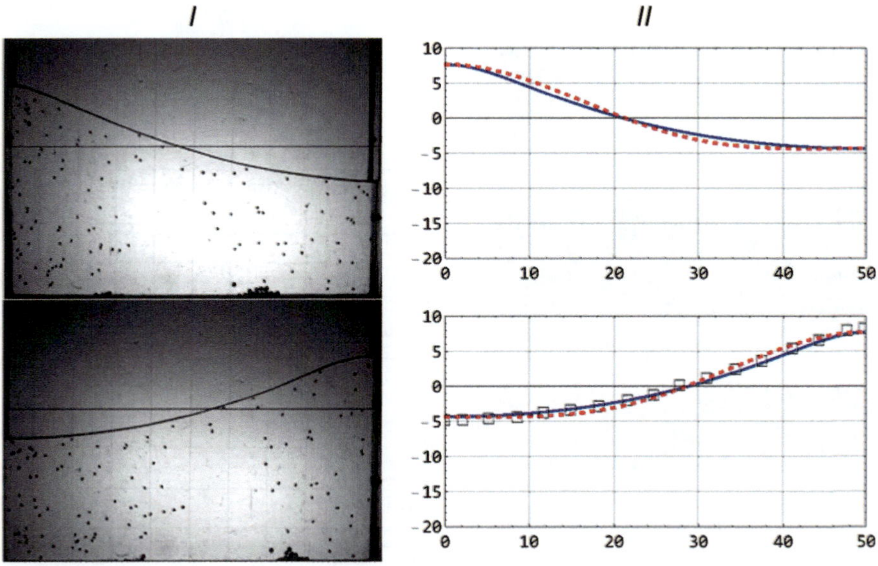

Fig. 2 First wave mode of maximum height: $n = 1$, $h = 20$ cm, $H = 12$ cm, $T = 0.924$ s ($t = 0$, $T/2$)

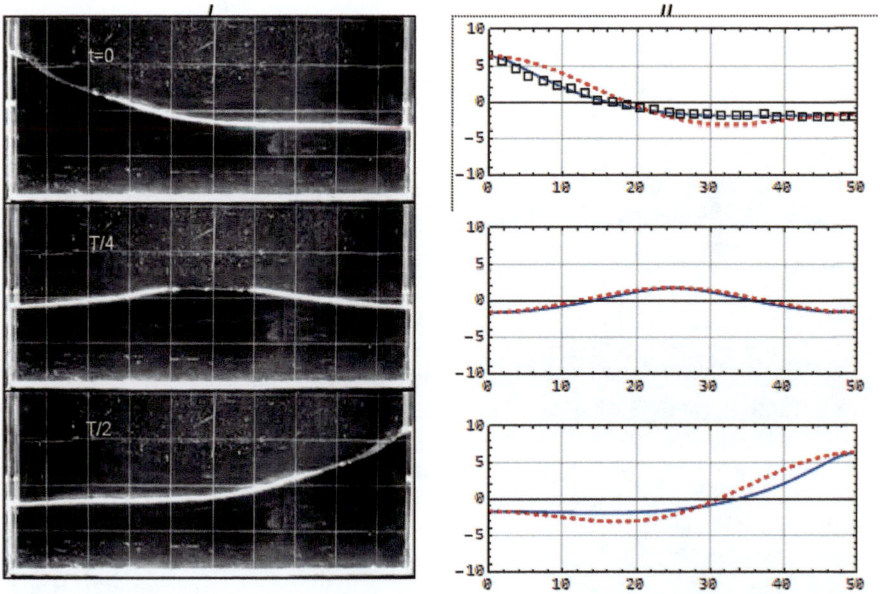

Fig. 3 First wave mode: $n = 1$, $h = 10$ cm, $H = 7.9$ cm, $T = 1.110$ s ($t = 0$, $T/4$, $T/2$)

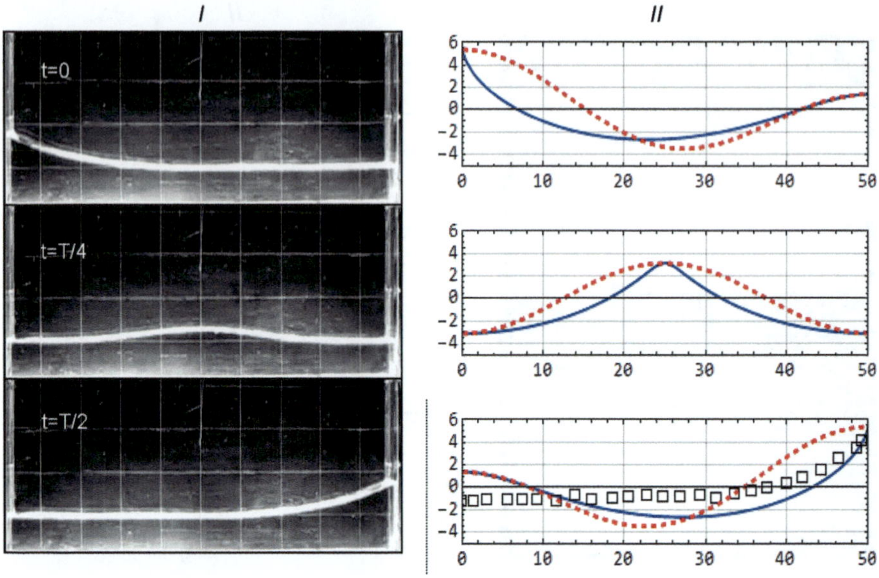

Fig. 4 First wave mode: $n = 1$, $h = 5$ cm, $H = 4$ cm, $T = 1.356$ s ($t = 0$, $T/4$, $T/2$)

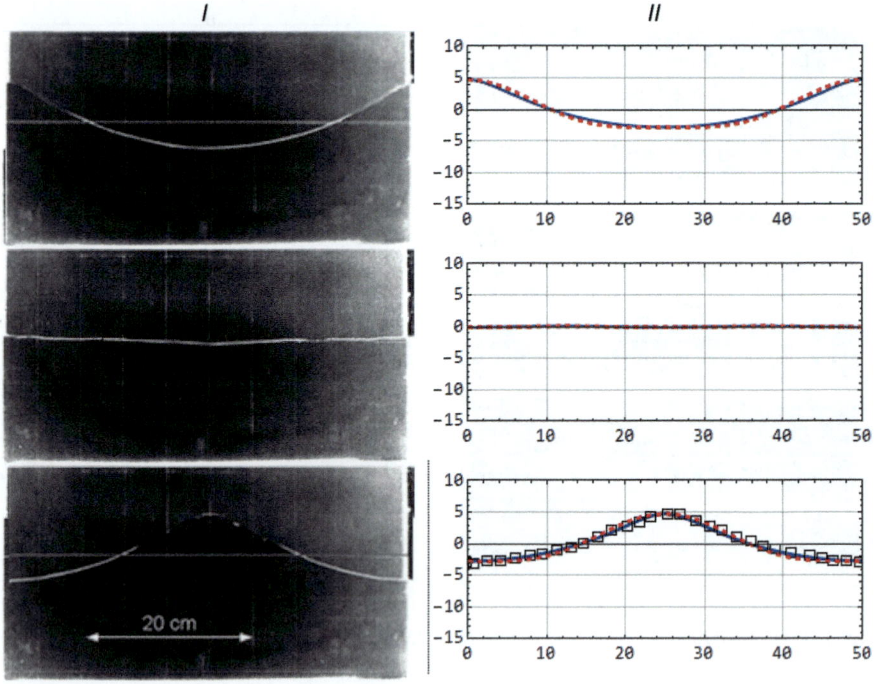

Fig. 5 Second wave mode: $n = 2$, $h = 15$ cm, $H = 7.5$ cm, $T = 0.940$ s ($t = 0$, $T/4$, $T/2$)

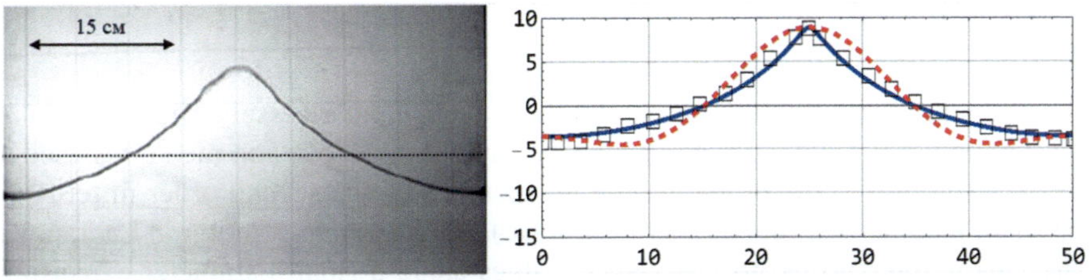

Fig. 6 Second wave mode of maximum height: $n = 2$, $h = 15$ cm, $H = 12.4$ cm, $T = 0.610$ s; experiment—the result of overlaying ten consecutive profiles of maximum development according to the results of the (Kalinichenko and Sekerzh-Zen'kovich 2007)

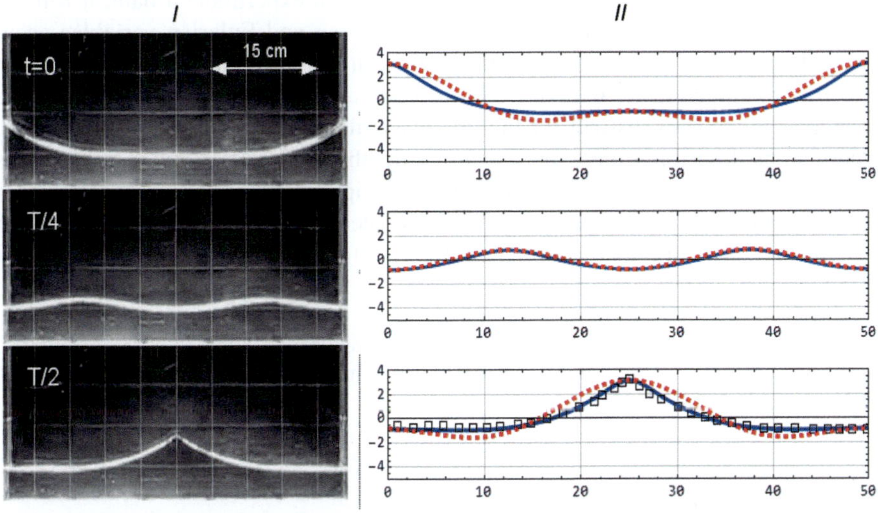

Fig. 7 Second wave mode: $n = 2$, $h = 5$ cm, $H = 4$ cm, $T = 0.814$ s ($t = 0$, $T/4$, $T/2$)

The values of Ur calculated from the experimental data are given in Table 3. The joint analysis of Figs. 2, 3, 4, 5 and 7 and Table 3 shows that the upper limit of applicability of the Eulerian variables is estimated by values of the Ursell number of about 20, and for the Lagrangian variables is about 100.

Table 3 Values of the Ursell parameter corresponding to the experimental conditions

Wave mode number	$n = 1$			$n = 2$		
λ, cm	100			50		
h, cm	20	10	5	15	15	5
H, cm	12	7.9	4	7.5	12.8	4
Ur	15	79	320	5.5	9.5	80

7 Conclusion

The problem of free oscillations of a liquid in a rectangular vessel of limited volume in Lagrangian variables is formulated. The asymptotic expansion method is used to obtain a solution to the problem with an accuracy of the second order in terms of the small parameter ε (steepness of the wave). The resulting solution in Lagrangian variables is transformed to Eulerian variables using a small parameter expansion. To verify the correctness of the solution, integral relations are calculated, both for parametric and explicit forms. It is shown that for the second mode, the accuracy of the formulas exceeds the integral value by an order of magnitude. This makes it possible to understand that the solutions were obtained mathematically correctly.

From a comparison of the theoretical model with experimental data, it follows that at great depths, the formulas in the Lagrangian and Eulerian variables accurately describe the wave process. The profiles in the Eulerian variables lose their effectiveness much faster with decreasing depth, and at heights below $h/\lambda \sim 0.1$, the discrepancy becomes distinct for the first wave mode. For the first wave mode, the theory stops describing the experiment at a depth of $h/\lambda \sim 0.05$ and a wave height of $H/\lambda \sim 4$. For the second wave mode, the description is more accurate. The upper limit of applicability of Lagrangian variables calculated using the Ursell parameter is almost an order of magnitude higher ($Ur \sim 100$) than the same value for Eulerian variables ($Ur \sim 20$).

Acknowledgements The work was supported by RSF grant No. 23-19-00,451. The experiments were carried out on the DSO bench (a unique scientific installation of the Ishlinsky Institute for Problems of Mechanics RAS).

References

Bridges TJ (1986) Cnoidal standing waves and the transition to the traveling hydraulic jump. Phys Fluids 29(9):2819–2827. https://doi.org/10.1063/1.865480

Kalinichenko VA, Nesterov SV, Sekerzh-Zen'kovich SY, Chaikovskii AA (1995) Experimental study of surface waves with Faraday resonance excitation. Fluid Dyn 30(1):101–107. https://doi.org/10.1007/BF02029933

Kalinichenko VA, Sekerzh-Zen'kovich SY (2007) Experimental investigation of Faraday waves of maximum height. Fluid Dyn 42(6):959–965. https://doi.org/10.1134/S0015462807060117

Kalinichenko VA (2024) Kinematics of the first wave Faraday mode on the side wall of a rectangular vessel. Fluid Dyn 59(5):1028–1037. https://doi.org/10.1134/S0015462824603802

Nesterov SV (1969) Parametric excitation of waves on the surface of a heavy liquid. Morskie Gidrofiz Issledovaniya 3(45):87–97. (in Russian)

Sekerzh-Zenkovich YI (1951) On the theory of standing waves of finite amplitude on the surface of a heavy fluid of finite depth. Izv Akad Nauk SSSR Ser Geogra Geojiz 15(1):57–73. (in Russian)

Mechanical Tests of Frozen Snow, Ice and Frozen Ground

D. M. Frolov, A. V. Koshurnikov, V. E. Gagarin, and G. A. Rzhanitsyn

Abstract This chapter examines the use of frozen snow, ice, frozen sand and clay as building materials in polar regions for construction, including underground storage facilities. Rapid climate change in the polar regions makes it necessary to maintain sustainable infrastructure, and snow, ice and frozen ground, due to its mechanical properties at low temperatures, become a promising option. For the reliable construction of these structures, calculations and modelling of the mechanical stability of these structures are important. To determine the calculation coefficients and modelling parameters, mechanical tests of the constituent frozen rocks are necessary. The paper presents the results of such laboratory mechanical tests.

Keywords Frozen snow · Ice · Frozen ground · Mechanical properties · Underground storage facilities

The polar areas are increasingly recognized for their unique geographical value and potential resources, making them a focal point of international scientific research. The indigenous peoples of these areas, particularly those in the Arctic, have demonstrated remarkable ingenuity in adapting to their harsh environments. One of the most notable examples is the construction of igloos by the Eskimos, which showcases their adept use of snow as a primary building material. These igloo huts are not only functional but also exemplify a deep understanding of the properties of snow and ice.

In the realm of modern engineering, M.M. Krylov's project from 1930 for cold warehouses made of ice represents an innovative application of these traditional techniques. By utilizing ice as a structural element, Krylov's design emphasizes sustainability and energy efficiency. The external insulation provided by materials such as sawdust or peat is crucial for maintaining low internal temperatures, allowing for the effective storage of perishable goods in extremely cold conditions.

This intersection of traditional knowledge and contemporary engineering highlights the potential for developing sustainable infrastructure in polar regions. As

---

D. M. Frolov (✉) · A. V. Koshurnikov · V. E. Gagarin · G. A. Rzhanitsyn
Faculty of Geography, Lomonosov Moscow State University, Moscow, Russia
e-mail: denisfrolovm@mail.ru

T. Chaplina (ed.), *Processes in GeoMedia—Volume IX*, Springer Geology,
https://doi.org/10.1007/978-981-96-9302-3_10

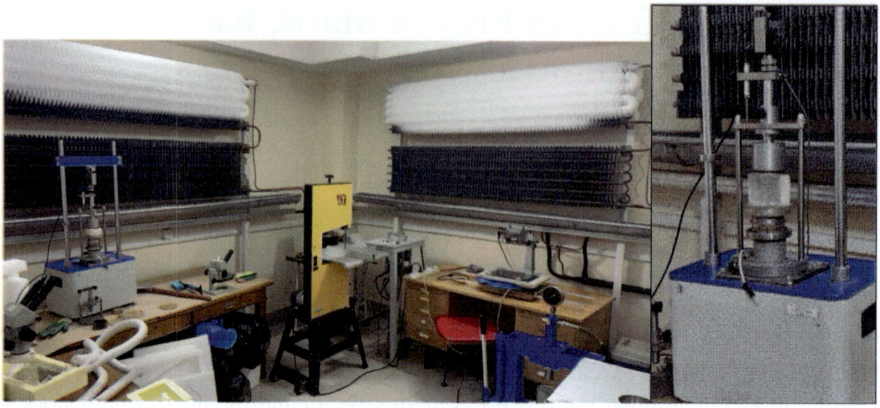

Fig. 1 Cold laboratory of the Geography Department of the Moscow State University. ASIS installation with one-sided compression device

climate change alters these environments, further research into these methods could yield valuable insights into enhancing resource management in challenging climates (Hughes et al. 2023; Biresselioglu et al. 2020; Ludescher et al. 2018; Stroeve and Notz 2018).

During the construction of underground storage facilities for petroleum products or spent waste in the frozen zone, and especially under climate change conditions, calculations and modelling of the mechanical stability of the structure are necessary. Mechanical testing of the ice and frozen rock constituents is required to determine the calculation factors and modelling parameters. We have carried out mechanical tests on frozen snow, ice and frozen sandy and clayey soil. The tests were performed in a cold laboratory on ASIS equipment for uniaxial compression (Fig. 1).

The cold laboratory was equipped in the middle of the last century. It can maintain negative temperatures from 0 degrees Celsius to—20 °C. Crystallographic, crystal-optical, mechanical and other tests of snow, ice and frozen ground are performed in the laboratory. The laboratory is equipped with a number of microscopes, a light table, a band saw, a compaction press and various thermometric equipment. There is also a thermostat at the laboratory's disposal, which allows setting the necessary temperature gradient on the investigated sample, as well as various scales and other devices. Research in the laboratory has been carried out for a number of years and includes determination of thermophysical, mechanical and crystallographic properties of snow, ice and frozen ground. The laboratory is also used for practical training of master's students of the Faculty of Geography.

Mechanical tests of frozen snow, ice, frozen clay and frozen sand to obtain values of strength model parameters play an important role in the design of permafrost facilities (underground storage facilities) (Shumsky 1955; Voytkovsky 1977, 1999; Abele 1990; Richter 1945; Wang et al. 2025; Gray et al. 1986; Engineering Glaciology 1971). Thus, the pressure–strain curves for frozen snow, ice, frozen clay and frozen sand obtained during laboratory tests are presented in Fig. 2.

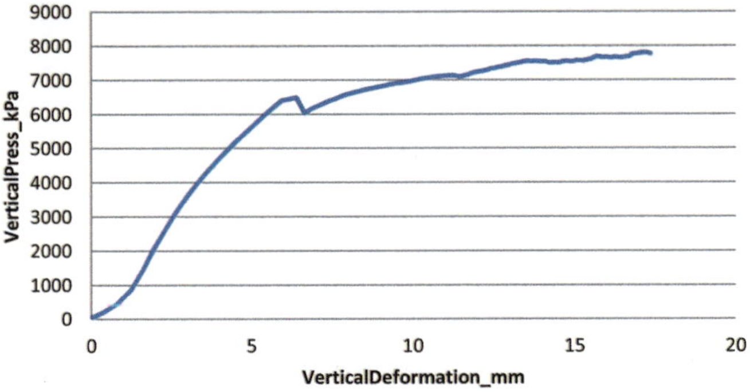

Fig. 2 Pressure–strain curves for frozen snow, ice, frozen clay and frozen sand

Thus, the figure shows that frozen sand (ice-cement) is practically incompressible, while frozen clay exhibits plastic properties under certain conditions. Also, frozen snow undergoes gradual destruction, while the destruction of the ice sample has a pronounced brittle character.

In general, the paper considers the use of frozen snow, ice and frozen ground as construction materials in polar regions for the construction of underground storage facilities. These materials have good mechanical properties and at low temperatures can be used in the construction of underground storage facilities. The aim of the study was to determine the physical and mechanical properties of frozen snow, ice and frozen ground, as well as to modify these properties for successful solution of applied problems. One of the main determinants of the properties of frozen snow, ice and frozen ground is temperature. Under the influence of temperature, frozen snow, ice and frozen ground changes its density, structure, plasticity and crystalline structure,

Fig. 2 (continued)

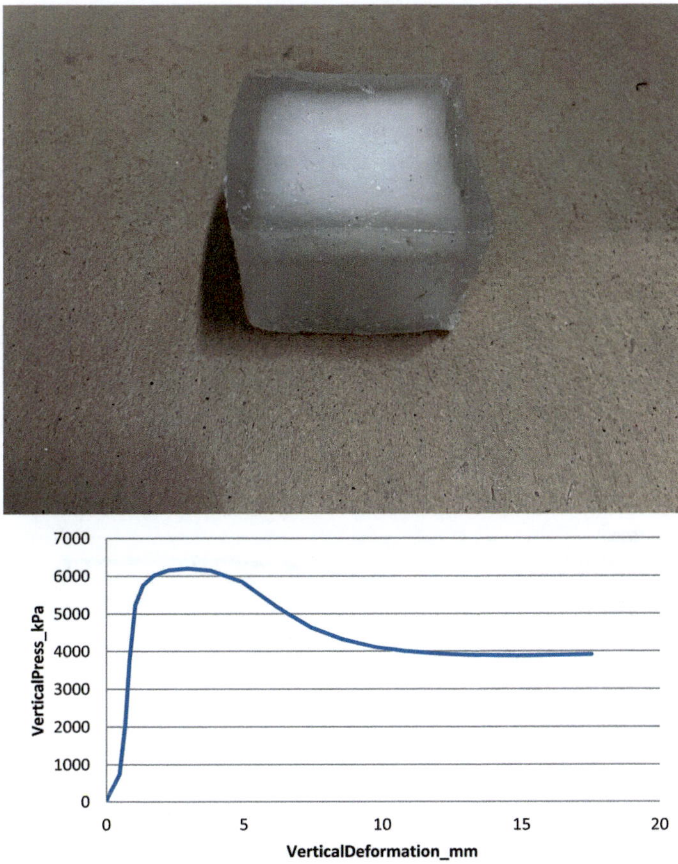

Fig. 2 (continued)

Fig. 2 (continued)

which in turn affects the other properties. Experience has shown that mechanical strength and other parameters increase with decreasing temperature.

Acknowledgements The work was carried out in accordance with the state budgetary themes of the Geography Faculty of MSU 'Danger and risk of natural processes and phenomena' (121051300175-4) and 'Evolution of the cryosphere under climate change and anthropogenic impact' (121051100164-0).

References

Abele G (1990) Snow roads and runways, vol 90, no 3. US Army Cold Regions Research and Engineering Laboratory

Biresselioglu ME, Demir MH, Solak B et al (2020) Investigating the trends in arctic research: the increasing role of social sciences and humanities. Sci Total Environ 729:139027. https://doi.org/10.1016/j.scitotenv.2020.139027

Engineering Glaciology – M (1971). Izd-vo Moscow University, 208 c

Gray DM, DH M, DH M (1986) Snow. Handbook

Hughes KA, Boyle CP, Morley-Hurst K et al (2023) Loss of research and operational equipment in Antarctica: balancing scientific advances with environmental impact. J Environ Manag 348:119200. https://doi.org/10.1016/j.jenvman.2023.119200

Ludescher J, Yuan N, Bunde A (2018) Detecting the statistical significance of the trends in the Antarctic sea ice extent: an indication for a turning point. Clim Dyn 53(1–2):237–244. https://doi.org/10.1007/s00382-0184579-3

Richter GD (1945) Snow cover, its formation and properties. Institute of Geography of the USSR, 121 p

Shumsky PA (1955) Fundamentals of structural ice science: petrography of fresh ice as a method of glaciological research. Izd-vo AS USSR

Stroeve J, Notz D (2018) Changing state of Arctic sea ice across all seasons. Environ Res Lett 13(10):103001. https://doi.org/10.1088/1748-9326/aade56

Voytkovsky KF (1977) Mechanical properties of snow. Nauka

Voytkovsky KF (1999) Fundamentals of glaciology. Nauka, Moscow, p 255

Wang Z, Zhang S, Zhang Z, Xie C, Zhou K, Yuan M (2025) Study on the hydrothermal effect in ice materials under the polar glacier runway dynamic load. Case Stud Constr Mater 22:e04214. https://doi.org/10.1016/j.cscm.2025.e04214

On the Existence of Hydrodynamic Connection Between the Outputs of Intensive Energy-Carrying Intramantle Vortexes and Its Influence on the Change of the Caspian Sea Level and Bottom Relief

S. Yu. Kasyanov

Abstract The chapter shows that in the Earth's mantle between the outlets of two hot energy-carrying intense vortices located beneath the Himalaya-Tian Shan Mountain Ring and beneath the area of the Mediterranean and Black Seas, an exchange of matter can occur. The hydrodynamic communication channel between these outlets passes under the Caspian Sea, the Caucasus, and the Black Sea. The flows of hot molten mantle matter between the two outlets of intense vortices are controlled by astronomical factors through the attraction of the planets of the solar system and cause changes in the pressure in the mantle under the Caspian Sea, leading to corresponding changes in the relief of the Caspian Sea floor and level, and also cause changes in the flows of endogenous heat into the waters of the Black and Azov Seas. All the above mentioned geological and climatic changes are controlled by astronomical factors.

Keywords Caspian sea level · Caspian sea floor topography · Endogenous heat · Intramantle vortices · Hydrodynamic coupling · Earth's mantle · Vortexes heat · Intramantle vortices · Hydrodynamic coupling · Earth's mantle · Vortex tubes · Vortex rings · Mud volcanism · Gravity anomalies · Deposits tubes · Vortex rings · Mud volcanism · Gravitational anomalies · Oil and gas deposits · Astronomical factors · Oil and gas · Astronomical factors of climate change · Black Sea climate

S. Yu. Kasyanov (✉)
N.N. Zubov's State Oceanographic Institute, Roshydromet, Moscow, Russia
e-mail: skas53@yandex.ru

T. Chaplina (ed.), *Processes in GeoMedia—Volume IX*, Springer Geology,
https://doi.org/10.1007/978-981-96-9302-3_11

1 Introduction

The system of hot intense vortices formed in the Earth's mantle due to the tangential fall of a temporary Earth satellite was considered earlier (Kasyanov 2012). The main role in it is played by the subsystem of three intensive large-scale energy-carrying intramantle vortices Himalaya-Tian-Shan—Kolyma (1), Kolyma—Yellowstone (2), Yellowstone—Mediterranean Sea (3) (Kasyanov and Samsonov 2021), in which the mixture of mantle matter and the invading body in the molten and gaseous state moves by inertia with high velocities. Secondary hot vortices appear around the large-scale energy-carrying vortices in the mantle. The system of three intense large-scale energy-carrying vortices is hydraulically connected to a chain of four hot intense vortices in the mantle beneath the western Pacific Ocean (Kasyanov 2023). These vortices were formed around the trajectory of the body during its partial immersion in the mantle and represent parts of vortex rings ending with outlets to the day surface (Kasyanov 2021). Further analysis of the system of intramantle vortices carried out in the present work supports the existence of a hydrodynamic connection between vortex (1) in the Himalaya-Tian Shan and vortex (3) in the area of the Mediterranean and Black Seas.

Part of the hot matter from the vortex outlet (1) in the Himalaya-Tian Shan can flow to the periphery of the Mediterranean vortex outlet (3) located under the Mediterranean and Black Seas. The channel through which the hot matter flows in the mantle passes under the Caspian Sea and is adjacent to the intramantle vortex located under it. The hot matter flowing under the Black Sea region forms a region of mud volcanoes around the Taman and Kerch Peninsulas.

The state of the system of intramantle vortices through the attraction of large bodies of the Solar System is determined by astronomical causes. This allows us to raise the question about the existence and mechanisms of connection of the Caspian Sea level and the Black Sea climate with astronomical factors and related global warming.

2 Secondary Vortices Between the Outputs of Two Energy-Carrying Intense Large-Scale Vortices

In the mantle, a complex system of secondary vortices in contact with each other appears in the space between the outlet of the intense energy-carrying vortex (1) in the Himalaya-Tian Shan region shown in Fig. 1 and the outlet of the intense energy-carrying vortex (3) in the Mediterranean and Black Sea region.

Analysis of the gravity anomaly maps shown in Fig. 2, Fig. 3, and Fig. 4, as well as the relief of the day surface shown in Fig. 5, allows us to speak about the existence of a region of decompressed hot matter, which begins at the western periphery of the Himalayan-Tian Shan vortex exit (1) and extends northwestward to the Mediterranean-Black Sea vortex exit (3).

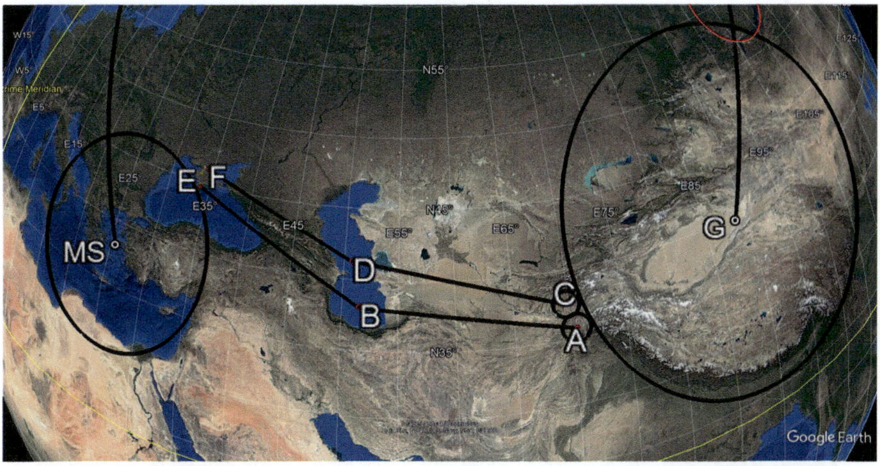

Fig. 1 Location of two vortex outlets from the system of three intense intramantle vortices (1), (2), (3) described in Kasyanov and Samsonov (2021) and location of traces in the relief of the day surface from the outlets of secondary intramantle vortices. G—Himalayan-Tian Shan vortex exit (1), MS—Mediterranean-Black Sea vortex exit (3). A, B, E, C, D, F—especially significant secondary vortices formed as a result of matter transfer from the Himalaya-Tien Shan exit of the intramantle vortex to the Mediterranean-Black Sea exit of the intense vortex. Cartographic basis is sourced from Google Earth (https://www.google.com/earth/about/versions/#download-pro)

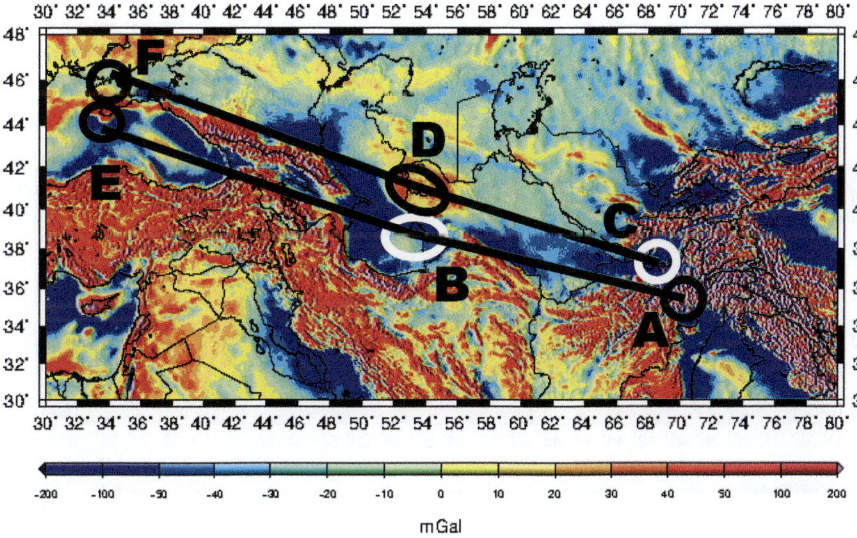

Fig. 2 Map of gravity anomalies in the free air of WGM2012 (Bonvalot et al. 2012). ABE and CDF—approximate schematic representation of the positions of the vortex tube axes of intramantle vortices linking the output of the intense intramantle vortex (1) under the Himalayas-Tian Shan with the output of the intense intramantle vortex (3) under the area of the Mediterranean and Black Seas. Latitude and longitude coordinates are marked along the borders

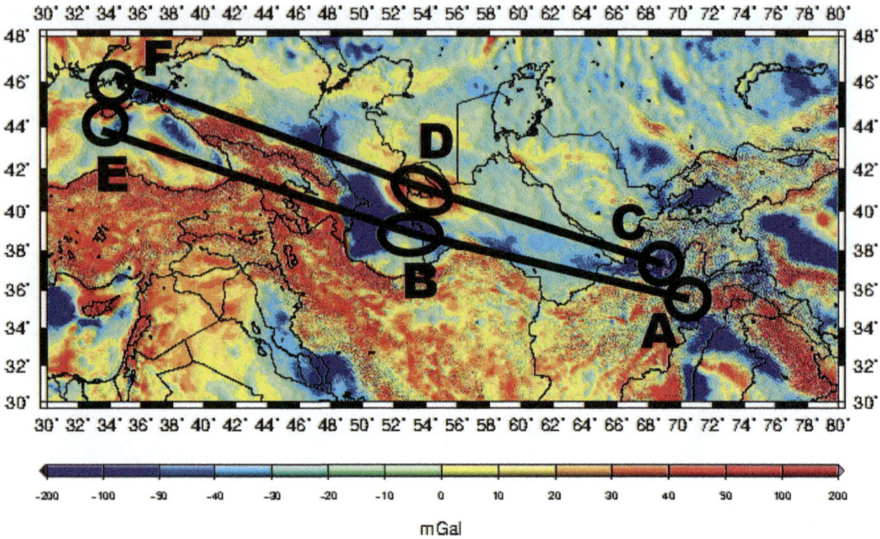

Fig. 3 Map of isostatic gravity anomalies of WGM2012 (Bonvalot et al. 2012). ABE and CDF—approximate schematic representation of the positions of the vortex tube axes of intramantle vortices linking the output of the intense intramantle vortex (1) under the Himalayas-Tian Shan with the output of the intense intramantle vortex (3) under the area of the Mediterranean and Black Seas. Latitude and longitude coordinates are marked along the borders

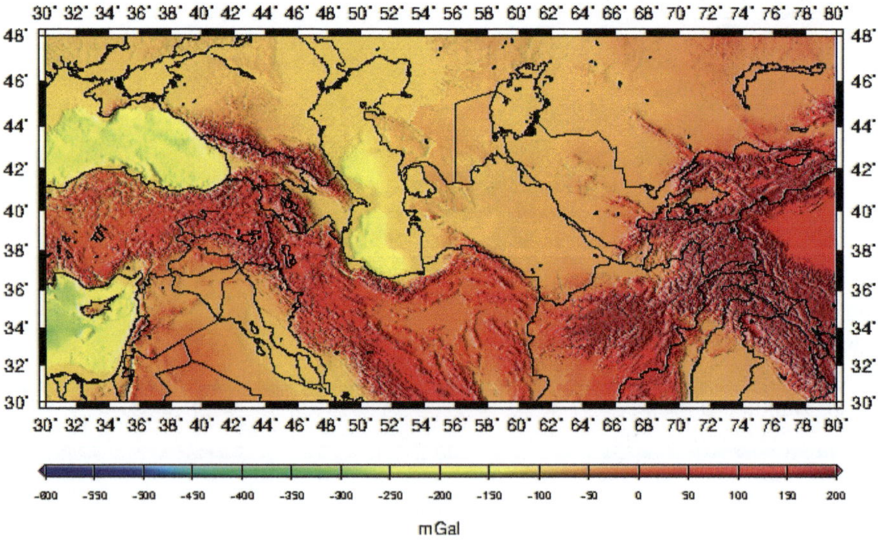

Fig. 4 Map of gravity disturbance WGM2012 (Bonvalot et al. 2012) in milligals (mGal). The map covers parts of Europe, the Middle East, and Asia, with a color scale at the bottom ranging from -600 to 200 mGal. Latitude and longitude lines are marked along the borders

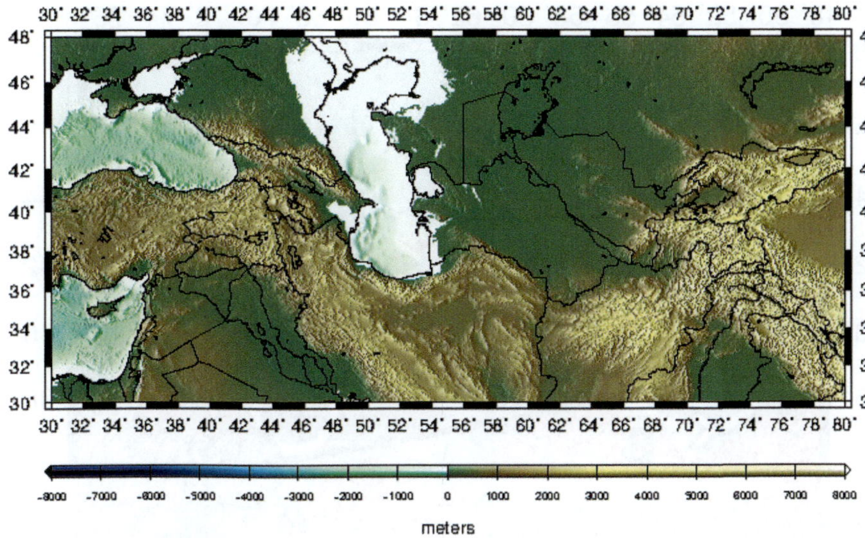

Fig. 5 Topographical map of the Middle East and Central Asia, showing elevation variations (Bonvalot et al. 2012). The map uses a color gradient from dark blue to yellow, representing elevations from -8000 to 8000 meters. Latitude and longitude markers are present along the borders

It will be shown below that the formation of a region of decompressed hot matter is a consequence of the existence of two vortex tubes of hot intramantle vortices linking the output of an intense intramantle vortex (1) beneath the Himalayas-Tian Shan with the output of an intense intramantle vortex (3) beneath the Mediterranean and Black Sea region. Figures 2 and 3 give an approximate schematic representation of the position of the vortex tube axes of these intramantle vortices. The refined position of the vortex tube axes of these intramantle vortices is shown in Figs. 6 and 7, where the position of the region of decompacted hot matter is presented in more detail.

Let us turn to Fig. 6. Consider the motion of the vortex jet at the ascending section of the annular motion in the Himalayan-Tien Shan outlet of vortex (1). At the extreme western point of the annular motion, the vortex jet (1) touches the meridional plane. Recall that the intense energy-carrying vortex (1) touches the Earth's liquid core under Baikal (Kasyanov and Samsonov 2021). This determines the position of the plane in which the matter motion occurs in the Himalayan-Tian-Shan outlet of the vortex (1).

Due to this location of the plane of motion of matter in the Himalayan-Tien Shan vortex outlet (1), the matter of vortex (1) near the westernmost point of its circular motion rises upward. At the same time, it shifts to the north, simultaneously moving away from the Earth's rotation axis. As a result, an anticyclonic vortex (1) appears at the extreme western point of the Himalayan-Tian-Shan outlet of the vortex (1), marked in the figures by the symbol A. Due to the anticyclonic rotation, the moving matter tends to gather to the center of vortex A, which explains the presence of a

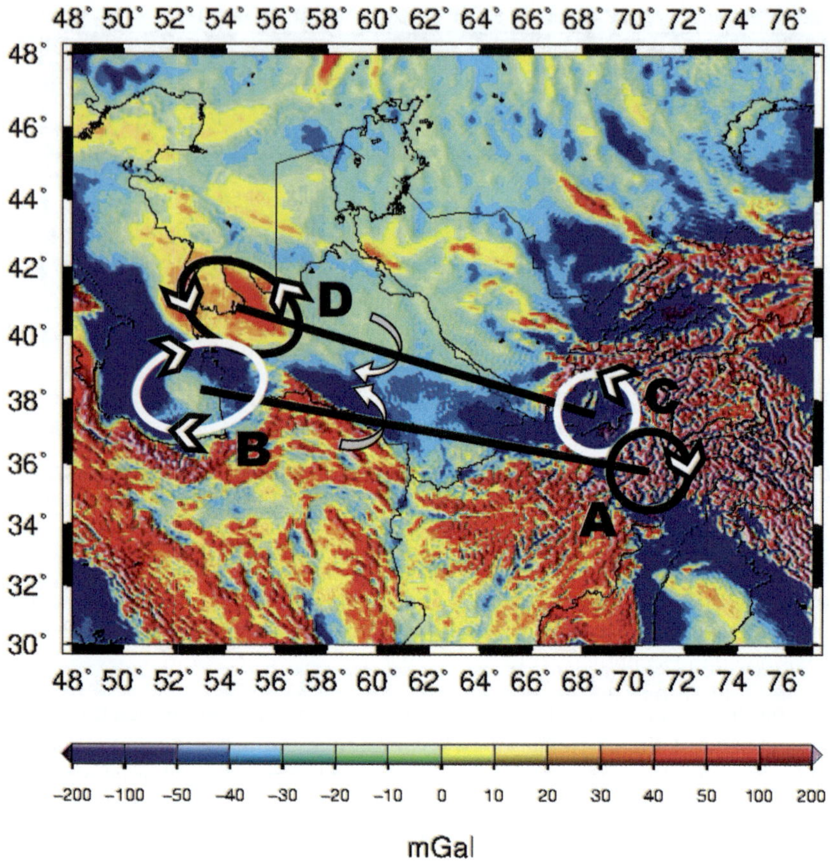

Fig. 6 Map of gravity anomalies in free air of WGM2012 (Bonvalot et al. 2012) for the region between the Caspian Sea and the Himalaya-Tian Shan Mountain Ring. Thick black lines show the direction of axes AB and CD of the swirling jets of intramantle vortices. Vortices A and B are anticyclonic, vortices C and D are cyclonic. Between the axes of vortices AB and CD, the hot matter of both vortices rises upward. Latitude and longitude coordinates frame the map. A color scale from blue to red at the bottom represents mGal values

positive gravity anomaly inside vortex A. Near the anticyclonic vortex A, a cyclonic vortex C is formed adjacent to it. Accordingly, due to the cyclonic rotation of vortex C, the moving matter tends to its periphery, which explains the presence of a negative gravity anomaly inside vortex C.

The matter in vortex A, which participates in the intense annular motion of matter in the Himalayan-Tien Shan vortex outlet (1), rises rapidly upward. This causes the upward motion of matter in the vortex C adjacent to vortex A. Due to the location of the plane of motion of matter in the Himalayan-Tien Shan outlet of vortex (1), in addition to the upward velocity component, both vortices A and C have a northward velocity component. Due to the Earth's rotation, the upwardly rising matter tends to

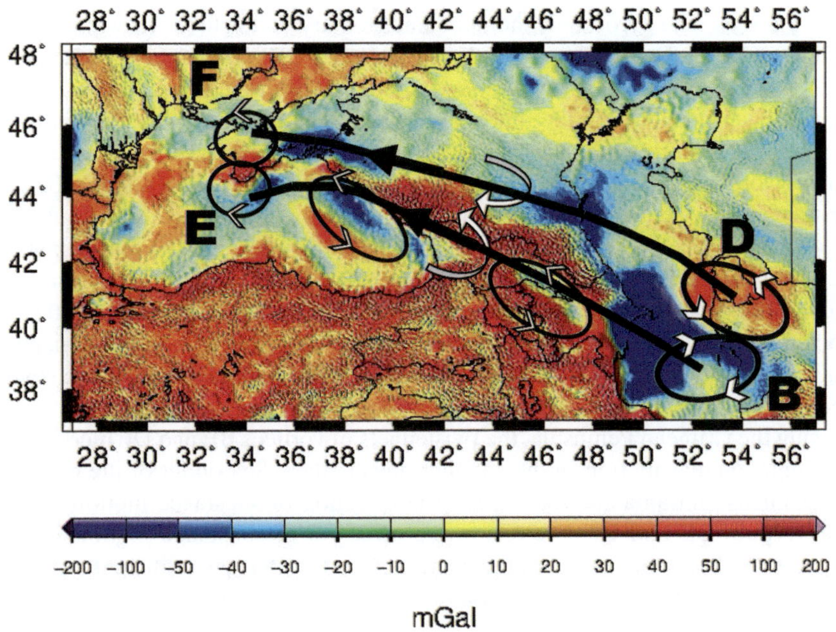

Fig. 7 Map of isostatic gravity anomalies of WGM2012 (Bonvalot et al. 2012) for the region between the Black Sea and the Caspian Sea, indicated by a color gradient from blue to red, representing values from -200 to 200 mGal. Thick black lines with black arrows show the directions of motion and the position of the BE and DF axes of the swirling jets of intramantle vortices. Vortices B and E are anticyclonic and vortices D and F are cyclonic. The ellipses show secondary vortices arising near the jet of vortex BE. The map includes latitude and longitude markers along the borders

be deflected to the west. Therefore, the vortex tubes departing upward from vortices A and C in Fig. 6 are deflected to the northwest and approach the lower boundary of the lithosphere beneath the Caspian Sea. Under the influence of the gyroscopic moment, the axes of the vortex tubes are turning. Traces of contact of a pair of contacting vortices with the lower boundary of the lithosphere under the Caspian Sea are visible in the field of gravity anomalies in Fig. 6 as contiguous ellipses B and D. Vortex B is anticyclonic and vortex D is cyclonic.

Oppositely rotating vortex tubes AB and CD are in contact. The hot substance in the area of their contact has an upward velocity component. It is still difficult to judge about the direction of the horizontal component of the velocity of motion. Nevertheless, it is clear that as a result of motion in the region between vortex tubes AB and CD, a region of hot decompacted matter appears, generating negative gravitational anomalies.

The matter released from vortices A and C rises to the lower boundary of the lithosphere beneath the Caspian Sea (see Fig. 6), forming anticyclonic vortex B and cyclonic vortex D. The gyroscopic moment tends to bury vortex tube AB against buoyancy forces. The gyroscopic moment tends to bury vortex tube AB while counteracting buoyancy forces. Therefore, the vortex tube AB shown in Fig. 6 gradually unfolds to a quasi-horizontal position and continues from the Caspian Sea region along the lower boundary of the lithosphere toward the Crimea. Its continuation is shown in Fig. 7 as vortex tube BE. Along the swirling jet of vortex tube BE, cyclonic secondary vortices, shown as ellipses next to the jet, occur. The axis of vortex tube BE is curved under the action of the gyroscopic momentum and approaches the southern coast of the Crimea. Near the southern coast of the Crimea, the motion of matter in the jet of vortex BE creates an intramantle anticyclonic vortex. The gyroscopic moment of this vortex creates negative gravity anomalies in the south and southeast and positive anomalies in the north and northwest, which are visible in the map of isostatic gravity anomalies in Fig. 7. Especially strong negative anomaly in the south can be seen in the map of gravity anomalies in free air in Fig. 2.

Passing the Crimean Peninsula, the twisted jets of vortices BE and DF merge into the cyclonic annular motion in the Mediterranean-Black Sea outlet of the energy-carrying intense intramantle vortex (3). The direction of substance motion in the jets of vortices BE and DF is close to the direction of substance motion in the Mediterranean-Black Sea outlet of the energy-carrying intensive intramantle vortex (3), so the jets flowing into vortex (3) intensify the cyclonic motion in it. Thereby, there appears a hydrodynamic connection between the outlet of the intense vortex (1) and the outlet of the intense vortex (3) in the system of three energy-carrying intense vortices (Kasyanov and Samsonov 2021).

Between the axes of vortices BE and DF, the hot matter of the vortices rises upward. The region of the hot matter rise corresponds to the position of the Main Caucasus Ridge and the location of the Crimean-Caucasian sutura shown in Fig. 8 (Aslanov et al. 2016).

Along the Crimean-Caucasian sutura, shown in Fig. 8, there are elevated flows of endogenous heat leading to the formation of areas of mud volcanism and oil and gas deposits. Around the Taman and Kerch Peninsulas there is an area of mud volcanoes and oil and gas deposits. These mud volcanoes are associated with a significant inflow of endogenous heat to seawater, which influences the climatic regime of the Black and Azov Seas. Oil fields are located in the Tersko-Sunzhensky region, highlighted in the figure by a light rectangle. The zone of developed mud volcanism and oil and gas deposits is located on the Apsheron Peninsula and continues on the coast and in the Caspian Sea waters to the south of the Apsheron Peninsula. From the Apsheron Peninsula, the Crimean-Caucasian Sutura line in Fig. 8 crosses the Caspian Sea to the Cheleken Peninsula. Mud volcanoes and rich oil deposits are located along this section along the sutura line, both in the sea and on the eastern shore of the Caspian Sea in Turkmenistan.

The location of marine mud volcanoes in the South Caspian Basin is shown in more detail in Fig. 9.

Within the framework of this study, we interpret the Crimean-Caucasian sutura line in Fig. 8 as the center line of the region where the converging upward flows of vortex tubes BE and DF come into contact. In the zone of convergence of the rising flows of vortexes BE and DF, the meeting vortex jets create additional dynamic

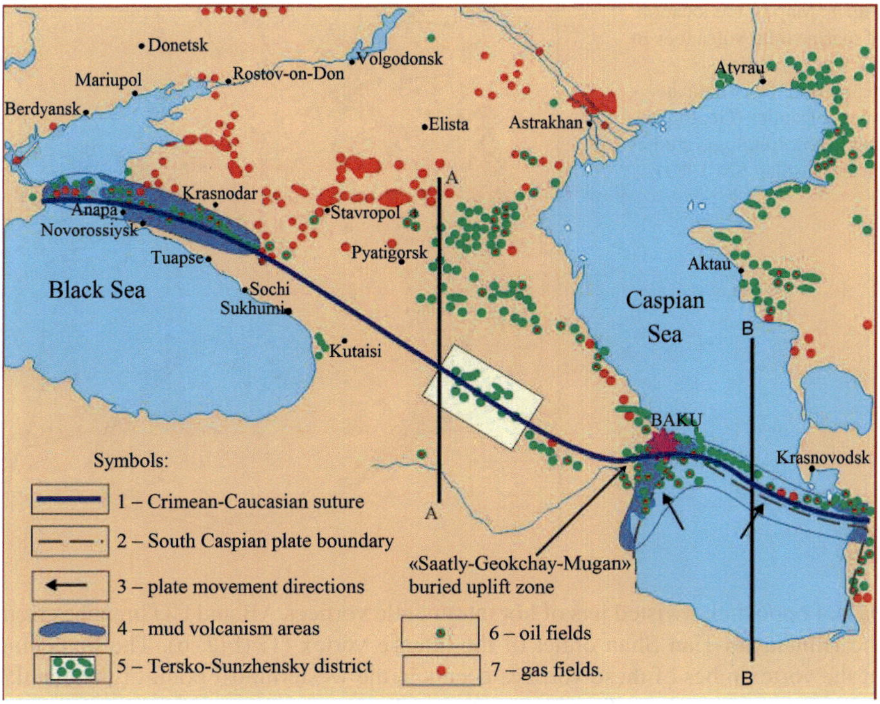

Fig. 8 Location map of the Crimean-Caucasian sutura, areas of mud volcanism and oil and gas deposits and tectonic features in the geographical region between the Black Sea and Caspian Sea (Aslanov et al. 2016). Key elements include the Greater Caucasus main thrust, South Caspian basin boundary, and seismic activity directions. Symbols indicate oil fields, gas fields, and tectonic zones. Major cities like Tbilisi, Baku, and Sochi are marked

pressure on the lower boundary of the lithosphere, which leads to elevation of the relief and formation of mountains (Fig. 5). Thus, an excess of mass is created here, which corresponds to positive gravity anomalies, both in free air and isostatic (Fig. 7). On both sides of the zone of convergence of upward flows of vortices BE and DF there are zones of quasi-horizontal motion in vortices BE and DF, which correspond to pressure minima and associated lowered values of terrain elevation (Fig. 5). In these areas, a mass deficit is created, which corresponds to negative gravity anomalies, both in free air and isostatic (Fig. 7).

3 Conclusions

A possible mechanism of hydrodynamic coupling between the output of the intense vortex (1) and the output of the intense vortex (3) in the system of three energy-carrying intense vortices has been established (Kasyanov and Samsonov 2021). A

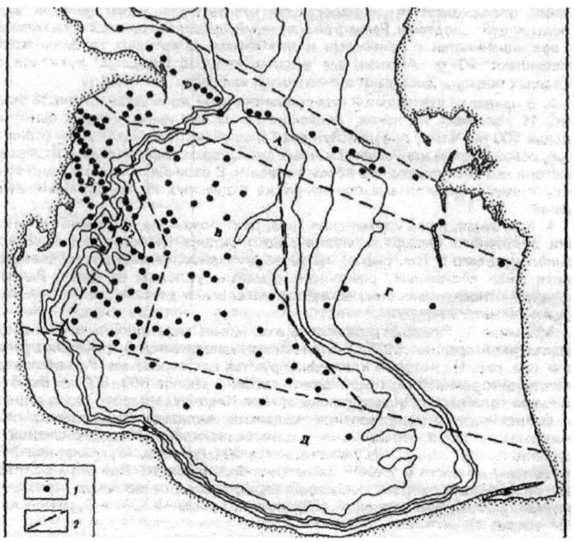

Fig. 9 Map of the location of marine mud volcanoes in the South Caspian depression. Dots indicate mud volcanoes (Geological and geomorphologic studies of the Caspian Sea 1983)

pair of oppositely twisted jets of hot intramantle vortices AB and CD branches from the Himalayan-Tian Shan outlet of the intense vortex (1) (Fig. 6). The branching of the vortex tubes of these vortices occurs at the westernmost point of the annular motion in the Himalaya-Tien Shan outlet of the intense vortex (1), at the site of the most rapid rise of the vortex substance.

The vortex tubes of vortices ABE and ADF (see Figs. 2 and 3) pass under the Caspian Sea, Caucasus, and Crimea and connect with the Mediterranean-Black Sea outlet of the intense intramantle vortex (3) (see Fig. 6), intensifying the cyclonic motion in it. The existence of these vortex tubes explains the presence and spatial structure of the area of negative gravity anomalies, divided in two by the zone of positive gravity anomalies. This zone starts from the western edge of the Himalayan-Tien Shan Mountain Ring and extends under the Caspian Sea, the Caucasus, and the Black Sea to the Crimea. This zone can be seen in Figs. 2, 3, 4, 6, and 7.

The vortex tubes of vortices ABE and ADF provide a possibility of hot matter transfer of intramantle vortices from the Himalayan-Tian Shan exit of the intense vortex (1) to the Mediterranean exit of the intense vortex (3).

The vortex tubes of the ABE and ADF vortices pass under the Caspian Sea. The movement of hot mantle matter in them changes the pressure field in the mantle beneath the Caspian Sea and the relief of the sea bottom, which leads to changes in the level of the Caspian Sea.

The movement of matter through the vortex tubes of vortices ABE and ADF between the outlets of two large-scale vortices belonging to the system of three energy-carrying intramantle vortices depends on the state of this vortex system and is determined by astronomical factors through the influence of the gravity of the solar system planets. Therefore, the level of the Caspian Sea also depends on astronomical factors.

The inflow of endogenous heat generated by mud volcanoes in the water area near the Kerch and Taman Peninsulas has an impact on the climatic regime of the Black and Azov Seas, which also depends on astronomical factors.

The location of the Crimean-Caucasian sutura line (Fig. 8), along which the areas of mud volcanism and oil and gas deposits are located, can be explained by the position of the convergence zone of the hot matter flows of vortex tubes ABE and ADF and coincides with the position of this zone.

References

Aslanov BS, Magerramov BI, Huduzade AI (2016) To the assessment hydrocarbon potential zone buried uplifts Saatli-Goychay-Mugan, SOCAR, 2:004–010

Bonvalot S, Balmino G, Briais A, Kuhn M, Peyrefitte A, Vales N, Biancale R, Gabalda G, Reinquin F, Sarrailh M (2012) World Gravity Map. Commission for the Geological Map of the World. Eds. BGI-CGMW-CNES-IRD, Paris

Geological and geomorphologic studies of the Caspian Sea. M., Nauka, 1983. p 71

Google Earth Pro. https://www.google.com/earth/about/versions/#download-pro

Kasyanov SY (2012) Modeling of the orbital motion of a large asteroid during a descent with a deepening into the lithosphere and mantle of the Earth. Phys Problems Ecology (Ecol Phys) (18):151–164. http://ocean.phys.msu.ru/pubs/ecophys/ecophys-18_pp150-246.pdf

Kasyanov SYu (2021) Formation of system of intense vortices in the mantle when a large temporary Earth satellite is immersed. In: Chaplina T (eds) Processes in GeoMedia, vol 3, Springer Geology, Springer, Cham, pp 217–232. https://doi.org/10.1007/978-3-030-69040-3_21

Kasyanov SY (2023) Hydrodynamic interaction between two systems of intense intra-mantle vortices as a possible cause of the connection between distant Earthquakes. In: Chaplina T (eds) Processes in GeoMedia, vol 7, Springer Geology, Springer, Singapore, pp 159–169. https://doi.org/10.1007/978-981-99-6575-5_15

Kasyanov SY, Samsonov VA (2021) About the modern system of three energy-carrying intensive vortices in the Earth's mantle. In: Chaplina T (eds) Processes in GeoMedia, vol 2, Springer Geology, Springer, Cham, pp 273–291. https://doi.org/10.1007/978-3-030-53521-6_31

Features of Seiche Oscillations in the System of Adjacent Bays

Yu. V. Manilyuk, S. V. Dovgaya, and V. V. Fomin

Abstract The expansion of the mode composition and amplification of long-wave oscillations in systems of adjacent bays due to their coupling through entrances can make a serious threat to the operation of ports and their coastal infrastructure. Based on the nonlinear numerical model *FUNWAVE* (*fully nonlinear WAVE model*), seiche oscillations in a system of two adjacent model bays of different sizes, excited by the action of a soliton, were studied. It was found that the large bay A has a significant effect on oscillations in the adjacent smaller bay B, in which, due to the coupling, an intense oscillation with a period corresponding to the Helmholtz mode of bay A is generated, dominating over all other eigenmodes. The condition under which the Helmholtz mode of bay A does not penetrate into bay B was obtained. Higher frequency eigenmodes are also excited in the bays under consideration. Their intensity is low and weakly depends on the distance between the bays.

Keywords Boussinesq equations · Numerical model FUNWAVE · Coupled bays · Seiches · Helmholtz mode

1 Introduction

Seiche oscillations in the bays (harbor oscillations) can represent a serious danger to moored vessels and coastal infrastructure. They may cause strong periodic horizontal currents that lead to the breaking of mooring lines, collisions of ships, and damage to quay walls (Rabinovich 2009, 1993). Usually the periods of such seiches do not exceed 10 min. The Helmholtz mode (zero mode) has the longest period in the bays. For most real bays, its period can be approximately calculated using the formula $T = 4L/\sqrt{gh}$, where L—length of the bay, g—acceleration of gravity, and h—average depth of the bay. This mode of oscillations has no nodal lines in the inner

Yu. V. Manilyuk (✉) · S. V. Dovgaya · V. V. Fomin
Marine Hydrophysical Institute of RAS, Sevastopol, Russian Federation
e-mail: uvmsev@yandex.ru

T. Chaplina (ed.), *Processes in GeoMedia—Volume IX*, Springer Geology,
https://doi.org/10.1007/978-981-96-9302-3_12

waters of the bay. It is easily excited by various types of disturbances and is usually always present in the spectrum of long-wave level oscillations, having high intensity.

A special feature of seiches in bays is that they are most often generated not by local fluctuations in atmospheric pressure and wind fluctuations, but by long-wave disturbances, coming from the open sea. (Rabinovich 1993). Of particular interest is the excitation of seiche oscillations in bays by a solitary wave, such as a tsunami. In works (Zheng et al. 2024; Ma et al. 2020; Dong et al. 2020) on the base of numerical and laboratory experiments, it was shown that this type of disturbance leads to the generation of a wide spectrum of seiche modes in bays, including transverse ones, which are difficult to excite by incident harmonic long waves. At the same time, these waves potentially can reach high intensity and represent a danger to ships in ports. Generation of transverse seiche modes by a solitary wave is confirmed by laboratory experiments (Dong et al. 2020). This may explain why extreme seiches, exacerbating the impact of the disaster, are generated in bays after the arrival of a tsunami.

In systems of connected bays, there are additional conditions for the occurrence of dangerous phenomena caused by the action of seiches. As a result of the effect of coupling of bays through their entrances, the modal composition of natural oscillations in them is expanded due to the penetration of natural modes of adjacent bays (Kovalev et al. 2024; Rabinovich et al. 1999; Medvedev and Arkhipkin 2015; Manilyuk et al. 2024a, 2020; Liu et al. 2003). When a system of connected bays contains a bay that is noticeably larger in size than the others, then its own modes penetrate into neighboring bays with high intensity. This effect occurs in the bays of Sevastopol (Manilyuk et al. 2024a, 2020), where the Helmholtz mode of Sevastopol Bay penetrates into Kruglaya and Quarantine Bays, or in the Balearic Islands, where the Helmholtz mode of the Bay of Ciutadella penetrates into the smaller Bay of Platja Gran, often causing destruction (Liu et al. 2003).

The interaction between bays may be of particular interest in the design and construction of artificial harbors, because with proper design it is possible to avoid serious problems that can occur in harbors, located in natural bays, where little can be changed. For example, if it is planned to build two adjacent harbors or to add an additional harbor to an existing bay, then the distance between the harbors and their geometric parameters can be selected in such a way so as to minimize the consequences of the interaction, arising from the connection.

The aim of this work is to study seiche oscillations in a system of two adjacent bays of different sizes, caused by the action of a solitary wave (soliton), and to establish the features of seiches in such a system.

2 Materials and Methods

Mathematical numerical model. In this work, the numerical hydrodynamic model of *Fully Nonlinear WAVE model* (*FUNWAVE*) (https:fengyanshi.github.io,build,html,basics.html; Kennedy et al. 2000; Shi et al. 2012) was used, which is based on the solution of nonlinear Boussinesq equations (Chen 2006). A detailed description of the modification of the model used in this work is given in Manilyuk et al. 2024b.

Scheme of numerical experiments. Figure 1 shows a schematic representation of the computational domain. It has a rectangular shape. Length of the domain is OG = 12,000 m and width of the domain is OD = 12,000 m. In its western part, there is a spongy layer of width OL = 1500 m, which is designed to dampen reflected waves, the distance LE equals 6800 m. The depth in the larger half of the computational domain ODIE is constant and it equals 50 m. On the shelf, with a width of EF = 1700 m, the depth changes linearly from 50–10 m. In the eastern part of the computational domain, there are two rectangular bays with a constant depth of 10 m. The length of the larger bay (A) is $l_1 = 1000$ m and the length of the smaller bay (B) is $l_2 = 500$ m. The widths of bays A and B in all numerical experiments were set equal to 500 m. In the calculations, the distance between bays (C) varied from 500–4000 m with a step of 500 m, while bay B moved, and points 20, 21, 23, 24, 26, and 27 moved synchronously with it. The distance on the normal from the line, passing through point F parallel to the y-axis, to points 25–27 is 1200 m. It is 200 m to points 22–24. The computational domain was covered with a uniform rectangular grid with a step 11.7 m, (1027 × 1025 nodes).

A solitary wave (Fig. 1, region C) with a height of $a_0 = 1$ m was used as the initial disturbance. It was set at a distance of 3000 m from the western boundary of the computational region in the form of a Korteweg–de Vries soliton moving along the normal to the bays and having the form (Kurkin 2005):

$$\zeta(x, t) = a_0 \cosh^{-2}\left(\frac{x - Ct}{\lambda}\right), \tag{1}$$

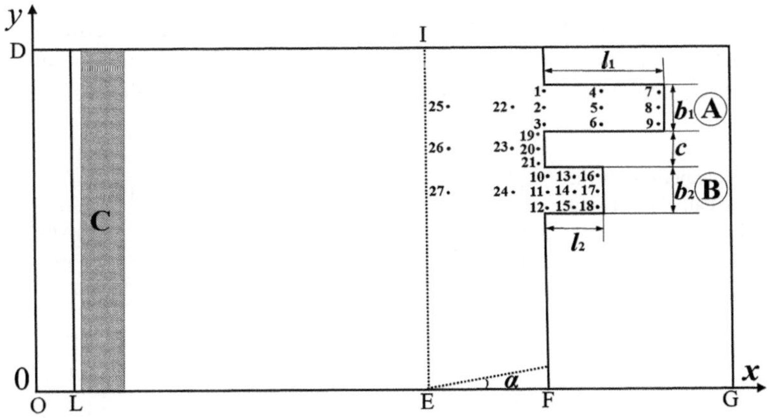

Fig. 1 Scheme of the computational domain. The numbers indicate the location of the control points

where a_0—wave height; t—time; $\lambda = 2H_0\sqrt{H_0/(3a_0)}$—horizontal scale of the soliton; $C = C_0(1 + a_0/(2H_0))$—phase velocity of the soliton; $C_0 = \sqrt{gH_0}$; H_0—depth of the basin.

The step of the integration on the time (Δt) was 0.1 s, the period of integration compiled 4000 s. After the calculations have been carried out, the time series containing intervals of the soliton running and completion of the transient process, caused by its impact, were excluded (1375 s). Then a spectral analysis of the obtained time series of water level fluctuations for the control points, shown in Fig. 1, was performed. As a result, the most energy-carrying periods of level fluctuations were identified.

Analytical estimates of eigenperiods of model bays. To estimate the values of the eigenperiods of model bays, we can use the following formula, obtained under the assumption that the bay has the rectangular shape and the constant depth, and the nodal lines of all eigenmodes are located at the entrance to it (Ivanov et al. 2018):

$$T_{km} = \frac{4ab}{\sqrt{gh}\sqrt{(1+2k)^2b^2 + 4m^2a^2}}, \tag{2}$$

where h—depth of the basin, a—length of the basin, b—width of the basin and its entrance, and k, m—non-negative integers, defining the mode number.

Table 1 presents the eigenperiods of the model bays, calculated using formula (2). Modes for which $m = 0$ are longitudinal, and those for which $k = 0$ are transverse. A feature of basins with an open entrance, in contrast to fully enclosed basins, is the presence of the Helmholtz mode (or the zero mode), in which $k = 0$ and $m = 0$. This mode has no nodal lines in the inner waters of the bay. In real bays, especially long and narrow ones, the Helmholtz mode is observed most frequently (Rabinovich 1993).

As can be seen from Table 1, the values of the periods of eigenoscillations in the considered system of connected bays do not exceed 7 min. It can be assumed that in the model calculations, the most intense level fluctuations should occur at periods

Table 1 Eigenperiods of the model bays, calculated using formula (2)

Number of mode		Eigenperiods (min)	
k	m	бухта A	бухта B
0	0	6.73	3.37
1	0	2.25	1.12
2	0	1.35	0.67
3	0	0.96	0.48
0	1	1.63	1.51
0	2	0.84	0.82
0	3	0.56	0.55
1	1	1.35	0.93

of about 6.5, 3.4, 2.3 min, corresponding to the eigenhead longitudinal modes of the bays of the system.

Shelf seiches. Considering that the model water area has the hypothetical "shelf"—the relatively wide area (1700 m) in the coastal part, where the depth decreases linearly as it approaches the solid boundary, it can be assumed that shelf seiches may arise as a result of the impact of the solitary wave. These are standing oscillations of the water level with periods corresponding to the resonant periods that depend on the slope of the bottom. In the computational domain (Fig. 1), the change in shelf depth at the distance of 1.7 km to the entrance into the bays is approximated by a linear dependence $h(x) = \alpha x$, where $\alpha = 0.0235$. The resonant periods of such the shelf are calculated using the formula (Rabinovich 1993):

$$T_n = 8\sqrt{L}\Big/ (n\sqrt{g\alpha}), \tag{3}$$

where L—shelf width, $n = 1, 3, 5\ldots$—number of the mode, g—acceleration of gravity.

Calculations, using formula (3) of the value of the period, for the first, third, fifth, and seventh modes correspondently give ~11.46 min, ~3.82 min, ~2.29 min, and ~1.64 min.

3 Results of Numerical Simulation and Discussion

Figure 2 shows the calculated mareograms for control points 9 and 18 located at the tops of the bays, for point 20 located in the coastal zone between the bays, and for point 24 located in the open part of the water area opposite the entrance to bay B (Fig. 1). They correspond to the time intervals for the completion of the transient process caused by the action of the disturbance and the free oscillation state. As can be seen from this figure, after the end of the transition process (t ~ 1375 s), the level oscillations have a pronounced periodic character, they are more intense in the smaller bay and fade over time. For the case when $c = 1000$ m, the level variation for points 20 and 24 is very close to that which occurs when $c = 500$ m. Therefore, the mareograms are given only for $c = 500$ m.

Figure 3 shows periodograms of level oscillations constructed for the tops of bays A (point 9) and B (point 18), as well as points 20 and 24, located in the open part of the basin.

It is clear from Fig. 3b that at point 9 (the top of Bay A) is an intense oscillation with a period of about 6.5 min, corresponding to the Helmholtz mode of bay A (Table 1), for all the considered values of parameter c (Fig. 1), which determines the distance between the bays. As the analysis of Fig. 3b and the calculation results shows, there is a decrease in the intensity of the Helmholtz mode of the given bay with an increase in the distance between bays from 500–2000 m. In this case, when $c = 1500$ m, the spectral energy density E decreases by 17% compared to the case when $c = 500$ m. If $c = 2000$ m, then the reduction is 37%. But then, with the increase

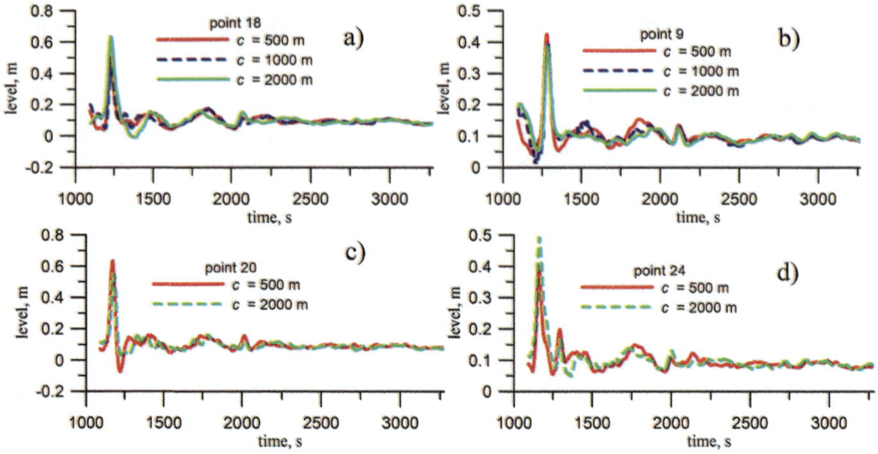

Fig. 2 Oscillations of the water level depending on the distance between bays: **a** point 18 (top of bay B); **b** point 9 (top of bay A); **c** point 20; **d** point 24

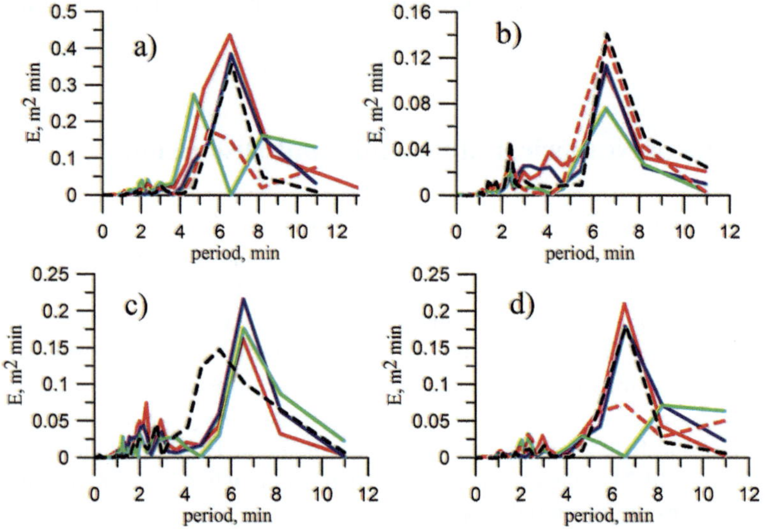

Fig. 3 Periodograms of water level oscillation depending on the distance between bays c. The red solid line corresponds to c = 500 m; the blue solid line—c = 1000 m; the green solid line—c = 2000 m; the red dotted line—c = 3000 m; the black dotted line—c = 4000 m: **a** point 9 (top of bay A); **b** point 18 (top of bay B); **c** point 20; **d** point 24

of c, the energy of oscillations also begins to increase over the period of 6.5 min. At $c = 2500$ m, E is already on 29% less than at $c = 500$ m. When the values of c reach 3000 and 4000 m, E becomes, respectively, 13 and 18% higher than at c = 500 m.

There are several other less significant peaks in bay A in addition to the peak at 6.5 min. The largest of them takes place at the period of 2.3 min and it corresponds to the single-node longitudinal mode of bay A (Table 1). At the same time with an increase in the distance between the bays, the intensity of the oscillations with the given period weakens, but it begins to increase when c>2000 m. This regularity is similar to the regularity that occurs for the 6.5 min period. It should be noted that the fifth mode of shelf seiches has the period of about 2.3 min that it also can increase the intensity of oscillations in the coastal zone on this period.

From the analysis of Fig. 3a, it is visible that if c = 500–1500 m then in bay B there is a strong peak at the period of 6.5 min, corresponding to the Helmholtz mode of bay A. This oscillation dominates over other modes excited in bay B, the periods of which are visible in Fig. 3a: above the Helmholtz mode of Bay B (period 3.4 min), the single-node longitudinal mode of Bay A, the fifth mode of shelf seiches (period 2.3 min), and the single-node transverse mode of Bay B (period 1.5 min). The intensity of oscillations with a period of 6.5 min weakens with increasing distance between bays. When c = 2000 m, the pattern of the level oscillations changes, and the peak at the period of 6.5 min is absent. But a fairly powerful peak appears at the period of about 4.0 min, which is close to the period of the second mode of shelf seiches (3.82 min). Further, with an increase in c, an intense oscillation with a period of 6.5 min appears again in the spectrum of level oscillations at the top of Bay B. At c = 4000 m, the periodogram takes the form that occurs when c is equal to 500 or 1000 m (Fig. 3a).

This transformation can be explained as follows. The eigenoscillations of bay A, excited by external influences, in particular, the Helmholtz mode, begin to spread into the open part of the basin area through the entrance to the bay in different directions. This is evidenced by Fig. 3c, d, which show periodograms for points 20 and 24, located in the open part of the basin. From these figures, it is clear that there is a strong peak at the period of 6.5 min. Waves are captured near the shore, as a result of which, they move parallel to the shore and, reaching the entrance to Bay B, waves penetrate into it. A coupled oscillation occurs between the bays. Considering that the Helmholtz mode of bay A is the standing wave, then its nodal lines and other elements do not move in space, but take a stable position that does not change over time. The magnitude of the deviation of the water level from the equilibrium state changes only. The wavelength of the Helmholtz mode of bay A is 4000 m. This can be explained as follows. There are no nodal lines of this mode in the bay, the first nodal line is at the entrance to the bay (Ivanov et al. 2018). Considering that the oscillations of the level in the bay occur according to the cosine law (Ivanov et al. 2018), the first zero of which is located at the point where $x = \lambda/4$, we can conclude that the quarter of the wavelength is placed in the bay. The cosine, according to its properties, has the following zero at the point corresponding to $(3\lambda)/4$. This value is 3000 m for bay A. Thus, when the distance between the bays $c = 2000$ m, at the entrance to bay B there will be the nodal line of the Helmholtz mode of bay A, captured by the coast (taking into account that the length of bay A is 1000 m). Therefore, this mode does not penetrate into bay B. The condition, when the Helmholtz mode of bay A does not penetrate into bay B, can be written as follows:

$$c + l_1 = \frac{3}{4}\lambda, \tag{4}$$

where c is the distance between the entrances to the bays (Fig. 1), l1 is the length of bay A (Fig. 1), and λ is the wavelength of the Helmholtz mode of bay A.

When the Helmholtz mode of bay A is captured by Bay B, its intensity in bay A is greater than in the case where capture does not occur (Fig. 3a, b). This can be explained as follows. When this mode is captured, the stable energy exchange occurs between bays A and B and its propagation in other directions into the open part of the water area decreases.

Note that when $c = 2000$ m, there is a powerful peak at point 20 for the period of 6.5 min (Fig. 3c), since the distance from this point to the top of bay A is 2000 m ($\sim\lambda/2$), where the cosine is 1. There is no peak at the entrance to bay B (point 24, Fig. 3d) for a period of 6.5 min, because the Helmholtz mode of bay A, captured by the coast, has a node in vicinity of bay B entrance.

4 Conclusion

Based on the *FUNWAVE* hydrodynamic model, the response of two adjacent bays of different sizes to the impact of the soliton is studied. The influence of the distance between the bays on the intensity of seiche oscillations in them is considered.

In the larger bay A, several modes of the eigenoscillations are excited, as a result of the action of the soliton. But the Helmholtz mode of this bay dominates over the other eigenmodes. The presence of the adjacent small bay B has little effect on seiche oscillations in bay A. The intensity of the Helmholtz mode of bay A can either increase or decrease depending on the distance between the bays *c*.

Large bay A has a significant influence on hydrodynamic processes in bay B. Due to the capture by the coast, the Helmholtz mode of bay A penetrates into bay B and dominates over all other modes of the seiches in this bay, including the Helmholtz mode of bay B. But under certain conditions, the Helmholtz mode of bay A does not penetrate into bay B. This happens when the given mode, captured by the shore, has the nodal line in the vicinity of the entrance to bay B. In this case, the intensity of the Helmholtz mode in bay A is minimal. This occurs, apparently, by the decrease in the amount of energy of the coupled oscillations of water between the bays and the increase of the propagation of the energy into the open area of the basin through the entrance to bay A.

The appearance of shelf seiches in both bays is insignificant. Presumably, this is due to the short duration of the disturbance.

The work was carried out within the state task of Federal Research Center Marine Hydrophysical Institute of RAS (MHI RAS), topic No. FNNN-2024-0016. The simulation was carried out on the computing cluster of the MHI RAS.

References

Chen Q (2006) Fully nonlinear Boussinesq-type equations for waves and currents over porous beds. J Eng Mech 132:220–230. https://doi.org/10.1061/(ASCE)0733-9399(2006)132:2(220)

Dong G, Zheng Z, Gao J, Ma X, Dong Y, Wu H (2020) Experimental investigation on special modes with narrow amplification diagrams in harbor oscillations. Coast Eng 159:103720. https://doi.org/10.1016/j.coastaleng.2020.103720

https://fengyanshi.github.io/build/html/basics.html

Ivanov VA, Manilyuk YuV, Sannikov VF (2018) Seiches in a basin with an open entrance. J Appl Mech Tech Phys 59:594–600. https://doi.org/10.1134/S0021894418040041

Kennedy AB, Chen Q, Kirby JT, Robert A, Dalrymple RA (2020) Boussinesq modeling of wave transformation. Breaking and runup. I: one dimension. J Waterw Port Coast Ocean Eng 126(1):726–726. https://doi.org/10.1061/(ASCE)0733-950X(2000)126:1(39)

Kovalev DP, Manilyuk Yu V, Kovalev PD (2024) Sea level oscillations in the adjacent bays—trade port and Kholmsk-Severny (Sakhalin Island). Phys Ocean 31(3):409–426

Kurkin AA (2005) Nonlinear and non-stationary dynamics of long waves in the coastal zone. N. Novgorod: NSTU, p 329 (In Russin)

Liu PL-F, Monserrat M, Macros M, Rabinovich AB (2003) Coupling between two inlets observation and modeling. J Geophys Res 108:3069:14–1–14–10. https://doi.org/10.1029/2002JC001478

Ma X, Zheng Z, Zhang X, Dong G (2020) Numerical investigation on special modes with narrow amplification diagram in harbor oscillations. Ocean Dyn 70(1):1–19. https://doi.org/10.1007/s10236-019-01325-8

Manilyuk Yu V, Lazorenko DI, Fomin VV (2020) Investigation of seiche oscillations in the adjacent bays by the example of the Sevastopol and the Quarantine bays. Phys Ocean 27(3):242–256. https://doi.org/10.22449/1573-160X-2020-3-242-256

Manilyuk Yu V, Fomin VV, Yurovsky Yu, Bagaev AV (2024) Sea level oscillations spectra of a shallow coastal bay: cost-effective measurements and numerical modelling in Kruglaya bay. Reg Stud Mar Sci 69:103326. https://doi.org/10.1016/j.rsma.2023.103326

Manilyuk YV, Dovgaya SV, Fomin VV, Lazorenko DI (2024) Numerical modeling of long-wave oscillations in two coupled bays. Process GeoMedia. V 5(8):285–295. https://doi.org/10.1007/978-981-97-6627-7_27

Medvedev IP, Arkhipkin VS (2015) Sea level fluctuations in Golubaya Bay (Gelendzhik). Bull Mosc Univ 5(3):70–78 (In Russin)

Rabinovich AB (1993) Long gravity waves in the ocean: capture. resonance. propagation. SPb: Hudrometeopubl, p 325 (In Russin)

Rabinovich AB (2009) Seiches and Harbor Oscillations (Chapter 9). Handbook of Coastal and Ocean Engineering. In: Kim YC (ed) World Science Publications, Singapore, pp 193–236

Rabinovich AB, Montserrat S, Fine IV (199) Numerical modeling of extreme seiche oscillations in the Balearic Islands region. Oceanology 39(1):16–24 (In Russin)

Shi F, Kirby JT, Harris JC, Geiman JD, Grilli ST (2012) A high-order adaptive time-stepping TVD solver for Boussinesq modeling of breaking waves and coastal inundation. Ocean Modelc 43–44:36–51. https://doi.org/10.1016/j.ocemod.2011.12.004

Zheng Z, Dong G, Ma X (2024) Special modes with narrow amplification diagrams in harbor oscillations: definition and parametric study. Ocean Dyn 74:511–523. https://doi.org/10.1007/s10236-024-01616-9

Thermal and Radiation Regime Modern Changes of the Asian Territory of Russia, as Well as the Movement of the North Magnetic Pole

A. V. Kholoptsev, R. G. Shubkin, and A. N. Baturo

Abstract The thermal and radiation regimes of any region of the world have a significant impact on the population life safety, as well as the development features of its economy and ecosystems. Therefore, the idea development about the reasons for their change for the Asian territory of Russia is a topical problem of physical geography, ecology, and emergency safety. The thermal regime of each area of the Earth's surface depends largely on its radiation regime, a significant factor of which is the change in the transparency coefficient of the atmosphere above it. Significant influence on the dynamics of the latter is exerted by variations in the state not only present over this area of aerosol (including cloud cover), but also the corresponding ozone layer. One of the reasons for the latest changes are the flow's variations of incoming rays to the atmosphere over the studied areas of the Earth's surface, which affect the intensity of its ionization, as well as the vorticity of the polar vortex. Therefore, a hypothesis has been put forward that the thermal and radiation regimes of the territory under consideration is linked for some spring months with the approach to the peninsular of Taimyr of the North Geomagnetic Pole. Its verification confirmed that the distribution changes of 2000–2024 of the monthly average values of total ozone for spring months over the region under study correspond to existing perceptions about the possible consequences of moving this pole. For many study sites, significant inter-annual changes in this indicator have been identified, as well as a number of characteristics of their radioactive and thermal regime with variations in solar activity that control the flows of incoming cosmic rays, also dependent on the position of the pole. The results suggest that if the pole moves further in the same

A. V. Kholoptsev
Doctor of Geography Sciences, The Siberian Fire and Rescue Academy of the State Fire Protection Service of EMERCOM of Russia, Zheleznogorsk, Krasnoyarsk Territory, Russia
e-mail: kholoptsev@mail.ru

R. G. Shubkin (✉) · A. N. Baturo
Candidate of Technical Sciences, The Siberian Fire and Rescue Academy of the State Fire Protection Service of EMERCOM of Russia, Zheleznogorsk, Krasnoyarsk Territory, Russia
e-mail: rshubkin@yandex.ru

A. N. Baturo
e-mail: baturo@sibpsa.ru

T. Chaplina (ed.), *Processes in GeoMedia—Volume IX*, Springer Geology,
https://doi.org/10.1007/978-981-96-9302-3_13

direction, the rate of warming of the climate of the Asian territory of Russia will increase, and the incoming streams of solar ultraviolet will raise. Ice conditions on the Northern Sea Route will be more favorable for shipping, but in the Arctic Zone of Russia a chance of accidents at man-made facilities and in its Forest Zone fire hazard will increase due to weather conditions.

Keywords The Asian territory of Russia · North geomagnetic pole · Cosmic ray flux · Solar activity · Ozone layer · Radiation regime · Thermal regime

1 Introduction

The life safety activities of population of territories of Russia, as well as the sustainability their ecosystems, economy, and transport development, primarily depend on the trends of further changes in the indicators of their thermal and radiation regimes. Therefore, the development of existing perceptions about the significant factors of these changes is a pressing issue in physical geography, ecology, climatology, and emergency safety.

Its solution is of considerable interest for the territories of the Asian Territory of Russia (here and after ATR), where the rate of climate warming is significantly higher than the average in the country. Many of these territories belonging to the Yamal-Nenets, Khanty-Mansiysk and Chuktskiy Autonomous Regions, Krasnoyarsk territory, and Republic of Sakha (Yakutia) have found promising mineral deposits, including hydrocarbons (Volkov 2019; Grigoriev 2019). Many of them are already operational and the development of others is a matter for the near future. The planned development of shipping along the Northern Sea Route and its land transport links can also make an important contribution to the socio-economic development of ATR (Plan of development of the Northern Sea Route n.d.; Schacha 2018; Sharonov 2018).

The thermal regime of any portion of the Earth's surface depends mainly on its radiation regime, a significant factor of which is the change in the transparency coefficient of the atmosphere above it (Ya 1965; Kurt 1986; Mahokhtkin et al. 2023). Significant influence on the dynamics of the latest is the state of both atmospheric aerosol (including tropospheric cloud) and ozone layer (Muller 2012; Larin 2000; Hoffman 2005; The ozone layer is on track to recover n.d.).

Among the cloud characteristics that significantly affect the transparency of the atmosphere is its optical density, which is monitored in Russia by the ground actinometric network Roshydrometa, which includes 77 stations carrying out observations in 6 standard times (Mahokhtkin et al. 2023).

The optical density of cloud depends on the distribution in the troposphere of temperature and humidity, as well as concentrations and characteristics of atmospheric condensation nucleus (Ya 1965). Among such nuclei, the most active ones are the ions present in air. Therefore, the ionization of humid air leads to an increase in the optical density of its cloud and coagulation of aerosol particles.

The state of the ozone layer is determined by the distribution over the Earth's surface of total ozone content (here and after TOC). Variations of this indicator over each site are determined by the dynamics of the physical ozone balance, the consumption of which depends on: among other things, and from the intensity of ionization of stratospheric air (Karol et al. 1982; Maurchev et al. 2019).

At the end of the twentieth century, the TOC decreased over many regions of our planet. Most rapidly, this process occurred over Antarctica and the Arctic (Database of changes in the solar radiation flux n.d.).

Since 2000, the decline in the average TOC of the Earth's atmosphere has slowed down and stabilized in recent years. In some regions of Russia belonging to the ATR, the TOC continued to decrease further (most rapidly over the Yakutsk town) (Bulletin n.d.).

In 2024, the average TOC in the latitude (60°S–60°N) was only 2% lower than in the 1980. Here, a significant increase in the average stratospheric ozone concentration after 2000 has been recorded at altitudes of 35–45 km. No tropical TOC changes detected (https:ozone.unep.org,sites,default,files, 2023-09,Ozone_and_UV_Bulletin_1_ru.pdf. n.d.).

For the high latitudes (85°N–60°N) of the northern hemisphere, the TOC in those years was lower than at the end of the twentieth century approximately about 3.5%. Here, the minimum values of the TOC are usually observed in February and the highest in September. Spring TOC increases over time (https:ozone.unep.org,sites,default,files, 2023-09,Ozone_and_UV_Bulletin_1_ru.pdf. n.d.).

The distribution changes of high and moderate latitudes in the northern hemisphere affect the location of the Arctic and Polar fronts, as well as the characteristics of their associated global circulatory cells. Therefore, they can influence the changes in the monthly average ground-air temperatures (here and after ATA) over many parts of Russia and the monthly total solar radiation flows to them (here and after SRF), and thus also on the fire hazard of the weather conditions for their landscapes.

The reasons for changes in TOC, ATA, and SRF over various sites of ATR have been studied for many years by domestic scientists (Svyagintsev et al. 2013; Kholoptsev 2012). It has been established that circulation changes in the troposphere and stratosphere as well as air intensity exchange between their different layers have a significant influence on these processes (Mokhanakumar 2011; Maurchev et al. 2019).

The decrease of TOC over some territories is usually observed when the above troposphere warms up and the stratosphere cools. Therefore the processes under consideration are interrelated [to the 24].

Over the territories located in the Polar Vortex Zone, the decrease of TOC also contributes to the increase of its turbulence, which activates air exchange between stratospheric layers and affects the transport of ozone-depleting substances (Maurchev et al. 2019).

The greater the concentration of ions in the stratosphere air carried by this vortex, the more strongly the Lorentz force influences its movement, the more intensive the exchange of substances between its layers. Therefore, one of the significant factors of TOC changes are intensity variations of air ionization of stratosphere by secondary particle downpours, formed during entering the Earth's atmosphere particles of cosmic rays. Therefore, some influence on the dynamics of TOC may be exerted by flux changes of such cosmic rays (here and after CRF) (Stozhkov et al. 2022).

Above the sites where the flux of stratospheric ozone-depleting substances from the troposphere is high enough, the activation of air exchange leads to a decrease in TOC. On areas where such flux is low, SRF increase leads to the opposite result.

Therefore, SRF changes entering the Earth's atmosphere over different parts of the Earth's surface are likely to affect both the SRF, ATA, and the TOC differently.

In the composition of cosmic ray's particles entering the terrestrial atmosphere, about 98% are protons with energies up to 10^{16} eV. Protons that arise from various astrophysical processes occurring outside the solar system usually possess such large energies. Therefore, their flux entering the heliosphere and reaching the Earth's orbit, due to the Forbush effect, is regulated by solar activity (Raspopov and Veretenko 2009; Golenkov et al. 1980; Jones Chris 2011; Ross 1834). The values of the correlation factor of the inter-annual changes of CRF over any portion of the Earth's surface with solar activity are close to -1.

Geomagnetic field hinders the entry of protons that do not have enough energy into the terrestrial atmosphere (Ross 1834). The minimum energy of the protons sufficient to overcome this obstacle over some section of ATP depends on the distance between it and the North Geomagnetic Pole (here and after NGP) (Stozhkov et al. 2022; Dolgorudnaya scientific station n.d.; Jones Chris 2011).

The NGP location was first established by James Ross in 1831 (Diagram of the movement n.d.). At that time, the pole was placed on Cape Adelaide, located on the west coast of the Butia Peninsula (Canadian Arctic Archipelago), which belongs to the Nunavut Territory (Canada). The further studies revealed that the geomagnetic poles of our planet are moving. Including NGP that moves in the direction of ATR (Bulletin n.d.). Since 2000, the speed has increased up to 55 km/year.

A diagram of the movement of the NGP in the period of 1900–2020 shown (according to [Diagram]) on Fig. 1

As Fig. 1 shows, during the period 1900 and 2020 the NGP approached the peninsula of Taimyr (Russia) on a trajectory that is almost straight.

Until 2017, it also approached the North Pole, moving away from the Polar Vortex zone. In 2017, the SGP crossed the Change of Dates line and began approaching the Polar Vortex zone located above the ATR.

According to the British Geological Service (BGS), the speed of movement of the NGP has decreased slightly, but its direction has not changed throughout recent years (North magnetic pole track n.d.).

Considering that the process that gives rise to the geomagnetic field is very inertial (Dzerdzeevskiy 1968), and the mentioned movement of NGP started in 1903 [Scheme] is highly likely will continue in the coming decades. As a result, the NGP can reach the territory of the Krasnoyarsk Territory.

It suggests that there are ATR sites where the decrease in monthly average TOC values, as well as the increase in SRF and ATA for the spring months was related to

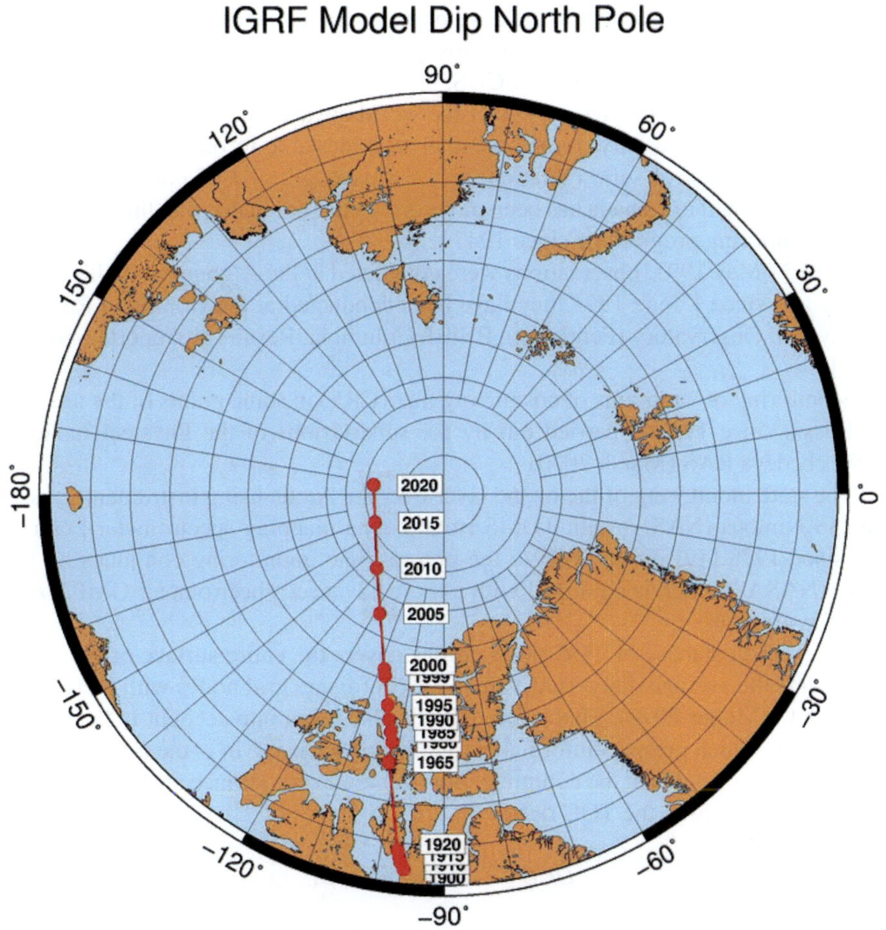

Fig. 1 A diagram of the movement of the NGP in the period of 1900–2020. (According to [Diagram])

the approach of the NGP. As a result, the characteristics mentioned could be changed in some areas synergistically with solar activity and in others in an antiphase manner.

This hypothesis is not trivial, because the thermal and radiative regime of different parts of ATR are also significantly influenced by changes in the atmospheric content of greenhouse gases [IPCC] and macro-circulation processes in the terrestrial atmosphere, which also cause changes in the distribution of aerosol and cloud (Mokhov 2020; Shakina 2010; Anisimov and Zhilitsova 2012). Therefore, its verification is not only of theoretical interest, but also practical.

The thermal and radiation regimes of many ATR sites has been monitored for decades by Roshydrometra hydrometeorological stations. It has been established that on any of its ATA sections (Third assessment report n.d.) are significantly raised. The results are taken into account in the development of a number of global atmospheric re-analyses. Among them, one of the most perfect is ERA 5 (The ozone layer is on track n.d.), supported by COPERNICUS.

Many astronomical observatories of the world observe the state of various characteristics of solar activity. In any weather, it is possible to observe such characteristics as the flux of solar radiation with a wavelength of 10.7 cm. The National Research Council of Canada has been monitoring changes in this flux through a solar radio-monitoring programme since 1947.

Until 31 May 1991, observations were conducted at the Algonquin Radio Observatory and since 1 June 1991 they have been conducted at the Dominion Radioastrophysical Observatory (Penticton), British Columbia. Results are reported in [Noa base].

Monitoring of variations of cosmic ray flux (CRF) in some points of the territory of Russia since 1957 is carried out by the subdivisions of the Physical Institute, P.N.Lebedeva RAS (Marsh 2000).

The state monitoring of the ozone layer over the Arctic using instruments placed on ISS Nimbus (NASA) with TOMS (total ozone mapping spectrometer), carried out since 1978 [49]. From 2005 to the present time, monitoring continues by ISS Aura (NASA), which is equipped with a more advanced spectrometer OMI (ozone meter instrument).

The world database of TOC observations covers the entire surface of the Earth. It presents the average daily TOC values for coordinate grid nodes with increments of $1^\circ \times 1^\circ$ (Muller 2012). This information is also taken into account in the ERA 5 re-analysis, which presents information on the «ozone reserve»—the total mass (M, in kg) of its molecules in the column of atmosphere over a surface area of 1m2. This indicator is related to the TOC ratio:

$$OCO = 100000 * M / \rho,$$

where ρ is the normal ozone density of 2.1445 g/dm3.

The NGP coordinates for a given year can be obtained from [track].

However, the hypothesis has not been tested before. The latter significantly limits the possibilities of applying the results of the above-mentioned monitoring in the long-term forecasting of the processes under study and planning of the activities of firefighters and other units of EMERCOM.

Checking the proposed hypothesis is the goal of this research. To achieve it, the following tasks have been solved:

1. Trends identification in TOC and SRF changes over the different ATR sites occurring during the spring months for the period 2000–2024.
2. The sites' identification of the region under study for which the inter-annual changes in the TOC, ATA, and SRF correspond to the spring months in 1979–2024 was significantly related to variations in solar activity.

2 Actual Material and Research Methodology

In order to identify the characteristics inherent to the trends of the studied processes over ATR, changes in their states over all areas of the Northern Hemisphere area limited by 80° northern latitude parallels were considered and 55° northern latitude.

As a factual material on the changes for any months of TOC, ATA, and SRF over all the sites of the designated area during the period of 1940–2024. ERA 5 re-analysis information is used [ERA 5 database].

As it follows from Marsh (2000), the number of points in the territory of Russia where CRF variations are monitored is small (at ATP observations were made only a few years ago in Norilsk). Therefore, in the test of the proposed hypothesis, inter-annual changes of TOC, ATA, and SRF were considered to be related to a single process controlling variations of CRF for all points of the Earth's surface-solar activity.

As a source of information about solar activity [Base Noa'a] was considered, where information on changes in the flow of solar radio radiation with 10.7 mm wavelength that occurred since 1947 is presented.

In solving the first task, as a characteristic of the trends of changes of TOC and SRF for one or another spring month, we considered the value of the difference between the average values of these indicators, which are calculated for 24 cycle (2009–2019) and 23 cycle (1997–2008).

The values of this indicator for all processes under consideration are calculated for all points in the Northern Hemisphere located between parallels 55 and 80° northern latitude, which made it possible to compare trends in the ATR and throughout Russia as well as in other parts of the zone.

When solving the second task, as a characteristic of the relationship between the processes studied, the value of the correlation coefficient of their time series was considered, which was estimated as for the entire period from 2000 to 2024, and for its part of 2012–2024. Therefore the method based on correlation analysis was applied. In the time series under consideration, the present linear trends are compensated.

The decision on the significance of the relationship between the series of the solar activity index under consideration as well as the series of characteristics under study was taken if its reliability was at least 0.95. The latest occurred when the correlation coefficient of the series to be compared was modularly above the 95% threshold of significance, which was determined by the t-criterion taking into account the number of degrees of freedom.

The values of the indicators under consideration are defined for all spring months and for ATR sites that correspond to a node of the ERA 5 re-analysis grid. For comparison, they are also estimated for February and June. Nodes within ATR were considered.

As a result, this method allows us to test the proposed hypothesis and determine the ATR sites on which during the period of approach to it GMP solar activity and therefore CRF changes, significant influence on the studied characteristics of their radiation and thermal regime.

However, it is clear that the direct influence on the state of the Earth's atmosphere is not caused by solar activity but by its controlled variations of CRF. The correlation between these processes is close to −1, but not equal. Therefore, the applied methodology allows to identify only qualitative patterns of studied processes.

3 Study Results and Analysis

In accordance with the described method of solving the first problem, for 23 and 24 solar activity cycles, the average values of TOC and SRF are determined, as well as their difference values, corresponding to the months from March to May.

The areas of the waters and territories of Russia, where the difference between the average values of the TOC for 24t cycle of solar activity in relation to 23 cycle, for the month from March to May took one or another value, are shown in Fig. 2.

As shown in Fig. 2a, there are many sites in Russia where the average monthly value of the TOC for 24 cycle was higher than or lower than that of 23 cycle.

On average across the northern hemisphere (55–80° northern latitude), the monthly average TOC increased by 4.23 Dobson units, confirming the validity of the (https:ozone.unep.org,sites,default,files, 2023-09,Ozone_and_UV_Bulletin_1_ru.pdf. n.d.) conclusion that the hemispheric ozone layer degradation has stalled.

The upward trend of TOC for March has been detected over the eastern part of the territory of Yakutia, Scandinavia, and many regions of Russia belonging to the North-West and Central Federal District (here and after NW and CFD). The rate of increase in TOC was maximized over the Canadian Arctic Archipelago from which the NGP was removed.

The TOC downward trend has been detected over the regions of Russia belonging to the Ural, Siberian Federal Districts (here and after U and SFD), as well as the western part of the territory of the republic of Sakha, Yakutia. At least 15 Dobson units of TOC has decreased over the territories of Krasnoyarsk Territory and Yakutia, to which the NGP is approaching with the greatest speed. The maximum decrease of the average value of TOC for March (more than 18 units of Dobson) was found in the Yakutsk territory (which corresponds to the conclusions [bulletin]).

As can be seen, the location of the Earth's surface areas on which some TOC changes have been identified for March suggests that it is possible to assume that the approach of the NGP to the ATR had a significant effect on them. As a result of the approach to NGP, the amplitude of the changes in CRF entering the atmosphere over the corresponding ATR areas increased during 24 solar cycle (relative to 23 cycle). This process caused an increase in the mean of the air ionization cycle intensity of the stratosphere and troposphere over such areas, many of which are located in the Polar Vortex zone. As a result, the NGP approach caused an increase in the 24-cycle mean of the destruction intensity over such tracts of stratospheric ozone and a decrease in the TOC. Over the area of the Canadian Arctic Archipelago from which the NGP was removed, the processes described have been developing in the opposite direction, which seems to have led to an increase in TOC occurring above it.

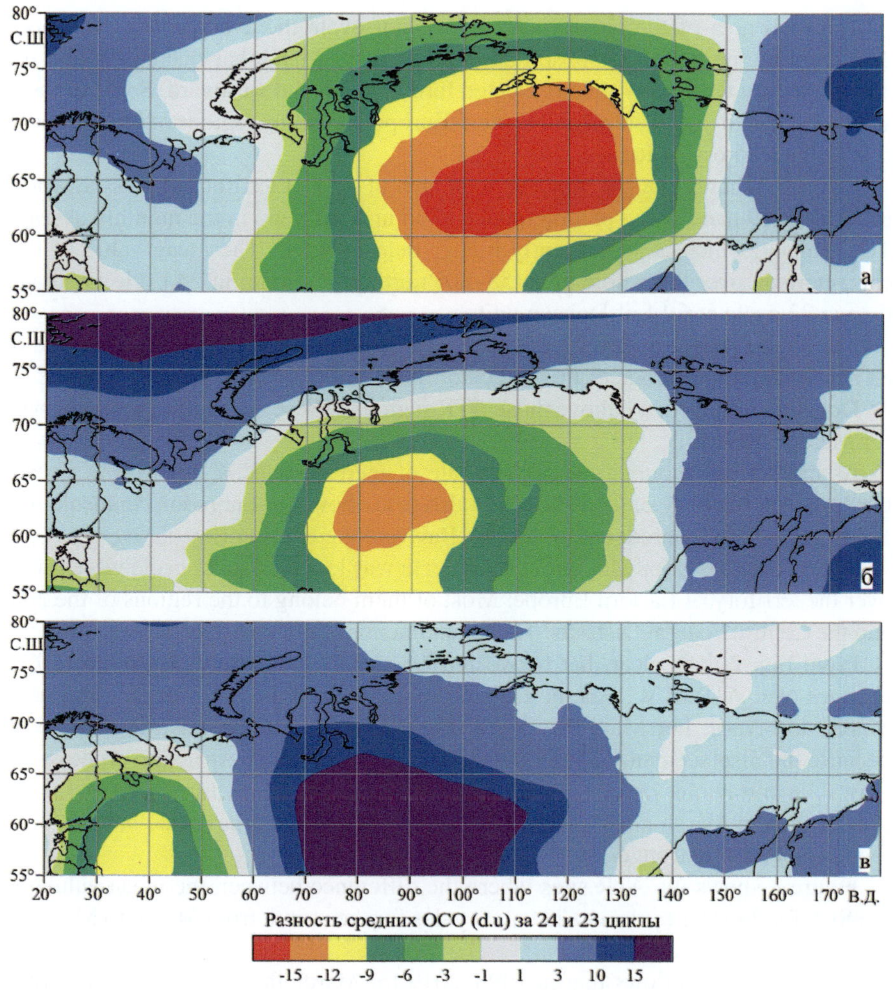

Fig. 2 Areas of waters and territories of Russia, where the difference between the average values of the TOC for 24 cycles of solar activity in relation to 23 cycles took one or another value for months: **a** March; **b** April; **c** May

Similar phenomena in the ozone layer over the territory of Russia have also been detected for April, which is confirmed by Fig. 2b. The average TOC over the whole (55–80° northern latitude) of 24 cycle increased by 7.27 Dobson units compared to 23 cycle.

As can be easily seen from Fig. 2, the total area of ATR sites where TOC has increased more than 15 Dobson units for April is slightly larger than for March.

The ATR sites, where the TOC for April decreased by more than 12 f Dobson units, are located mainly in the Krasnoyarsk territory, Khanty-Mansiysk Autonomous District, Tomsk, Tyumen, and Sverdlovsk territories.

The total area of the sites where in 24 solar cycle the average TOC for April decreased by more than 1 Dobson unit compared to March has decreased, and the entire territory where it is sited is located further south and westward. Some of these sites are situated over Central and Eastern Europe, including regions of the European part of the territory of Russia.

Figure 2c shows that for May, the total area of sites of territory of Russia, where in 1998–2019 there was a predominance of an upward trend in the monthly average TOC values, which is significantly higher than in April. The mean value of this characteristic over the whole area (55–80° northern latitude) in 24 cycle is greater than in 23 cycle at 6.14 of Dobson units.

The most significant feature of the TOC distribution for May is the presence of an area where the average of this indicator in 24 cycle is higher than in 23 cycle on 15 Dobson units, directly above those ATR regions, where the decline was observed for March and April.

As in previous months, the highest rate of increase in TOC for May was over the Nunavut Territory of Canada. They also rose above the Labrador Peninsula, the Baffin Island, the Labrador Sea, the Fox Bay, and the Hudson Bay.

Sites where TOC for May in 24 solar cycle was less than in 23 cycle were found over the territory of Eastern Europe. Most of them belong to the regions of the NW and the Central Federal Districts. This phenomenon was also detected over Moscow.

From Fig. 2, it is clear that between 1998 and 2019 the most significant TOC changes over ATR as well as the entire zone (55–80 northern latitude) occurred in March and April. For May, they were much weaker. The latter is not surprising, because the displacement of NGP can affect only the ozone destruction process and may solar ultraviolet radiation streams involved in the ozone formation are much greater than in March and April, as a result their impact on the ozone balance (and therefore TOC) is stronger.

Figure 3 shows the ATR sites where the difference between the mean values of the SRF for 24 solar cycle and 23 solar cycles for the month from March to May was different.

From Fig. 3, it follows that in 1998–2019 for March there is a downward trend of the SRF predominated over the western part of the territory of the Republic of Sakha (Yakutia), the Khabarovsk and the Amur territories, the Yamal-Nenetskiy Autonomous District, and the northern part of the Krasnoyarsk territory.

The opposite trend in the same years occurred over the eastern part of the territory of the Republic of Sakha (Yakutia), the Sverdlovsk and Magadan regions, as well as the territories of the Chukotsky and Khanty-Mansiysky Autonomous Districts, and Southern Siberia.

Figure 3b demonstrates that the total area of ATR sites for April, where the trend to increase the SRF was dominant during the period under review, is much larger than for April. The boundary of the territories where this trend prevailed in Western Siberia is located behind the 60th parallel and in Eastern Siberia slightly south of it. In the eastern part of the Yakutia territory and within the Chukotskiy Autonomous territory, the total area of such plots in April is less than in March.

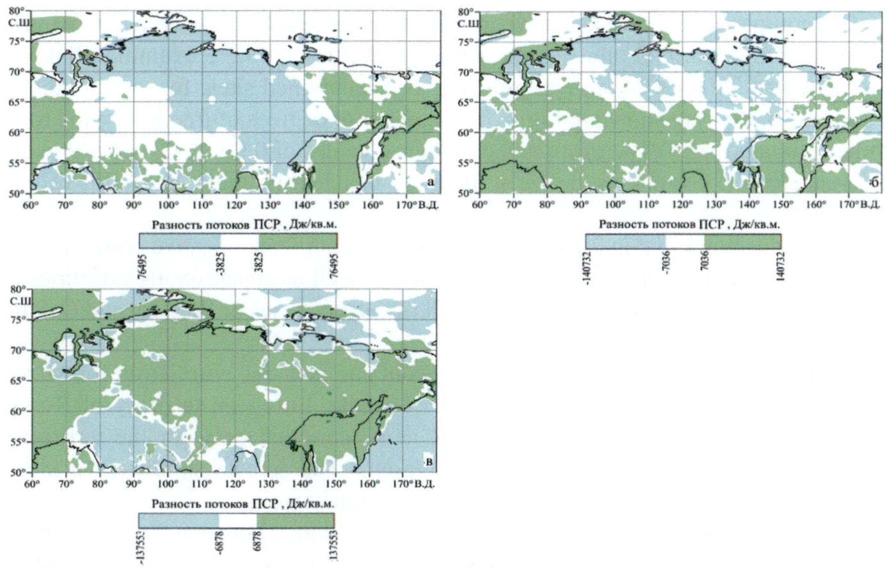

Fig. 3 ATR sites where the difference between the mean values of the SRF for 24 cycle and 23 cycle of solar activity took one or more values for months **a** March; **b** April; **c** May

Significant downward trend of the SRF for April was observed only in the northern and western parts of Yakutia, in the north of the Krasnoyarsk territory, as well as in the Yamalo-Nenets Autonomous District.

As shown in Fig. 3b, the upward trend of SRF for May has been prevalent in almost all ATR. The opposite trend has been observed in southern Western Siberia, the Transbaikal region, as well as in some coastal areas of the Yamal-Nenets Autonomous District and the Republic of Yakutia.

The comparison of Fig. 3a, c demonstrates that from March to May the total area of the plots where the SRF in 1998–2019 significantly decreased and the areas where the SRF increases become larger. The most important changes in the trends of the process under study occur in the territories to which the NGP is approaching.

The observed pattern may be due to the gradual destruction of the Polar Vortex during the spring months, as a result of which the impact of the movement of the NGP on stratospheric ozone destruction is reduced. At the same time, the influence of this process on the intensity of ionization of air, both stratosphere and troposphere, during spring remains unchanged. As a result, the resulting change in the optical density of tropospheric cloud and aerosol in May is more significant than the change in the TOC over many ATR sites.

During objective 2 solution, inter-annual changes in TOC have been analyzed with variations in solar activity for the months under review and all parts of the Northern Hemisphere located in an area bounded by parallels of 80° north latitude and 55° north latitude.

For comparison, the links between the same processes for February have also been studied. Identified areas of the territory of Russia, where examined links in the period of 2000–2024 were significant. The mentioned sites, for the months of February to May, are shown in Fig. 4. The number of degrees of freedom of the rows to be compared is 25. Therefore, the confidence values of 0.99, 0.95, and 0.9 in the relevance of the correlation between them are matched by their thresholds of 0.5, 0.4, and 0.3.

Figure 4a shows that in 2000–2024 the area of significant negative correlation of inter-annual changes in the TOC for February, as well as synchronous variations in solar activity, the territory of Russia, as well as the territories of Western and Central Siberia.

Figure 4b proves that there was a significant and negative correlation for March of the processes studied in the same years over the European territory of Russia and countries of Eastern Europe. In ATR, it is identified only for the territories of Chelyabinsk, Kurgan, and Sverdlovsk regions. The area of significant positive correlation is located only above the Kamchatka territory.

Figure 4c indicates that a significant positive correlation between the variations of TOC and solar activity for April has been identified over a large part of the Eastern Sector and Central Arctic Basin of the Russian Arctic. The identified area directly surrounds the area within which in 2000–2024 moved NGP. No area of search was found over the territory of Russia.

Figure 4 allows us to establish that for May, as well as for April, the area of significant positive correlation of the processes under study is located over the eastern sector of the Russian Arctic, as well as the Central Arctic Basin. The area is slightly smaller, but includes some parts of the territory of Allaikhovsk District of Yakutia.

Thus, Fig. 4 a-g proves that the changes in solar activity of the period 2000–2024 have had a significant impact on the variations of the OSB over many regions of the Northern Hemisphere, some of which were ATR.

The most significant links between the processes under consideration were for February and March, as well as for the areas to which NGP approached in the same years.

Practically the same location features of significant correlation areas of the same processes were identified for the period 2012–2024.

Similar researches for the territories and waters of Canada have shown that they were met for all months by a significant positive correlation of solar activity and TOC. The strongest correlation was for the waters and territories of the Canadian Arctic Archipelago belonging to the Nunavut Territory (Canada), from which the NGP was removed at maximum speed in the same years.

Therefore, the location of the areas of significant TOC correlation and solar activity is related to the direction of motion of the NGP.

ATR sites were also identified, where significant links of solar activity with variations of SRF and ATA in the periods of 2000–2024 and 2012–2024.

It has been established that for the months of March and April such sites are almost absent, but for May and June there are many. Therefore, Fig. 5 shows the ATR sites for which the second task was solved in 2012–2024 the links between the

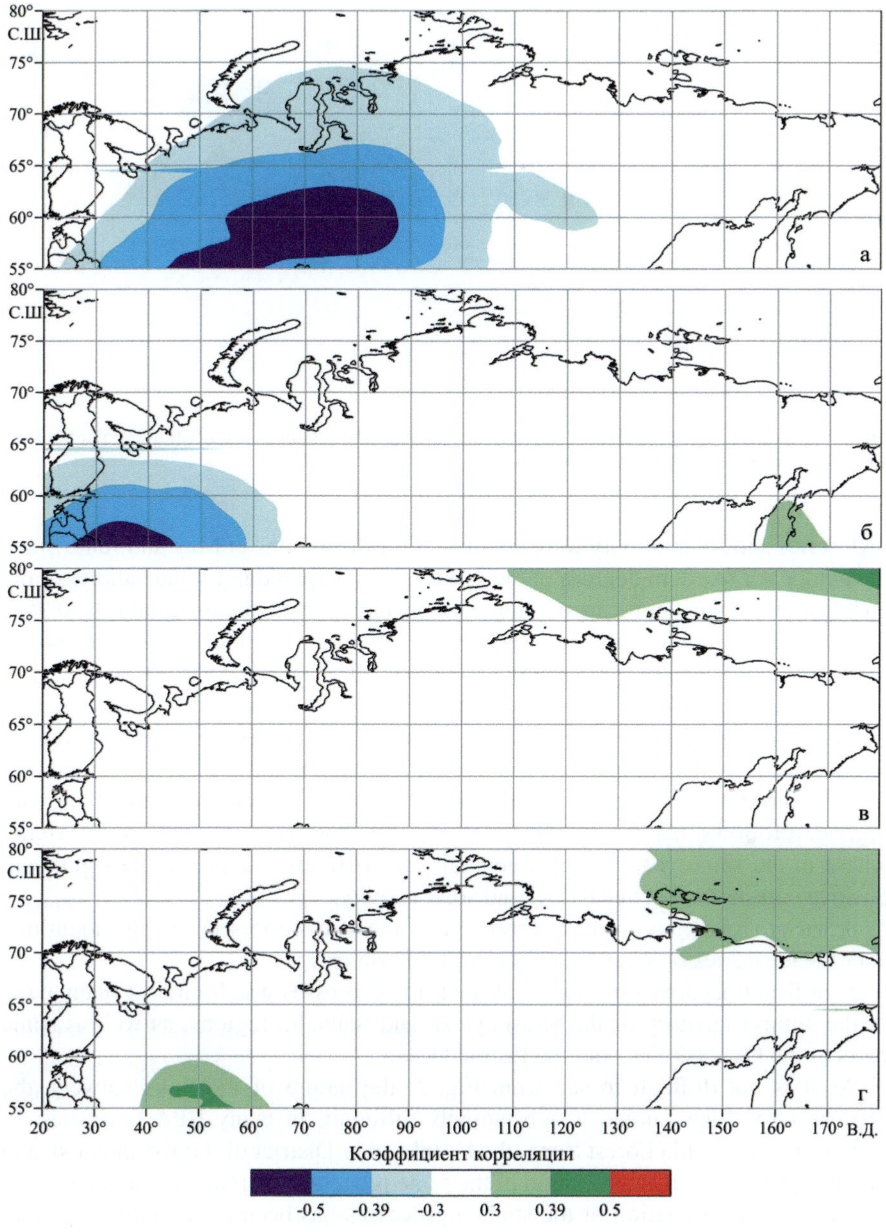

Fig. 4 Areas of the territory of Russia, where the significant links between the series of TOC and CRF, corresponding to the period 2000–2024 for months **a** February; **b** March; **c** April; **d** May

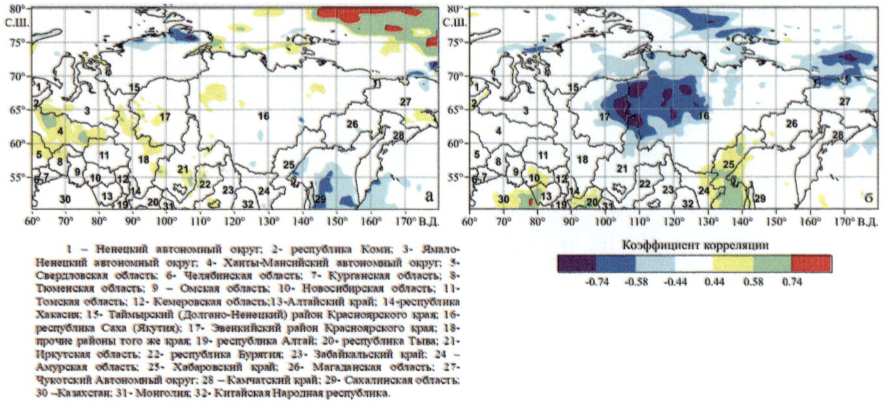

Fig. 5 ATR sites, where in 2012–2024 the inter-annual SRF changes are related to solar activity, are significant for months: **a** May; **b** June

SRF series and solar activity were significant for the months of May and June. Since the number of freedom degrees of the series to be compared is 13, the validity of the conclusions on the significance of the relationship between them was at least 0.95 if their correlation coefficient was greater than 0.58. Their confidence 0.9 corresponds to the threshold level of 0.44, and the confidence 0.99 is the level of 0.74.

Figure 5a shows one of the significant positive correlations of inter-annual changes in the SRS for May, with solar activity is over the East Siberian Sea and the Central Arctic Basin. By the comparison with Fig. 1, it is clear that it is directly surrounding the area through which the NGP moved in 2012–2024. From the comparison with Fig. 4c, it follows that this area includes many of those areas of the Russian Arctic, where in the same years, and, in general, for 2000–2024, there was a significant positive correlation between TOC and solar activity.

Areas of significant positive correlation of the same processes are also identified over the territories of Western and Central Siberia regions.

Significant negative correlation of the same processes was found for many parts of the Taimyr territory of the Krasnoyarsk and Sakhalin regions, as well as some parts of the Okhotsk Sea, and the Pacific Ocean.

As it is not difficult to see from Fig. 5, the nature of the links between the processes for June studied is substantially different. In many BPA territories, as well as in the Yakutia Forest Zone, the Evenkiyskiy District of the Krasnoyarsk and the Irkutsk regions, the correlation of the same processes is significant and negative. No significant correlation of the studied processes has been identified in the NGP area.

The ATR sites are shown in Fig. 6, where the inter-annual ATA changes for the months from February to May with variations of solar activity corresponding to the period 2000–2024 were significant. Since the number of degrees of freedom of the series to be compared is 25, the confidence values of 0.99, 0.95, and 0.9 in the

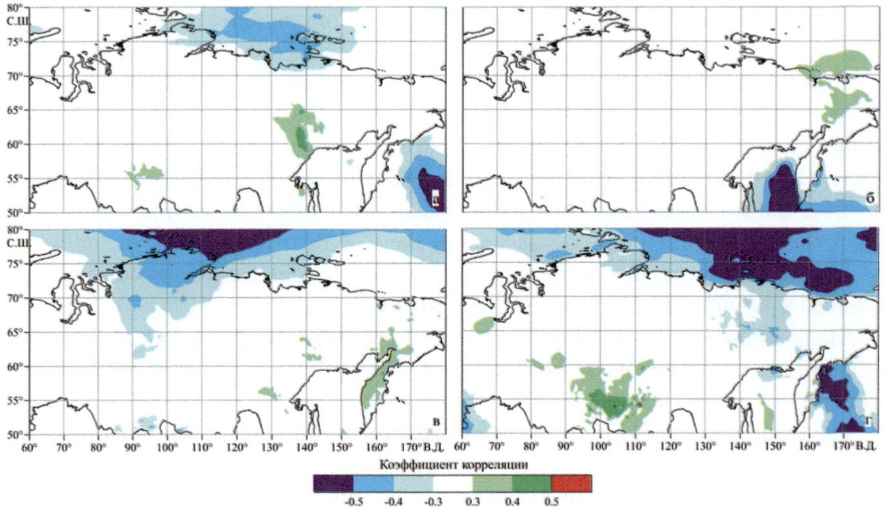

Fig. 6 ATR sites, where inter-annual ATA changes are related to variations of solar activity, corresponding to the period 2000–2024 were significant for months **a** February; **b** March; **c** April; **d** May

significance of the correlation between them correspond to its thresholds of 0.5, 0.4, and 0.3.

As Fig. 6a demonstrates ATR sites, where for the period of 2000–2024 correlation of the inter-annual changes of the ATA for February, with variations in solar activity, with a confidence of at least 0.9 considered significant and negative, identified in the north of the peninsula of Taymyr as well as in the coastal territories of Yakutia, which belong to Bulunskoe, Ust-Yanskoe, and Allaikhovsky Ouolums (districts). Significant and positive correlation sites found in the territories of the Khabarovsk and the Magadan territories, as well as Oymekonskiy and Ust-Mayskiy Districts (uluses) of Yakutia.

Figure 6b indicates that for March, only those areas of ATP have been identified, where for the period of 2000–2024, the correlation of inter-annual changes in ATA, with variations in solar activity, with a confidence of at least 0.9 is considered to be significant and positive. All these plots are located in the territory of the Chukotskiy Autonomous Region.

From Fig. 6c, it follows that for April there are numerous ATR plots for which the correlation of inter-annual changes in the years under consideration by ATA for April, with variations in solar activity, is considered significant and negative. Such areas are identified in the territories of Krasnoyarsk region (Tyimyr (Dolgan-Nenetskiy), Evenikiiskiy, and Turuhanskiy Districts), as well as The Republic of Yakutia (Anabarskiy and Olenekskiy National Districts (uluses).

It is important to note that the strongest link between the processes under consideration was found over the waters of the Russian Arctic, surrounding the area within which in 2000–2024 moved by NGP.

Significant and positive correlations were found mainly in the territory of Kamchatka territory and Chukotskiy Autonomous District.

Figure 6d allows us to conclude that the studied areas of ATR, where the correlation of the processes in question for May was significant and negative, belonged to the Taymirskiy (Dolgan-Nenetskiy) District of Krasnoyarsk territory, as well as Abayyye, Allaikhovskiy, Momskiy, and Ust-Yanskiy Ululums (districts) Yakutia. As in April, the strongest link between the studied processes was over the waters of the Laptev and East Siberian Seas, as well as the Central Arctic Basin, including the NGP area. This is practically the same area where for May significant correlation of solar activity with variations of TOC and SRF was found.

Areas where the correlation of the same processes is significant and positive, found in the territories of the Irkutsk region, the republics of Tyva and Buryatia, as well as the Zabaikalsky region.

The comparison from Figs. 4, 5 and 6 demonstrates that for April and May the significant causes of ATA changes over many parts of the Eastern Sector of the Russian Arctic were variations of their corresponding SRF, as well as the TOC. On the ATR sites belonging to the Krasnoyarsk and Yakutia territories, variations of the ATA depended only on changes in the SRF, and they were not related to the dynamics of the TOC.

It allows us to argue that in these months the main mechanism causing SRF variations over such areas was ionization of tropospheric air and its aerosol by secondary particle showers from cosmic rays entering the atmosphere above them. The location of the identified areas (Taimyr Peninsula, North-Siberian and Yano-Indigirskiy lowlands) indicates that they are all close to the direction where NGP is approaching the territory of Russia.

Similarly, it has been found that over the parts of the waters and territories of the Canadian Arctic archipelago from which the NGP is removed, the correlation of the processes in question is also significant, but the sign is the opposite.

As noted above, the response of the atmospheric transparency factor over some ATR sites to changes in solar activity depends on both their phase and location relative to the Polar Vortex and its removal relative to the NGP.

The obtained results confirm the importance of the influence of changes in distance between the NGP and some section of the Polar Vortex Branch, on the relationship of the intensity of ionization of its air with variations of CRF, as well as solar activity. It follows that the smaller the distance of the CRF from the section of the beam of this air current under consideration, the greater the amplitude of the effects on the corresponding area of the atmosphere of the CRF caused by changes in solar activity, the more they affect the intensity of ionization of its air.

In the period of 1998–2019, the distance from NGP to the North Pole was reduced. At the same time, the distance from the NGP to the Polar Vortex over Canada has been increased, and the distance from the NGP to the Polar Vortex over ATR has been reduced. As a result, the average monthly CRF entering this vortex area over Canada decreased and increased above ATR.

As shown in Fig. 2, these differences in the CRF trends dynamics entering the atmosphere above Canada and ATR led to opposite trends of changes in their air ionization intensity and, consequently, variations in the TOC.

The effect on TOC of CRF variations is manifested because the latter control the rate of stratospheric ozone depletion. Partially or completely compensates for this effect by the process of ozone formation in the reactions of the Chapman cycle, whose intensity is greater the more flux of ultraviolet radiation from the Sun enters the atmosphere. For this reason, the opposite trends in TOC over ATR and Canada are most pronounced in February and March, and are almost negligible in May.

As we move from winter to summer, the flux of ultraviolet radiation from the Sun, involved in the reactions of the Chapman cycle, increases. As a result, the relationship between decomposition and stratospheric ozone formation in the spring is shifting in favor of the latter, and the influence of stratospheric air ionization changes on TOC and SRF variations in the spring months is declining. The optical density of tropospheric clouds and other types of aerosols is increased. The changes in the SRF due to the effects of these mechanisms may be reversed over many areas of the Earth's surface. As a result, they may be partially or fully compensated. The latter explains the changes identified for the spring months in the total area of ATP plots, where in the period of 1998–2019 the trends toward an increase in the SRF are predominant, as shown in Fig. 3.

As is well known, variations in solar activity cause antiphase changes of CRF (Erlykin 2011). At the same time, the phases of the mentioned changes coincide both over Russia and over Canada. As a result, the intensity of ionization of their stratospheric and tropospheric air is changed in both regions. These changes cause corresponding variations in both TOC and the optical density of tropospheric clouds and other types of aerosols. However, the above-mentioned effect of the mutual compensation of SRF responses on these processes affects the relationship between changes in TOC as well as SRF and ATA over different ATR sites with solar activity in different months differently.

As shown in Fig. 4, the mentioned links are only relevant for April and May. The latter corresponds to the conclusions [of 14, Holopists], from which it follows that for these months the influence on the dynamics of TOC processes of stratospheric ozone destruction is less significant than the processes of its formation. Because of this, changes in the intensity of air ionization by secondary particle showers are more pronounced in variations of the optical density of tropospheric cloud and aerosol.

An important result is the identification of ATR sites where, in some spring months, SRF and ATA connections with solar activity changes (and therefore also with variations of SRF) in the studied period were significant. As can be seen from Fig. 5, there are both sites where the significant correlation between SRF and solar activity is negative and those where it is positive. This is possible if on the first sections in the respective months ionization influences the TOC, and, on the second, changes in the optical density of cloud and aerosol.

For March, as well as February, the inter-annual CRF changes, caused not only by solar activity but also by the displacement of the NGP, control the changes of the SRF and ATA mainly through the influence of CRF on the variations of the TOC.

For the May and summer months, their influence on the optical characteristics of tropospheric clouds and aerosols predominates.

For May, a significant feature is the existence in the Arctic of a region comprising an area within which the NGP moved during the years under review, where the correlation of solar activity with the variations of the TOC, SRF, and ATA was significant positive. Increased solar activity, due to the Forbush effect, leads to a decrease in CRF (Erlykin 2011), which causes a decrease in the intensity of ionization of air, both troposphere and stratosphere (Stozhkov et al. 2022). In the NGP area, amplitude and the average intensity of these atmospheric impacts are maximum, but in May stratospheric ozone formation occurs more intensively than its destruction, as a result of which the correlation between TOC and solar activity is positive, but the impact of TOC changes on the SRF is not significant. However, the CRF influence on tropospheric cloudiness and aerosol is significant in the same area, so the correlation of solar activity with variations of SRF and ATA is considerable and positive (the higher the solar activity, the less the CRF, more the SRF and ATA).

It is clear that all the features identified indicate a significant link between the movement of the NGP and the variations of the TOC, SRF, and CRF not only over the ATR but also over the entire relevant zone of the Northern Hemisphere.

Therefore, it is reasonable to assume that as the NGP continues to shift toward the Tymyr Peninsula, the distance from it to the North Pole will begin to increase and the Polar Vortex above the ATR will decrease. At the same time, the influence of CRF variations on air ionization, both stratosphere and troposphere, over ATR will be increased. This will lead to a decrease in TOC over the identified regions of the ATR, which may pose a danger to their populations.

The links between solar activity and not only the variations of the TOC in February and March, but also the changes of the SRF and the ATA for the months from March to May and summer will be strengthened.

The latter will increase the risk of accidents at ATP on man-made objects caused by the destruction of long-standing ice rocks under their foundations, to extend the time frame of the fire season for many of its territories and reduce the periods of safe operation of motor vehicles, as well as increase their fire safety due to weather conditions. From this point of view, it is relevant to develop a plan of activities to protect the population and economy of ATR regions from dangerous consequences of NGP approaching to the Russian territory, as well as its impact on the processes considered.

In general, the results confirm the validity of the hypothesis. Thus, the aim of this research has been achieved and its objectives met.

4 Discussion of the Results

As it follows from the obtained results, they fully correspond to the existing understanding of the reasons for the change in thermal regime of various ATR sites (Third assessment report n.d.), the role of variations in their radiation regime (Ya 1965), the transparency factor of the Earth's atmosphere (Kurt 1986; Mahokhtkin et al. 2023) and TOC (Karol et al. 1982; Muller 2012; Larin 2000; Hoffman 2005; The ozone layer n.d.). However, some of the findings have significant new elements. These include.

1. The sections of ATR and other regions of the Northern Hemisphere, located in the Polar Vortex zone, where in the period of 1998–2019 some NGP changes, corresponding to different spring months, predominated.
2. Trends of changes in TOC and SRF over the identified sites correspond to their distribution assumptions for the case where the impact of NGP movement on them was significant.
3. ATR sites have been established, for which the inter-annual changes of the TOC, SRF, and ATA are related to variations in solar activity, and therefore also to the CRF, between 2000 and 2024 were significant. For May, many of them form an area directly surrounding the area within which the NGP moved in those years. In this area, the correlation of solar activity with variations of TOC, SRF, and ATA is significant and positive.

The characteristics of the studied statistical relations between the processes under consideration suggest that these relations are causal.

5 Conclusions

Thus, it is established that

1. In the Asian territory of Russia there are numerous sites, where inter-annual change trends of the average monthly values of total ozone content and monthly flux of total solar radiation for spring months between 1998 and 2024 qualitatively corresponded to the presented ideas about the consequences of approaching to it the North Geomagnetic Pole.
2. For many of them, trends in the total ozone content have been observed to decrease during the months of March, April, and winter, despite the fact that the average values increased over the same period for the whole area of the northern hemisphere.
3. There are many areas of the Asian territory of Russia for which monthly flows of total solar radiation for spring months, from 1998 to 2024, significantly increased. Their total areas from March to May steadily increase.
4. Areas of the study area identified for which the statistical relationships between the inter-annual changes in the average monthly values of total ozone content, surface air temperatures, and monthly cumulative solar radiation flows for spring months, with synchronous variations of solar activity, are significant. For May, many of them form an area directly surrounding the North Geomagnetic Pole where the correlation between all these processes is significant and positive.

5. It is reasonable to assume that the approach of the North Geomagnetic Pole to Russia is one of the significant factors of climate warming, as well as fire hazard for many parts of its Asian territory.

References

Anisimov OA, Zhilitsova EL (2012) On the estimates of climate change in the regions of Russia in the 20th and early 21st centuries according to observations. Meteorol Hydrol 6:95–107

Atmospheric ozone and global climate change/Alexander EL, Karol IL, Rakipova LR, Sedunov US, Khrgian AKh- L (1982) Hydrometeoroizdat, p 165

Database of changes in the solar radiation flux with a wavelength of 10.7 mm [Electronic resource]. Access Mode: ftpprefftp.ngdc.noaa.gov/STP/space-weather/solar-data/solar-features/solar-radio/noontime-flux/penticton/penticton_observed/listings/listing_drao_noontime-flux-observed_monthly.txt

Diagram of the movement of the north magnetic pole towards russia [Electronic resource]. Access https://travelask.ru/blog/posts/13546-kuda-dvizhetsya-severnyy-magnitnyy-polyus-i-kogda-polyusa-sn

Dolgorudnaya scientific station named S.N. Vernova. Laboratory of the physics of the sun and cosmic rays. Physical institute named P.N. Lebedev RAN. [Electronic Resource]. Access https://sites.lebede3v.ru/DNS_FIAN

Dolgorudnaya scientific station named S.N. Vernova. Laboratory of the physics of the sun and cosmic rays. Physical institute named P.N. Lebedev RAS. [Electronic Resource]. Access https://sites.lebede3v.ru/DNS_FIAN

Dzerdzeevskiy BL (1968) Circulatory mechanisms in the atmosphere of the Northern Hemisphere in the XX century [Text]. - Moscow : [b. and 1968]. p 240

Erlykin AD (2011) Cosmic ray effects on cloud cover and their relevance to climate change/Erlykin AD, Wolfendale AW//J Atmos Sol -Terr Phys 77:1681–1686

Filipova NA (2019) Ensuring efficient and reliable delivery of northern cargo to the regions of the Far North and the Arctic zone of Russia: monograph/Filipova NA, Vlasov VM, Bogumil VNM (2019) Techpolygraph Center p 224

Golenkov AE, Ochlopkov VP, Svirezhevskaya AK, Svirezhevsky NS, Stozhkov UI (1980) Planetary distribution of cosmic rays intensity by measurements in the stratosphere. Works of FIAN. M.: Science 122:3–14

Grigoriev GA (2019) Prospects for the development of hydrocarbon resources of the Russian Arctic shelf—strategic pause is imminent/Mineral resources of Russia. Econ Manag 2(165):37–45

Hoffman MJ (2005) Ozone depletion and climate change/Hoffman MJ New-York: State Univ. press, p 276

Hólm E, Janisková M, Keeley S et al (2020) The ERA5 global reanalysis/Q J R Meteorol Soc 146:1999–2049

https://ozone.unep.org/sites/default/files/2023-09/Ozone_and_UV_Bulletin_1_ru.pdf

Ionov DV, Timofeev UM, Shalamyansky AM (2002) Comparison of satellite (equipment GOME, TOMS) and ground measurements of total ozone content/Ions DW, Timofeev UME, Shalamyansky AM/Earth Survey from space 3:10–19

Jones Chris A (2011) Planetary magnetic fieldsand fluid dynamos (англ.)/Annual review of fluid mechanics[англ.]—Annual Reviews 43:583—614

Kholoptsev AV (2012) World Ocean and the Ozone Sphere/Kholoptsev AV, Nikiforova MP, Bolshikh AV (2014) Saarbrucken, Deutschland: LAP Lambert Academic Publishing, p 522

Kurt VG (1986) Clearness of the atmosphere/Physics of space small encyclopedia. Under the ed. Shuniaeva YA 2nd edn 2nd reworked and supplemented/Moscow.: Soviet encyclopedia, p 783

Larin IK (2000) Ozone layer chemistry and life on Earth/Larin IK/Chemistry and life 21st century 7(10):15

Mahokhtkin AN, Mahokhtkina EL, Plachina IN (2023) Transparency of the atmosphere on the territory of Russia: current measurements in the twenty-first century/Works of the Main Geophysical Observatory named after A.I. Voyekova 609:103–119

Marsh ND (2000) Cosmic rays, clouds and climate/Masrsh ND, Svenmark H//Space Sci Rev 94:215–230

Maurchev EA, Balabin YuV, Germanenko AV, Mikhalko EA, Gvozdevsky BB (2019) Calculating the ionization rate induced by GCR and SCR protons in Earth atmosphere/Solar-Terrestrial Physics 5(3):68–74

Mokhanakumar K (2011) Interaction stratospheric and troposphere/Mohanakumar K, pr. c. R. Y. Lukyanova. - M.: Fizmatlite, p 451

Mokhov II (2020) Features of contemporary climate change in the arctic and its consequences/ Problems of the Arctic and Antarctica 66(4):446–462. https://doi.org/10.30758/0555-2648-2020-66-4-446-462

Muller R (2012) Stratospheric ozone depletion and climate change/Muller R (2012) RSC Publishing, p 346

Nikiforova MP, Svyagintsev AM, Vargin PN (2017) Abnormally low levels of total ozone above Siberia in late January 2016/Optics and the atmosphere 30(1):12–19

North magnetic pole track [Electronic resource]. Access https://web.archive.org/web/20220427002623/; https://www.ngdc.noaa.gov/geomag/data/poles/NP.xy

Ozone Bulletin. [Electronic resource]. Access mode:

Plan of development of the Northern Sea Route until 2035. Accessed http://government.ru/news/46171

Polyakov A, Kriukovskikh EYa, Virolainen G, Nerobelov D, Kozlov Yu, Timofeyev (2024) Determining the total ozone column from spectral measurements of IKFS-2 in 2015–2022, Izvestiya, Atmospheric and Oceanic Physics 60(6);689–698

Raspopov OM, Veretenko SV (2009) Solar activity and cosmic rays: influence on cloudiness and processes in the lower atmosphere (memory and by the 75th anniversary of MI Pudovkin)/ Geomagnetism and aerometry 49(2):147 155

Reshetnyak MJ (2018) Geodimo models/Izv. universities. Radiophysics 61(8–9):605–613

Database Results of reanalysis ERA5 hourly data on pressure levels from 1979 to present. [Электронный ресурс]. Режим доступа: https://cds.climate.copernicus.eu/cdsapp!/dataset/reanalysis-era5-pressure-levels?tab=form. https://doi.org/10.24381/cds.bd0915c6

Ross JC (1834) On the position of the north magnetic pole (англ.)/Philos Trans R Soc Lond: I 124:47–52

Schacha M (2018) Impacts of an ice-free Northeast passage on LNG markets and geopolitics/ Schacha M, Madlenerb R (2018) Energy policy 122:438–448. https://doi.org/10.1016/j.enpol.2018.07.009

Shakina N (2010) Blocking anticyclones/Shakina N, Ivanova AR/Meteorology and hydrology 11:5–18

Sharonov AY (2018) Hydrometeorological tasks of year-round navigation in the East Siberian Sea/ Sharonov AY, Shmatkov VA (2018) Bulletin of the State University of Marine and River Fleet named after Admiral SO Makarova 10(1):170–182. https://doi.org/10.21821/2309-5180-2018-10-1-170-182

Stozhkov UI, Mahmutov VC, Svirezhevsky NS (2022) Research of comical rays on cylinders at the Physical Institute named after P.N. Lebedev RAN/Successes of Physical Sciences T.192(9):1054–1063

Svyagintsev AM, Kuznetsov GI, Kuznetsova IN (2013) Spring anomalies of the ozone layer over Russia/Meteorology and hydrology 5:5–13

The movement of the north magnetic pole to the territory of Russia has slowed down [Electronic resource]. Access https://www.rbc.ru/rbcfreenews/67394e8b9a794716d3a3cb81

The ozone layer is on track to recover, which will avoid global warming by 0.5 o. [E resource]. Access https://www.unep.org/ru/novosti-i-istorii/press-reliz/ozonovyy-sloy-nakhoditsya-na-puti-k-vosstanovleniyu-chto-pomozhet
Third assessment report on climate change and its consequences in the territory of the Russian Federation. General summary. SPB.: High-tech technologies, p 124
Todd MC, Kniveton DR (2001) Changes in cloud cover associated with Forbush decreases of galactic cosmic rays. J Geophis Res 106(23):32031–32041
URL: https://www.wmo.int/pages/prog/arep/WMOArcticOzoneBulletins2016.html
Veretenko SV (1996) Effects of forbush-decreases of galactic cosmic rays in variations of general cloudiness/Veretenko SV, Podovkin MI/Geomagnetism and aeronomy 36(1):153–156
Volkov AV (2019) Deposits of strategic metals in the Arctic region/ Volkov AV, Bortnikov NS (2019) Scientific support for implementation of priorities of scientific and technological development of the Russian Federation. Moscow: Russian Academy of Sciences. Scientific sessions of the General Meetings of RAS Divisions, p 104–114
WMO Arctic ozone bulletin. 2016. N 1. [Электронный ресурс]. Режим доступа:
Kondratiev K Ya (1965) Actinometry. Hydrometeorological publishing house, Leningrad, p 692

Method of Assessing the Influence of Snow and Vegetation Cover on the Thermal Regime of Soils Using the Example of the Msu Weather Station

V. E. Gagarin, A. V. Koshurnikov, D. M. Frolov, T. A. Sidorova, and Ozoda Akhmedova

Abstract The chapter describes the methodology for assessing the influence of snow and vegetation cover on the thermal regime of soils. These works were carried out at the MSU meteorological station. There thermometric boreholes were drilled: one on the site with vegetation and the other on the site without vegetation and snow cover. The thermometric measurements obtained were used to construct thermoisopets, which allow analyzing the effect of cover on heat fluxes in the ground and are essential for understanding the freezing and thawing processes. The results of the research can be useful for engineering decisions in the design of foundations under conditions of climate change. The research is carried out within the framework of state budgetary themes at the Geography Department of MSU.

Keywords Freezing · Thermal regime · Soil · Snow and vegetation cover

The year 2024 turned out to be one of the warmest years in the history of instrumental observations according to the World Meteorological Organization, and in the current conditions of climate change and changes in glacier reserves, the role of geocryological research cannot be overstated. The study of the regularities of formation of the temperature regime of rocks is primarily connected with the study of the regularities of formation of the layer of seasonal freezing and the layer of seasonal thawing, as it is in these layers that the main part of thermal processes is realized, which primarily include annual heat turnover (conductive heat exchange) and heat transfer (convective heat exchange). Annual heat turnover in rocks is a component of the radiation-thermal balance of the Earth's surface and can be represented as a function of the surface temperature regime, thermophysical properties of rocks and their humidity. The experimental site on the basis of the MSU meteorological station was organized to assess the influence of vegetation and snow cover on the surface temperature regime.

V. E. Gagarin · A. V. Koshurnikov · D. M. Frolov (✉) · T. A. Sidorova · O. Akhmedova
Moscow State University Named After M.V. Lomonosov, Moscow, Russia
e-mail: denisfrolovm@mail.ru

T. Chaplina (ed.), *Processes in GeoMedia—Volume IX*, Springer Geology,
https://doi.org/10.1007/978-981-96-9302-3_14

For this purpose, two boreholes were drilled to the depth of zero annual temperature fluctuation amplitudes, in which thermocones with temperature sensors and loggers for recording temperature readings were placed. One well was drilled in undisturbed landscape conditions, the other in the area where vegetation cover is removed and snow is removed during winter. Some results of our research are presented in the Chapter.

In general, the thickness of the seasonally thawed layer (STL) reflects the state of permafrost soils (PFS) and fluctuates from year to year depending on climatic conditions. The information on its long-term changes and peculiarities of recent years published annually in Roshydromet Reports is based on the data of measurements of this value obtained according to a standardised methodology within the framework of the International Monitoring Programme CALM (Circum Polar Active-Layer Monitoring). More than 60 CALM sites have been established in the Russian cryolithozone at different times, but only a small part of them has a long series of observations. It is important to note that these measurements are of non-governmental nature and are carried out within the framework of separate scientific projects, carried out on an initiative basis, mainly by specialists from the Institute of Earth Cryosphere of the Russian Academy of Sciences, Institute of Permafrost Science of the Siberian Branch of the Russian Academy of Sciences with the participation of researchers from other scientific institutions.

Thus, in this chapter we considered the methodology of organizing thermometric boreholes and determining the influence of snow and vegetation cover on the thermal regime of soils on the example of the meteorological station of Moscow State University. For this purpose, boreholes were drilled, logger measurements were obtained and thermoisopleths were constructed both for the borehole with vegetation and for the second borehole with no vegetation and snow cover. The thermometric borehole on the territory of the observation site of the MSU meteorological observatory with remote natural cover is presented in Fig. 1.

As a result, logger measurements were obtained and thermal isopleths were plotted for both the well with vegetation and the second well with no vegetation and snow cover (Fig. 2).

These studies allow us to draw reasonable conclusions about the influence of snow and vegetation cover on the thermal regime of soils on the example of the MSU meteorological station by comparing the thermal regime of the soil on the site with open ground and on the site with the presence of vegetation and snow cover based on the results of thermal measurements in boreholes. The results of the work can be applied to assess the warming effect of cover in engineering and thermophysical calculations for foundations of buildings and linear structures in conditions of perennial and seasonal permafrost and changing climate.

The work is carried out in accordance with the state budgetary themes of the Geography Faculty of MSU ‘Danger and risk of natural processes and phenomena’ (121051300175–4) and ‘Evolution of the cryosphere under climate change and anthropogenic impact’ (121051100164–0).

Fig. 1 Thermometric wells on the territory of the MSU meteorological observatory

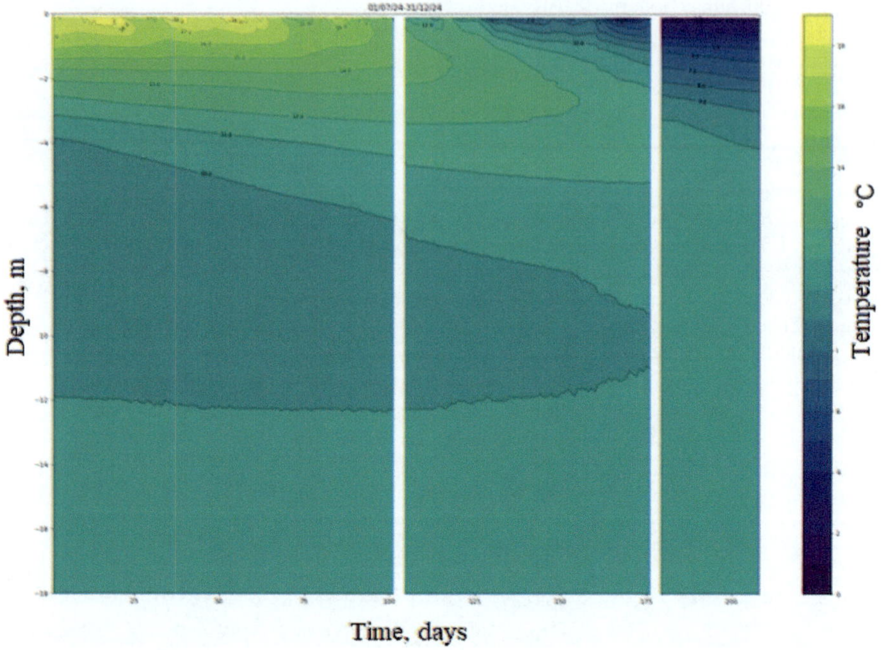

Fig. 2 Results of thermometric measurements of loggers in boreholes and constructed thermoisopleths for the territory of the MSU meteorological observatory

References

Electronic resource of the intergovernmental panel on climate change URL:http://www.ipcc.ch/. Accessed 18 Feb 2025

Electronic resource of the National Snow and Ice Data Centre URL: http://nsidc.org/data/docs/noaa/g02135_seaice_index/. Accessed 18 Feb 2025

Frolov DM, Komarov AYu, Seliverstov YuG et al (2019) Study of spatial and temporal heterogeneity of the snow column at the MO MSU site in winter 2018/2019/Ecological and climatic characteristics of the atmosphere of Moscow in 2018 according to data from the Lomonosov Moscow State University Meteorological Observatory. - Moscow: MAKS Press LLC, pp 225–230

Frolov DM, Seliverstov YG, Sokratov SA et al (2023) Cryological studies at the meteorological site of MSU in winter 2022/23/Arctic and Antarctic 1:1–13. https://doi.org/10.7256/2453-8922.2023.1.40448

Frolov DM, Seliverstov YG, Koshurnikov AV et al (2023) Investigation of evolution and thermal state of the snow cover thickness at the observation site of the lomonosov msu in winter of 2022/2023/Processes in GeoMedia, vol 7, Springer Geology, Singapore: Springer Singapore, pp 135–138. https://doi.org/10.1007/978-981-99-6575-5_12

https://wmo.int/media/news/wmo-confirms-2024-warmest-year-record-about-155degc-above-pre-industrial-level

https://www.un-glaciers.org/

Komarov AYu, Seliverstov YuG, Sokratov SA, Grebennikov PB (2016) Study of spatial and temporal heterogeneity of the snow column using a high-resolution penetrometer snowmicropen on the territory of the MSU Meteorological Observatory/Ecological and climatic characteristics of the atmosphere in 2015 according to the data of the Lomonosov Moscow State University Meteorological Observatory/Edited by Shilovtsev OA, MAKS Press Moscow, pp 201–210

Komarov AYu, Seliverstov YG, Sokratov SA, Grebennikov PB (2017) Study of spatial and temporal heterogeneity of the snow column at the site of the MSU Meteorological Observatory in winter 2016/2017/Ecological and climatic characteristics of the atmosphere in 2016 according to the data of the Lomonosov Moscow State University Meteorological Observatory / Edited by E. I. Nezval, I. V. Soshinskaya. - Moscow: MAKS Press LLC, pp 190–202

Komarov AYu, Seliverstov YG, Sokratov SA, Grebennikov PB (2017) Study of spatial and temporal heterogeneity of the snow column using snowmicropen penetrometer/III International Symposium Physics, Chemistry and Mechanics of Snow: Collection of Papers, Part I, Yuzhno-Sakhalinsk, 2–6 October 2017. - Sakhalin branch of FGBUN FGBUN Far Eastern Geological Institute FEB RAS Yuzhno-Sakhalinsk, pp 64–68

Komarov AYu, Seliverstov YuG, Sokratov SA et al (2017) Study of spatial and temporal heterogeneity of the snow column at the MSU Meteorological Observatory site in winter 2017/2018/ Ecological and climatic characteristics of the atmosphere of Moscow in 2017 according to the Lomonosov Moscow State University Meteorological Observatory/Edited by Lokoschenko MA, Moscow: MAKS Press LLC, pp 199–203

Komarov AYu, Seliverstov YuG, Grebennikov PB, Sokratov SA (2018) Spatial and temporal heterogeneity of the snow column according to the snowmicropen penetrometer data/Ice and Snow 58(4):473–485. https://doi.org/10.15356/2076-6734-2018-4-473-485

Site weather schedule [electronic resource] https://rp5.ru/. Accessed 22 Feb 2025

Sokratov SA, Seliverstov YG, Komarov AY et al (2018) Simultaneous use of different techniques in assessment of spatial-temporal variability of the characteristics of snow cover. In: Proceedings of the International Snow Science Workshop ISSW 2018, Innsbruck, Austria, 7–12 October 2018/Fischer J-T, Adams M, Dobesberger P, Fromm R, Gobiet A, Granig M, Mitterer C, Nairz P, Tollinger C, Walcher M (eds) Innsbruck ISSW, pp 373–376

Sokratov SA, Komarov AY, Vasilchuk YK et al (2023) Spatial and temporal heterogeneity of δ18o values and the structure of the snow column on the territory of the MSU Meteoobservatory. Ice and Snow 63(4):569–582. https://doi.org/10.31857/S2076673423040154

Website weather and climate [electronic resource] http://www.pogodaiklimat.ru/. Accessed 22 Feb 2025

Influence of Technogenesis on the Hydrosphere Condition of the Eastern Orenburg Region

I. V. Kudelina, T. V. Leontyeva, and M. V. Fatyunina

Abstract Depending on the methods and systems of deposit development, technogenic disturbances of the geological environment have different effects on its pollution. Pyrite complex deposits of the Orenburg region, developed by open-pit mining, have a technogenic impact on the environment. The ores of these deposits are a natural concentration of toxic elements. The main agent that ensures the oxidation of elements, removes reaction products, regulates the migration process of heavy metals in the conditions of external waste rock dumps is water. The purpose of the study is to assess the impact of technogenesis on the hydrosphere conditions of the Eastern Orenburg region. As a result, it was concluded that there are periodic excesses of such indicators as iron, manganese, lead, zinc, oil products at the "Autumn" and "Summer" copper-pyrite deposits. The hydrochemical situation at the copper-pyrite deposits of the Dombarovsky industrial hub is generally satisfactory. Regular monitoring and implementation of new technologies and, in particular, the use of complex barriers are recommended.

Keywords Anthropogenic pollution · Groundwater · Heavy metals · Complex barriers

1 Introduction

The Dombarovsky ore district is located in the southern part of the East Magnitogorsk zone. In metallogenic terms, it belongs to the ore-bearing zone of the same name, and in structural terms, it corresponds to the Dombarovsky structural-formational zone. The confinement of the district to the mobile junction zone of two largest megastructures of the Southern Urals—the Magnitogorsk subsidence and the East Ural uplift—determined the complexity of its geological structure, the widespread development of folded deformations, intrusive magmatism and metamorphism (Plotnikov 1987; Kosarev 2010).

I. V. Kudelina (✉) · T. V. Leontyeva · M. V. Fatyunina
Orenburg State University, Orenburg, Russia
e-mail: kudelina.inna@mail.ru

T. Chaplina (ed.), *Processes in GeoMedia—Volume IX*, Springer Geology,
https://doi.org/10.1007/978-981-96-9302-3_15

Copper pyrite deposits are located within the central zone of the Dzhailgan volcanotectonic structure. In geological terms, the structure is represented by intrusive, volcanogenic metamorphic rocks of the Silurian and Ordovician, the Mesozoic weathering crust of the Mesozoic and the Cenozoic cover of loose sediments (Seravkin 2007; Prokin 2011). The structure contains the Letneye, Osennee and Levoberezhnoye deposits and is estimated as one of the most promising in the region for the discovery of new industrial copper pyrite objects. The area of the Osennee deposit, which is a local feeding area, is a slope of the watershed. In the area of the deposit, an aquiferous Quaternary alluvial horizon, an aquiferous locally weakly aquiferous Mesozoic weathering crust complex and a locally aquiferous zone of Proterozoic-Paleozoic metamorphic, intrusive, effusive and volcanogenic-sedimentary rocks are widespread. In this zone, water-bearing rocks are represented by diabases, diabase porphyrites, their pillow lavas, gabbro-diabases, dacite and liparite-dacite porphyries, tuffs of different compositions and constitute a single aquiferous zone of volcanogenic formations of bedrock.

At the deposit, according to the results of hydrochemical testing, the hydrocarbonate-chloride type of groundwater is most widespread, hydrocarbonate-sulphate and chloride-hydrocarbonate are present. Sodium waters predominate in terms of cation composition.

In the area of the Letneye deposit, the following aquifers and complexes are distinguished: alluvial deposits; fissure waters of volcanogenic-sedimentary rocks of the Middle Devonian; fissure waters of diabase-porphyrite rocks of the Upper Silurian of the Middle Devonian, the Kiembaevskaya suite (Kovalevsky 2001; Goldberg 1984).

The aquifer complex of volcanogenic-sedimentary rocks of the Middle Devonian is confined to the wings and core of the Kurmansay structure and to the south-western part of the deposit area. The water content of the complex is associated with weathering fracturing with a thickness of 45.0–80.0 m in tuffs, tuffites, tuff breccias, tuff sandstones. The aquiferous complex of diabase-porphyry rocks of the Upper Silurian—Middle Devonian is widely developed and is represented by fissure and fissure-vein unconfined waters of diabases, diabase and andesite-basalt porphyrites, spilites, flooded in contact zones with intrusions and in zones of tectonic disturbances.

During visual examination of the core, intense fracturing is noted to a depth of 50 m; deeper, the fracturing decreases and in most cases the cracks have a very weak degree of opening. These are mainly ground cracks and only in some of them is an opening of up to 1–2 mm sometimes observed. The degree of waterlogging of bedrock is weak and decreases sharply with depth. At a depth of 150 m, only individual cracks are found, the circulation of groundwater practically ceases, and therefore this boundary is taken as an aquiclude (Borevsky 2014; Gaev 2019).

Changes in the chemical composition and mineralization of groundwater over time are insignificant and are related to the amount of precipitation.

Replenishment of natural groundwater resources occurs mainly due to infiltration of melt water and, to a lesser extent, in autumn, due to rain. In summer, when evaporation is maximum, despite the large share of precipitation (62% of the annual amount), infiltration is very small and the graphs of regime observations show almost no pronounced rise in the water level during this period.

Groundwater at the deposit is hydrocarbonate and hydrocarbonate-chloride, less common are hydrocarbonate-sulphate and chloride-sulphate. The pH value fluctuates from 6.7–8.4, the water is predominantly alkaline and only in single wells is pH<7. Water mineralization mostly does not exceed 1 g/dm^3. With depth, it usually increases, sometimes up to 2.0–4.1 g/dm^3. There is an increased iron content (from 0.4–0.6 to 9.6–15.5 mg/dm^3), the amount of metals is small Cu (0.16–0.5 mg/dm^3), Zn (0.6–1.4, sometimes up to 2.3 mg/dm^3), Mo (0.008–0.014 mg/dm^3), U—up to 6.5 $\times 10^{-7}$ mg/dm^3), As, Pb, Hg, as a rule, are absent. In the area of the ore loading site, one aquifer is developed, confined to the Mesozoic weathering crust of porphyrites and to fractured porphyrites. The waters are unconfined, the groundwater depth is 13 m, during pumping with a decrease of 32.5 m, a flow rate of 1.0 l/s was obtained. Specific flow rate is 0.03 l/s. The waters are fresh, dry residue 0.32 g/dm^3, moderately hard (hardness 6 mg-eq/dm^3), magnesium-calcium hydrocarbonate, alkaline (pH 7.4).

2 Results and Discussion

The quality of groundwater at the copper-pyrite deposits of the Dombarovsky industrial hub, namely the "Osennee" and "Letnee" deposits, is characterized by 251 chemical analyses of samples. Of these, 132 analyses account for the analysis of groundwater at the "Letnee" deposit, and 119—at the "Osennee" copper-pyrite deposit.

The indicator of groundwater pollution is given in comparison with waters suitable for domestic and drinking water supply according to regulatory documents (On the state of sanitary 2018; On approval of the regional program 2019).

Observations of quality indicators at the "Osennee" and "Letnee" deposits were carried out for dry residue, hardness, hydrogen index, copper, zinc, iron (total), calcium, magnesium, cobalt, chlorides, sulfates, lead, manganese, oil products and nitrates for 5 years. Observations of underground waters and study of the qualitative composition were carried out 3 times a year. Water sampling was accompanied by threefold sampling of the well column in order to determine reliable values of the studied parameters. The following results were obtained for the "Osenniy" deposit (Leontyeva 2019; Cherepansky 2018).

Dry residue (mineralization). The dry residue value in observation wells varies from 49.0 (well No. 10n) to 492.0 (well No. 8) mg/dm^3, with average dominant values from 150.0–280.0 mg/dm^3. The waters are fresh, with hydrocarbonate anions.

Hardness. The hardness value was not determined directly, although this value is one of the main ones for characterizing groundwater. However, judging by the dry residue value and the magnesium and calcium content, we can conclude that its quantities are insignificant. The quantitative characteristics of the magnesium and calcium content directly indicate the hardness value.

Hydrogen index (pH). The value of the hydrogen index in the groundwater of the "Osennee" deposit varies from 5.93 (well No. 2n) to 9.11 (in well No. 2n), with average dominant values of 6.6–8.7. The waters are classified as slightly alkaline to slightly acidic. The maximum permissible concentrations by the value of the hydrogen index are 6–9. From the above, it can be concluded that the waters are generally healthy by the hydrogen index, with the exception of isolated cases exceeding the MAC by 0.1 units.

Nitrates. The value of nitrates in the groundwater of the deposit for the observation period from 2008 to 2012 fluctuates from 0.1–30.6 mg/dm^3, with average dominant values of 1.5–3.0 mg/dm^3. An exception is the nitrate value in well 7–50.1–153.4 mg/dm^3, which is about 1.1–3.2 MAC for drinking water. Most likely, the pollution is directly related to the storage pond, near which the well is located. Nitrate levels are exceeded in May–June, during the post-flood period. Melt and waste waters are drained into groundwater due to oversaturation in one way or another. In connection with the above, it is recommended to take into account the drainage of melt water from storage ponds by barriers (low-permeability embankments).

Iron (total). The iron content in the deposit as a whole does not exceed the MAC in groundwater, however, systematic excesses were recorded during the observation period. Iron is a priority indicator of groundwater pollution from the development of copper-pyrite deposits, the ores of which contain iron in high concentrations. The iron content varied from 0.01–2.44 mg/dm^3, with average dominant values of 0.03–0.1 mg/dm^3 (Fig. 1). Single excesses of the MPC for iron were noted in almost all observation wells (except well No. 10n). Iron is a natural pollutant of groundwater, characteristic of most sediments within the Orenburg region.

Manganese. Manganese concentrations vary from 0.01–2.15 mg/dm3 (with a MAC of 0.1 mg/dm^3). Exceeding the MAC by 20 times was noted for one well, exceeding the MAC by 3 or more times—for 11 wells (Fig. 2). Increased content is typical for groundwater of copper-pyrite deposits, since this element is most mobile in groundwater. It is recommended to increase water sampling from 3 times a year to 12 times

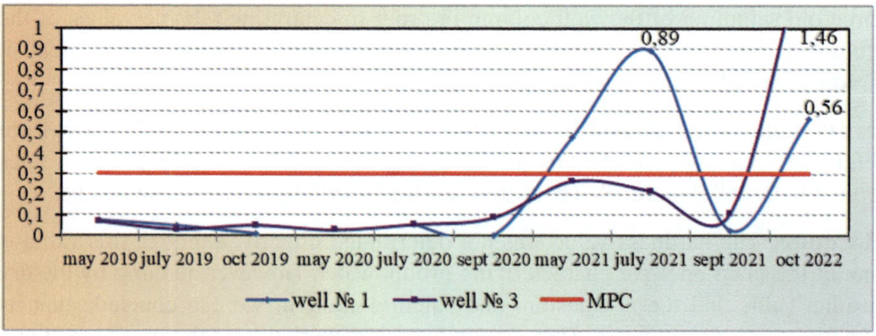

Fig. 1 Change in iron content in wells of the "Osenniy" deposit

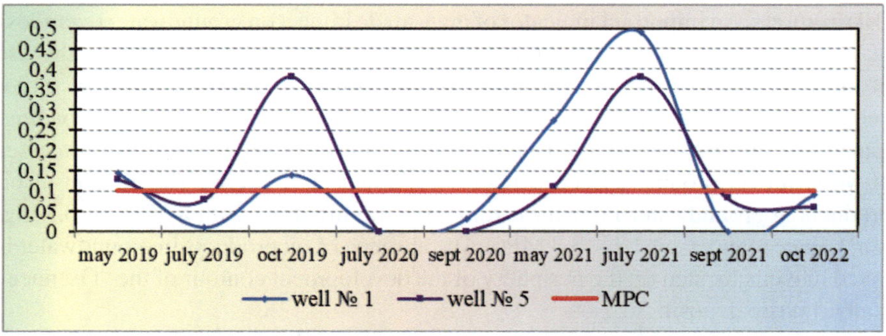

Fig. 2 Change in manganese content in wells of the "Osenniy" deposit

(once a month) during further observations in order to determine whether the excess of manganese is seasonal.

Zinc. The zinc content in the groundwater of the "Osennee" copper-pyrite deposit does not exceed the MAC. Values fluctuate from 0.003–0.07 mg/dm^3, with average dominant values of 0.02–0.03 mg/dm^3. Zinc is a priority indicator of groundwater pollution from copper-pyrite deposit development activities. Its insignificant content in groundwater apparently indicates its weak migration in groundwater (heavy metal).

Copper. Along with zinc (included in the same group of pollutants), it is a priority indicator of groundwater pollution from copper-pyrite deposit development. Copper content in groundwater fluctuates from 0.02–0.397 mg/dm^3, with average values of 0.01–0.1 mg/dm^3. Copper content in groundwater of the "Osennee" copper-pyrite deposit does not exceed 0.3 MAC (MAC 1.0 mg/dm^3).

Lead. For unknown reasons, lead determinations were carried out irregularly. According to the given indicators, lead concentration varied from 0.0001–0.022 mg/dm^3 (exceeding MAC by 2.2 times in well 7). Most likely, pollution is directly related to the storage pond. For a number of wells, this indicator was not determined at all. It is necessary to recommend organizing observations of lead in groundwater in the above-mentioned wells, since this heavy metal is highly toxic.

Cobalt. Analysis of observations of this component indicates its insignificant quantities in groundwater. Its content during the study period varies from 0.001–0.019 mg/dm^3, with average dominant values of 0.003–0.007 mg/dm^3. However, during the entire study period, a single case of exceeding 3.6 MAC (0.36 mg/dm^3) was noted. Most likely, a single excess is random (error or laboratory error).

Macrochemical indicators (chlorides, sulfates, calcium, magnesium). All macrocomponents of groundwater quality are significantly below the MAC for drinking water. Thus, chlorides were contained in quantities from 3.0–85.2 mg/dm^3 (MAC 350 mg/dm^3), sulfates—from 2.5–98.8 mg/dm^3 (MAC 500 mg/dm^3), magnesium—from 2.4–43.6 (MAC 50 mg/dm^3), calcium—from 4.0–76.0 mg/dm^3.

Oil products. An important indicator of man-made impact on groundwater regardless of the type of production. Pollution with oil products is associated with the use of motor vehicles at the site. At the "Osennee" deposit, a large number of motor vehicles are used throughout the mining and processing cycle: ore extraction, loading, transportation and unloading. The observation network regularly shows excesses of MPC for oil products. Exceedances were noted in 12 wells. The content of oil products in groundwater fluctuates from 0.04–4.6 mg/dm^3 (with a MPC of 0.3 mg/dm^3). Exceedances are 1.1–15.3 MPC. The absence of oil products in groundwater is noted in wells located on the periphery of the development contour of the "Osennee" copper-pyrite deposit.

As a result of the study at the "Letnee" deposit, the following results were obtained.

Dry residue (mineralization). The dry residue value in the observation wells of the regime network of the Letnee field varies from 62.0–648.0 mg/dm^3 (at the MAC of 1000 mg/dm^3), the dominant average values from 150.0–280.0 g/dm^3. The waters are fresh, sulfate-hydrocarbonate, hydrocarbonate, hydrocarbonate-sulfate.

Hardness. The hardness value is determined irregularly. Its value is low and varies across the regime network of wells from 0.4–7.6 mg-eq/dm^3.

Hydrogen index (pH). The value of the hydrogen index in the groundwater of the Letnee field varies from 6.25–9.52 (in well No. 8), with average dominant values of 7.0–8.6. Moreover, in well No. 8, excess of MAC is observed regularly. Waters are classified as from slightly alkaline to slightly acidic. Maximum permissible concentrations by the value of the hydrogen index are 6–9. It can be concluded that the waters by the hydrogen index are generally healthy, with the exception of numerous isolated cases exceeding MAC by 0.1–0.52 units.

Nitrates. The nitrate content in the groundwater of the deposit for the observation period from 2018 to 2022 fluctuates from 0.08–61.83 mg/dm^3, with average dominant values of 1.0–8.0 mg/dm^3. Increased nitrate values were noted in wells No. 11, 12. Most likely, the pollution is directly related to the main pond and the storage pond, near which the wells are located. The excess of nitrates occurs in May–June, in the post-flood period. Melt and waste waters are drained into the groundwater in one way or another due to oversaturation. In connection with the above, the Subsoil User is recommended to take into account the drainage of melt water from the storage ponds with barriers (low-permeability embankments). Iron (total). The iron content in the deposit as a whole does not exceed the MAC in groundwater, however, systematic excesses were recorded during the observation period. Iron is a priority indicator of groundwater pollution from the development of copper-pyrite deposits, the ores of which contain iron in high concentrations. The iron content varied from 0.01–1.753 mg/dm^3, with average dominant values of 0.05–0.13 mg/dm^3 (Fig. 3). Single excesses of the MAC for iron were noted practically in observation wells. 7, 8, 9, 10, 11, 12, 13, 14, 17. Moreover, excesses in well 12 were constantly noted for the period 2020–2022, which can be associated with the leakage of wastewater that has undergone treatment (apparently insufficient to neutralize the iron content in the water).

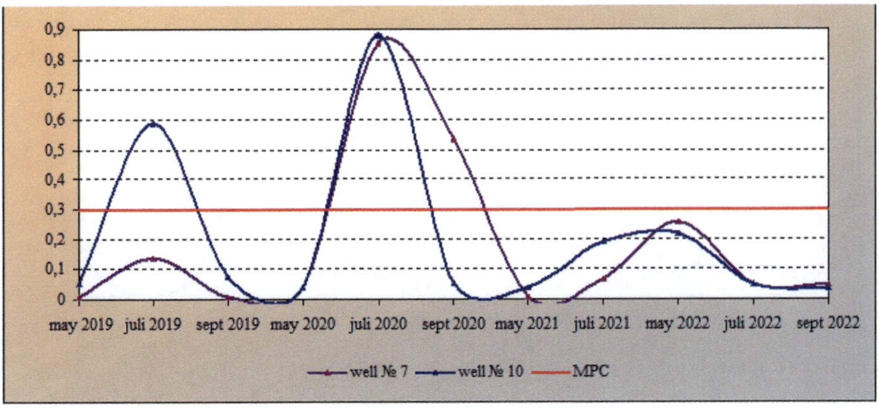

Fig. 3 Change in iron content in boreholes of the Letneye deposit

Manganese. Manganese content readings were not determined at the Letneye copper pyrite deposit. By analogy with the Osennee deposit, we can assume elevated concentrations of this component and recommend regular measurement of this indicator.

Zinc. Zinc content in groundwater at the Letneye copper pyrite deposit does not exceed the MAC (as well as at the Osennee deposit). Zinc valuesin groundwater fluctuate from 0.001–0.14 mg/dm^3, with average dominant values of 0.01–0.05 mg/dm^3. Zinc is a priority indicator of groundwater pollution from copper pyrite deposit development activities. Its insignificant content in groundwater apparently indicates its weak migration in groundwater (heavy metal).

Copper. It is a priority indicator of groundwater pollution from copper-pyrite deposit development activities. Copper content in groundwater of the Letnee deposit fluctuates between 0.002 and 0.086 mg/dm^3, with average values of 0.01–0.05 mg/dm^3. Copper content in groundwater of the Letnee copper-pyrite deposit does not exceed the MAC, which indicates healthy groundwater.

Lead. As in the case of the "Osenniy" deposit, lead determinations were carried out irregularly for unknown reasons. It can be assumed that the Subsoil User is concealing the true contamination of groundwater with lead, an element hazardous to human and animal health (let's not forget about plants). According to the given indicators, lead concentrations varied from 0.006–0.15 mg/dm^3 (exceeding the MAC by 15 times in well 12). Most likely, the contamination is directly related to the storage pond, near which well 12 is located. MAC excesses are also noted in a number of observation wells 1, 4, 7, 8, 9, 10, 11, 12, 13, 14, 16, 17. It is necessary to recommend that the Subsoil User organize lead observations in groundwater in the above-mentioned wells, since this heavy metal is highly toxic. Cobalt. The analysis of observations of this component indicates its insignificant quantities in groundwater. Its content

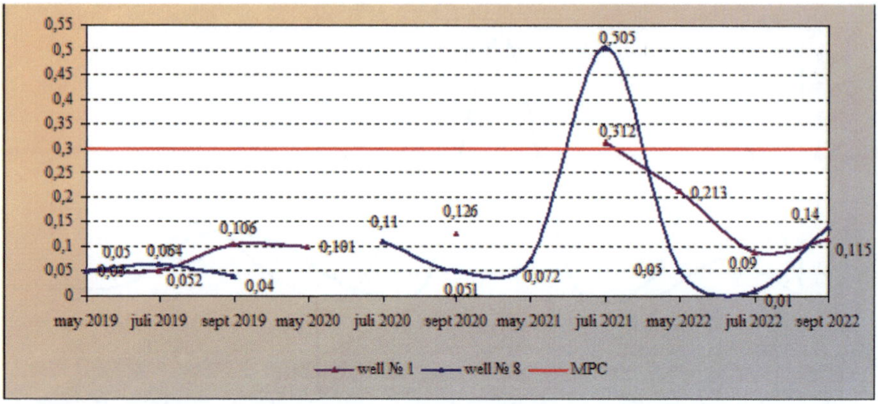

Fig. 4 Change in the content of oil products in the wells of the "Letney" field

during the study period varies from 0.001–0.0066 mg/dm^3, with average dominant values of 0.003–0.005 mg/dm^3.

Macrochemical indicators (chlorides, sulfates, calcium, magnesium). All macrocomponents of groundwater quality are significantly lower than the MAC imposed on drinking water. Thus, chlorides were contained in quantities from 3.4–81.54 mg/dm^3 (MAC 350 mg/dm^3), sulfates—from 1.7–224.68 mg/dm^3 (MAC 500 mg/dm^3), magnesium—from 2.4–40.6 (MAC 50 mg/dm^3), calcium—from 2.0–64.0 mg/dm^3.

Petroleum products. An important indicator of man-made impact on groundwater regardless of the type of production. Pollution with oil products is associated with the use of motor vehicles at the site. Both at the Osenneye field and at the Letneye field, a large number of motor vehicles are used throughout the mining and processing cycle: ore extraction, loading, transportation and unloading. Regular excesses of MPC for oil products have been noted in the observation network. Exceedances have been noted in wells 1, 7, 8, 9, 10, 11, 12, 13. The exceptions are wells 2, 4, 14, 15, 16, 17. The content of oil products in groundwater fluctuates from 0.009–2.38 mg/dm^3 (with a MPC of 0.3 mg/dm^3). The excess amounts to 1.1–7.9 MAC (Fig. 4). The absence of oil products in groundwater is noted in wells located on the periphery of the development contour of the Letneye copper-pyrite deposit.

3 Conclusions

From the analysis of the situation of the state of groundwater, it is evident that according to the data of water samples from the observation network of wells at the "Osenney" and "Letney" copper-pyrite deposits, there are periodic excesses of such indicators as: iron, manganese, lead, zinc, oil products.

In conclusion, it should be noted that, in general, the hydrochemical situation at the copper-pyrite deposits of the Dombarovsky industrial hub ("Osenney" and "Letney") is generally satisfactory. It is possible to prevent the development of negative processes by introducing new technologies and, in particular, the technology of complex barriers.

References

Borevsky BV (204) Typification of combined water intake systems for the joint use of groundwater and surface water. Borevsky BV, Kozak NS (2014) Exploration and protection of mineral resources 05:51–55

Cherepansky MM (2018) The influence of natural and man-made factors on the underground component of water resources in the western part of the Moscow artesian basin. Cherepansky MM, Karimova OA, Dzyuba AV, Vyazkova OE (2018) Min J 11:29–34. https://doi.org/10.17580/gzh.2018.11.04

Gaev AYa (2019) Water management problems of water-deficient territories on the example of the Southern Urals/Gaev AYa, Kudelina IV, Leontyeva TV (2019) Science, new technologies and innovations of Kyrgyzstan 4:218–222

Goldberg VM (1984) Hydrogeological foundations for protecting groundwater from pollution. Goldberg VM, Gazda SM (1984) Science 262

Kosarev AM (2010) Geochemical features of volcanogenic formations of the southern urals and their productivity for pyrite mineralization. Kosarev AM (2010) Lithosphere 3:177–184

Kovalevsky VS (2001) Combined use of surface and groundwater resources. Kovalevsky VS (2001) Scientific World 332

Leontyeva TV (2019) On the analysis of the water supply system of the population in the Eastern Orenburg region. Science, new technologies and innovations of Kyrgyzstan 4:240–243

On approval of the regional program "Improving the quality of water supply in the Orenburg region" for 2019–2024: Resolution of the Government of the Orenburg region dated 10.10.2019 No. 775-pp (as amended on 01.02.2022). URL: https://docs.cntd.ru/document/561580593. Accessed 24 Feb 2025

On the state of sanitary and epidemiological well-being of the population in the Orenburg region in 2018: State report. Orenburg 241

Plotnikov NI (1987) Hydrogeology of ore deposits. Plotnikov NI, Roginets II, Nedra M 287

Prokin VA (2011) Geological conditions of placement and prospects for identifying large copper-pyrite deposits in the urals. Prokin VA, Seravkin IB, Vinogradov AM (2011) Lithosphere 6:123–133

Seravkin IB (2007) Endogenous metallogeny of the southern urals and a general assessment of the prospects of its Orenburg part for pyrite, copper-porphyry and gold mineralization. Seravkin IB, Znamensky SE (2007) Geological collection No. 6:24

Uncertainties in Chlorophyll-A Concentration Estimation from OLCI Satellite Data Over the Sea of Japan in April 2023 Due to the Presence of Dust in the Atmosphere

A. S. Papkova, P. A. Salyuk, and D. M. Shukalo

Abstract The occurrence of a high concentration of absorbing dust aerosol over the sea surface significantly complicates satellite measurements of the Ocean Color characteristics. This study presents an analysis of the variability of satellite estimates of the chlorophyll-a (Chl-a) concentration during the transfer of mineral dust over the Sea of Japan in April 2023. This transfer is confirmed not only by visual analysis of satellite images, but also by the calculation results of the SILAM software package. Statistical processing of OLCI satellite data for the central and northeastern parts of the Sea of Japan shows that on the days of a high concentration of mineral dust over the region, the Chl-a concentration sharply increased by 3–4 times compared to the average monthly values. This is a clear consequence of the atmospheric correction error, since even if mineral dust affects an increase in the Chl-a due to the deposition of mineral substances into the upper layer of the sea, this effect cannot occur immediately. This study examines possible causes of this error, namely physically incorrect values of the remote sensing reflectance (Rrs(λ)) of the Sea of Japan obtained during standard processing of OLCI data.

Keywords Sea of Japan · Mineral dust · Chlorophyll-a · Remote sensing reflectance · SILAM · OLCI

A. S. Papkova (✉) · D. M. Shukalo
Marine Hydrophysical Institute of the RAS, Sevastopol, Russian Federation
e-mail: hanna.papkova@gmail.com

P. A. Salyuk
Far Eastern Branch, RAS, V.I. Il'ichev Pacific Oceanological Institute, Vladivostok, Russian Federation

T. Chaplina (ed.), *Processes in GeoMedia—Volume IX*, Springer Geology,
https://doi.org/10.1007/978-981-96-9302-3_16

1 Introduction

Every year, visible plumes of atmospheric mineral dust transported eastward from the Gobi and Taklamakan deserts are recorded over the Sea of Japan using remote sensing methods (Song et al. 2016). The appearance of a large amount of absorbing dust aerosol over the sea surface significantly complicates satellite measurements of Ocean Color characteristics in the study region (Fukushima and Toratani 1997; Shanmugam and Ahn 2007; Park et al. 2022), especially considering that this region is subject to significant river runoff (Cui et al. 2014; Salyuk et al. 2024; Aleksanin and Kachur 2016). In this paper, we propose to assess the effect of mineral dust on the error in calculating chlorophyll-a (Chl-a) based on satellite data from the Ocean and Land Colour Instrument (OLCI) scanner for the Sea of Japan. A similar study was previously conducted to assess the impact of Saharan dust on the quality of the atmospheric correction of satellite data for the Black Sea (Suetin et al. 2004; Papkova et al. 2024; Shybanov and Papkova 2022). The results showed that the standard atmospheric correction method proposed by Gordon and Wang (Gordon and Wang 1994), in the presence of dust atmospheric aerosol, does not work properly and provides physically incorrect values of the remote sensing reflectance and, as a consequence, this leads to further errors in the calculation of the concentration of Chl-a and yellow matter (Suetin et al. 2004).

To analyze satellite data on Chl-a, it is necessary to know the regional features of its distribution. In the study (Zvalinsky et al. 2022), based on in situ data, it is noted that at the beginning of the spring period, the maximum values of Chl-a are observed in the surface layer. The work (Bukin et al. 2007) described an increase in satellite estimates of Chl-a after the entry of aerosol matter into the upper layer of the ocean in the Sea of Japan in the spring. The paper (Lipinskaya et al. 2023) considered various physical factors that influence the spatial distribution of Chl-a based on satellite optical data. In such studies, it is important to take into account the fact that the obtained satellite estimates of Chl-a can be distorted by the presence of absorbing aerosol in the atmosphere. This influence can be partially reduced through the use of empirical bio-optical algorithms that do not use the 412 nm and 443 nm spectral channels (Salyuk et al. 2010).

In this study, we propose to consider a specific dust transport episode recorded during the period of April 13–21, 2023 over the study area and to assess the variability of the Chl-a concentration (https://modis.gsfc.nasa.gov/gallery/individual.php?db_date=2023-04-17). This work will allow us to evaluate the quality of the provided satellite estimates of the Chl-a concentration during an intense multi-day transport of mineral dust over the Sea of Japan in combination with the presence of strong cloudiness over the region.

2 Tools and Methods

Satellite measurements of the second processing level (level2) of the OLCI optical radiometer installed on the Sentinel-3A (S3A) and Sentinel-3B (S3B) satellites were used as the main source of data. The technical characteristics of OLCI and its calibration stages are described in detail in (https://sentiwiki.copernicus.eu/web/s3-olci-instrument). At the moment, the OLCI scanner is one of the most reliable and accurate remote sensing instruments of Ocean Color. Its advantage is high spatial resolution and good validation results with in situ data, as well as taking into account the contribution of solar illumination (highlight) and, as a consequence, its rare appearance in images (NASA Goddard Space Flight Center 2022). The second-level processing data were taken from the Ocean Color website (https://oceancolor.gsfc.nasa.gov). The Chl-a estimates were calculated using the standard OCI algorithm (Hu et al. 2012) with framework (NASA Goddard Space Flight Center 2022).

An additional source of data on the concentration of mineral dust over the study region and its movement in the atmosphere was the SILAM (System for Integrated modeling of Atmospheric coMposition) model. The SILAM software package is widely used to study the impact of forest fires, volcanic eruptions, dust transport, etc., and is a modern powerful tool for modeling the scattering properties of aerosols, gas components, dust particles, radionuclides and natural allergens in the atmosphere. SILAM calculations are based on the Lagrange-Euler model. One of the SILAM parameters used in this work is the content of suspended particles of 2.5 and 10 μm in size (PM2.5 and PM10) (Sofiev et al. 2006). Since mineral dust most often represents coarse particles of PM10, an increase in their concentration can serve as an indicator of its presence in the atmosphere.

Visualization of satellite images of the OLCI scanner, allocation of the study area, as well as statistical analysis for the second-level processing data were performed in the SeaDAS program (https://seadas.gsfc.nasa.gov/).

3 Results and Discussions

Figure 1 shows True Color satellite images for the central and northeastern parts of the Sea of Japan obtained from the OLCI scanner. In this paper, all images for April 2023 were considered and, unfortunately, dust identification was complicated by the presence of clouds over the studied region. From April 12 to 22, 2023, yellow plumes are visible on satellite images, which are the places of concentration of mineral dust in the atmosphere. From visual analysis, it is clear that the cloudiest days are April 10 and 27, but the image for the 10th is characterized by the presence of highlight, while in the image for the 27th it is noted only in the northeastern part.

Confirmation that the place of origin of dust aerosol is the Gobi and Taklamakan deserts is not only a visual analysis of satellite data, but also SILAM maps for the distribution of dust particles in the studied region, where yellow dust plumes at a

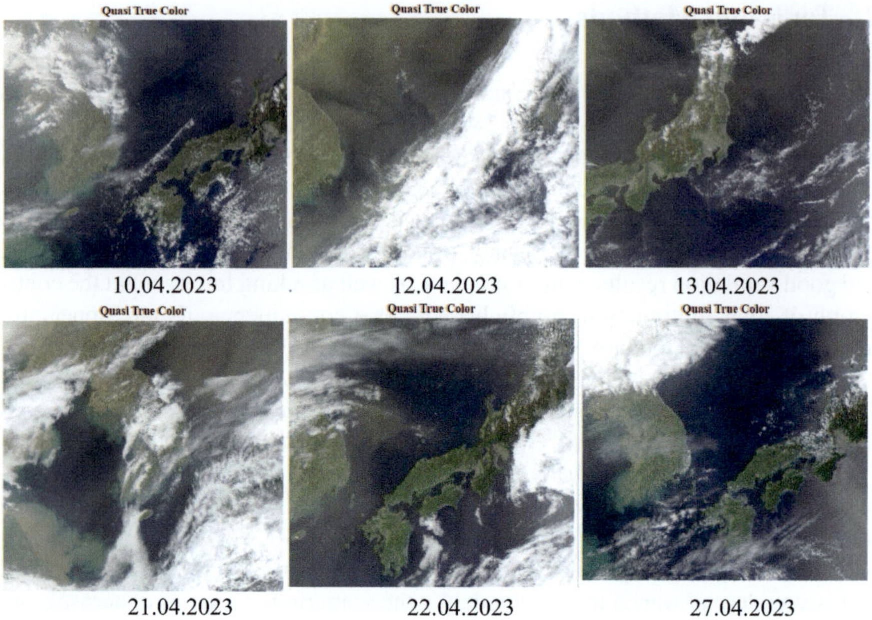

Fig. 1 A series of six satellite images showing quasi true color views of Japan and surrounding regions on different dates in April 2023. Each image displays varying cloud cover and atmospheric conditions. The dates are 10.04.2023, 12.04.2023, 13.04.2023, 21.04.2023, 22.04.2023, and 27.04.2023. The images highlight changes in weather patterns and visibility over time(*Source* https://oceancolor.gsfc.nasa.gov/)

concentration of 100–150 $\mu g \cdot m^{-3}$ were observed over the Sea of Japan on 12.04.2023 (Fig. 2).

The analysis of the concentration of coarse aerosol particles PM10 for 13.04.2023 and 21.04.2023 is presented in Fig. 3, which confirms the presence of mineral

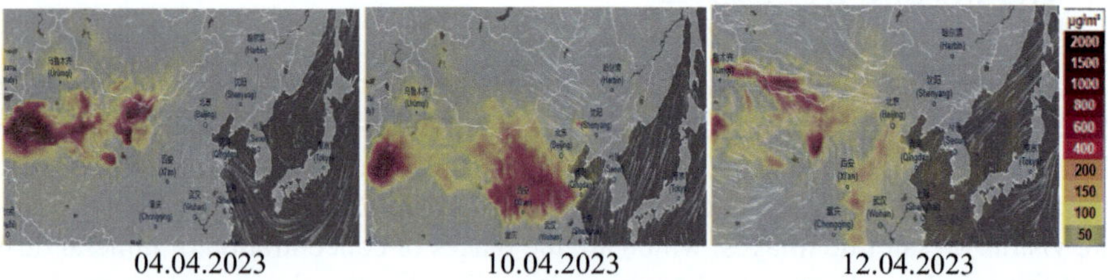

Fig. 2 Three-panel geographical map showing air pollution levels over parts of China on different dates: April 4, April 10, and April 12, 2023. Each map uses a color gradient from yellow to purple to indicate pollution concentration in micrograms per cubic meter ($\mu g/m^3$), with a scale ranging from 50 to 2000 $\mu g/m^3$. Key cities like Beijing, Shanghai, and Chongqing are labeled. The maps depict changes in pollution distribution over time, with notable concentrations in central and eastern regions (*Source* https://www.ventusky.com/)

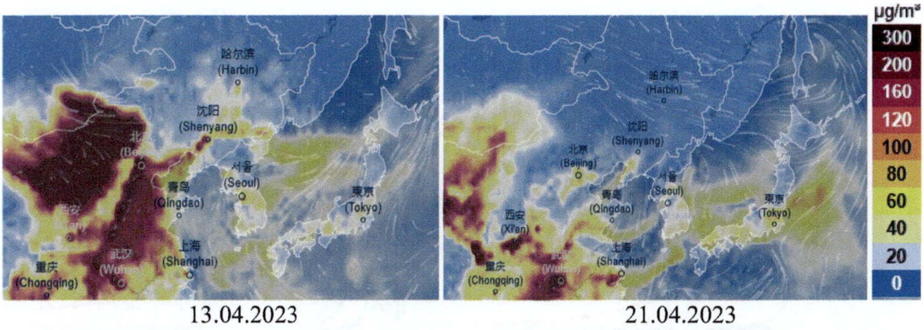

Fig. 3 Two geographical maps showing air pollution levels in East Asia on April 13 and April 21, 2023. The maps highlight regions including Beijing, Harbin, Shenyang, Seoul, Tokyo, Qingdao, Shanghai, Wuhan, Chongqing, and Xi'an. A color scale on the right indicates pollution levels in micrograms per cubic meter, ranging from blue (low) to dark red (high). The map on April 13 shows higher pollution levels, particularly around Beijing, compared to April 21 (*Source* https://www.ventusky.com/)

dust identified in the OLCI satellite images at an approximate concentration of 60–100 $\mu g \cdot m^{-3}$ (Fig. 3).

In Ref. (Li et al. 2024), this episode of dust transport over the Sea of Japan was confirmed by calculated back trajectories using the Hybrid Single-Particle Lagrangian Integrated Trajectory model (HYSPLIT) kinematic model and lidar data on stratification and atmospheric aerosol typing from the Cloud-Aerosol Lidar and Infrared Pathfinder Satellite Observations (CALIPSO), where according to forecast data, the transfer was carried out at an altitude of 2–5 km.

High concentrations of mineral dust in the atmosphere, based on visual analysis of OLCI satellite images for April 2023, were observed over the Sea of Japan on April 13 and 21. Based on this, the study areas were selected, namely, pixel subset in the northeastern Sea of Japan (subset boundaries: (N: (41.032–41.39); E: (138.033–138.403)) and in the central part of the Sea of Japan (subset boundaries: (N: (38.106–39.901); E (134.217–135.238))). The main criteria for constructing the subsets were: visual homogeneity of the Chl-a distribution and aerosol optical thickness (AOT), the presence of absorbing aerosol, and the absence of clouds for the selected subset. For comparative analysis, all cloud-free images for the study polygons for April 2023 were selected to study the temporal variability of the Chl-a distribution. In total, data arrays for the following dates were selected for the northeastern Sea of Japan: (01.04; 09.04; 13.04; 23.04 and 27.04); and for the central part of the Sea of Japan: (02.04; 10.04; 21.04; 23.04 and 27.04). In the SeaDAS program, the statistics of the distribution of Chl-a and Rrs(λ) by pixels for the selected subsets were calculated. In addition, the probability distribution functions (pdf) of Chl-a were constructed, which are shown in Fig. 4a for the northeastern part and in Fig. 4b for the central part of the Sea of Japan.

Based on Fig. 4, it can be seen that on April 13 and 21, maximum concentrations of Chl-a are recorded, reaching abnormally high values of more than 10 $mg \cdot m^{-3}$. Further

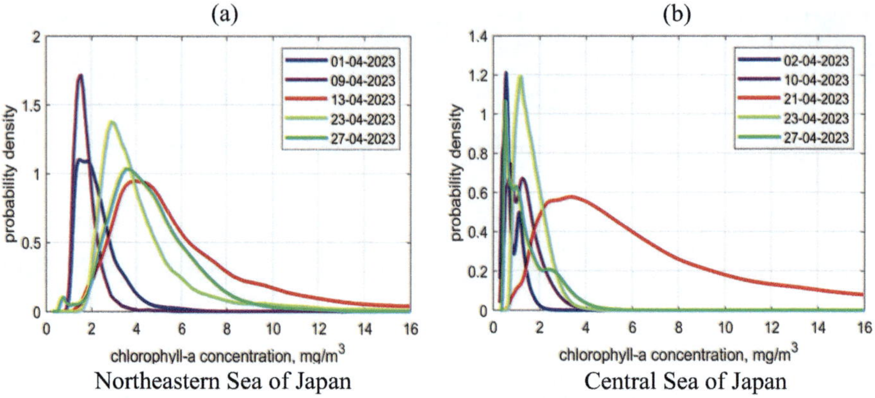

Fig. 4 Two X-Y charts comparing chlorophyll-a concentration probability density in the Sea of Japan. Chart (a) represents the Northeastern Sea of Japan, while chart (b) represents the Central Sea of Japan. Both charts display multiple lines for different dates in April 2023, with varying peaks and distributions. The x-axis shows chlorophyll-a concentration in mg/m^3, and the y-axis shows probability density. Legends indicate specific dates for each line

statistical processing of the data in SeaDAS showed that on April 13, the chlorophyll concentration averaged over the pixels of the selected area sharply increased to 5.23 mg·m^{-3}, although before that on April 10, 2023, it varied on average within 1.58 mg·m^{-3}. In this study, it is assumed that this is an obvious consequence of the atmospheric correction error, since even if mineral dust affects the increase in the Chl-a concentration, this effect cannot occur immediately. On April 21, the highest concentration of Chl-a content in the Sea of Japan is recorded in its central part with the highest standard deviation, even exceeding the average value itself, which may be an additional sign of errors. For the dates before the dust event (02.04 and 10.04) and after it (23.04 and 27.04), the average value of the Chl-a concentration is 4 times lower than for 21.04. For the northeastern part of the Sea of Japan, the effect of increased Chl-a values is preserved on the dates after the intensive removal of dust aerosol, namely, April 23 and 27, 2023, which can be explained by the deposition and entry of mineral matter that limited the development of phytoplankton cells into the upper layer of the Sea of Japan, which is typical for the spring period (Bukin et al. 2007).

In the work (Li and Wang 2023), the authors obtained a similar result using the Himawari-8/9 satellite by conducting generalized statistics of Chl-a in the study area (37 N–42 N and 145 E–165 E) in the period from April 10–18, 2023, and also obtained a peak increase in Chl-a for April 13–14, 2023. However, the analysis of the remote sensing reflectance values was not carried out and atmospheric correction errors were not noted. In this paper, the variability of the remote sensing reflectance of the Sea of Japan was analyzed. On April 13, 2023, negative Rrs(λ) values were identified in the northeastern part at 400 and 412 nm, which is an anomaly. In turn, on April 21, 2023, the Rrs(λ) values in the central part of the Sea of Japan were

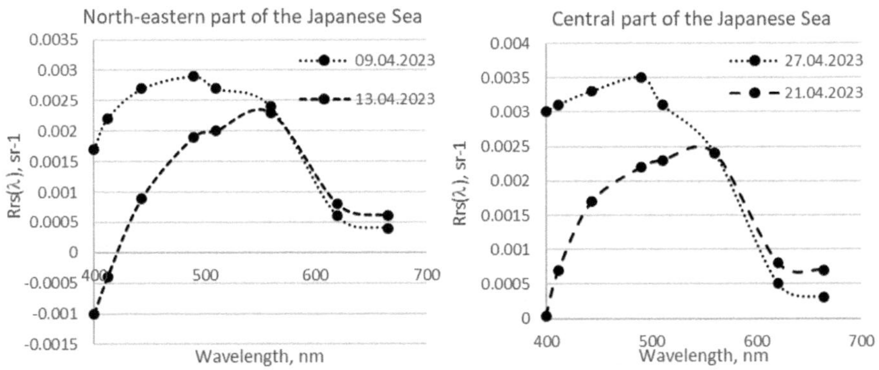

Fig. 5 Two X-Y charts comparing remote sensing reflectance (Rrs) in the Japanese Sea. The left chart shows data for the north-eastern part with two lines: a dotted line for April 9, 2023, and a dashed line for April 13, 2023. The right chart represents the central part with a dotted line for April 27, 2023, and a dashed line for April 21, 2023. Both charts plot Rrs(λ) in sr−1 against wavelength in nm, ranging from 400 to 700 nm

greatly underestimated without negative values. The spectral behavior of the Rrs(λ) values averaged over the subsets on a day with mineral dust and the nearest day with a clear atmosphere (the presence of background aerosol, the AOT value is less than the monthly average) for April 13 and 21 are shown in Fig. 5

It is known that within the framework of reprocessing (NASA Goddard Space Flight Center 2022) when analyzing OLCI information, the Chl-a values are calculated using the standard OCI algorithm (Hu et al. 2012), which uses Rrs(λ) measurements in the following spectral channels: 443, 490, 510 and 560 nm. From Fig. 5 it is evident that the spectral variation of the Rrs(λ) up to 560 nm on days with the presence of mineral dust in the atmosphere is greatly underestimated, and therefore so are the estimates of the values included in the calculation of Chl-a, namely Rrs(443), Rrs(490) and Rrs(510). With a decrease in the corresponding Rrs(λ) according to the OCI formulas, the Chl-a value will increase.

4 Conclusions

The present study examines the effect of mineral dust aerosol on the Chl-a calculation error based on OLCI satellite scanner data for the Sea of Japan in a specific case for April 2023. Statistical processing of satellite data in the SeaDAS program shows that during the days of the dust storm, the Chl-a concentration increased sharply up to 10 mg·m^{-3}, which is a clear consequence of the error, since even if mineral dust affects the increase in the Chl-a concentration, this effect cannot occur instantly. In this paper, it is assumed that this error is associated with underestimated and even negative values of the remote sensing reflectance up to 560 nm and, as a consequence, incorrect calculations using the OCI algorithm.

In the future, it is planned to conduct a more detailed analysis of this dust episode with the calculation of the spectral course of the atmospheric correction error, an expanded analysis of the atmospheric parameters and additional involvement of a source of in situ measurements.

Acknowledgements The work was carried out within the framework of the topics of the state assignment of MHI RAS FNNN-2024-0012 "Analysis, diagnosis and operational forecast of the state of hydrophysical and hydrochemical fields of marine waters based on mathematical modeling using data from remote and contact measurement methods ("Operational Oceanology") and the state assignment of POI FEB RAS No. 124022100080-0" Integrated use of Earth remote sensing methods, as well as ground-based and underwater measuring systems in studies of marine weather formations, hydrodynamic and biogeochemical processes in the upper ocean and seas. "

References

Aleksanin AI, Kachur VA (2016) Features of atmospheric correction of satellite ocean color data in the Far Eastern region. Earth Research from Space 6:56–67. https://doi.org/10.7868/S020596141605002X

Bukin OA, Pavlov AN, Salyuk PA, Kulchin Yu N, Shmirko KA, Stolyarchuk S Yu, Bubnovsky A Yu (2007) Features of the aerosol altitude distribution during the passage of dust storms over Peter the Great Bay in 2006 and their impact on phytoplankton communities of the Sea of Japan. Opt Atmos Ocean, 20 pp 341–349

Cui T, Zhang J, Tang J et al (2014) Assessment of satellite ocean color products of MERIS, MODIS and SeaWiFS along the East China Coast (in the Yellow Sea and East China Sea). ISPRS J Photogramm Remote Sens 87:137–151. https://doi.org/10.1016/j.isprsjprs.2013.10.013

Fukushima H, Toratani M (1997) Asian dust aerosol: Optical effect on satellite ocean color signal and a scheme of its correction. J Geophys Res: Atmos 102:17119–17130

Gordon HR, Wang W (1994) Influence of oceanic whitecaps on atmospheric correction of SeaWiFS. Appl Opt 33:7754–7763

Hu C, Lee Z, Franz B (2012) Chlorophyll-a algorithms for oligotrophic oceans: A novel approach based on three-band reflectance difference. J Geophys Res, 117 https://doi.org/10.1029/2011JC007395

Li Y, Wang W (2024) Long-Range transport of a dust event and impact on marine chlorophyll-a concentration in April 2023. Remote Sens. 16, https://doi.org/10.3390/rs16111883

Li J, Wong M, Shi G (2024) Multi-faceted analysis of dust storm from satellite imagery, ground station, and model simulations, a study in China. Atmos Res, 299, https://doi.org/10.1016/j.atmosres.2023.107195

Lipinskaya NA, Salyuk PA, Golik IA (2023) Variations and depth of formation of submesoscale eddy structures in satellite ocean color data in the southwestern region of the peter the great bay, Remote Sens., 15, 5600. https://doi.org/10.3390/rs15235600

NASA Goddard Space Flight Center. Ocean ecology laboratory, ocean biology processing group. ocean and land colour imager (OLCI) Ocean Color Data, 2022 Reprocessing; NASA OB.DAAC: Greenbelt, MD, USA, (2022)

Papkova AS, Kalinskaya DV, Shybanov EB (2024) The effect of dust aerosol on satellite data from different color scanners, physical oceanography. No 3:720–735

Park M, Lee S, Ahn J et al (2022) Decadal measurements of the first geostationary ocean color satellite (GOCI) compared with MODIS and VIIRS data, remote sens., 14, https://doi.org/10.3390/rs14010072.

Salyuk PA, Stepochkin IE, Shmirko KA, Golik IA (2024) Using satellite multi-angle polarization measurements to characterize atmospheric aerosol above Bohai Bay. Adv Space Res 73:514–522. https://doi.org/10.1016/j.asr.2023.10.007

Salyuk P, Bukin O, Alexanin A et al (2010) Optical properties of peter the great bay waters compared with satellite ocean colour data. Int J Remote Sens, 31, https://doi.org/10.1080/01431161.2010.485219

Shanmugam P, Ahn Y (2007) New atmospheric correction technique to retrieve the ocean colour from SeaWiFS imagery in complex coastal waters, Journal of Optics A: Pure and Applied Optics. No 9(5):511–530. https://doi.org/10.1088/1464-4258/9/5/016

Shybanov EB, Papkova AS (2022) (2022a) Algorithm for additional correction of remote sensing reflectance in the presence of absorbing aerosol: case study. Phys Oceanogr 29:688–706. https://doi.org/10.22449/1573-160X-2022-6-688-706

Sofiev M, Siljamo P, Valkama I, Ilvonen M, Kukkonen J (2006) A dispersion modelling system SILAM and its evaluation against ETEX data. Atmosph. Environ. 40:674–685. https://doi.org/10.1016/j.atmosenv.2005.09.069

Song H, Zhang K, Piao S, Wan S (2016) Spatial and temporal variations of spring dust emissions in northern China over the last 30 years. Atmos Environ 126:117–127. https://doi.org/10.1016/j.atmosenv.2015.11.052

Suetin VS, Korolev SN, Suslin VV, Kucheryavy AA (2004) Manifestation of features of the optical properties of atmospheric aerosol over the Black Sea in the interpretation of data from the SeaWiFS satellite instrument. Phys Oceanogr 1:69–79

Zvalinsky VI, Lobanova PV, Tishchenko P Ya et al (2022) Estimates of primary production in the northern part of the Sea of Japan in different seasons based on ship and satellite data, Oceanology, 22 pp 726–742, https://doi.org/10.31857/S0030157422050215

Multifractal Analysis of Seismic Energy Spatial Distributions in the Process of Preparation of the Pedro Bay Earthquake (Alaska, 2016, $M = 7.1$)

I. R. Stakhovsky

Abstract The paper proposes the method of structural analysis of the seismic flux in the focal area of a strong earthquake, applied to study the preparation process of the Pedro Bay earthquake (Alaska, 24.01.2016, $M = 7.1$). The method includes the use of the sliding time window and multifractal analysis of seismic data. 19 spatial distributions of seismic energy, characterizing the development of the Pedro Bay earthquake preparation process in time over 20 years, were studied. It is shown, that as the seismogenerating system approaches the moment of strong earthquake, the structure of the seismic flux experiences significant changes. The quantitative characteristics of the detected effects, as well as their physical interpretation are proposed.

Keywords Pedro bay earthquake · Seismic flux structure · Multifractal analysis · Earthquake preparation process

1 Introduction

Studies of the scale-invariant structure of spatial distributions of seismicity, intensively conducted since the 1980s (Kagan and Knopoff 1980; Hooge et al. 1994), have formed a conceptually new paradigm of seismogenesis, closely related to the perception of the seismogenerating media as a self-organized dynamical system whose properties are subject to various forms of scaling (Telesca and Lovallo 2009; Sornette and Werner 2009). Thus, the functional space of seismological studies has gone far beyond the applicability of the laws of classical (equilibrium) physics (Turcotte 1999; Bak et al. 2002; Stakhovsky 2016a). The very fact of the appearance of scale invariance in complex systems, including seismogenerating systems, is interpreted by the theory of nonequilibrium processes as a consequence of their transition to a

I. R. Stakhovsky (✉)
Schmidt Institute of Physics of the Earth, Russian Academy of Sciences, Moscow, Russia
e-mail: stakhov@ifz.ru

T. Chaplina (ed.), *Processes in GeoMedia—Volume IX*, Springer Geology,
https://doi.org/10.1007/978-981-96-9302-3_17

nonequilibrium state. The constructive role in the formation of dissipative structures in nonequilibrium systems belongs to the energy of fluctuations (Prigogine 1980).

In the epicentral areas of forthcoming large-scale earthquakes the growth of fluctuation energy of small-scale seismicity induced by the forming source of a strong lithospheric event leads to changes in the structure of seismic energy spatial distributions, which can be detected by the methods of multifractal analysis of seismic data. In the following, multifractal measures that model the seismic energy spatial distributions will be referred to by the term «seismic energy fields» for brevity. Variations in the structure of seismic energy spatial distributions before strong earthquakes in their epicentral areas lead to the widening and the change in the shape of asymmetry of $f(a)$-spectra of seismic energy fields (Stakhovsky 2023a, 2016b). The physical cause of these structural variations is the transition of the seismogenerating system into a strongly nonequilibrium state before the main shock of earthquake. The effect of seismic energy fields $f(a)$-spectra widening allows us to detect, to study and to describe quantitatively the evolution of strong nonequilibrium regions in the lithosphere, i.e., in purely seismological application—regions of preparation of large-scale seismic events. Therefore, the study of variations of seismic energy fields before strong earthquakes should be considered as an urgent task of seismophysics, oriented to the physical interpretation of basic phenomenology of the seismic process.

This paper presents the results of multifractal analysis of seismic energy fields before the Pedro Bay earthquake (Pedro Bay, Alaska, USA, 24.01.2016, M = 7.1) in its epicentral area. This earthquake occurred near the southern coast of Alaska at the latitude of St. Petersburg, but its epicenter turned out to be placed in a remote and almost uninhabited terrain with subarctic climate. The earthquake was named after the census-designated place with 40 inhabitants in 55 km to the west from epicenter. Using local toponyms this event is also called the Old Iliamna earthquake after the name of the abandoned Eskimo camp on the river of Iliamna or the Iniskin earthquake after the name of the peninsula near which the epicenter was located. The Pedro Bay earthquake had happened in winter, when a layer of snow and ice covered the neighborhood of epicenter. As it is extremely difficult to carry out geomorphologic and seismotectonic studies under such conditions, seismic data are actually the only source of information about the process of preparation of this event.

This study used data from the seismic catalog of the United States Geological Survey (USGS) (Geological Survey (USGS). The flux of seismic events accompanying the process of preparation of the Pedro Bay earthquake in its epicentral area was analyzed by the methods of theory of multifractals with the help of a sliding time window. Within the scope of the study, 19 $f(a)$-spectra of seismic energy fields, featuring in chronological order the development phases of preparation process of this earthquake, were obtained. It is shown that as the preparation process of the Pedro Bay earthquake approached the moment of the main shock, $f(a)$-spectra of seismic energy fields demonstrated significant changes in their width and asymmetry. A physical interpretation of these effects is presented. The results of the study allow us to consider multifractal analysis of seismic energy fields as a method of seismic monitoring, as well as a method of quantitative characterization of the process of preparation of a strong earthquake.

2 Initial Data

The Pedro Bay earthquake had happened on January 24, 2016 at 10:30:29 UTC at the southern coast of Alaska Peninsula about 260 km to the southwest from Anchorage in one of the bays on the western shore of Cook inlet, which is a part of the Pacific Ocean Gulf of Alaska. Its magnitude was $M = 7.1$, the epicenter was located near the coast at the coordinates $59.62°N$, $153.34°W$ (Fig. 1). The hypocenter was situated beneath the sea floor at the depth of $H = 125.6$ km. Since there is no any urbanized environment in the vicinity of the Pedro Bay earthquake epicenter, this event did not cause significant destructions of engineering infrastructure. The highway connecting Anchorage with the settlement of Kenai was damaged on the eastern side of Cook inlet. In Anchorage, power outages, water main ruptures, as well as cracks in the structures of the city bridge were reported. Gas leaks led to local flare-ups. Fires were also observed in the settlement of Kenai, but without any human casualties.

The southern coast of Alaska is located above the zone of subduction of the Pacific plate under the North American plate and is among the most seismically active regions of the Earth (Daly et al. 2021). About 200 km south of the coast, the Aleutian Trench, 3400 km long and up to 8 km deep, extends over the ocean floor, along which the

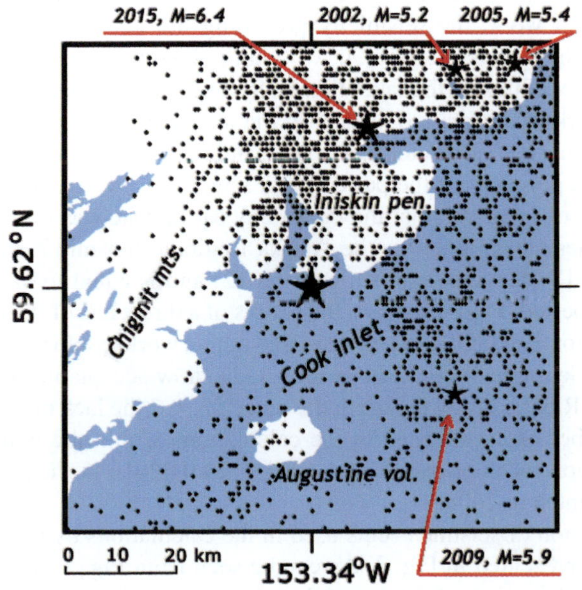

Fig. 1 Map of the investigated polygon of 90 × 90 km in the vicinity of the epicenter of the Pedro Bay earthquake (large star in the center, M = 7.1). The water area of Cook inlet is shown by blue color, the coast of Alaska in the north-west of the polygon is shown by white color. The locations of epicenters of earthquakes with M $\geq$ 2.0 for the period 1996 ÷ 2016 are marked with dots. The locations of epicenters of moderate earthquakes (5.2 $\leq$ M $\leq$ 6.4) that occurred during the same period are indicated by small stars and red arrows. Abbreviations used on the map: pen. → peninsula, vol. → volcano, mts → mountains

Pacific plate dips into the mantle in the northwestern direction at a rate of ~ 60 mm per year, resulting in frequent and strong earthquakes. In particular, at the southern coast of Alaska in 120 km to the east from Anchorage and 370 km to the northeast from the epicenter of the Pedro Bay earthquake there is a point of the epicenter of the Great Alaska earthquake on March 27, 1964, one of the strongest earthquakes in the history of instrumental observations of seismicity, which magnitude is estimated as $M = 9.3$.

The Alaska Regional Seismic Network has been deployed in Alaska after the Great Alaska Earthquake of 1964. Today it consists of almost 500 digital automated seismic stations (D'Alessando and Ruppert 2012), many of which, due to the climatic conditions of Alaska, are accessible for technical support only by helicopters (the seismometric equipment may be damaged by snow drifts and polar bears). The highest density of seismic stations is on the southern coast of the Alaska Peninsula. Seismic information is received by the Alaska Earthquake Center of the Geophysical Institute of the University of Alaska in Fairbanks, which cooperates with seismological institutions in Canada and Russia. The Alaska Earthquake Center, in particular, performs seismic monitoring of the Trans-Alaska Oil Pipeline and dams of hydraulic structures. Seismic data are collected from seismic stations in digital format via satellite communications, dedicated telephone lines, the Internet, etc., then systematized and transmitted to the U.S. Geological Survey.

In the present work, the seismicity preceding the Pedro Bay earthquake was studied within the boundaries of a 90×90 km polygon with the center coinciding with the epicenter of this earthquake (Fig. 1). Several seismic stations of the Regional Seismic Network of Alaska are located in the territory of the polygon. In calculations we used the USGS catalog data on seismic events within the boundaries of the investigated polygon for the period from January 1, 1996 to January 24, 2016, i.e., the process of preparation of the Pedro Bay earthquake was traced for 20 years. The completeness magnitude for the Alaska Regional Seismic Network data was investigated in D'Alessando and Ruppert (2012) and is equal to $M \approx 2.0$ for the specified time period. Figure 2 shows the results of a direct test of the representativeness threshold for the USGS catalog data used in this work, confirming this estimate. In the magnitude range $M \geq 2$, the seismic data show acceptable compliance with the Gutenberg-Richter law, whereas in the range $M < 2$ the lack of data is observed. The total number of events used in the calculations within the boundaries of the specified polygon for the period 01.01.1996 $\div$ 24.01.2016 in the magnitude range $M \geq 2.0$ amounted to 7608.

The distribution of seismic events used in the calculations by the depths of their hypocenters is presented in Fig. 3. As can be seen from Fig. 3, 84% of the events have depths in the range of 80 $\div$ 140 km, i.e., the process of preparation of the Pedro Bay earthquake took place mainly below the Earth crust at medium focal depths. According to the modern seismotectonic reconstruction of the region (Doser and Rodriguez 2011), the mechanism and depth of the hypocenter of the Pedro Bay earthquake correspond to the location of the source of this event in the massif of the Pacific plate, bending and sinking in the mantle. The shift of the fault surfaces in the source occurred almost vertically. The fault plane was ~ 50 km along strike and

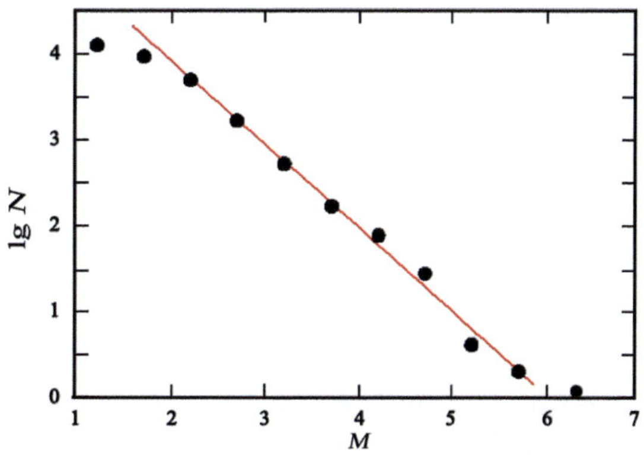

Fig. 2 The recurrence plot for the seismic data used in the calculations. The points are calculated in steps of 0.5 magnitude units; the straight red line is a linear regression graph constructed by these points. In the range M < 2 the USGS catalog data become unrepresentative

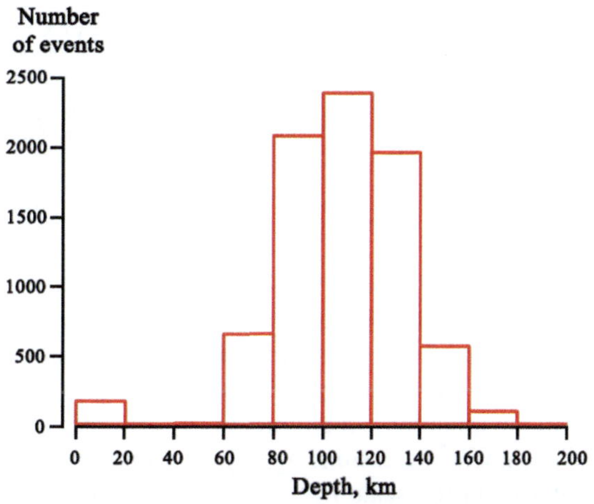

Fig. 3 Histogram of distribution of seismic events used in calculations by depths of their hypocenters

$\sim$ 20 km tall. It can be reasonably assumed that the dominant cause of seismic activity preceding the Pedro Bay earthquake was a significant deformation of the Pacific plate, supporting a strongly non-equilibrium state of its material in the process of subduction.

3 Counting Technique

The mathematical technique of constructing $f(a)$-spectra of seismic energy fields used below was described earlier in Stakhovsky 2016b. To convert magnitudes into energies of seismic waves, the formula of Bath (Bath 1973) was used:

$$\lg E = 5.24 + 1.44M\,, \tag{1}$$

where E is a seismic energy (J), M is a magnitude. For the purposes of the scaling analysis, the investigated polygon was covered by a two-dimensional renormalizable scale grid of non-intersecting square boxes numbered by index $i (i = 1,2,3\ldots)$. Let us introduce a measure P modeling the spatial distribution of seismic energy. The content of the measure P in the i-th box of the grid p_i was estimated by means of normalization:

$$p_i = E_i/E_0, \tag{2}$$

where E_i is the value of seismic energy in the i-th box, E_0 is the total value of seismic energy of all events in the catalog data sample under study.

The global self-similarity of seismic energy fields was investigated using the algorithm for direct calculation of the $f(a)$-spectrum (Chhabra et al. 1989):

$$\tilde{p}_i(q, r) = p_i^q(r)/\sum_j p_j^q(r), \tag{3}$$

$$a(q) = \lim_{r\to 0}(\ln r)^{-1}\sum_{i=1}^{N}\tilde{p}_i(q, r)\ln p_i(r), \tag{4}$$

$$f(a(q)) = \lim_{r\to 0}(\ln r)^{-1}\sum_{i=1}^{N}\tilde{p}_i(q, r)\ln\tilde{p}_i(q, r), \tag{5}$$

where q is the moment order, r is the size of the grid box (i.e., scale), $a(q)$ is the singularity index defining the local self-similarity of the measure, $f(a(q))$ is the spectrum of singularities defining the global self-similarity of the measure, N is the total quantity of non-blank grid boxes over which the summation is performed. In practical computations, the limits in (4)-(5) were estimated numerically as coefficients of the linear regressions $\sum_{i=1}^{N}\tilde{p}_i(q, r)\ln p_i(r)$ and $\sum_{i=1}^{N}\tilde{p}_i(q, r)\ln\tilde{p}_i(q, r)$ on $\ln r$. The changes in the scale r ($r = r_1, r_2, r_3\ldots$) were realized by the grid renormalization. The q values were iterated discretely with the step $\Delta q = 0.05$ in the range of values $-10 \le q \le 10$.

Let us also consider the quantitative characteristics of $f(a)$-spectra. In (Stakhovsky 2002), it is proposed to estimate the unfoldment of the $f(a)$-spectrum by the help of the integral S:

$$S = \int_{a_{min}}^{a_{max}} f(a)\mathrm{d}a, \tag{6}$$

where S denotes the region of coordinate space enveloped by the $f(a)$-spectrum, while a_{max} and a_{min} represent the maximum and minimum values of the singularity indices of the studied field, respectively. The unfoldment parameter S can always be numerically estimated. The width of the $f(a)$-spectrum is often approximated simply as the difference $\Delta a = a_{max} - a_{min}$ (with Δa sometimes loosely referred to as the «multifractality level»). However, the integral unfoldment parameter S is more informative, as it accounts not only for the width but also for the shape of the $f(a)$-spectrum.

The asymmetry of $f(a)$-spectra was evaluated by the help of the following relation:

$$A = \frac{2a_{ex} - a_{max} - a_{min}}{a_{max} - a_{min}}, \tag{7}$$

Here, $\mathbb{A}$ is the asymmetry parameter of the $f(a)$-spectrum, and a_{ex} is the singularity index corresponding to the extremum of the $f(a)$-spectrum. The asymmetry parameter $\mathbb{A}$, defined in this manner, can take values within the range $-1 < \mathbb{A} < 1$. Negative values of $\mathbb{A}$ correspond to the $f(a)$-spectrum extremum displacement toward lower singularity index values from the center of the range Δa, positive values of $\mathbb{A}$ correspond to the $f(a)$-spectrum extremum displacement toward higher singularity index values from the center of the range Δa. In the case of spectrum symmetry $\mathbb{A} = 0$.

The analysis of seismic data was conducted by the help of a sliding time window that encompassed a fixed quantity of seismic events, successively occurring within the boundaries of the investigated polygon, totaling 1200 events. This quantity of events ensured the obtaining of sufficiently complete singularity spectra of equal representativeness. For reasons of the seismic process heterogeneity, the time intervals covered by the window were not constant, ranging from 2.25 to 4.5 years. The window was shifted in increments of 1 year. Each count (window position) corresponded to a distinct data sample, for which a full cycle of computations was carried out by the help of Eqs. (1–7).

The first day of the initial sample of data (the first count) corresponded to January 1, 1996, as earlier catalog data were sparse. After processing the first 1200 events that formed the initial sample, the window was shifted so that the first day of the second sample (the second count) aligned with January 1, 1997. The computations were then repeated for the next 1200 events from this day forward. Succeeding counts corresponded to January 1, 1998, January 1, 1999, and so on. An exception was made for the final sample, which began on November 1, 2013, instead of January 1, 2014, due to an insufficient volume of data for the period 01.01.2014÷24.01.2016 in the USGS catalog. The sets of events in the samples of neighboring counts partially intersected, but no data stayed unprocessed. Henceforth, the numeration of counts and their appropriate samples of data will be assigned in chronological order.

The described data processing algorithm, based on the sliding time window shifts, enabled the construction of 19 $f(a)$-spectra, characterizing the temporal variations of seismic energy fields during the preparation of the Pedro Bay earthquake. This approach provided a quantitative description of the earthquake preparation process to the degree that this information is included in the scale-invariant characteristics of the seismic energy spatial distributions. The physical interpretation of the parameters S and $\mathbb{A}$ will be presented below.

4 Calculation Results

The calculation results can be seen from Figs. 4 and 5. The numbers next to the plots in Fig. 4 indicate the sequence number of a count (a sample). For brevity, we will also avail subscripts in the notations S_k, $\mathbb{A}_k$, $f(a)_k$ to associate them with the relevant counts and samples, where k denotes the number of a count.

Figure 4a illustrates the general trend of variations of seismic scaling during the preparation process of the Pedro Bay earthquake. The spectrum $f(a)_2$ (black curve, constructed from the second sample) exhibits the lowest unfoldment value, $S_2 = 5.82\ldots$ among all the constructed spectra. The spectrum $f(a)_3$(blue curve) has the lowest asymmetry parameter, $\mathbb{A}_3 = -0.20\ldots$. Meanwhile, $\mathbb{A}_2 = -0.13\ldots$ and $S_3 = 6.50\ldots$. The left branches of the spectra $f(a)_2$ and $f(a)_3$ practically coincide. At the beginning of the observation period ($17 \div 18$ years before the mainshock of the Pedro Bay earthquake), the $f(a)$-spectra of seismic energy fields can be characterized as "narrow" in comparison with the $f(a)$-spectra constructed by chronologically subsequent samples. The asymmetry of the spectra $f(a)_2$ and $f(a)_3$ is characterized by a displacement of their extrema toward lower singularity index values, which in accordance with definition (7) can be referred to as «negative» asymmetry.

The spectrum $f(a)_6$ (green curve) represents the seismic process 14 years before the mainshock of the Pedro Bay earthquake. It is nearly symmetric, with $\mathbb{A}_6 =$

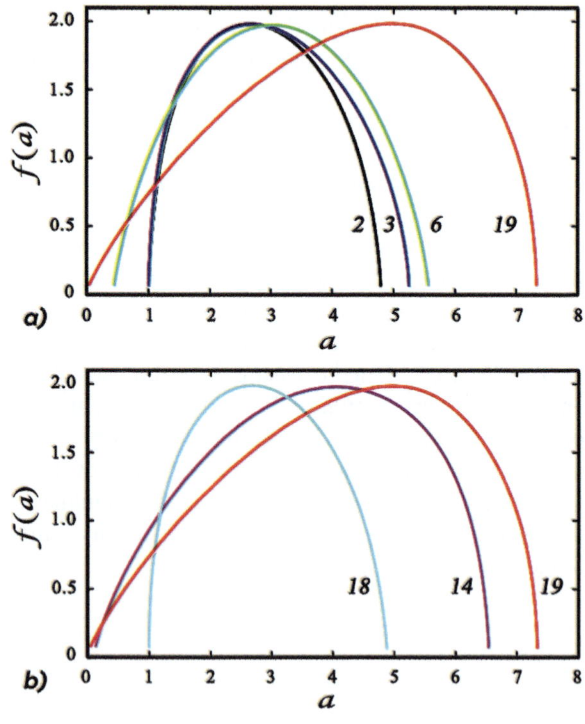

Fig. 4 Singularity spectra of seismic energy fields constructed from USGS catalog data for six counts. The numbers next to the plots indicate the numbers of the counts (samples). **a**—Singularity spectra constructed from the 2nd, 3rd, 6th, and 19th samples, illustrating the general structural changes in the seismic flux during the preparation of the Pedro Bay earthquake. b—Singularity spectra constructed from the 14th, 18th, and 19th samples, highlighting the structural changes in the seismic flux during the final stage of the Pedro Bay earthquake preparation process

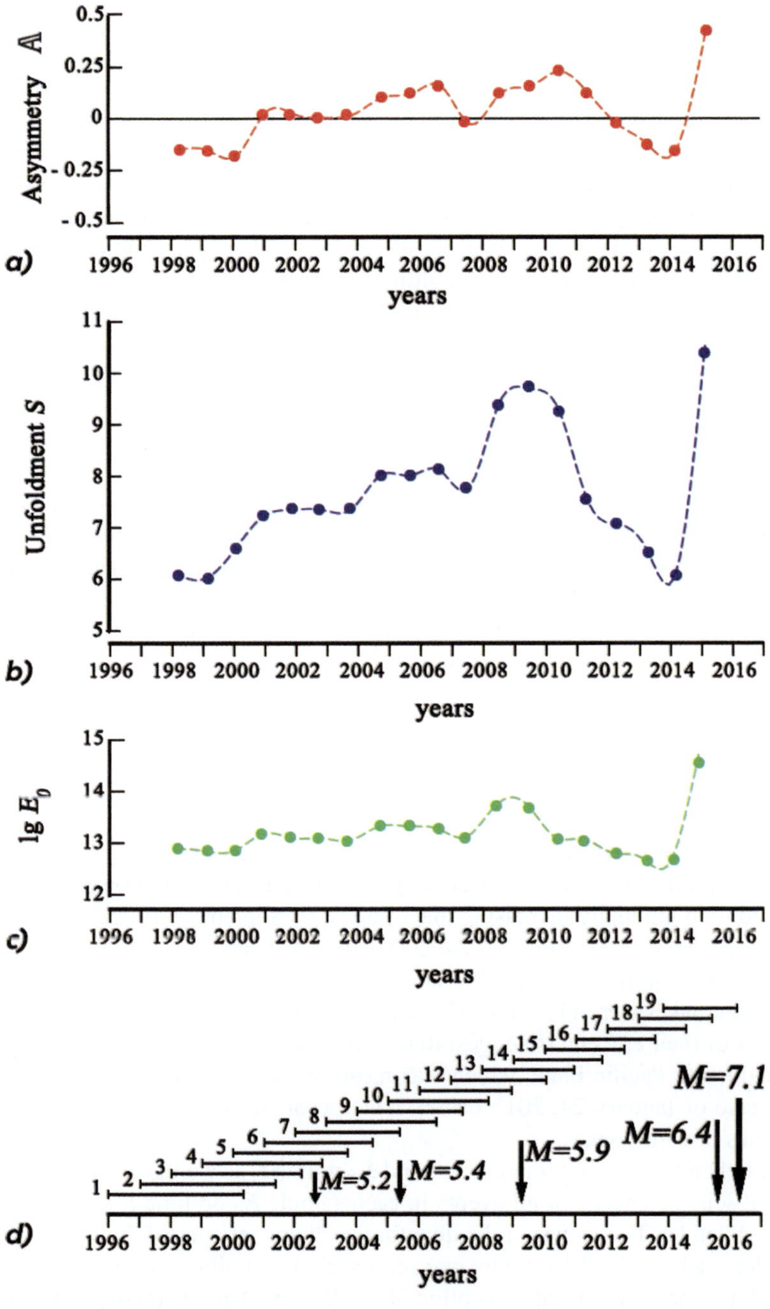

Fig. 5 **a**—The dependence of the asymmetry parameter of the $f(a)$-spectra on time, $\mathbb{A}(t)$, based on all counts for the period 01.01.1996 ÷ 24.01.2016. **b**—The dependence of the unfoldment parameter of the $f(a)$-spectra on time, $S(t)$, based on all counts. **c**—The dependence of the total seismic energy of all events in each sample on time, $E_0(t)$. **d**—Time intervals covered by each sample. The dashed lines in Figs. **a**, **b**, and **c** represent graphs of functions interpolating the computed dependencies $\mathbb{A}(t)$, $S(t)$ and $E_0(t)$, provided only to «guide an eye»

0.01 ..., and its unfoldment is $S_6 = 7.40\ldots$. Over the course of four years of seismic process evolution, the asymmetry of the $f(a)$-spectrum of the seismic energy field changed significantly, while its unfoldment increased by approximately 30%.

The spectrum $f(a)_{19}$ (red curve) was constructed by the last sample of data and characterizes the seismic process during the last two years before the Pedro Bay earthquake (the Pedro Bay earthquake itself was not included in the sample). The spectrum $f(a)_{19}$ exhibits the highest unfoldment value among all computed spectra, $S_{19} = 10.19\ldots$, as well as the highest value of asymmetry parameter, $\mathbb{A}_{19} = 0.36\ldots$. Its extremum is significantly displaced toward higher singularity index values, which, in accordance with definition (7), can be referred to as "positive" asymmetry. Over the analyzed period of seismic process evolution in the epicentral area of the Pedro Bay earthquake, the unfoldment parameter S of the $f(a)$-spectra of seismic energy fields increased by 75%, while the asymmetry parameter $\mathbb{A}$ changed its sign, with its variation amounting to 28% of the possible range $-1 < \mathbb{A} < 1$ as the formula (7) defines.

As we can see, the increase in fluctuation energy of small-scale seismicity induced by the forming source of a large-scale earthquake can be considered as a direct provider of information about the earthquake preparation process. This information manifests in the form of the effect of seismic energy fields $f(a)$-spectra widening. Figure 4a clearly illustrates this effect, i.e., the general trend of structural changes in the seismic energy spatial distributions before the Pedro Bay earthquake in its epicentral area. The preparation process of this earthquake was accompanied by both a significant widening and a transformation in the asymmetry of the $f(a)$-spectra of seismic energy fields in accordance with the chronology of counts. However, the widening and the shape variations of the $f(a)$-spectra exhibited a distinctly nonlinear and non-monotonic character.

During the period from 2002÷2015, against the background of weak diffused seismicity within the boundaries of studied polygon, four moderate-power earthquakes occurred with gradually increasing magnitudes: on September 10, 2002 ($M = 5.2$), May 19, 2005 ($M = 5.4$), January 24, 2009 ($M = 5.9$), and July 29, 2015 ($M = 6.4$). The epicenters of these earthquakes are marked in Fig. 1. The hypocenter depths of these earthquakes were 117.2 km, 95.5 km, 97.9 km, and 119.3 km, respectively. The locations of their epicenters suggest that their sources were situated in those part of the subducting Pacific Plate, which was most stressed by bending. The Pedro Bay earthquake of January 24, 2016 ($M = 7.1$) appears to be merely the final event in this series of earthquakes.

The dependencies of parameters $\mathbb{A}$ and S on time t, as well as the dependence of the total seismic energy of all events in each sample E_0 on time t, are presented in Fig. 5a, b, and c. The values of the quantitative characteristics of the seismic energy fields for each of the 19 samples are represented as points on the $\mathbb{A}(t)$, $S(t)$ and $E_0(t)$ plots. For better visual perception of the $\mathbb{A}(t)$, $S(t)$ and $E_0(t)$ dependencies, the graphs of interpolating functions constructed using cubic splines are shown as dashed lines in Fig. 5a, b, and c. The time intervals covered by each sample are displayed in Fig. 5d, where their numbering corresponds to the numbering of counts and samples. The calculated values of the parameters $\mathbb{A}$, S, and E_0 (points), obtained from seismic data for these time intervals, are aligned with the centers of the corresponding time intervals in Fig. 5a, b, and c. Arrows on the time axis in Fig. 5d indicate the dates of the strongest earthquakes over the 15 years preceding the Pedro Bay earthquake. The computed dependencies $\mathbb{A}(t)$, $S(t)$ and $E_0(t)$ are, strictly speaking, non-uniformly discrete. However, due to the significant number of the counts under the sliding time window shifting, the distribution of computed points in time allows for a relatively detailed track of changes in the parameters $\mathbb{A}$, S, and E_0 during the preparation process of the Pedro Bay earthquake.

The first three counts (1996 ÷ 2000) are characterized by «negative» asymmetry and low values of unfoldment of the $f(a)$-spectra (the spectra $f(a)_2$ and $f(a)_3$ are shown in Fig. 4a). This period can be considered as a period of a steady or background seismic regime. The counts 4 through 6 demonstrate a surge in seismic activity associated with the preparation process of the September 10, 2002 earthquake ($M = 5.2$). This surge led to an increase in the value of unfoldment parameter S and a change of the asymmetry form from «negative» to «positive» for the $f(a)$-spectra, derived from the corresponding samples. The counts 7 ÷ 10 are characterized by $f(a)$-spectra influenced by the preparation process of the May 19, 2005 earthquake ($M = 5.4$), which resulted in a further growth of value of the asymmetry parameter $\mathbb{A}$ in the positive range, along with another increase in value of the unfoldment parameter S. At the count 11, a slight decline in seismic activity is observed, accompanied by another sign reversal of the asymmetry parameter $\mathbb{A}$. However, the counts 12 ÷ 14 again exhibit a strong increase in values of both the unfoldment parameter S and the asymmetry parameter $\mathbb{A}$, associated with the preparation of the January 24, 2009 earthquake ($M = 5.9$).

During the counts 15 to 18, the opposite process occurs - a sharp decline in the total seismic energy E_0 of events corresponding to these counts and samples. The values of the unfoldment parameter S and the asymmetry parameter $\mathbb{A}$ nearly return to the levels observed during the steady seismic regime. The evolution of the seismic flux structure during this period (2010 ÷ 2014) is illustrated in Fig. 4b. As seen in Figs. 4a and 4b, the spectrum $f(a)_{18}$ (aquamarine-colored curve) is quite similar in shape to the spectrum $f(a)_2$. However, at the count 19, a sudden surge in the unfoldment parameter S occurs, along with a transition in the asymmetry of the $f(a)$-spectrum from «negative» to «positive». These changes are attributed to the preparation process of the July 29, 2015 earthquake ($M = 6.4$), during which the values of parameters S and $\mathbb{A}$ reach their maximums. The time interval between this earthquake and the Pedro Bay earthquake was six months, meaning the event on July 29, 2015, can be interpreted either as a foreshock of the Pedro Bay earthquake or as a part of a broader sequence of events culminating in the Pedro Bay earthquake. Herewith, unlike the highly variable parameters S and $\mathbb{A}$, the monofractal dimensions of seismic energy fields (corresponding to the extremum values of the $f(a)$-spectra) did not exhibit significant variations in this study (Figs. 4a, b).

As can be seen in Fig. 5, changes in the parameters S and A can occur quite quickly, so that the two-to-three-year data samples required to construct f (a)-spectra (under the existing representativeness of seismic statistics), inevitably smooth out the sought variations of these parameters. Nevertheless, the unfoldment parameters

S and asymmetry parameters A of the f (a)-spectra of seismic energy fields turned out to be sensitive not only to the process of preparing an earthquake with a magnitude of M = 7.1, but also to the processes of preparing weaker events. Almost before all the earthquakes of moderate power marked in Fig. 5d, the unfoldment parameter S sharply increased its value, and the asymmetry parameter A changed sign, i.e., tended to return to the negative values, but the process of preparation of the next earthquake of moderate power returned it to positive values. The study of spatial distributions of seismic energy (instead of spatial distributions of epicenters) allows us to identify a new property of the structure of seismic flux—the inversion of the sign of the asymmetry parameter A of seismic energy fields f (a)-spectra before strong earthquakes.

5 Physical Interpretation of the Study Results

The effect of seismic energy fields $f(a)$-spectra widening provides a general statistical description of the rather complex process of evolution of a seismogenerating system in a strongly nonequilibrium state. This effect can be defined as a reflection of changes in the structure of the seismic flux caused by the physical processes of earthquake preparation.

The reason for the appearance of the effect of seismic energy fields $f(a)$-spectra widening is the transition of the seismogenerating system to a strongly nonequilibrium state before the main shock of the earthquake, accompanied by an increase in fluctuations of seismic activity. It is rather complicated to detach these fluctuations as an independent object of study against the background of high intermittency of the seismic process, however, in the structure of multifractal measures that model the spatial distributions of seismic energy, the growth of fluctuations of seismic activity manifests itself cumulatively, in the form of an increase in the integral unfoldment of the $f(a)$-spectrum of the seismic energy field in contrast to preceding fields.

The asymmetry of the $f(a)$-spectrum is explained by statistical distribution of relatively weak and relatively strong events in the data sample. The right branch of the $f(a)$-spectrum is defined by the statistics of singularities that focus on themselves the minimal content of the measure P. In a steady seismic regime, the portion of events with relatively low magnitudes in the total flux is predominant, which is expressed in the shape of the $f(a)$-spectrum as an «elongation» of the right branch compared to the left one. The left branch of the $f(a)$-spectrum is defined by the statistics of singularities that focus on themselves the maximal content of the measure P. During the transition of the seismogenerating system to a strongly nonequilibrium state, the portion of events with relatively high magnitude values in the total flux continuously increases, which is expressed in the shape of the $f(a)$-spectrum as an «elongation» of the left branch compared to the right one. The increase in the amplitude of fluctuations, predicted by the theory (Prigogine 1980), makes these changes of unfoldment and asymmetry of the seismic energy field $f(a)$-spectra a natural phenomenon that may be expected before any lithospheric earthquakes.

The sign inversion of the asymmetry parameter $\mathbb{A}$ with the succeeding steady increase in the value of $\mathbb{A}$ in the range of positive values can be assumed as the start-up of the final phase of the process of preparing a strong earthquake. If strong earthquakes form a time sequence, then, as the results of the study show, an inversion of the sign of the asymmetry parameter $\mathbb{A}$ may be observed before each such earthquake, i.e., in the process of earthquake preparation a change of the $f(a)$-spectra asymmetry from «negative» to «positive» may mean that the moment of the main shock is approaching. Note also that under the conditions of scale invariance of the seismic process, the concepts of «weak» and «strong» earthquakes should be interpreted not in terms of magnitudes, but in terms of scale symmetry, which determines the scale of events by the scale of the seismogenerating system under consideration.

Change of the state of a seismogenerating system from weakly to strongly nonequilibrium is physically associated with the build-up of discontinuities in the lithosphere stuff due to energy fluctuations (Turcotte et al. 2014; Johansen and Sornette 2000). An aggregation of discontinuities at any scale levels gives rise to the development of fractal dissipative structures, which is shown by computer modeling methods (Stakhovsky 1995; Borodich 1997), but primarily by direct experimental studies of the process of rock fracture (Johansen and Sornette 2000; Velde et al. 1991; Sen et al. 2002; Jouini et al. 2011). Under the conditions of lithosphere, the aggregation of discontinuities leads to the generation of seismic waves. The effect of seismic energy fields $f(a)$-spectra widening is explained by the impact of these processes on the generation of seismic events, or, more precisely, this effect tracks these physical processes in the lithosphere regions undergoing a transition to a strongly nonequilibrium state.

It is well known that the theory of multifractal measures has principal analogies with nonequilibrium thermodynamics (Mandelbrot 1989; Multifractals 2001). These analogies are not just formal, but they reflect the general orientation of both theories toward the statistical description of complex (and in physical implementation—multicomponent) nonequilibrium systems. As a consequence, the description of the spatial distributions of seismic energy in the terms of multifractal fields corresponds to the physical essence of the earthquake preparation process not only at the conceptual, but also at the structural levels. The effect of seismic energy fields $f(a)$-spectra widening turns out to be a practical method that allows monitoring the processes of strong earthquake preparation. Accordingly, the numerical characteristics of this effect S and $\mathbb{A}$ turn out to be one of the possible means of quantitative description of these processes.

6 Discussion of Results and Conclusions

The conducted research shows that the results of multifractal analysis of seismic energy fields are closely related to the physical nature of the processes of lithospheric earthquake preparation and make it possible to carry out their quantitative description. The transition of the seismogenerating system to a strongly nonequilibrium state in the epicentral area of forthcoming strong earthquakes is manifested in temporal variations of the topological characteristics of the spatial distributions of seismic energy. Structural changes in small-scale seismicity before a large-scale earthquake lead to an increase in the unfoldment and transformations of the asymmetry forms of $f(a)$-spectra of seismic energy fields.

At the same time, the monofractal dimensions of seismic energy fields were not subject to considerable changes in this study. Apparently, this is explained by the fact that seismic events did not so much perform random walks across the field, but occurred repeatedly at the same points (points of energy dissipation associated, for example, with the fault network of the lithosphere), plotted on the map in Fig. 1. However, it is also necessary to take into account that the series of observations used in this study is relatively short, and the USGS catalog data used in the calculations is limited by the representativeness threshold (i.e., in fact, they contain an artificial scaling break). Under these conditions, detecting differences between structurally close measures (seismic energy fields based on the data of neighboring counts) by the help of such a «rough» method as calculating their monofractal dimensions becomes problematic.

Physically justified nature of the behavior of the seismic energy field monofractal scaling in the process of a strong earthquake preparation has not been studied enough till now. In existing structural models of the seismic process, the multifractal fields with both localized singularities (Stakhovsky 2019, 2023b) and delocalized ones (Lovejoy and Schertzer 2007; Sotolongo-Costa et al. 2000) are investigated, which can lead to either constant values of monofractal dimensions or variable ones. Physical interpretation of temporal variations of the spatial seismic distribution monofractal scaling requires a significant increase in the level of representativeness of seismic data and an increase in the duration of observation series.

The effect of the seismic energy fields $f(a)$-spectra widening is defined by the demeanor of the higher moments of multifractal measures that model the structure of the real spatial distributions of seismic energy. Changes in the state of the system, reflected in changes in the organization of the seismic flux, are manifested in the natural increase in fluctuations of seismic activity. The ability to describe the effect of seismic energy fields $f(a)$-spectra widening by scalar characteristics (for example, $\mathbb{A}$ and S) allows us to speak of the seismic scaling as a macroparameter of a seismogenerating system, which makes it possible, to a certain extent, to mensurate the process of the system approaching the moment of global stability loss.

During the seismic process the total divergence between the values of seismic energies generated by the points of the field with the highest and lowest activity (according to the form of $f(a)$-spectrum) defines the value of unfoldment parameter S, while the distribution of seismic energy values throughout the whole field defines the value of asymmetry parameter $\mathbb{A}$. The asymmetry of the $f(a)$-spectrum of the seismic energy field is associated with the absence of strong energy fluctuations in the steady (background) seismic regime and the nature of their distribution under approaching the main shock of earthquake. Energy fluctuations lead to the variations in the value or the inversion of the sign of parameter $\mathbb{A}$, which can deliver an auxiliary information about the last phase of the process of a strong earthquake preparation.

Note that earlier, when analyzing the spatial distributions of earthquake epicenters, the asymmetry of $f(a)$-spectra could not be studied, because the lack of data usually leads to the impossibility of constructing complete right branches of $f(a)$-spectra (Stakhovsky 2002), which justifies the feasibility of studying the spatial distributions of seismic energy.

Thus, the effect of seismic energy field $f(a)$-spectrum widening, which is closely related to the physics of the seismic process, can be used both for identification of the very fact of existence of the processes of strong lithospheric earthquake preparation in the seismogenerating medium, and for obtaining the quantitative characteristics of these processes. Variations in the scaling structure of seismic energy fields can be considered as a tool for analyzing the fine structure of the seismic process, which is not yet available for analysis by other methods. Let us note also that the $f(a)$-spectra of multifractal fields are scale-invariant functions that do not contain direct information about the scale-dependent parameters of the seismogenerating system, for example, about the magnitude of the forthcoming earthquake. However, structural analysis of seismicity can be used to identify areas of strong nonequilibrium in the lithosphere, including the identification of the geometric size of such areas. Due to the increase in the amplitude of fluctuations under conditions of strong nonequilibrium, temporal variations of dissipative seismic structures distinguish the area of strong nonequilibrium from the surrounding weakly nonequilibrium medium that allows for a natural transition to estimates of the scale-dependent characteristics of specific events in a scale-invariant seismic process.

The occurrence of the effect of seismic energy field $f(a)$-spectrum widening before strong earthquakes is explained by the generic evolution laws of nonequilibrium systems, first of all, by the growth of energy fluctuations in strongly nonequilibrium systems before fracture, regardless of the physical nature of the systems. Thus, the effect of seismic energy field $f(a)$-spectrum widening has no direct connection with the geology or tectonics of the seismogenerating system, i.e., has a universal character. At the same time, scaling analysis of seismic energy fields is carried out on the basis of strictly formalized mathematical procedures and does not imply subjective expert assessments of the data, as well as their simplifications or idealizations. This allows us reasonably to suppose that the proposed algorithm for studying the spatial distributions of seismic energy can be efficient in systems for monitoring the processes of preparation of strong earthquakes in the current time, and not only when analyzing seismic catalogs.

References

Bak P, Christensen K, Danon L, Scanlon T (2002) Unified scaling law for earthquakes. Phys Rev Lett. 88 pp 178501–178504. https://doi.org/10.1103/PhysRevLett.88.178501

Bath M (1973) Introduction to seismology. Basel and Stuttgart. Birkhauser Verlag p 350

Borodich FM (1997) Some fractal models of fracture. J Mech Phys Solids. 45(2). pp 239–259. https://doi.org/10.1016/S0022-5096(96)00080-4

Chhabra AB, Meneveau C, Jensen RV, Sreenivasan KR (1989) Direct determination of the f(a)-Singularity spectrum and its application to fully developed turbulence. Phys Rev A 40(9):5284–5294

D'Alessando A, Ruppert NA (2012) Evaluation of location performance and magnitude of completeness of the alaska regional seismic network by the SNES method. Bull Seismol Soc Am 102(5):2098–2115. https://doi.org/10.1785/0120110199

Daly KA, Abers GA, Mann ME, Roecker S, Christensen DH (2021) Subduction of an oceanic plateau across southcentral Alaska: High-resolution seismicity. J Geophys Res: Solid Earth. 126. e2021JB022809. https://doi.org/10.1029/2021JB022809

Doser DI, Rodriguez H (2011) A seismotectonic study of the Southeastern Alaska Region. Tectonophys. 497. 1–4. pp 105–113. https://doi.org/10.1016/j.tecto.2010.10.019

Geological Survey (USGS), Department of the Interior/USGS https://earthquake.usgs.gov/earthquakes/search/

Harte D (2001) Multifractals: theory and applications. New York. Chapman and Hall/CRC. p 264. https://doi.org/10.1201/9781420036008

Hooge C, Lovejoy S, Pecknold S, Malouin F, Schertzer D (1994) Universal multifractals in seismicity. Fractals 2(3):445–449

Johansen A, Sornette D (2000) Critical ruptures. Eur Phys J B 18:163–181

Jouini MS, Vega S, Mokhtar EA (2011) Multiscale characterization of pore spaces using multifractals analysis of scanning electronic microscopy images of carbonates. Nonlinear Process Geophys 18(6):941–953. https://doi.org/10.5194/npg-18-941-2011

Kagan YY, Knopoff L (1980) Spatial distribution of earthquakes: the two-point correlation function. Geophys J R Astr Soc 62 P 303–320

Lovejoy S, Schertzer D (2007) Scaling and multifractal fields in the solid earth and topography. Nonlinear Processes in Geophysics. 14(4). pp 465–502. https://hal.archives-ouvertes.fr/hal-00331103

Mandelbrot B (1989) Multifractal measures, especially for geophysicist. Fractals in Geophysics. Basel: Springer, pp 5–42. https://doi.org/10.1007/978-3-0348- 6389–6_2

Prigogine I (1980) From being to becoming. W.H. Freeman and Co., San Francisco, p 200

Sen D, Mazumder S, Tarafdar S (2002) Pore morphology and pore surface roughening in rocks: a small-angle neutron scattering investigation. J Mater Sci. 37 pp 941–947. https://doi.org/10.1023/A:1014391629262

Sornette D, Werner MJ (2009) Statistical physics approaches to seismicity. In: Encyclopedia of complexity and systems science (ed. Meyers R.A.). New York. Springer. P 7872–7891. https://doi.org/10.48550/arXiv.0803.3756

Sotolongo-Costa O, Antoranz JC, Posadas A, Vidal F, Vazquez A (2000) Levy flights and earthquakes. Geophys Res Lett 27(13):1965–1968. https://doi.org/10.1029/2000GL011394

Stakhovsky IR (1995) Modeling of crack aggregation in nonequilibrium medium. Mathematical Modeling. T. 7. 6. C. 54–64 (in Russian)

Stakhovsky IR (2002) Widening of the *f(a)* Spectra of seismic fields in preparation Zones of strong earthquakes. Izv, Phys Solid Earth 38(2):156–160

Stakhovsky IR (2016a) Attractor reconstruction from the time series of information entropy of seismic kinetic process. Izvestiya, Physics of the Solid Earth. 52(5):740–753. https://doi.org/10.1134/S1069351316040091

Stakhovsky IR (2019) Structural model of the relation between seismic scalings and the generalized scaling law of seismicity. Lett J Exp Theor Phys (JETP Lett). 109(12):816–820. https://doi.org/10.1134/S0370274X19120105

Stakhovsky IR (2024) Variations of multifractal scaling of spatial distributions of seismic energy before the Preci earthquake (Central Italy, 2016, Mw = 6.6). Processes in GeoMedia—Volume VIII, Springer Geology. Ed. by T. Chaplina. pp 103–115. Springer Nature Singapore. https://doi.org/10.1007/978-981-97-6627-7_10

Stakhovsky IR (2023) The influence of strong nonequilibrium on multifractal scaling of two-dimensional distributions of seismic energy. In: Physical and Mathematical Modeling of Earth and Environment Processes—2022. Springer Proceedings of 8th International Scientific Conference. Ed. by Karev V. Cham (Switzerland). Springer. P. 431–442. https://doi.org/10.1007/978-3-031-25962-3_42

Stakhovsky IR (2023) Generalized scaling law of seismicity in relation to the seismic process in Coso Range Region (Eastern California). Processes in GeoMedia—V. VI. (Ed. by T. Chaplina). Cham (Switzerland). Springer Geology. pp 1–8. https://doi.org/10.1007/978-3-031-16575-7_1

Telesca L, Lovallo M (2009) Non-uniform scaling features in central Italy seismicity: A non-linear approach in investigating seismic patterns and detection of possible earthquake precursors. Geophys Res Lett 36(1):L01308. https://doi.org/10.1029/2008GL036247

Turcotte DL (1999) Seismicity and self-organized criticality. Phys Earth Planet Inter. 111. P 275–293. https://doi.org/10.1016/S0031-9201(98)00167-8

Turcotte DL, Schubert G (2014) Geodynamics (3-d ed.). Cambridge University Press. Cambridge. p 636. https://doi.org/10.1017/CBO9780511843877

Velde B, Dubois J, Moore D, Touchard G (1991) Fractal patterns of fractures in granites. Earth Planet Sci Lett 104(1):25–35. https://doi.org/10.1016/0012-821X(91)90234-9

Assessment of the Hydrosphere Conditions in the Hydrocarbon Production Area of Western Orenburg Region

I. V. Kudelina, T. V. Leontyeva, and M. V. Fatyunina

Abstract The territory of the west of the Orenburg region is the southeastern part of the Volga-Kama artesian basin and borders on the Caspian artesian basin. Due to the intensification of oil and gas production, the hydrosphere of the Southern Urals is undergoing major changes. Within the western Orenburg region, the territory of the Kurmanaevskoye and Tananykskoye oil fields is allocated, located in one of the oldest oil production areas (Mukhanovo-Erokhovsky oil and gas region) with a developed oil field and social infrastructure. Its rapid development is especially relevant for maintaining the optimal structure of the mineral resource base of the Orenburg region. The purpose of the study is to assess the hydrosphere condition in the hydrocarbon production area of the Western Orenburg region. As a result, it was concluded that there are periodic excesses of such indicators as: iron, manganese, ammonia, copper, oil products in the influence zone of the Kurmanaevskoye and Tananykskoye oil fields. The hydrochemical situation at these oil fields is generally satisfactory. Regular monitoring, well repair and implementation of new technologies, in particular, the use of complex barriers, are recommended.

Keywords Underground water · Mineralization · Oil products · Maximum permissible concentrations · Technogenesis

1 Introduction

The Kurmanaevskoye and Tananykskoye oil fields are located in the Kurmanaevskiy district of the Orenburg region, southwest of the city of Buzuluk.

The fields have been in operation for over 40 years and during their long-term operation, emergency situations have repeatedly occurred, which resulted in inevitable spills, leaks and filtration of the extracted liquid into the soil. The maximum national economic effect can be achieved with the complete extraction of hydrocarbon

I. V. Kudelina (✉) · T. V. Leontyeva · M. V. Fatyunina
Orenburg State University, Orenburg, Russia
e-mail: kudelina.inna@mail.ru

T. Chaplina (ed.), *Processes in GeoMedia—Volume IX*, Springer Geology,
https://doi.org/10.1007/978-981-96-9302-3_18

reserves from the subsoil and in compliance with the requirements of environmental legislation, which allows meeting the region's needs for mineral resources on the basis of reasonable sufficiency and environmental and social feasibility.

Stationary observations of the level and temperature of groundwater are carried out both to determine general patterns of changes in the state of groundwater in the territory of the oil field and to identify the first signs of man-made changes in the environment. Due to the high degree of level and piezoconductivity of the environment, groundwater reacts much earlier to man-made impacts by changing the quality of groundwater. When placing the network of observation wells, the following factors were taken into account: the location of potential pollution sources, their proximity to populated areas, gully networks, surface water courses, artificial reservoirs, lithology and permeability of rocks of the geological section. The location of observation wells is confined mainly to places of the greatest accumulation of oil producing, injection wells, in the area of the head structures of complex oil preparation, preliminary water discharge units, automated group metering units, etc.

2 Results and Discussions

In the area of the Kurmanaevskoye oil field, the aeration zone is subject to pollution, represented by an anhydrous permeable Quaternary alluvial-deluvial horizon (adQ), the first from the surface Upper Neopleistocene-Holocene alluvial horizon (aQIII-H) and the main aquifer Lower Triassic complex (T1), which is widespread and of primary economic and drinking importance, so observations are organized for these aquifers and complexes (Environmental impact assessment 2002). The results of the measurements were compared with the standard indicators of maximum permissible concentrations (MPC).

Below, the qualitative state of groundwater in each of these complexes is characterized separately.

Anhydrous permeable Quaternary alluvial-deluvial horizon (adQ).

According to the chemical composition, groundwater mixed with a mineralization of 1.7 mg / dm3. Of the microcomponents, a significant excess of the norm was observed for iron—1.95 mg/dm3 (with a MAC of 0.3 mg/dm3) and bromine—1.20 mg/dm3 (with a MAC of 0.2 mg/dm3), petroleum products were found in an insignificant amount—0.37 mg/dm3 (with a MAC of 0.1 mg/dm3).

Analyzing the data on the qualitative composition of groundwater in the observed horizon, it is evident that in all observation wells, dry residue exceeded 1.2 MPC to 3 MPC (from 1213 mg/dm3 to 3014 mg/dm3) with MPC of 1000 mg/dm3. The chemical composition of the water is hydrocarbonate-calcium, and in well 5k, where increased mineralization is noted (from 1299 to 1622 g/dm3), the groundwater is hydrocarbonate-chloride magnesium-calcium. With an increase in mineralization, the content of calcium and magnesium increases, and, accordingly, the hardness of the groundwater in the aquifer under consideration increases. Observation well 5k is located in a zone of increased technogenic load, downstream from the central pumping station and the freight yard, and apparently, the recorded pollution is associated with this. In observation well 5k, the chloride content was found to be in excess of 568.6 to 847.4 mg/dm3 (with a MAC of 350 mg/dm3), iron from 0.955 to 6.99 mg/dm3 (with a MAC of 0.3), and oil products were found in insignificant quantities of 0.11 mg/dm3.

An excess of manganese content was found in observation wells 1k, 2k, 4k and 5k from 0.22 to 1.46 mg/dm3 (with a MAC of 0.1).

High turbidity and color content up to 3.4 to 57.0 mg/l (at MAC 1.5), which may be either the result of poor pumping before sampling, which is impossible due to the small water column in shallow wells 3k, 5k, or is the result of oxidation of divalent iron, the content of which exceeds the MAC. The latter is possibly associated with the natural background (Gaev 2014; Kudelina 2018).

Exceeding the MAC for mineralization and water hardness by 1.5–2.3 times may be a consequence of anthropogenic impact. The increase in hardness is observed mainly due to the increase in the calcium content in groundwater, respectively, the removable hardness increases.

In accordance with the general requirements for the composition and properties of water in water bodies, the pH value should not go beyond the range of 6.5–8.5. The pH indicator in the water of the assessed wells varies from 7.2 to 8.14. The waters are classified as neutral.

The aquiferous Upper Neopleistocene-Holocene alluvial horizon (aQIII-H) is distributed mainly in the Buzuluk River valley within the low, high floodplain and in the Tarpanovka River valley (the right tributary of the Buzuluk River), as well as in the mouths of numerous gullies and ravines, in the gully-ravine network.

Monitoring wells 4k, 10k, 11k are equipped for this aquifer. The depth of the monitoring wells is 18.0 m. The water-bearing rocks are fine-grained, fractional sands, with the inclusion of gravel-pebble and sand-gravel deposits. The thickness of the water-bearing rocks is 10.0–15.0 m. The water is unconfined. The groundwater level is revealed at depths of 7.0–7.3 m.

In terms of chemical composition, groundwater is variegated; along with hydrocarbonate waters, there were hydrocarbonate-sulphate, hydrocarbonate-chloride and mixed waters. Sodium and calcium ions predominated among the cations. Groundwater mineralization was 0.3–0.8 g/dm3 and in isolated cases exceeded 1.0 g/dm3. In observation well 4k, located downstream from the booster pumping station (BPS), water mineralization was 1.2 g/dm3. In terms of water hardness, the water was mostly moderately hard, and in well 4k the water was very hard—14.8 mmol/dm3 (with a MAC of 7 mmol/dm3). Of the microcomponents, the excess of the norm was recorded for iron from 1.92 to 6.50 mg/dm3 (with the MAC of 0.3 mg/dm3), manganese from 0.43 to 1.11 mg/dm3 (with the MAC of 0.1 mg/dm3) and bromine from 0.21 to 1.0 mg/dm3 (with the MAC of 0.2 mg/dm3). Total hardness varies from 2.36 to 14.8 mmol/dm3, most often it was 3.0–4.0 mmol/dm3.

When assessing the available information on the qualitative composition of groundwater in the Upper Pleistocene-Holocene alluvial aquifer, it is clear that the groundwater in the aquifer is of mixed composition. Mineralization in observation well 4k varies within 0.949–1.382 g/dm3. The chemical composition of the water is hydrocarbonate-chloride magnesium-calcium. With increasing mineralization, an increase in the calcium and magnesium content is observed, and the water hardness increases accordingly. Observation well 4k reflects an obvious excess of the MAC for almost all observed components of groundwater. According to the value of total hardness, equal to 12.48–18.04 mg-eq/dm3, groundwater is characterized as very hard (with a MAC of 7 mg-eq/dm3). In the remaining observation wells, where water mineralization does not exceed 1.0 g/dm3, the water is moderately hard. Exceeding the MAC for mineralization and hardness of groundwater by 1.5–2.3 times in individual wells may be a consequence of anthropogenic impact. Observation well 4k is located in the zone of increased technogenic load, downstream from the booster pumping station, and this is probably associated with the pollution of the observed aquifer. For example, the content of oil products in almost all samples in observation well 4k is above the permissible MAC limits, from 0.15 to 0.36 mg/dm3. The content of iron and manganese in groundwater varies from 1.5 to 12.8 MAC, which may be the result of poor pumping before sampling or is a natural background.

Excess of turbidity and color is recorded in all wells of the regime network of observation wells of the Kurmanaevskoye oil field, which may be the result of poor pumping before sampling.

The Lower Triassic aquifer (T1) is widespread in the area of the described field. Depending on the depth of immersion, it is the second or third aquifer from the surface.

The aquifer complex was tested by observation wells of the regime network 2k, 7k, 9k. The depth of the wells is 85.0 m. An increased content of oil products was observed in the groundwater in wells 2k, 7k, 9k from 0.16 to 0.5 mg / dm3 with a MAC of 0.1 mg / dm3. In wells 7k, 9k, an increased iron content was found from 0.59 to 1.37 mg / dm3 with a MAC of 0.3 mg / dm3. Well 2k recorded an increased dry residue content of 2839 mg/dm3 with a MAC of 1000 mg/dm3. With increased mineralization, the chemical composition of groundwater is hydrocarbonate-chloride magnesium-calcium. With increased mineralization, the content of calcium and magnesium increases and, accordingly, the hardness of the water increases, which is 33.5 mg-eq/dm3 and the water is characterized as very hard. Exceeding the MAC for mineralization and water hardness by 1.5–2.3 times can be a consequence of anthropogenic impact, and it is in observation wells 2k and 9k that the impact of the development of neighboring oil fields is felt. These two observation wells were drilled by the project as background ones, to control the qualitative composition of groundwater from the possible impact of the development of the Bobrovskoye and Shulaevskoye oil fields.

When analyzing the chemical composition of water in wells 2k, 7k and 9k, it is evident that excesses are recorded in all observation wells, mainly in color and turbidity. Water in wells 4k and 5k in terms of dry residue content fluctuates from 1053 to 1544 mg/dm3, respectively, mineralization varies within the range of 1.0–1.54 g/dm3, in spring an increase in dry residue is observed in the wells. The chemical composition of the water is hydrocarbonate-chloride magnesium-calcium. With an increase in mineralization, the content of calcium and magnesium increases and, accordingly, the hardness of the water increases. According to the value of total hardness, equal to 10.4–15.52 mg-eq/dm3, the water is characterized as very hard. The excess of total hardness is 1.5–2.5 MAC. In the remaining wells, where the mineralization of the water does not exceed 1.0 g/dm3, the water is moderately hard. Exceeding the MAC for mineralization and water hardness by 1.5–2.3 times may be a consequence of anthropogenic impact. The described aquifer complex is the most promising for organizing both domestic and drinking water supply and technical water supply of the deposit. In erosional cuts, i.e. where the complex lies close to the daylight surface, it can be used for drinking water supply in its upper and middle parts. In the lower parts of the aquifer complex, where mineralization increases above 1.0 g/dm3, it is recommended for use for technical water supply.

In the process of the planned activity it is necessary to control the limit of use of water of domestic and technical quality and forecast the development of possible pollution and depletion of groundwater.

In the zone of possible influence of oil production of the Kurmanaevskoye oil field facilities there are the villages of Kurmanaevka, Petrovka, Skvortsovka, Kandaurovka of the Kurmanaevskiy district. Monitoring of groundwater is also carried out at wells 4486 and 5964 of domestic and drinking water intakes located - 1.8 km southeast of the village of Petrovka and north of the village of Kurmanaevka. At hydroposts on the Buzuluk River (2 posts) and on the Tarpanovka River (1 post).

In modern domestic and drinking water supply, groundwater of the Lower Triassic deposits is exploited by the Kurmanaevskoye domestic and drinking water intake, which is located within the mining allotment of the oil field. The water intake consists of one well located 1.8 km southeast of the village of Petrovka in the Kurmanaevsky district. The well depth is 120.0 m.

Analysis of the quality of drinking water in rural settlements of the Kurmanaevsky district revealed that the standards for such indicators as total hardness, iron, petroleum products, ammonia, and copper were exceeded.

Thus, in the village of Romashkino, the standards for general hardness were exceeded by 1.56 MAC, in the village of Kurmanaevka for ammonia nitrogen—2.1 MAC and oil products 5.1 MAC. Exceeding the standards of MAC of household and drinking water supply for general hardness is noted in the villages of Semenovka, Petrovka, Danilovka and Volzhsky from 1.1 to 1.3 MAC. Exceeding the standards for iron was noted in the villages of Petrovka - 6.1 MAC and Kandaurovka—1.4 MAC, for copper—in the village of Volzhsky—3.5 MAC, total mineralization—in the village of Danilovka—1.3 MAC. All components of the excesses can be influenced not only by the Kurmanaevskoye field, but also by other nearby oil fields. Pollution control of surface and ground waters included systematic sampling of water from observation wells, operating water intakes and watercourses along the regime network of the Kurmanaevskoye oil field. Generalization and analysis of analytical data, identification of stable trends (positive or negative) in changes in the state of the aquatic environment, which are recorded by the content of pollutant components and general indicators. During the local monitoring of the qualitative composition of groundwater, the content of iron and manganese in the groundwater of the Kurmanaevskoye oil field varies from 1.5 to 12.8 MAC, there is an excess, which may be the result of poorly performed pumping before sampling or is a natural background of this area of work.

In accordance with the general requirements for the composition and properties of water in water bodies, the pH value should not exceed the range of 6.5–8.5. The pH value in the water of the assessed wells varies from 7.2 to 8.14. The waters are neutral.

The content of oil products in almost all samples is within the MAC (0.1 mg / dm3). The exception is the observation well 2k, drilled to the Lower Triassic deposits, where the content of oil products was 0.5 mg / dm3, i.e. 5.0 MAC, there is also an excess of oil products, which is 0.48 mg / dm3 (with a MAC of 0.1). In wells 4k, 5k there is also an excess of 1.5 to 3.6 MAC. Wells 4k, 5k, 7k, 9k, 10k, 11k show turbidity and color exceeding 1.4–57.0 MAC, which may be the result of poor pumping before sampling or the possibility of pumping before sampling. In the territory of the Tananyk oil field, the observed groundwater of the Lower Triassic aquifer complex is chloride-sulfate, chloride, sulfate–chloride, hydrocarbonate-chloride, chloride-hydrocarbonate in chemical composition according to the results of local monitoring (Environmental Impact Assessment 2002; Kovalevsky 2001). Excessive chlorine ion content in the groundwater of the observed Lower Triassic aquifer complex is present everywhere according to the analysis results. The MAC of chlorine ions is 350 mg/dm3. The presented chemical analyses show that the highest chloride content was recorded in those wells with the highest dry residue content, i.e. in wells 24, 30, 32.

In well 24, the chloride content is from 25 to 45 MAC, in well 30, the chloride content is observed in the amount of 11–13 MAC, in well 32 from 0.0.3 to 5 MAC.

In observation wells 25, 29, 33, 34, 35, isolated excesses of the MAC of chlorides by 1.1–1.6 times are noted. In the provided analyses for the remaining observation wells, the content of chlorine ions is in the range from 0.02 to 0.4 MAC.

An increased content of dry residue was found in observation wells 24, 30. In well 24, 25.0 m deep, the maximum dry residue content is 24,600 mg/dm3 (24.6 MAC). Excess dry residue content is also observed in wells 30 and 32, 25.0 m deep. In well 30, the dry residue excess was 7 times the MAC, in well 32 it was 3 times the MAC.

Monitoring wells 31 and 33, 100,0 m deep, equipped for the lower part of the Lower Triassic aquifer, do not exceed 400 mg/dm3 in dry residue content. It can be assumed that the increased dry residue content is the result of possible contamination "from above". Well 24 is located in the zone of influence of the Tananyk complex oil treatment plant (COP), well 30 is located in the zone of influence of the production oil well, well 32 is in the zone of influence of the automated group metering unit and injection well 224. In the remaining wells 33, 34, 36, isolated excesses of the maximum permissible concentration of dry residue by 1.1 times were recorded.

The observed groundwater of the Lower Triassic complex varies in hardness. In the territory of the Tananyk oil field, there are very hard groundwaters, more than 9 mg-eq/dm3, in wells 24, 30, 32, 34. The value of total hardness, expressed in mg-eq/dm3, depends on the content of calcium and magnesium salts. The maximum permissible concentration of hardness is 7 mg-eq/dm3. The assessment of groundwater by the degree of total hardness is given according to the classification of O. A. Alekin (Goldberg 1984; Borevsky 2014). Increased hardness values were recorded in well

24 from 21 to 45 MPC. Hard waters from 6 to 9 mg-eq/dm3 are found in well 38. Throughout the entire observation period, the hardness value in well 38 varied in the range from 6.6 to 8.7 mg-eq/dm3. Moderately hard waters from 3 to 6 mg-eq/dm3 and soft waters from 1.5 to 3 mg-eq/dm3 are found in wells 25, 28 and 29, 31, 33, 35, 36, 37. In wells 29, 33, the hardness index varies from 1.3–1.6 to 8.3–9.36 mg-eq/dm3. No regularity in the change in the hardness index is observed. The forecast of the graphs-diagrams in the observation wells shows an increase in hardness in groundwater, only in wells 24, 33 is there a tendency to a decrease in the hardness value.

Groundwater of the Lower Triassic aquifer complex contains iron ions at almost all observed depths with varying concentrations. The MPC of iron for groundwater is 0.3 mg/dm3. It should be noted that in wells 25, 29, 31, 33, 35, 37, 38 with a depth of 100.0 m, the concentration of iron ions is higher than in shallow wells. The maximum permissible content of oil products in groundwater according to is 0.1 mg / dm3. Oil products in groundwater of the Lower Triassic aquifer complex are present in each observed interval. In wells 24, 30, 32, 34 with a depth of 25.0 m, the content of oil products is in the range from 0 to 0.42 mg / dm3, i.e. the maximum excess of MAC is 4.2 times. In wells 32, 34, oil products were found in every sample. In well 34 there are single excesses of MAC by 2.2 and 1.6 times. In observation wells with a depth of 100.0 m, oil products were found in almost every sample. The content of petroleum products varies from 0 to 0.8 mg/dm3.

Manganese is one of the priority standardized indicators of groundwater quality. The maximum permissible level of manganese content in groundwater is 0.1 mg/dm3. In groundwater of the Lower Triassic aquifer complex, sampled to a depth of 25.0 m, manganese is found everywhere. The manganese content in samples taken from wells 24,30,32,34 is found both below the MAC and above from 2 to 70 times.

Manganese is found in almost all water samples taken from observation wells 100.0 m deep. The concentration of manganese in the samples varies, the lowest content of manganese is in samples taken from wells 35, 37.

The color of groundwater refers to the standardized physical properties, the determination of which is required by regulatory documents. Colour is a secondary organoleptic indicator of groundwater quality and cannot be formed without certain geochemical processes. Formation of increased colour of groundwater is possible when it is contaminated with micro-quantities of specific organic compounds. In observation wells 25.0–50.0 m deep, groundwater often has increased colour. In wells 24.30, over the entire observation period, the colour of groundwater is recorded above the MAC from 1.25 to 16 times. In observation wells 50.0 m deep, the colour of groundwater varies within the MAC. In observation wells 100.0 m deep, there is an increased increase in the colour index of groundwater in the range from 50.0 to 100.0 m.

Increased color of groundwater is possible with increased concentration of ferrous iron. When determining the color of groundwater in the presence of iron hydroxide using optical instruments, there is an increase in light absorption due to suspended particles and, accordingly, the color is artificially inflated. This color is "apparent" and regulatory documents before measuring this indicator require ultramembrane filtration of a water sample through a filter with a pore diameter of 0.42 microns, which is recommended to be carried out in the presence of suspended and colloidal particles in the water (Gaev 2019; On the state of sanitary 2019). The turbidity indicator refers to the standardized physical properties of groundwater. Turbidity of groundwater is formed in the same way as color in cases where suspended, colloidal particles are present in them (On approval of the regional program 2022; Leontyeva 2019). Groundwater of the Lower Triassic aquifer complex at all observed levels has an increased turbidity indicator. In wells 25.0 m deep, the turbidity index exceeds the MAC by 3.4 to 76.5 times. In wells 50.0 m deep, the turbidity exceeds the permissible norms by 3 to 40 times. In wells 25, 29, 31, 33, 35, 37, 38 with a depth of 100.0 m, the turbidity index increases by 1.4 to 93 times.

The turbidity value also depends on the amount of suspended particles, a large content of which is found in the unpumped aquifer complex. Therefore, before collecting water samples from observation wells, it is necessary to pump the water according to the existing methodology. The results of monitoring observations at observation points on the Bobrovka River were used to assess the state of surface watercourses. The quality of water bodies is assessed by abbreviated chemical analysis, pollutants, including oil products (Cherepansky 2018). Water samples were collected twice a year during periods of high spring standing of groundwater levels and in the autumn low water. The regime network includes three hydrological posts located on the Bobrovka River and one sampling site from a stream flowing into the Bobrovka River near the village of Semenovka. Hydrological post 6 is located on the southwestern border of the mining allotment of the deposit near the bridge over the Bobrovka River, on the northern outskirts of the village of Volzhsky. Hydropost 7 is located in the central part of the mining allotment, near the bridge over the Bobrovka River, between the settlement of Volzhsky and the village of Semenovka. Hydropost 8 is located near the bridge over the Bobrovka River below the village of Semenovka, south of the border of the mining allotment of the Tananyk deposit.

When analyzing the current state of surface waters based on the results of chemical analyses, surface waters in the Bobrovka River range from fresh to slightly brackish in composition with a dry residue of 271–1773 mg/dm3. The chemical composition of the water is varied: hydrocarbonate, mixed with a predominance of hydrocarbonates, hydrocarbonate-chloride, rarely hydrocarbonate-sulphate. Magnesium-calcium and calcium waters predominate in cations. With a dry residue of 271–289 mg/dm3, the water is soft (the total hardness is 2.7–3.0 mg-eq/dm3), with an increase in mineralization to 0.910 g/dm3, the water becomes hard (8.3 mg-eq/dm3) and with a mineralization of 0.997–1.773 g/dm3, the total hardness changes within the range of 11–23.2 mg-eq/dm3 and the water is characterized as very hard. The active reaction of the environment is neutral and slightly alkaline—the hydrogen index (pH) is 7.7–8.68. The content of petroleum products during the observation period is within the MAC for rivers of fishery importance (0.5 mg/dm3) or was not detected in water samples. Exceedances over the standards in the studied water samples were found for the following indicators: nitrites (1.15–10 MAC), mineralization (1.2–1.8 MAC) and, accordingly, chlorides (2.5 MAC), calcium (1.2–1.4 MAC), magnesium (1.3–1.9 MAC). With an increase in water mineralization, as a rule, there is an increase in the content of calcium and magnesium, that is, an increase in hardness. Nitrogen compounds can enter the water with fertilizers and detergents. As a result of examining the content of chemicals in the surface waters of the rivers according to the qualitative analysis of monitoring studies, it can be concluded that the composition of the water in the Bobrovka River does not correspond to the original natural composition.

Generalization and analysis of analytical data, identification of stable trends (positive or negative) in changes in the state of the aquatic environment, control over possible pollution of surface and groundwater, which are recorded by the content of pollutant components and general indicators are based on systematic sampling of water from observation wells, existing water intakes and watercourses along the regime network of the Tananyk oil field. The location of observation wells, according to the project, is confined mainly to places of the largest accumulation of oil producing, injection wells, in the area of head structures. In terms of chemical composition, groundwater in the area of the Tananyk oil field at the beginning of the monitoring, according to the qualitative composition, was observed to be of varying quality. In observation wells located upstream from the head structures and in places of the smallest accumulation of oil producing wells, mainly hydrocarbonate composition with a mineralization of 0.3–0.9 g / dm3 is observed. In observation wells, located mainly downstream from the head structures and in places where oil producing wells are most concentrated, sulfate–chloride or chloride-sulfate waters with mineralization from 1.1 to 4.8 g/dm3 are observed.

During the analysis it can be noted that the underground water in observation wells 24,25,27,28,29,31,32,33,34,35,36,37,38 is mainly fresh with mineralization of 0.216–0.721 g/dm3. According to the chemical composition, the underground water is hydrocarbonate magnesium-calcium, calcium-magnesium. According to the value of total hardness, varying within the range of 1.4–7.4 mg-eq/dm3, the water is from very soft to hard. In wells 32,34, isolated excesses of mineralization (1.08–1.7 MAC) and hardness (2.7–2.8 MAC) are noted. The water is characterized as very hard, the chemical composition changes to hydrocarbonate-chloride magnesium-calcium and sulfate–chloride calcium. Isolated cases of increased mineralization and hardness may be the result of anthropogenic impact.

Analysis of the results of chemical studies of water in observation wells shows that the hydrochemical situation in most wells has not changed during comparison. An exception is well 24, where the water does not meet the requirements of standards for drinking water in terms of mineralization, hardness, chloride content, ammonium nitrogen and physical properties. The water in the well is very brackish with a mineralization of 12.3–18.9 g / dm3, according to the value of total hardness equal to 148.4–208 mg-eq / dm3, the water is very hard. The unsatisfactory composition of the water in the well is apparently associated with the close location of the Tananyk oil field treatment plant.

In accordance with the general requirements for the composition and properties of water in water bodies, the pH value should not go beyond the range of 6–9. The pH indicator in the water of all the assessed wells varies from 6.8 to 9.4. The reaction of the aqueous environment is from neutral to alkaline.

The content of oil products in almost all samples (73%) is within the MAC (0.1 mg/dm3). The exception is observation wells 27,32,33,34, where the content of oil products was 0.15–0.24 mg/dm3, i.e. 1.5–2.4 MAC. Almost all samples show increased turbidity (5.8–93 MAC) and color (1.05–16 MAC). This indicates that water samples were most often taken without mandatory preliminary pumping of observation wells. Water pollution in reservoirs in the Tananyk oil field is mainly due to runoff from agricultural fields and discharges from livestock farms. This pollution is seasonal and occurs mainly in the warm season. The results of laboratory monitoring of water in reservoirs in the territory of the Tananyk oil field indicate the absence of samples exceeding sanitary standards, both for bacteriological and chemical indicators do not exceed the MAC.

It should be noted that water samples were taken in observation wells at least twice a year based on the results of monitoring and the monitoring regulations were mainly carried out based on the qualitative composition of groundwater and surface water.

According to the qualitative composition of groundwater, the content of iron and manganese in groundwater varies from 1.5 to 12.8 MAC, it is clear that there is an excess and this may be the result of poorly performed pumping before sampling or is a natural background in this area of work.

In accordance with the general requirements for the composition and properties of water in water bodies, the pH value should not go beyond the range of 6.5–8.5. The pH indicator in the water of the wells being assessed varies from 7.2 to 8.14. The waters are neutral, which may be the result of poor pumping before sampling, or the impossibility of pumping before sampling due to the small water column in shallow wells.

3 Conclusions

Based on the results of monitoring observations, several pollution sources were identified on the territory of the fields, related to the leakage of formation waters. In the zone of influence of the Kurmanaevskoye and Tananykskoye oil fields, such indicators as iron, manganese, ammonia, copper, oil products, and mineralization are periodically exceeded. The hydrochemical situation at these oil fields is generally satisfactory. Regular monitoring, well repair, and the introduction of new technologies, in particular, the use of complex barriers, are recommended.

References

Borevsky BV (2014) Typification of combined water intake systems for the joint use of groundwater and surface water. Borevsky, BV, Kozak NS. Exploration and protection of mineral resources. 05. pp 51–55

Cherepansky MM (2018) The influence of natural and man-made factors on the underground component of water resources in the western part of the Moscow artesian basin. Cherepansky MM, Karimova OA, Dzyuba AV, Vyazkova OE. Min J. 11 pp 29–34. https://doi.org/10.17580/gzh.2018.11.04

Environmental impact assessment for the "Analysis of the development status of the Kurmanaevskoye oil field in the Orenburg region". OOO "Center for Geoecological Research". Orenburg, (2002)

Environmental Impact Assessment for the "Technological Scheme for the Development of the Tananyk Oil Field". OOO "Center for Geoecological Research. Orenburg, (2002)

Gaev A Ya (2014) On the protection of water intakes for domestic and drinking purposes using the example of Orenburg. Gaev AYa, Kudelina IV, Malikova ON. Regional problems of water use in changing climatic conditions: Materials of MNPK, 11–12. Ed.: Gareev AM. Ufa: Aeterna, pp 277–280

Gaev A Ya (2019) Water management problems of water-deficient territories on the example of the Southern Urals A Ya Gaev, Kudelina IV, Leontyeva TV. Sci, New Technol Innov Kyrg. 4. pp 218–222

Goldberg VM (1984) Hydrogeological foundations for protecting groundwater from pollution. Goldberg VM, Gazda S. Mosc: Sci, p 262

Kovalevsky VS (2001) Combined use of surface and groundwater resources. Kovalevsky VS. Mosc: Sci World, p 332

Kudelina IV (2018) Ways to stabilize the regime of alluvial water intakes in a semi-arid climate. Kudelina IV. Science, new technologies and innovations of Kyrgyzstan. Bishkek, 2 pp 82–86

Leontyeva TV (2019) On the analysis of the water supply system of the population in the Eastern Orenburg region. Sci, New Technol Innov Kyrg. 4:240–243

On approval of the regional program. Improving the quality of water supply in the Orenburg region for 2019–2024: Resolution of the Government of the Orenburg region dated 10.10.2019 775-pp (as amended on 01.02.2022). https://docs.cntd.ru/document/561580593

On the state of sanitary and epidemiological well-being of the population in the Orenburg region in 2018: State report. Orenburg, (2019). p 241

Conditions for the November Storms in the Black Sea

T. E. Danova and T. M. Bayankina

Abstract The paper discusses the conditions for the rare, non-standard processes in the ocean–atmosphere system that lead to catastrophic storms in the Black Sea region. We consider the weather patterns of storms on November 11, 2007, November 11, 2009 and November 26–27, 2023. The data from pressure topography maps, satellite imagery in the visible and infrared (IR) spectral ranges, as well as data from a surface network of weather stations on the Crimean Peninsula show the movement of Mediterranean cyclones to the Black Sea region, which then intensify to severe storms. We show the effect of orographic features of the Crimean Peninsula, temperature fluctuations, in the month of November, of the underlying land and sea surface on the intensity of latitudinal mixing of air masses. According to 3-h observations at coastal weather stations, at the time when storms pass, an excess of the heat threshold is recorded at the level of the 95% percentile of maximum air temperature values.

Keywords Storm in the black sea · Mediterranean cyclone · Periods of warming in the winter season on the crimean peninsula

1 Introduction

For the Black Sea region, the most significant weather-forming pressure patterns are cyclones, which are associated with the transfer of moisture in the atmosphere, the formation of powerful cumulonimbus clouds and various types of precipitation, dangerous weather phenomena (Bukharov et al. 2010; Gippius et al. 2013; Dotsenko and Ivanov 2010; Divinskii et al. 2003). The area where the cyclone is formed and the trajectory of its movement are of great importance, for example, the trajectories of southern cyclones, with abundant precipitation for both cold and warm seasons, run along the western-northwestern coast of the Black Sea, skirting the Carpathians from the East (Danova and Bayankina 2018; Bayankina and Danova 2020).

T. E. Danova (✉) · T. M. Bayankina
FRC Marine Hydrophysical Institute of RAS, Sevastopol, Russian Federation
e-mail: danova8@mail.ru

T. Chaplina (ed.), *Processes in GeoMedia—Volume IX*, Springer Geology,
https://doi.org/10.1007/978-981-96-9302-3_19

As a rule, the process of cyclogenesis depends on the geographical features of the region and is due to complex orography (Speranza 2001), the arrival of moisture-saturated Mediterranean air, which contributes to increased baroclinic instability, intense release of latent heat and an increase in temperature contrasts (Fantini and Davolio 2001). Latitudinal mixing of air masses with different thermal and pressure characteristics (Lionello et al. 2002; Trigo et al. 2002) with powerful humidity-dynamic parameters leads to unstable stratification of the atmosphere and the emergence of cyclones (Danova and Bayankina 2018; Nesterov 2018).

Since the beginning of the twenty-first century, a large number of works have been devoted to the study of the conditions for dangerous weather phenomena associated with cyclones passing over the Black Sea. For example, modeling of catastrophic weather phenomena in the Black Sea region associated with cyclones is discussed in Diansky and Fomin (2017); Shokurov 2012); the possibility of using satellite data to diagnose and forecast hazardous weather phenomena using modern technology is discussed in Bayankina and Danova (2020); Nesterov 2018; Efimov et al. 2008).

A topical issue in the field of hydrometeorological problems and the most important scientific and practical task for the diagnosis and forecast of dangerous weather phenomena is the analysis of rare, non-standard processes in the ocean–atmosphere system, leading to catastrophic storms in the Black Sea region. These include the severe storms of November 11, 2007, November 11, 2009 and November 26–27, 2023, which resulted in the loss of life, shipwrecks and significant financial damage to the coastal zone. The twentieth century also saw severe storms on November 10–12, 1981 and November 14–16, 1992, which resulted in extensive material damage, damaged oil and gas platforms, destroyed concrete booms and washed away part of the beaches in Odessa and Crimea. Historical documents reference the strongest November storms of the nineteenth century. Arseny Markevich, one of the most famous Crimean scientists at that time, a historian, local historian, chairman of the Taurida Scientific Archival Commission, mentioned that on November 1, 1824, “there was a terrible hurricane in Crimea” (Efimov et al. 2015). The most famous November storm was the storm during the Crimean War, which took place on November 14, 1854, which destroyed numerous foreign ships anchored in the Balaklava Bay.

As we can see, all these storms, which can be characterized as catastrophic, were regularly observed in November. In this regard, the purpose of the paper is to analyze the conditions for the catastrophic storms in November and uncover their features.

2 Materials and Methods

The climate database “Reliable Prognosis” (http://rp5.ru) was used to obtain baseline data to determine the surface weather; data was obtained for each station of the Crimean Peninsula on the dates when the storms passed over the Crimean Peninsula: November 11, 2007, November 11, 2009 and November 26–27, 2023. The direction of movement of the cyclones was tracked using synoptic surface maps (https://www.wetter3.de) and satellite imagery MODIS\Terra (https://worldview.earthdata.nasa.gov/) (Moderate Resolution Imaging Spectroradiometer).

Considering that strong storms were observed in November, the variability of the surface thermal regime on the Crimean Peninsula is of primary interest. To this end, we used the results of studies of warming and cooling periods in the seasonal distribution of average daily values of surface air temperature on the Crimean Peninsula. The method used to determine warming and cooling periods in the seasonal distribution of average daily values of surface air temperature used well-known approaches presented in the study (Markevich 1905). According to this method, the cold threshold is set at the 5% percentile of the temperature distribution in the cold season from November to March, which is close to the 3% percentile of the year-round distribution. To determine the warming and cooling periods during the cold season of the year (from November to March), average daily air temperatures were used.

A warming period is a sequence of ≥ 3 days with temperatures $\geq 95\%$ percentile in the cold season (November to March) of average daily temperatures for the period from 2006 to 2022; a cooling period is a sequence of ≥ 3 days with temperatures $\leq$ 5% percentile in the cold season of average daily temperatures for the same months and years.

3 Results

A number of works are devoted to the weather pattern associated with the formation of the cyclone on November 9, 2007, which consider the specifics of synoptic processes, as well as the possibilities of remote observation and modeling of such situations (Wang et al. 2016; Lopatukhin et al. 2007). The cyclone formed over Italy with a pressure of 1001.0 hPa in the center. On November 10 the cycle began to quickly spread across the Balkan Peninsula towards North-East, to the Azov-Black Sea basin. The cloud mass shown in the satellite image has a classic spiral shape (Fig. 1a).

On November 11, in the center of the cyclone, located over the western part of the Black Sea, the pressure dropped to 983.0 hPa, which caused an increase in the radius of the zone of maximum wind speeds and their increase over the sea area. On the morning of November 11, the speed of the westerly wind near the southwestern Crimea increased to 25–32 m/s, the height of the waves spreading from the southwest reached 3–5 m at Cape Chersoneses (Fig. 1b). With a south-southwest wind speed of 25 m/s, the wave height in the southern part of the Kerch Strait reached 4–5 m. On November 11, from 05:00 AM to 01:00 PM, a frontal zone of increased air temperature gradients at a level of 850.0 hPa, caused by a cold atmospheric front, passed through the Kerch Strait region. The rate of pressure drops over 3 h reached 5.0 hPa. The cyclone was not very deep, but was characterized by large pressure gradients, which resulted in wind speeds of up to 30–35 m/s (Efimov et al. 2008).

The storm was accompanied by a squally wind, which resulted in numerous shipwrecks, a catastrophic environmental situation and human casualties; more than a

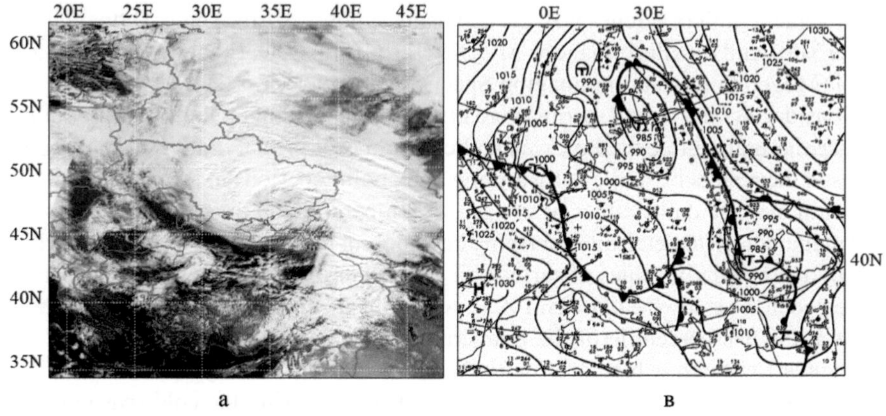

Fig. 1 Storm of November 11, 2007; **a**—satellite image MODIS\Terra (11/11/2007), **b**—synoptic map of surface weather analysis during the storm of November 11, 2007

thousand tons of fuel oil spilled into the water, enormous material damage was caused to Crimea and Sochi.

The weather pattern on November 8, 2009 showed that a cyclone with a pressure of 995.0 hPa in its center began to form in the area of the Gulf of Genoa; after that, the cyclone moved through the Balkan Peninsula to the North-East, while the associated cold front moved through the Black Sea (Fig. 2a, b). During the storm, the height of the sea waves reached 9–10 m. The storm resulted in the sinking of several ships, in addition to human casualties. As the front passed Sevastopol, the weather on the ground was characterized by heavy rainfall and storm winds.

The so-called storm of the century hit the Black Sea coast on the night of November 26–27, 2023. The storm was caused by a cyclone that formed in the Gulf of Genoa

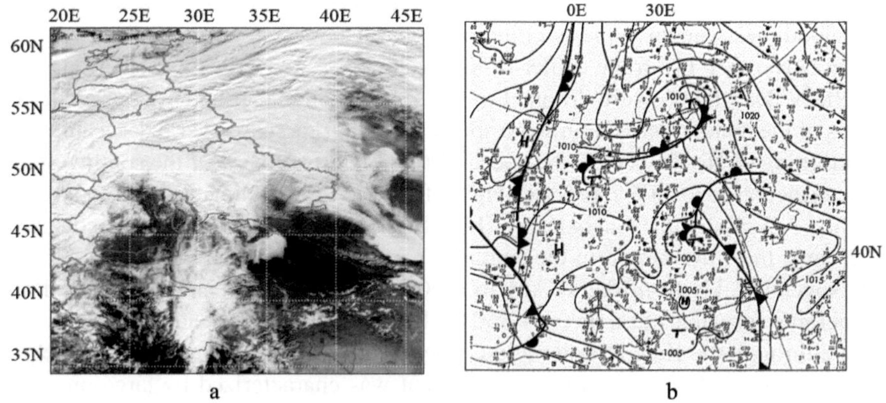

Fig. 2 Storm of November 11, 2009; **a**—satellite image MODIS\Terra (11/11/2009), **b**—synoptic map of surface weather analysis during the storm of November 11, 2009

on November 20 with a pressure of 1010.0 hPa in the center. On November 22, the cyclone intensified over the Mediterranean Sea and was characterized by a system of fronts. On November 23, the storm began to occlude and on November 24, a cold-type occluded front was observed.

During this period, a large cyclone with a developed frontal system and a pressure of 972.0 hPa in the center was active in the Northern Europe, which originated in the North Atlantic and moved to the East. On November 25, these two cyclones merged into a huge area of low pressure with two centers. In the center of the cyclone over the Mediterranean Sea, the pressure dropped abruptly from 1010.0 hPa to 992.4 hPa per day, and it began to spread over to the Black Sea. By November 26, the cyclone constituted a huge low-pressure area (980.0 hPa in the center) with large gradients; the associated frontal system was oriented from the South-West to the North-East. At the same time, on the MODIS\Terra satellite image from November 26, 2023, the cloud system covers the entire Black Sea and part of the Eurasian territory (Fig. 3a).

On November 27, 2023, the front began to occlude, horizontal pressure gradients increased significantly, the pressure in the center was 980.0 hPa (Fig. 3b). The height of the sea waves reached 7–10 m, the wind speed in the coastal zone reached 40 m/s. The entire coastal infrastructure was significantly damaged. In the southern Black Sea, sea levels in the coastal zone rose by approximately 1 m. According to specialists from the "Phobos" weather center, up to 33 mm of precipitation fell in Crimea in some places during the day, and up to 43 mm in Kuban. In the last 12 h, wind speed recordings at weather stations have ranged from 20 to 31 m/s. In Simferopol, the wind speed reached 29 m/s; at the Chernomorskoye station, the wind speed reached 40 m/s.

The damage from the storm that hit Sevastopol amounted to about 6 billion rubles. Most of this amount is needed to restore the damaged berthing infrastructure. The

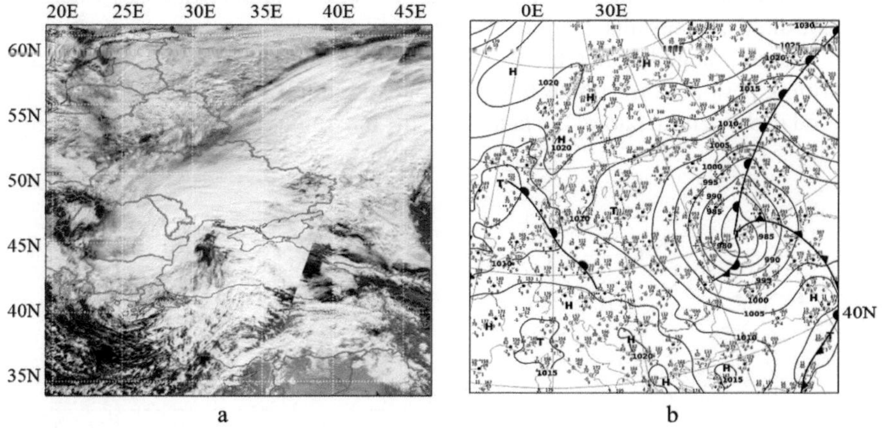

Fig. 3 Storm of November 26–27, 2023; **a**—satellite image MODIS\Terra (26/11/2023), **b**—synoptic map of surface weather analysis during the storm on 27/11/2023

so-called storm of the century that hit Crimea resulted in the death of more than 800 marine animals in the aquarium museum in Sevastopol.

4 Discussion

Dangerous weather phenomena in the Black Sea region are due to cyclonic activity and are associated primarily with the humidity and thermodynamic characteristics of the ocean–atmosphere system. One of the reasons for unstable stratification of the atmosphere is the latitudinal mixing of air masses with different thermal and pressure characteristics. An analysis of the synoptic situation showed that November storms in the Black Sea region are due to Mediterranean cyclones, which bring warm air saturated with moisture to the region.

The time of year plays an important role in the strongest storms in the Black Sea region: November is the last autumn month, which is characterized by abrupt and significant changes in air temperature, i.e. intense latitudinal mixing of air masses. In addition, the orographic features of the Crimean Peninsula also play a significant role: in November, there are pronounced temperature differences between the underlying land and sea surfaces. In this regard, a study was conducted of the surface thermal regime on the Crimean Peninsula for the period from 2006 to 2023. Warming and cooling periods in the cold season of the year (from November to March) were calculated for average daily values of surface air temperature at stations on the Crimean Peninsula. The cold threshold is set at the 5% percentile of the temperature distribution, while the heat threshold is set at the 95% percentile.

The calculations showed that on the Crimean Peninsula, during the cold period of the year, in November, only warming periods are observed annually. At stations on the Crimean Peninsula for the specified period, the number of cases where the heat threshold was sustained at the 95% percentile level, characterized as periods of warm weather, ranged from 6 to 13. The spatial distribution of the number of cases of warming periods corresponds to the region's orography, the minimum values are observed in the central part of the peninsula, the maximum values in the coastal zone (Fig. 4a).

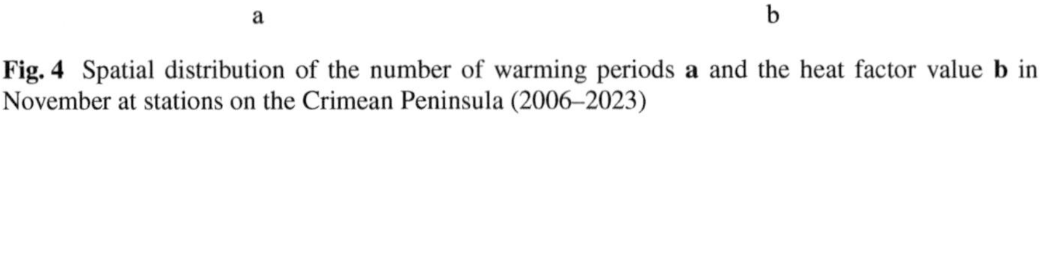

Fig. 4 Spatial distribution of the number of warming periods **a** and the heat factor value **b** in November at stations on the Crimean Peninsula (2006–2023)

Of all the months of the cold period, it is in November that a significant warming trend is observed at all stations of the Crimean Peninsula. It is November that is characterized by the maximum values of incoming heat during the cold period of the year, which are interpreted as a heat factor and were determined as the average total excess of the calculated threshold value over the warming period. The spatial distribution of the heat factor also bears the features of the region's orography; the maximum values are characteristic of coastal stations (Fig. 4b).

Analysis of surface air temperature values showed that, according to 3-h observations at coastal stations during the storms on November 11, 2007, November 11, 2009 and November 26–27, 2023, air temperature maxima exceeded the 95% percentile heat threshold. In the case of the storm on November 11, 2009, all coastal stations recorded an excess of the heat threshold at the level of the 95% percentile for average daily air temperatures.

5 Conclusion

The research into the conditions for November catastrophic storms in the Black Sea region on the Crimean Peninsula allow us to draw the following conclusions:

the strongest November storms in the Black Sea region are due to Mediterranean cyclones, which bring warm air saturated with moisture to the region;
November is characterized by abrupt and significant changes in air temperature, a significant role belongs to the orographic features of the Crimean Peninsula, the temperature differences of the underlying land and sea surface are pronounced, which contributes to intensive latitudinal mixing of air masses with different thermal and pressure characteristics;
long-term series of surface air temperature at all stations of the Crimean Peninsula indicate a significant warming trend in November; the month is characterized by maximum values of incoming heat in the cold period of the year;
according to 3-h observations at coastal meteorological stations during the storms of November 11, 2007, November 11, 2009 and November 26–27, 2023, the maximum air temperature values exceeded the heat threshold at the 95% percentile.

The results obtained have allowed to identify the features of catastrophic storms in the Black Sea region, which can be used in forecasting.

Acknowledgements The research was conducted under the theme of the Marine Hydrophysical Institute **No. FNNN-2024-0014 "Fundamental Research into the Interaction Processes in the Ocean-Atmosphere System that Shape the Variability of the Physical State of the Marine Environment on Various Spatio-Temporal Scales." ("Ocean-Atmosphere Interaction"** code).

References

Bayankina TM, Danova TE (2020) Cyclogenesis in the Mediterranean-Black Sea region from satellite data. Sovrem Probl Distantsionnogo Zondirovaniya Zemli Iz Kosm. T. 17(3). C. pp 231–239. https://doi.org/10.21046/2070-7401-2020-17-3-231-239 (In Russian)

Bukharov MV, Golovlev KN, Mironova NS, Sizenova EA (2010) Analysis of the reasons for the increase of squall winds to storm force in the Kerch Strait on November 11, 2007. Sovrem Probl Distantsionnogo Zondirovaniya Zemli Iz KosmA 7(3):9–15 (In Russian)

Danova TE, Bayankina TM (2018) Thermodynamic characteristics of a super-storm over the Black Sea. In: Proc. SPIE, 10833, 24th international symposium on atmospheric and ocean optics: atmospheric physics, 108337H. https://doi.org/10.1117/12.2503960.

Diansky NA, Fomin VV (2017) Modeling of the Sea of Azov circulation and features of the reproduction of extreme surges in the Taganrog Bay. Trudy GOIN. 218:74–104 (In Russian)

Divinskii BV, Kos'yan RD., Podymov IC, Pushkarev OV (2003) Extreme waves in the north-eastern Black Sea in February 2003. Oceanology. T. 43(6). pp 891–893. (In Russian)

Dotsenko SF, Ivanov VA (2010) Natural disasters of the Azov-Black Sea region. Sevastopol': NAN Ukrainy, 174 s. (In Russian)

Efimov VV, Stanichnyi SV, Shokurov MV, Yarovaya DA (2008) Observation of a quasi-tropical cyclone over the Black Sea. Russ Meteorol Hydrol 33(4):233–239. https://doi.org/10.3103/S1068373908040067.(InRussian)

Efimov VV, Volodin EM, Anisimov AE, Barabanov VS (2015) Regional projections of climate change for the Black Sea – Caspian Sea area in late XXI century. Phys Oceanogr 5:49–66. https://doi.org/10.22449/1573-160X-2015-5-49-66

Fantini M, Davolio S (2001) Instability of neutral eady waves and orography. J Atmos Sci. 58 pp 1146–1154

Gippius FN, Arkhipkin VS, Surkova GV (2013) Assessment of modern characteristics and long-term variability of extreme waves in the Black Sea. Ecol Saf Coast Shelf Zones Compr Use Shelf Resour. 27:92–96 (In Russian)

Lionello P, Dalan F, Elvini E (2002) Cyclones in the Mediterranean region: the present and the doubled CO2 climate scenarios. J Clim Res. 22 pp 147–159

Lopatukhin LI, Bukhanovsky AV, Chernysheva ES, Kolesov AM (2009) About the storm on the Black Sea in November 2007. M.: Rossijskij Rechnoj Registr, pp 35. (In Russian)

Markevich A (1905) Taurida province during the crimean war (Based on archival materials). Izvestiya Tavricheskoj uchenoj arxivnoj komissii. Simferopol'. Tipog. Tavrich. Gubern. Zemstva., 37. pp 1–260. (In Russian)

Nesterov ES (2018) Extreme cyclones over the seas of the European part of Russia. Gidrometeorologicheskie issledovaniya i prognozy'. 1 (367). pp 97–115. (In Russian)

Peskov BE, Dmitrieva TG (2009) Synoptic processes that caused dangerous winds and waves in the north-east of the Black Sea, in the Kerch Strait and in the Sea of Azov on November 11, 2007. Meteospektr. 1:91–96 (In Russian)

Shokurov MV (2012) Numerical simulation of tsunami in the Black Sea region. Ecological safety of coastal and shelf zones and comprehensive use of shelf resources. 26(2). pp 301–320. (In Russian)

Speranza A (2001) The physical mechanisms of secondary cyclogenesis. Recent Res Dev Atmos Sci. V. 1. P. 57–68

Trigo IF, Bigg GR, Davies TD (2002) Climatology of cyclogenesis mechanisms in the Mediterranean. J Mon Weather Rev. 130. pp 549–649

Wang L, Liu T, Hu M et al (2016) The impact of cold spells on mortality and effect modification by cold spell characteristics, Sci Rep 6, 38380 https://doi.org/10.1038/srep38380

Variational Algorithms for Identification of Transfer Model Parameters and Experimental Planning

S. V. Kochergin and V. S. Kochergin

Abstract The aim of the work is to build and implement variational algorithms for identifying the capacities of a series of pollution sources in a passive impurity transfer model. The paper considers an example of variational assimilation of impurity concentration data in the upper layer based on minimizing the prediction quality functional and an approach based on variational filtering of linear systems of equations. In both cases, an essential element of the algorithms is the solution of related problems, which are also used in the construction of information matrices. The paper considers an example of variational assimilation of data on the concentration of suspended matter in the upper layer of the Sea of Azov in the area of the Dolgaya Spit, which is of great importance for navigation in this area. When solving the problem of identifying the power of pollution sources in the passive impurity transfer model based on measurement data, the question arises of building optimal plans to improve the computational properties of algorithms. The computational properties of the algorithms used in this case can be significantly improved by choosing the most optimal measurement scheme.

Keywords Concentration of suspended matter · Variational algorithm · Assimilation · Adjoint problem · Sea of Azov · Pollution source capacities · Experiment planning

1 Introduction

To solve the environmental problems of the Azov-Black Sea basin, systems are required that allow obtaining operational information about the ecological state of the reservoir. Such systems require numerical dynamic models (Marchuk and Sarkisyan 1988; Ivanov and Fomin 2008), models for the transport of various suspended solids (Marchuk 1982), and procedures for assimilating information about its concentration in the studied basin (Marchuk and Penenko 1979; Kochergin et al. 2020; Kochergin

S. V. Kochergin · V. S. Kochergin (✉)
Marine Hydrophysical Institute RAS, Sevastopol, Russia
e-mail: vskocher@gmail.com

T. Chaplina (ed.), *Processes in GeoMedia—Volume IX*, Springer Geology,
https://doi.org/10.1007/978-981-96-9302-3_20

and Kochergin 2017a). In this paper, a variational assimilation algorithm is used to identify the power of pollution sources. When implementing the procedure, the solution of the main, conjugate tasks and tasks in variations are used to build a functional gradient and organize the iterative process. A comprehensive overview of the methods of assimilation of measurement data is presented in Shutyaev (2019). The model in the algorithm used acts as a space–time interpolant (Penenko 1981) and the resulting solution to the problem in the time interval used is consistent not only with the model, but also with the measurement data. In addition, an algorithm based on filtering systems of linear equations is presented. The construction of which is based on the evaluation method and the solution of related tasks.

2 Building an Optimal Plan

Optimal planning algorithms (Gorsky 1984; Ermakov and Zhiglyavsky 1987) are based on studying the properties of the Fisher information matrix:

$$M(\varepsilon_N) = \sum_{i=1}^{N} F_i p_i F_i^T, \tag{1}$$

where $\varepsilon_N = \begin{cases} x_1, x_2, \ldots, x_N \\ p_1, p_2, \ldots, p_N \end{cases}$ is the normalized plan, $0 \le p_i \le 1$, $\sum_{i=1}^{N} p_i = 1$, F_i is the Jacobi matrix, which determines the variations of the impurity concentration field depending on the variations of certain parameters of the problem. There are different approaches to finding optimal plans. The simplest one is as follows:

- a non-degenerate initial plan is being built;
- an additional point is determined, the addition of which to the plan
- increases the conditionality of the task being solved;
- the least informative point of the plan is being searched for, which is discarded.

The determinant of matrix (1) characterizes the conditionality of the solved problem of identifying the parameters of numerical modeling. Therefore, it is possible to compare several a priori defined plans. The comparison is based on the value of the determinant of the Fisher information matrix (1).

3 Transport Model

The model of passive admixture transfer-diffusion used for the water area of the Sea of Azov in the σ–coordinates has the form

$$\frac{\partial DC}{\partial t}+\frac{\partial DUC}{\partial x}+\frac{\partial DVC}{\partial y}+\frac{\partial WC}{\partial \sigma}=\frac{\partial}{\partial x}DA_H\frac{\partial C}{\partial x}+\frac{\partial}{\partial y}DA_H\frac{\partial C}{\partial y}+\frac{\partial}{\partial \sigma}\frac{K}{D}\frac{\partial C}{\partial \sigma} \tag{2}$$

with boundary conditions:

$$\Gamma:\frac{\partial C}{\partial n}=0,$$

$$\sigma=0:\frac{\partial C}{\partial \sigma}=0$$
$$\sigma=-1:\frac{\partial C}{\partial \sigma}=Q=\sum_i Q_i^B\cdot\delta(x-x_i)\cdot\delta(y-y_i),\, i=1,I. \tag{3}$$

and the initial data:

$$C(x,y,\sigma,0)=C_0(x,y,\sigma), \tag{4}$$

where $t\in[0,T]$—time; D—dynamic depth; x, y—is horizontal coordinates; σ—dimensionless vertical coordinate varying from -1 (at the bottom) to 0 (on the sea surface); U, V, W—components of the velocity field; C—impurity concentration; A_H and K—coefficients of horizontal and vertical turbulent diffusion; n is the normal to the lateral boundary, Γ—the boundary of the model integration area M; $M_t = M\times[0,T]$, I—the number of pollution sources at the bottom with a given capacity Q_i^B.

4 A Variational Algorithm for Assimilation of Measurement Data

Let's consider a variational assimilation algorithm. The problem is solved by minimizing the convex quadratic prediction quality functional of the following form:

$$I_0=\frac{1}{2}\big(P\big(RC-C\big),P\big(RC-C\big)\big)_{M_t}, \tag{5}$$

where is the operator for zero–filling the forecast discrepancy field in the absence of measurement data, and is the projection operator at the observation points. The functional (5) under linear constraints (2–4) is written as follows:

$$I = I_0 + \left[\frac{\partial DC}{\partial t} + \frac{\partial DUC}{\partial x} + \frac{\partial DVC}{\partial y} + \frac{\partial WC}{\partial \sigma} - \frac{\partial}{\partial x} DA_H \frac{\partial C}{\partial x} - \frac{\partial}{\partial y} DA_H \frac{\partial C}{\partial y} - \right.$$
$$\left. - \frac{\partial}{\partial \sigma} \frac{K}{D} \frac{\partial C}{\partial \sigma}, C^* \right]_{M_t} + \left(\frac{\partial C}{\partial n}, C^* \right)_t + \left(-_0, C^* \right)_M + \left(\frac{\partial C}{\partial \sigma}, C^* \right)_{\sigma_t^0} + \tag{6}$$
$$+ \left(\frac{\partial C}{\partial \sigma} - \sum_i Q_i^B \cdot \delta(x - x_i) \cdot \delta(y - y_i), C^* \right)_{\sigma_t^{-1}},$$

where the dot product in L_2 is defined in the standard way. Let us integrate the corresponding (6) expression for the variation of the functional in parts, taking into account the analogue of the continuity equation and boundary conditions. In (6) $\sigma_t^0 = \sigma^0 \times [0, T], \sigma_t^{-1} = \sigma^{-1} \times [0, T]$, σ^0 – surface of the sea, and σ^{-1}—the bottom. By choosing the solution of the following conjugate problem as Lagrange multipliers:

$$- \frac{\partial DC^*}{\partial t} - \frac{\partial DUC^*}{\partial x} - \frac{\partial DVC^*}{\partial y} - \frac{\partial WC^*}{\partial \sigma} - \frac{\partial}{\partial x} DA_H \frac{\partial C^*}{\partial x} - \frac{\partial}{\partial y} DA_H \frac{\partial C^*}{\partial y} - \frac{\partial}{\partial \sigma} \frac{K}{D} \frac{\partial C^*}{\partial \sigma} = 0,$$
$$\Gamma : \frac{\partial C^*}{\partial n} = 0, \sigma = 0 : \frac{\partial C^*}{\partial \sigma} = 0, \sigma = -1 : \frac{\partial C^*}{\partial \sigma} = 0,$$
$$t = T : C^* = P(C - R). \tag{7}$$

When identifying constant power sources operating over the entire time period, we have:

$$\nabla_{Q_i^B} I = \int_0^T C^*(-1, x_i, y_i, t) dt. \tag{8}$$

When determining the power of an instantaneous point source, we have:

$$\nabla_{Q_i^B} I = C^*(-1, x_i, y_i, 0). \tag{9}$$

The value of the source capacities is determined during the iteration process.

$$Q_{i_{n+1}}^B = Q_{i_n}^B + \tau \cdot \nabla_{Q_i^B} I, \tag{10}$$

where is the iterative parameter, which is found taking into account the solution of the problem in variations. It should be noted that the test calculations performed showed good convergence of the iterative process and sufficient accuracy in identifying the power of sources using complete information about the concentration field. When specifying a small number of measurement points, the result fully depends on their location, and then the task of planning the experiment arises naturally.

5 The Assessment Method. Filtering Algorithm

Multiplying (2–4) by C^* and integrating piecemeal, taking into account the boundary conditions and the analogue of the continuity equation in σ–coordinates, we choose C^* as the solution of the following adjoint problem:

$$-\frac{\partial DC^*}{\partial t}-\frac{\partial DUC^*}{\partial x}-\frac{\partial DVC^*}{\partial y}-\frac{\partial WC^*}{\partial \sigma}-\frac{\partial}{\partial x}DA_H\frac{\partial C^*}{\partial x}-\frac{\partial}{\partial y}DA_H\frac{\partial C^*}{\partial y}-\frac{\partial}{\partial \sigma}\frac{K}{D}\frac{\partial C^*}{\partial \sigma}=0, \tag{11}$$

$$\Gamma:\frac{\partial C^*}{\partial n}=0, \sigma=0:\frac{\partial C^*}{\partial \sigma}=0, \sigma=-1:\frac{\partial C^*}{\partial \sigma}=g, \tag{12}$$

$$t=T:C^*=0. \tag{13}$$

As a result, from the integral identity we get

$$\sum_i \int_{B_t} Q_i^B \cdot C^* \cdot dB_t = \int_{B_t} C \cdot g \cdot dB_t, \tag{14}$$

where B—the bottom of the sea, $g=\delta(t-T)\cdot\delta(x-x_n)\cdot\delta(y-y_n), n=1,...,N, N$ – the total number of measurement points at a finite time. Given the representations for Q and g from formula (14), we have:

$$\sum_i Q_i^B \cdot \int_0^T C_n^*(t,x_i,y_i,-1)dt = C(T,x_n,y_n,-1), n=1,...,N. \tag{15}$$

Thus, taking into account the information about the concentration at N points for a finite moment of time, a system of N linear equations is constructed. In the case of an overridden system, i.e., when the number of equations is greater than the number of sources, one of the well-known variational filtering methods is used, for example, values Q_i^B are determined from system (15) based on the method (Strakhov 1991), which is successfully used to solve similar problems. For the redefined system (15), in case of measurement errors, data filtering is necessary, taking into account all the information available in N equations. The initial system of equations is transformed to some equivalent system by means of orthogonal transformation of plane rotations. The rotation angles are found from the maximum of the functional characterizing the ratio of the coefficients of the system, a priori given information about the solution and the right-hand sides. The algorithm is performed iteratively. The coefficients of such a transformation are sought based on the maximum of a quadratic functional of a special kind. As a result, we obtain another system of algebraic equations equivalent to (15), taking into account all available information.

As a result of this procedure, the equations of the new system are sorted by rank. As a result, the equations with the best information content appear in the upper lines of the system.

6 Construction of the Jacobi Matrix

Let be C_{u3m} present at N points, and Q_B^i given at I points. The Jacobi matrix, which characterizes the variation of the solution depending on the variations in the parameters, is written as:

$$F = \begin{pmatrix} \frac{\partial C_1}{\partial Q_B^1} & \frac{\partial C_2}{\partial Q_B^1} & \cdots & \frac{\partial C_N}{\partial Q_B^1} \\ \frac{\partial C_1}{\partial Q_B^2} & \frac{\partial C_2}{\partial Q_B^2} & \cdots & \frac{\partial C_N}{\partial Q_B^2} \\ \cdots & \cdots & \cdots & \cdots \\ \frac{\partial C_1}{\partial Q_B^I} & \frac{\partial C_2}{\partial Q_B^I} & \cdots & \frac{\partial C_N}{\partial Q_B^I} \end{pmatrix}. \tag{16}$$

Considering the representation of the estimation method (16), we have

$$F = \begin{pmatrix} \int_0^T C_1^*(t, x_1, y_1, -1)dt & \int_0^T C_2^*(t, x_1, y_1, -1)dt & \ldots & \int_0^T C_N^*(t, x_1, y_1, -1)dt \\ \int_0^T C_1^*(t, x_2, y_2, -1)dt & \int_0^T C_2^*(t, x_2, y_2, -1)dt & \ldots & \int_0^T C_N^*(t, x_2, y_2, -1)dt \\ \cdots & \cdots & \cdots & \cdots \\ \int_0^T C_1^*(t, x_I, y_I, -1)dt & \int_0^T C_2^*(t, x_I, y_I, -1)dt & \ldots & \int_0^T C_N^*(t, x_I, y_I, -1)dt \end{pmatrix}. \tag{17}$$

Similarly, for an instantaneous point source, we have:

$$F = \begin{pmatrix} C_1^*(0, x_1, y_1, -1) & C_2^*(0, x_1, y_1, -1) & \ldots & C_N^*(0, x_1, y_1, -1) \\ C_1^*(0, x_2, y_2, -1) & C_2^*(0, x_2, y_2, -1) & \ldots & C_N^*(0, x_2, y_2, -1) \\ \cdots & \cdots & \cdots & \cdots \\ C_1^*(0, x_I, y_I, -1) & C_2^*(0, x_I, y_I, -1) & \ldots & C_N^*(0, x_I, y_I, -1) \end{pmatrix}, \tag{18}$$

where C_n^*—the solution of the adjoint problem for the nth dimension at the corresponding points in space with the coordinates of the pollution sources.

In the derivation (18), the ratio of the estimation method (14) for an instantaneous point source is taken into account.

Thus, having solutions to the conjugate problems (11–14), we can immediately evaluate all the terms of the Jacobi matrix, and hence Fischer (1). Next, the optimal plan is selected based on the determinant of the corresponding matrix. Next, solving the system (15), we estimate the values of the input parameters. Considering the results of Kochergin and Kochergin 2017b, it can be concluded that the best result is obtained when measuring concentrations in the region of maximum values.

7 Results of Numerical Experiments

Numerical experiments were carried out using a barotropic version of the model (Ivanov and Fomin 2008) for the waters of the Sea of Azov with a real bottom relief and wind action, which is characterized mainly by eastward winds with values of the order of 10 m/s. The calculated velocity fields and coefficients of the model were the input information for the integration of the passive impurity transfer-diffusion model for a period of 5 days. In the model, the values of steps in time $\Delta t = 240$ sec and space $\Delta x = 0, 78$ km, $\Delta y = 1, 125$ km were set. Vertically, the model uses a calculated grid in σ–coordinates with 15 horizons.

Numerical experiments have shown the sensitivity of the variational assimilation algorithm to the measurement plan used. For example, Table 1 shows the found power values of five sources of suspended matter concentration in the area of Dolgaya Spit. The true values of the source capacities were set to -1, i.e. agitation was simulated. In Table 1, the first row shows the values obtained when using a non-optimal measurement plan, and the second row shows the values when using an optimal plan based on the results of the work (Strakhov 1991).

The table shows that when using a more optimal measurement network, the process of searching for known power values of impurity sources is better. According to the results of Kochergin and Kochergin 2017b, the optimal measurement points were located in the areas of maximum concentration values. When implementing the filtering algorithm, in the case of the redefined system (15), from the top five equations with the best conditionality in the equivalent system of equations, we find the known power values with sufficient accuracy. The result of such identification is presented in Table 2.

Table 1 Values Q_i^B at the fifth iteration

Q_1	Q_2	Q_3	Q_4	Q_5
– 1.0182	– 0.916	– 1.2634	– 0.8959	– 0.5948
– 1.0108	– 1.0177	– 0.9910	– 0.9562	– 1.0258

Table 2 Values Q_i^B when implementing the filtering algorithm

Q_1	Q_2	Q_3	Q_4	Q_4
– 1.0105	– 1.0167	– 0.9921	– 0.9478	– 1.0312

8 Conclusions

In the work, variational assimilation of distributed information was carried out, which made it possible to obtain the initially set values of the power of suspended matter sources. A similar result was obtained by applying a variational algorithm for filtering systems of linear algebraic equations, which is constructed using the estimation method and solving adjoint equations. Identification procedures can be used to determine the parameters of marine pollution sources.

Acknowledgements The work was performed within the framework of the state assignment on the topic FNNN-2024-0016 "Investigation of the spatial and temporal variability of oceanological processes in the coastal, coastal and shelf zones of the Black Sea under the influence of natural and anthropogenic factors based on contact measurements and mathematical modeling" (code "Coastal research").

References

Ermakov SM, Zhiglyavsky AA (1987) Mathematical theory of optimal experiment—Moscow: Nauka, p 320

Gorsky VG (1984) Planning kinetic experiments. Nauka, Moscow, p 240

Ivanov VA, Fomin VV (2008) Mathematical modeling of dynamic processes in the sea-land zone. Sevastopol: ECOSI–hydrophysics, p 363

Kochergin VS, Kochergin SV (2017) Identification of the parameters of an instantaneous point source of pollution in the Sea of Azov based on the method of conjugate equations. Mar Hydrophys J. 1 pp 66–71. EDN XSWBBB. https://doi.org/10.22449/0233-7584-2017-1-66-71

Kochergin VS, Kochergin SV (2017) Using the solution of conjugate problems in identifying the input parameters of the transfer model and planning the experiment. Ecol Bull Sci CentS Black Sea Econ Coop. 2 pp 42–47

Kochergin VS, Kochergin SV, Stanichny SV (2020) Variational assimilation of satellite data on the surface concentration of suspended matter in the Sea of Azov. Mod Probl Remote Sens Earth Space. 17(2) pp 40–48. https://doi.org/10.21046/2070-7401-2020-17-2-40-48

Marchuk GI (1982) Mathematical modeling in the problem of the environment. Nauka M, p 320

Marchuk GI, Penenko VV (1979) Application of optimization methods to the problem of mathematical simulation of atmospheric processes and environment. Model Optim Complex Syst. Ed. G. I. Marchuk. Berlin : Springer, pp 240–252. https://link.springer.com/chapter/10.1007%2FBFb000416

Marchuk GI, Sarkisyan AS (1988) Mathematical modeling of ocean circulation. Nauka, Moscow, p 302

Penenko VV (1981) Methods of numerical modeling of atmospheric processes. L.: Gidrometeoizdat p 350

Shutyaev VP (2019) Methods for observation data assimilation in problems of physics of atmosphere and ocean. Izv, Atmos Ocean Phys 55 pp 17–31. https://doi.org/10.1134/S0001433819010080

Strakhov VN (1991) Method of filtering systems of linear algebraic equations—the basis for solving linear problems of gravimetry and magnetometry. Dokl. ANSSR. 320(3), pp 595–599

Analytical Model of Wind Circulation Under Variable Wind Action

S. V. Kochergin, V. S. Kochergin, and S. N. Sklyar

Abstract The paper provides an analytical solution for a three-dimensional model of wind currents under the influence of wind of complex configuration. The availability of analytical solutions for barotropic velocity components, additional velocities, and vertical components is important when testing and analyzing difference schemes and algorithms in the process of constructing hydrodynamic models of reservoir dynamics.

Keywords Dimensionless problem · Wind currents · Test problem · Analytical solution

1 Introduction

Numerical modeling of the dynamics of reservoirs is an important tool in solving urgent problems of environmental monitoring and climatic problems related to weather forecasting. Significant development of computer technology makes it possible to use an increase in the discretization of solved problems to increase the accuracy of describing the dynamic processes taking place in the ocean. However, the direction related to the construction and correction of the models themselves and the methods of their numerical implementation is successfully developing, because when using the available computer power together, new computing schemes and algorithms can lead to a significant joint effect. When testing numerical models of ocean dynamics, the results obtained from different models are most often compared with each other, which is often subjective. Therefore, comparing the numerical solution with the exact analytical one allows us to choose a more correct scheme for the numerical implementation of the ocean dynamics models used, which are quite

S. V. Kochergin · V. S. Kochergin (✉)
Marine Hydrophysical Institute RAS, Sevastopol, Russia
e-mail: vskocher@gmail.com

S. N. Sklyar
American University of Central Asia, Bishkek, Kyrgyzstan

T. Chaplina (ed.), *Processes in GeoMedia—Volume IX*, Springer Geology,
https://doi.org/10.1007/978-981-96-9302-3_21

complex. Analytical solutions for dynamic models exist only for the simplest statements, for example, the Stommel model (1948, 1965), Stommel and Gol'fstrim (1965), Eremeev et al. (2002). In Eremeev et al. (2002), such a task is implemented using the method of dynamic operator inversion to study the applied computational schemes of a special type for calculating velocity fields. In this paper, we consider a more complex problem, the solution of which allows us to obtain an analytical solution for the barotropic velocity component, its three-dimensional additional part and the vertical component. In Kochergin et al. (2019), this problem was solved under the condition that the wind was specified in Stommel's formulation. In this paper, analytical solutions are obtained when specifying a sufficiently general wind effect.

2 A Problem in a Dimensionless Form

Let's consider the problem of wind circulation for the reservoir in question, which has the shape of a rectangle in the $x0y$ plane:

$$\Omega_0 = [0, r] \times \left[0, q\right]$$

and its depth $H > 0$. The axes of the Cartesian coordinate system have the following directions: $0x$—east, $0y$—north, $0z$—vertically downward. In the three-dimensional domain $\Omega = \{(x, y, z)|(x, y) \in \Omega_0,\ 0 \le z \le H\}$, we consider the system of equations of motion in a dimensionless (Kochergin and Kochergin 2019) form:

$$\begin{cases} -\ell v = -\dfrac{\partial P^s}{\partial x} + \dfrac{\partial}{\partial z}\left(k\dfrac{\partial u}{\partial z}\right), \\ \ell u = -\dfrac{\partial P^s}{\partial y} + \dfrac{\partial}{\partial z}\left(k\dfrac{\partial v}{\partial z}\right), \qquad t > 0,\ \ (x, y, z) \in \overset{0}{\Omega}, \\ \dfrac{\partial u}{\partial x} + \dfrac{\partial v}{\partial y} + \dfrac{\partial w}{\partial z} = 0, \end{cases} \tag{1}$$

with boundary conditions:

$$\left\{t > 0,\ \ z = 0,\ \ (x, y) \in \overset{0}{\Omega}_0\right\}: \quad k\frac{\partial u}{\partial z} = -\tau_x,\ \ k\frac{\partial v}{\partial z} = -\tau_y,\ \ w = 0; \tag{2}$$

$$\left\{t > 0,\ \ z = H,\ \ (x, y) \in \overset{0}{\Omega}_0\right\}: \quad k\frac{\partial u}{\partial z} = -\tau_x^b,\ \ k\frac{\partial v}{\partial z} = -\tau_y^b,\ \ w = 0; \tag{3}$$

$$\{t > 0,\ \ 0 \le z \le H,\ \ (x, y) \in \partial\Omega_0\}: \ \ U \cdot n_x + V \cdot n_y = 0; \tag{4}$$

and the initial data:

$$\{t = 0,\ (x, y, z) \in \Omega\} :\ u = u^o,\ \ v = v^o.\ w = w^o. \tag{5}$$

In (4), the integral velocities are defined in the standard way:

$$U(t, x, y) = \int_0^H u(t, x, y, z)\mathrm{d}z,\quad V(t, x, y) = \int_0^H v(t, x, y, z)\mathrm{d}z,$$

and in (3), a variant of parameterization of bottom friction is used:

$$\tau_x^b = \mu \cdot U,\quad \tau_y^b = \mu \cdot V,\quad \mu \equiv \text{const} > 0. \tag{6}$$

As in the Stommel model, let:

$$\ell = \ell_0 + \beta \cdot y,\ k \equiv \text{const} \tag{7}$$

and the components of the tangential stress of wind friction will be given by the formulas:

$$\begin{cases} \tau_x = [F_1 \cdot \cos(r_l x) + F_2 \cdot \sin(r_l x)] \cdot \cos(q_m y), \\ \tau_y = [G_1 \cdot \cos(r_s x) + G_2 \cdot \sin(r_s x)] \cdot \sin(q_p y); \end{cases} \tag{8}$$

in which the designations are accepted:

$$\begin{aligned} &r_l = \frac{\pi l}{r};\ r_s = \frac{\pi s}{r};\ q_m = \frac{\pi m}{q};\ q_p = \frac{\pi p}{q}; \\ &l,\ s = 0,\ 1,\ 2,\ \ldots;\quad m,\ p = \ 1,\ 2,\ \ldots \end{aligned} \tag{9}$$

Thus, the wind model contains a number of real: $F_1,\ F_2,\ G_1,\ G_2$ and whole ones: $l,\ m,\ s,\ p$ numerical parameters, the choice of which allows us to describe fairly general situations of wind forcing.

The horizontal components of the velocity vector are written as:

$$u = U \cdot H^{-1} + \hat{u},\qquad v = V \cdot H^{-1} + \hat{v} \tag{10}$$

where the first terms are called barotropic, and the second terms are called the additional components of velocity.

3 Analytical Solutions. Barotropic Components

In the system of Eqs. (1)–(5), we integrate both equations in "z" from 0 to H, taking into account the boundary conditions. As a result, we get the problem for integral velocities:

$$\begin{cases} \mu U - \ell V = -H\dfrac{\partial P^s}{\partial x} + \tau_x, \\ \ell U + \mu V = -H\dfrac{\partial P^s}{\partial y} + \tau_y, \\ \dfrac{\partial U}{\partial x} + \dfrac{\partial V}{\partial y} = 0, \quad (x, y) \in \overset{0}{\Omega}_0; \\ U \cdot n_x + V \cdot n_y = 0, \quad (x, y) \in \partial\Omega_0. \end{cases} \tag{11}$$

From the first two equations in (11), we exclude pressure gradients using cross-differentiation and introducing the current function using $\Psi = (x, y)$ the formulas:

$$U = \frac{\partial \Psi}{\partial y}, \quad V = -\frac{\partial \Psi}{\partial x}$$

to find the current function, we obtain the problem:

$$\begin{cases} \mu\left(\dfrac{\partial^2\Psi}{\partial x^2} + \dfrac{\partial^2\Psi}{\partial y^2}\right) + \beta\dfrac{\partial \Psi}{\partial x} = \dfrac{\partial \tau_x}{\partial y} - \dfrac{\partial \tau_y}{\partial x}, \quad (x, y) \in \underset{0}{\overset{0}{\Omega}}, \\ \Psi = 0, \quad (x, y) \in \partial\Omega_0. \end{cases} \tag{12}$$

The solution of Eq. (12) is represented as the sum of partial solutions:

$$\Psi(x, y) = \Psi_1(x, y) + \Psi_2(x, y)$$

satisfying the following tasks:

$$\begin{cases} \mu\left(\dfrac{\partial^2\Psi_1}{\partial x^2} + \dfrac{\partial^2\Psi_1}{\partial y^2}\right) + \beta\dfrac{\partial \Psi_1}{\partial x} = \dfrac{\partial \tau_x}{\partial y}, \quad (x, y) \in \underset{0}{\overset{0}{\Omega}}, \\ \Psi_1 = 0, \quad (x, y) \in \partial\Omega_0, \end{cases} \tag{13}$$

$$\begin{cases} \mu\left(\dfrac{\partial^2\Psi_2}{\partial x^2} + \dfrac{\partial^2\Psi_2}{\partial y^2}\right) + \beta\dfrac{\partial \Psi_2}{\partial x} = -\dfrac{\partial \tau_y}{\partial x}, \quad (x, y) \in \underset{0}{\overset{0}{\Omega}}, \\ \Psi_2 = 0, \quad (x, y) \in \partial\Omega_0. \end{cases} \tag{14}$$

To solve each of the problems (13), (14), we find partial solutions, and then the general solution of the corresponding homogeneous equations, summing them up and satisfying the boundary conditions for these problems, due to the arbitrariness contained in the general solution of the homogeneous equation, we end up with:

$$\Psi_1(x,y) = \left[C_1e^{Ax} + C_2e^{Bx} + D_1\cdot\cos(r_lx) + D_2\cdot\sin(r_lx)\right]\cdot\sin(q_my),$$
$$\Psi_2(x,y) = \left[\overline{C}_1e^{\bar{A}x} + \overline{C}_2e^{\bar{B}x} + \overline{D}_1\cdot\cos(r_sx) + \overline{D}_2\cdot\sin(r_sx)\right]\cdot\sin(q_py), \qquad (15)$$

$$A = -\frac{\beta}{2\mu} + \sqrt{\left(\frac{\beta}{2\mu}\right)^2 + q_m^2};\quad B = -\frac{\beta}{2\mu} - \sqrt{\left(\frac{\beta}{2\mu}\right)^2 + q_m^2},$$
$$D_1 = q_m\cdot\frac{\mu\cdot(r_l^2+q_m^2)\cdot F_1 + \beta\cdot r_l\cdot F_2}{\mu\cdot(r_l^2+q_m^2)^2 + \beta^2\cdot r_l^2},\ D_2 = q_m\cdot\frac{\mu\cdot(r_l^2+q_m^2)\cdot F_2 - \beta\cdot r_l\cdot F_1}{\mu\cdot(r_l^2+q_m^2)^2 + \beta^2\cdot r_l^2},$$
$$C_1 = D_1\cdot\frac{e^{Br}-(-1)^l}{e^{Ar}-e^{Br}},\ C_2 = -D_1\cdot\frac{(-1)^l - e^{Ar}}{e^{Ar}-e^{Br}},$$

$$\overline{A} = -\frac{\beta}{2\mu} + \sqrt{\left(\frac{\beta}{2\mu}\right)^2 + q_p^2};\quad \overline{B} = -\frac{\beta}{2\mu} - \sqrt{\left(\frac{\beta}{2\mu}\right)^2 + q_p^2},$$
$$\overline{D_1} = r_s\cdot\frac{\mu\cdot(r_s^2+q_p^2)\cdot G_2 + \beta\cdot r_s\cdot G_1}{\mu^2\cdot(r_s^2+q_p^2)^2 + \beta^2\cdot r_s^2},\ \overline{D_2} = r_s\cdot\frac{\mu\cdot(r_s^2+q_p^2)\cdot G_1 + \beta\cdot r_s\cdot G_2}{\mu^2\cdot(r_s^2+q_p^2)^2 + \beta^2\cdot r_s^2},$$
$$\overline{C_1} = \overline{D_1}\cdot\frac{e^{\bar{B}r}-(-1)^s}{e^{\bar{A}r}-e^{\bar{B}r}},\ C_2 = -\overline{D_1}\cdot\frac{(-1)^s - e^{\bar{A}r}}{e^{\bar{A}r}-e^{\bar{B}r}}.$$

Finally, the integral velocities are found by the formulas:

$$U(x,y) = q_m\left[C_1\cdot e^{Ax} + C_2\cdot e^{Bx} + D_1\cdot\cos(r_lx) + D_2\cdot\sin(r_lx)\right]\cdot\cos(q_my) +$$
$$+ q_p\left[\overline{C_1}\cdot e^{\bar{A}x} + \overline{C_2}\cdot e^{\bar{B}x} + \overline{D_1}\cdot\cos(r_sx) + \overline{D_2}\cdot\sin(r_sx)\right]\cdot\cos(q_py)$$
$$V(x,y) = -\left[A\cdot C_1\cdot e^{Ax} + B\cdot C_2\cdot e^{Bx} + r_l\cdot D_1\cdot\sin(r_lx) + r_l\cdot D_2\cdot\cos(r_lx)\right]\cdot\sin(q_my) -$$
$$-\left[\bar{A}\cdot\overline{C_1}\cdot e^{\bar{A}x} + \bar{B}\cdot\overline{C_2}\cdot e^{\bar{B}x} - r_s\cdot\overline{D_1}\cdot\cos(r_sx) + r_s\cdot\overline{D_2}\cdot\sin(r_sx)\right]\cdot\sin(q_py),$$

4 Analytical Solution. Additional Components

Let us substitute the horizontal components of the velocity vector written in the form (10) into the first two equations of system (11). Considering the equations for integral velocities:

$$\begin{cases} -\ell V H^{-1} = -\dfrac{\partial P^s}{\partial x} + \dfrac{\tau_x - \tau_x^b}{H}, \\ \ell U H^{-1} = -\dfrac{\partial P^s}{\partial y} + \dfrac{\tau_y - \tau_y^b}{H}, \end{cases}$$

we obtain the problem for the additional components of horizontal velocities:

$$\begin{cases} -\ell \hat{v} = \dfrac{\partial}{\partial z}\left(k\dfrac{\partial \hat{u}}{\partial z}\right) + \dfrac{\tau_x^b - \tau_x}{H}, \\ \ell \hat{u} = \dfrac{\partial}{\partial z}\left(k\dfrac{\partial \hat{v}}{\partial z}\right) + \dfrac{\tau_y^b - \tau_y}{H}, \end{cases} \quad (x, y) \in \Omega_0,\ 0 < z < H. \tag{16}$$

$$\{z = 0,\ (x, y) \in \Omega_0\}: \quad k\frac{\partial \hat{u}}{\partial z} = -\tau_x,\ k\frac{\partial \hat{v}}{\partial z} = -\tau_y; \tag{17}$$

$$\{z = H,\ (x, y) \in \Omega_0\}: \quad k\frac{\partial \hat{u}}{\partial z} = -\tau_x^b,\ k\frac{\partial \hat{v}}{\partial z} = -\tau_y^b. \tag{18}$$

Let's write the problem (16)–(18) in a complex form and introduce the following functions:

$$\begin{aligned} &\theta = \hat{u} + i \cdot \hat{v},\ \ \tau^b = \tau_x^b + i \cdot \tau_y^b = \mu(U + i \cdot V), \\ &\tau = \tau_x + i \cdot \tau_y = [F_1 \cdot \cos(r_l x) + F_2 \cdot \sin(r_l x)] \cdot \cos(q_m y) + \\ &\quad + i \cdot [G_1 \cdot \cos(r_s x) + G_2 \cdot \sin(r_s x)] \cdot \sin(q_p y). \end{aligned}$$

Then we have:

$$\begin{cases} i\ell\theta - k\dfrac{\partial^2 \theta}{\partial z^2} = \dfrac{\tau^b - \tau}{H}, & 0 < z < H; \\ z = 0: & k\dfrac{\partial \theta}{\partial z} = -\tau, \\ z = H: & k\dfrac{\partial \theta}{\partial z} = -\tau^b. \end{cases} \tag{19}$$

The general solution of the equation from (19) is found as the sum of the partial solution of this equation and the general solution of the corresponding homogeneous equation. Let we have:

$$\begin{aligned} C(z) &= \cos(H - z)\eta \cdot ch(H + z)\eta - \cos(H + z)\eta \cdot ch(H - z)\eta, \\ S(z) &= \sin(H - z)\eta \cdot sh(H + z)\eta - \sin(H + z)\eta \cdot sh(H - z)\eta, \end{aligned}$$

where:

$\eta = \sqrt{\frac{\ell}{2k}}$,

$$L(z) = \frac{C(z)}{C(H)} - \frac{z}{H}, \quad M(z) = \frac{S(z)}{C(H)},$$

$$Cs(z) = \cos(H-z)\eta \cdot sh(H+z)\eta + \cos(H+z)\eta \cdot sh(H-z)\eta,$$
$$Sc(z) = \sin(H-z)\eta \cdot ch(H+z)\eta + \sin(H+z)\eta \cdot ch(H-z)\eta.$$

Taking into account the boundary conditions on the surface and bottom, one can obtain from (19):

$$\begin{aligned}\hat{u} = {} & \frac{\mu V - [G_1 \cdot \cos(r_s x) + G_2 \cdot \sin(r_s x)] \cdot \sin(q_p y)}{\ell H} + \\ & + \frac{\eta}{\ell C(H)} \cdot \{([G_1 \cdot \cos(r_s x) + G_2 \cdot \sin(r_s x)] \cdot \sin(q_p y) + \\ & + [F_1 \cdot \cos(r_l x) + F_2 \cdot \sin(r_l x)] \cdot \cos(q_m y)) \cdot Cs(H-z) + \\ & + ([G_1 \cdot \cos(r_s x) + G_2 \cdot \sin(r_s x)] \cdot \sin(q_p y) - \\ & - [F_1 \cdot \cos(r_l x) + F_2 \cdot \sin(r_l x)] \cdot \cos(q_m y)) \cdot Sc(H-z) - \\ & - \mu(U+V) \cdot Cs(z) + \mu(U-V) \cdot Sc(z)\},\end{aligned}$$

$$\begin{aligned}\hat{v} = {} & -\frac{[G_1 \cdot \cos(r_s x) + G_2 \cdot \sin(r_s x)] \cdot \sin(q_p y) - \mu V}{\ell H} + \\ & + \frac{\eta}{\ell C(H)} \cdot \{([G_1 \cdot \cos(r_s x) + G_2 \cdot \sin(r_s x)] \cdot \sin(q_p y) - \\ & - [F_1 \cdot \cos(r_l x) + F_2 \cdot \sin(r_l x)] \cdot \cos(q_m y)) \cdot Cs(H-z) + \\ & + ([G_1 \cdot \cos(r_s x) + G_2 \cdot \sin(r_s x)] \cdot \sin(q_p y) + \\ & + [F_1 \cdot \cos(r_l x) + F_2 \cdot \sin(r_l x)] \cdot \cos(q_m y)) \cdot Sc(H-z) + \\ & + \mu(U-V) \cdot Cs(z) - \mu(U+V) \cdot Sc(z)\}.\end{aligned}$$

5 Analytical Solution. The Vertical Component of the Velocity Vector

The vertical component "w" of the velocity vector is determined from the continuity equation. To do this, we integrate the continuity equation for the variable z from 0 to z, taking into account the boundary condition (2) and the continuity equation for integral velocities.

$$\frac{\partial U}{\partial x} + \frac{\partial V}{\partial y} = 0.$$

You can get:

$$w(z) = D + C, \tag{20}$$

where

$$\begin{aligned}
D = &-(q_p \cdot [G_1 \cdot \cos(r_s x) + G_2 \cdot \sin(r_s x)] \cdot \cos(q_p y)- \\
&- r_l \cdot [F_1 \cdot \cos(r_l x) + F_2 \cdot \sin(r_l x)] \cdot \cos(q_m y)) \cdot \frac{M(H-z)}{l}+ \\
&+ (\mu \cdot (q_m \cdot [A \cdot C_1 e^{Ax} + B \cdot C_2 e^{Bx} - r_l \cdot D_1 \cdot \sin(r_l x)+ \\
&+ r_l \cdot D_2 \cdot \cos(r_l x)] \cdot \cos(q_m y)) \cdot \frac{M(z)}{l}+ \\
&+ (\mu \cdot (q_p \cdot [\overline{A} \cdot \overline{C_1} e^{\overline{A}x} + \overline{B} \cdot \overline{C_2} e^{\overline{B}x} - r_s \cdot \overline{D_1} \cdot \cos(r_s x)+ \\
&+ r_s \cdot \overline{D_2} \cdot \sin(r_s x)] \cdot \cos(q_p y)) \cdot \frac{M(z)}{l}+ \\
&+ (r_s \cdot [G_1 \cdot \sin(r_s x) + G_2 \cdot \cos(r_s x)] \cdot \sin(q_p y)- \\
&- q_m \cdot [F_1 \cdot \cos(r_l x) + F_2 \cdot \sin(r_l x)] \cdot \sin(q_m y)) \cdot \frac{L(H-z)}{l}- \\
&- (\mu \cdot ([A^2 \cdot C_1 e^{Ax} + B^2 \cdot C_2 e^{Bx} - r_l^2 \cdot D_1 \cdot \cos(r_l x)+ \\
&+ r_l^2 \cdot D_2 \cdot \sin(r_l x)] \cdot \sin(q_m y)) \cdot \frac{L(z)}{l}+ \\
&+ (\mu \cdot ([\overline{A} \cdot \overline{C_1} e^{\overline{A}x} + \overline{B} \cdot \overline{C_2} e^{\overline{B}x} - r_s^2 \cdot \overline{D_1} \cdot \cos(r_s x)+ \\
&+ r_s^2 \cdot \overline{D_2} \cdot \sin(r_s x)] \cdot \sin(q_p y)) \cdot \frac{L(z)}{l}.
\end{aligned} \tag{21}$$

To calculate the value of C, we define the necessary functions:

$$P(z) = (H+z)[\cos(H-z)\eta \cdot sh(H+z)\eta + \sin(H+z)\eta \cdot ch(H-z)\eta],$$
$$Q(z) = (H+z)[\sin(H-z)\eta \cdot ch(H+z)\eta - \cos(H+z)\eta \cdot sh(H-z)\eta];$$
$$CY(z) = \frac{\beta}{4k\eta \cdot C(H)} \cdot [P(z) - P(-z)], \quad SY(z) = \frac{\beta}{4k\eta \cdot C(H)} \cdot \left[Q(z) - Q(-z)\right].$$

Then:

$$\begin{aligned}
C = &\frac{[F_1 \cdot \cos(r_l x) + F_2 \cdot \sin(r_l x)] \cdot \cos(q_m y)}{\ell} \cdot \\
&\left\{CY(H-z) - CY(H) \cdot \left[L(H-z) + \frac{H-z}{H}\right]\right\}- \\
&- \frac{\mu U}{\ell} \cdot \left\{CY(z) - CY(H) \cdot \left[L(z) + \frac{z}{H}\right]\right\} - \frac{\mu V}{\ell} \cdot [SY(z) - CY(H) \cdot M(z)].
\end{aligned}$$

6 Conclusions

An analytical solution of the wind circulation model with complex wind effects has been constructed for use as a test solution (Kochergin et al. 2022, 2023a, b) in the analysis of schemes (Il'in 1969; Samarskij 1983) and algorithms for constructing dynamic models of reservoirs. Solutions to the problem for determining barotropic, additional velocities (variables in space) and velocity components in the vertical direction have been found, comparison with which will allow testing various difference discretizations and computational algorithms for constructing dynamic models of reservoirs.

Acknowledgements The work was performed within the framework of the state assignment on the topic FNNN-2024-0016 "Investigation of the spatial and temporal variability of oceanological processes in the coastal, coastal and shelf zones of the Black Sea under the influence of natural and anthropogenic factors based on contact measurements and mathematical modeling" (code "Coastal research").

References

Eremeev VN, Kochergin VP, Kochergin SV, Sklyar SN (2002) Matematicheskoe modelirovanie gidrodinamiki glubokovodnyh bassejnov. Sevastopol: Ekosi-Gidrofizika, 238 p

Il'in AM (1969) Raznostnaya skhema dlya differencial'nogo uravneniya s malym parametrom pri starshej proizvodnoj. Matematicheskie zametki 6(2):237–248

Kochergin VS, Kochergin SV (2019) Perekhod k bezrazmernoj zadache v modeli vetrovoj cirkulyacii. Ekologicheskij vestnik nauchnyh centrov CHernomorskogo ekonomicheskogo sotrudnichestva 4(16):25–30. https://doi.org/10.31429/vesnik-16-4-25-30

Kochergin VS, Kochergin SV, Sklyar SN (2019) Analiticheskaya testovaya zadacha vetrovyh techenij. Processy v geosredah—Moskva 2(20):193–198

Kochergin VS, Kochergin SV, Sklyar SN (2022) Testirovanie raznostnyh skhem pri reshenii uravneniya dlya funkcii toka na osnove resheniya zadachi vetrovyh techenij. Ekologicheskij vestnik nauchnyh centrov CHernomorskogo ekonomicheskogo sotrudnichestva 19(2):53–61. https://doi.org/10.31429/vesnik-19-2-53-61

Kochergin VS, Kochergin SV, Sklyar SN (2023a) Aspekty chislennogo resheniya nestacionarnoj zadachi vetrovogo dvizheniya zhidkosti. Ekologicheskij vestnik nauchnyh centrov CHernomorskogo ekonomicheskogo sotrudnichestva 20(1):12–18. https://doi.org/10.31429/vesnik-20-1-12-18

Kochergin VS, Kochergin SV, Sklyar SN (2023b) Calculation of the total flow components in the models of wind fluid motion. Phys Oceanogr 30(3):274–287. https://doi.org/10.29039/1573-160X-2023-3-274-287

Samarskij AA (1983) Teoriya raznostnyh skhem M.: Nauka, 616 p.

Stommel H (1948) The westward intensification of wind-driven ocean currents. Trans Am Geoph Un 29:202–206

Stommel H (1965) The gulf stream. A physical and dynamical description. University of California Press

Stommel H, Gol'fstrim M (1965) IL, p 227

Quantum Kinetic Theory and the Quantum Image Particle Method for Kinetic Simulation of Wakes Including Particles with Spin

Maxim G. Ponomarev

Abstract Availability of the quantum mechanical model with particle spin is demonstrated in addition to the previous classical physics, classical mechanics models. The spin is introduced using a spin transform, or Q-transform. Quantum kinetic theory with particle spin and the quantum Image Particle Method are formulated for simulations of charged particle fluxes, which allow to include effects of generated electric and magnetic fields in addition to all previous results for charged particle flows simulations in external electromagnetic fields. A spin transform, or Q-transform, and gauge invariant Wigner function have been introduced in this paper for kinetic simulation of wakes, disturbances in plasma with formulation in terms of general electromagnetic fields. These spin transform and Wigner formulation are the natural ways to describe quantum transport, and to monitor decoherence of macroscopic quantum states in interaction with the environment. According to this new quantum Image Particle Method with particle spin, the collision integrals, corresponding simulated collision operators for these "image particles" describe the effects from the generated electric and magnetic fields. In addition to quantum generalization these models show obvious comparison and correspondence with previous both analytical, and numerical Boltzmann and Vlasov models.

Keywords Spin · Spin transform · Q-transform · Quantum transport · Interaction with the environment · Gauge invariant Wigner function · Wigner model · Quantum kinetic theory · Wakes · Disturbances · Weak coupling · Intermediate coupling · Strong coupling · Weak magnetization · Strong magnetization · Extreme magnetization · Geomagnetic fields · Magnetic fields · Charged particle flows · Magnetic field effect on ionized flows · Plasma · Geomagnetism · Charged particle sources · Explosive charged particle sources · Charged particle fluxes · Electric and magnetic field effects · Effect of induced electric and magnetic fields · Electromagnetic fields

M. G. Ponomarev (✉)
Cambridge Research and Consultancy, Cambridge, UK
e-mail: maxim.pv@gmail.com

Department of Mathematics and Physics, Russian State Hydrometeorological University, Saint-Petersburg, Russia

T. Chaplina (ed.), *Processes in GeoMedia—Volume IX*, Springer Geology,
https://doi.org/10.1007/978-981-96-9302-3_22

1 Introduction

This paper extends directly analytical results of previous paper by Ponomarev (2024b) by addition of influence of quantum particle spin. Furthermore, this publication gives yet another example from extensive number of innovative physical models with applications to real world problems published in Ponomarev (2022a, b, c, 2023a, b, c, 2024a, b), Ponomarev et al. (2010, 2011, 2012a, b, 2013a, b, 2014, 2022), Iovea et al. (2012, 2013), Webb and Ponomarev (2007), Ponomarjov and Carati (2003, 2004, 2004a, b, c, d, 2006), May et al. (2006), Carati and Ponomarjov (2003), Ponomarjov (1995, 1996, 1997a, b, 1998, 1999, 2000a, b, c, d, e, f, 2001a, b, c, d, 2002), Ponomarjov and Gunko (1995), Gun'ko and Ponomarev (1993, 1995a, b).

Quantum Particle Spin and Magnetic Moment

It is shown by experiments that particles with spin $\overrightarrow{s_e}$ carry intrinsic magnetic moments, which is proportional to their spins. For electrons, the Landé g-factor g_e is described by the following relation between the magnetic moment $\overrightarrow{\mu_e}$ and spin $\overrightarrow{s_e}$:

$$\overrightarrow{\mu_e} = \frac{g_e}{2}\frac{e}{m_e}\overrightarrow{s_e} \tag{1}$$

And it is estimated both theoretically and experimentally as having a value close to -2, the negative sign is from the electric charge $-|e|$ of the electron (National Institute of Standards and Technology 2025), electron g factor g_e

$$g_e \simeq -2.00231930436092 \tag{2}$$

This most recent estimation has standard uncertainty, 0.00000000000036 and relative standard uncertainty 1.8×10^{-13}.

2 Kinetic Simulation of Disturbances and Wakes of Bodies with Quantum Particle Spin

Furthermore, the consideration based on the Schrödinger equation will be extended to add the spin dynamics. The higher order contributors in the BBGKY-hierarchy (Bogolyubov 1946) will be neglected. And, therefore, the evolution equation will be derived from the single-particle density matrix, based on the von Neumann equation, which then proceeded with a Wigner transform. Important addition here is that the density matrix now depends on the spin state.

For this reason, the single component in the Scrödinger case is replaced by a two by two matrix, where the different components correspond to different spin states.

Pauli Hamiltonian for Evolution Equations Taking into Account Quantum Spin

To add the spin dynamics, in the non-relativistic approximation, the Pauli Hamiltonian for evolution equations is introduced in the following form:

$$\widehat{H} = m_i c^2 + e_i \varphi + \frac{\left(\widehat{\vec{p}} - e_i \vec{A}\right)^2}{2m_i} - \frac{e_i \hbar \vec{\sigma} \cdot \vec{B}}{2m_i} \tag{3}$$

Here $\vec{\sigma}$ is a vector with the 2×2 Pauli matrices as components. A gauge-invariant scalar kinetic theory is developed utilising a density matrix description for a spin-1/2 particle (Zamanian et al. 2010; Curtright et al. 2018; Weinbub and Ferry 2018; Brodin and Zamanian 2022).

The von Neumann equation is used here, therefore the evolution equation for the density matrix is written as follows:

$$i\hbar \frac{\partial \widehat{\rho_{\alpha\beta}}}{\partial t} = \left[\widehat{H}, \widehat{\rho_{\alpha\beta}}\right] \tag{4}$$

Furthermore, the Wigner–Stratonovich transform (Stratonovic 1970) has the following form:

$$W_{\alpha\beta}\left(\vec{r}, \vec{p}, t\right) = \left(\frac{1}{2\pi\hbar}\right)^3 \int d\vec{\xi} \exp\left(\frac{-i\vec{\phi} \cdot \vec{\xi}}{\hbar}\right) \rho_{\alpha\beta}\left(\vec{r} + \vec{\xi}/2, \vec{r} - \vec{\xi}/2, t\right) \tag{5}$$

with the phase

$$\vec{\phi} = \vec{p} - e_i \int_{-1/2}^{1/2} d\theta \vec{A}\left(\vec{r} + \vec{\xi}\theta, t\right) \tag{6}$$

This phase $\vec{\phi}$ is introduced to have gauge invariance of the resulting distribution function.

We see that Eqs. (4) and (5) can be considered as kinetic evolution equations, but there is no physical meaning for the individual components of this Wigner matrix $W_{\alpha\beta}\left(\vec{r}, \vec{p}, t\right)$. For this reason, more functions with clear physical meaning are required.

Spin Transform or Q-transform

The one way to introduce a function with clear physical meaning is to introduce a spin transform, or Q-transform according to Scully and Wódkiewicz (1994).

The scalar function $f\left(\vec{r}, \vec{p}, \vec{s}, t\right)$ is defined from the Wigner matrix $W_{\alpha\beta}\left(\vec{r}, \vec{p}, t\right)$ in the extended phase space in the following expression:

$$f(\vec{r}, \vec{p}, \vec{s}, t) = \frac{1}{4\pi} \sum_{\alpha\beta=1}^{2} (1 + \vec{s} \cdot \vec{\sigma})_{\alpha\beta} W_{\beta\alpha}(\vec{r}, \vec{p}, t) \tag{7}$$

here spin $\vec{s}$ is a vector of unit length. The spin variable $\vec{s}$ can be represented in spherical coordinates as well, as presented at Figs. 1 and 2:

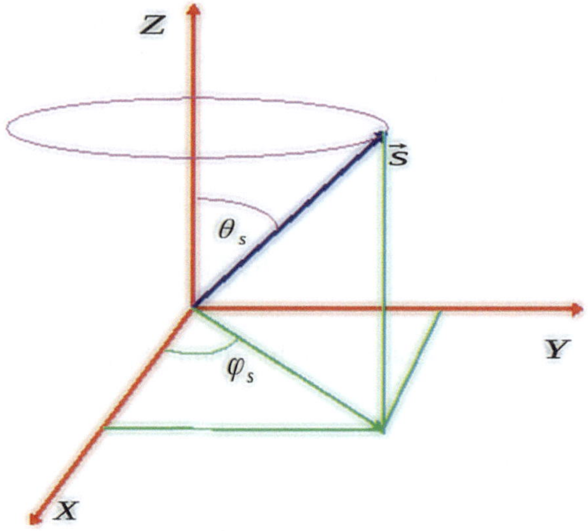

Fig. 1 Spin $\vec{s} = (\cos\varphi_s \sin\theta_s; \sin\varphi_s \sin\theta_s; \cos\theta_s)$ in spherical coordinates

Fig. 2 Spin $\vec{s} = (0; 0; 1)$ in spherical coordinates

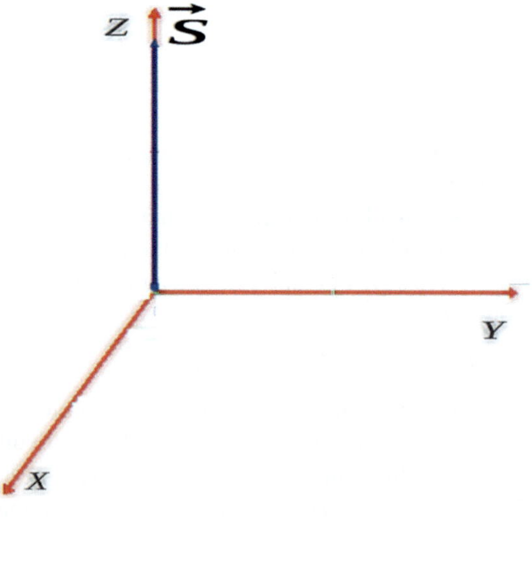

$$\vec{s} = (\cos\varphi_s \sin\theta_s;\ \sin\varphi_s \sin\theta_s;\ \cos\theta_s) \tag{8}$$

Reduced distribution function $f\left(\vec{r}, \vec{s}, t\right)$ can be obtained after integration over all space of momentums $\vec{p}$ as follows:

$$f(\vec{r}, \vec{s}, t) = \int f(\vec{r}, \vec{p}, \vec{s}, t) \mathrm{d}\vec{p} \tag{9}$$

It shows that this reduced distribution function $f\left(\vec{r}, \vec{s}, t\right)$ is proportional to the probability to have the particle close to position $\vec{r}$ with spin-up in the direction of $\vec{s}$. .

By analogy, introduce

$$f(\vec{p}, \vec{s}, t) = \int f(\vec{r}, \vec{p}, \vec{s}, t) \mathrm{d}\vec{r} \tag{10}$$

This Eq. (10) is proportional to the probability to have the particle close to momentum $\vec{p}$ with spin-up in the direction of $\vec{s}$.

Furthermore, these equations with velocity $\vec{v}$ as an independent variable are outlined.

For time dependent gauge-independent Wigner function, according to Brodin and Zamanian (2022), we have

$$\begin{aligned} &\frac{\partial f_W(\vec{r}, \vec{v}, \vec{s}, t)}{\partial t} + \left(\vec{v} + \delta\widehat{\vec{v}}\right) \cdot \frac{\partial f_W(\vec{r}, \vec{v}, \vec{s}, t)}{\partial \vec{r}} + \frac{e_i}{m_i}\left(\left[\vec{v} + \delta\widehat{\vec{v}}, \widehat{\vec{B}}\right] + \widehat{\vec{E}}\right) \\ &\cdot \frac{\partial f_W(\vec{r}, \vec{v}, \vec{s}, t)}{\partial \vec{v}} + \frac{\mu}{m_i}\frac{\partial}{\partial \vec{r}}\left(\left(\vec{s} + \frac{\partial}{\partial \vec{s}}\right) \cdot \widehat{\vec{B}}\right) \cdot \frac{\partial f_W(\vec{r}, \vec{v}, \vec{s}, t)}{\partial \vec{v}} \\ &\frac{+2\mu}{\hbar}\left(\vec{s} \times \left(\widehat{\vec{B}} + \delta\widehat{\vec{B}}\right)\right) \cdot \frac{\partial f_W(\vec{r}, \vec{v}, \vec{s}, t)}{\partial \vec{s}} = Q_i(\vec{r}, \vec{v}, \vec{s}, t) \end{aligned} \tag{11}$$

and we have the following operators here:

$$\delta\widehat{\vec{v}}(\vec{r}, t) = ie_i \frac{\hbar}{m_i^2} \frac{\partial}{\partial \vec{v}} \times \int_{-1/2}^{1/2} \xi d\xi\, \vec{B}\left(\vec{r} + i\frac{\hbar}{m_i}\xi\frac{\partial}{\partial \vec{v}}, t\right) \tag{12}$$

$$\widehat{\vec{E}}(\vec{r}, t) = \int_{-1/2}^{1/2} d\xi\, \vec{E}\left(\vec{r} + i\frac{\hbar}{m_i}\xi\frac{\partial}{\partial \vec{v}}, t\right) \tag{13}$$

$$\widehat{\vec{B}}(\vec{r}, t) = \int_{-1/2}^{1/2} d\xi\, \vec{B}\left(\vec{r} + i\frac{\hbar}{m_i}\xi\frac{\partial}{\partial \vec{v}}, t\right) \tag{14}$$

The new added terms in Eq. (11) are Stern and Gerlach force and the torque of the spin (respectively). Furthermore, the new operator $\delta\widehat{\vec{B}}(\vec{r},t)$is introduced in the following form:

$$\delta\widehat{\vec{B}}(\vec{r},t) = -i\frac{\hbar}{m_i}\int_{-1/2}^{1/2}\xi d\xi\,\vec{B}\left(\vec{r}+i\frac{\hbar}{m_i}\xi\frac{\partial}{\partial\vec{v}},t\right)\overleftarrow{\nabla}_{\vec{r}}\cdot\vec{\nabla}_{\vec{v}} \quad (15)$$

The added spin terms are to describe the splitting into two of the particle beam in Stern and Gerlach's experiments (Gracia-Bondá et al. 1993).

Here μ is the magnetic moment of the particle, for electrons are defined by the Bohr magneton

$$\mu_e = \frac{-g_e}{2}\frac{e\hbar}{2m_e} \quad (16)$$

where the estimation of this Land´e g-factor g_e is in Eq. (2).

Here $\vec{E} = \vec{E}(\vec{r},t)$ and $\vec{B} = \vec{B}(\vec{r},t)$ are time-dependent electric and magnetic fields (respectively).

$Q_i(\vec{r},\vec{v},\vec{s},t)$ simulates the sources of particles.

Furthermore, the Eq. (11) looks similar to the Boltzmann equation. Nethertheless, there are following differences:

(a) the operators $\widehat{\vec{E}}$ and $\widehat{\vec{B}}$ replace the electromagnetic fields $\vec{E}$ and $\vec{B}$ in accordance to Eqs. (13) and (14);
(b) added operator of quantum mechanical perturbation$\delta\widehat{\vec{v}}$ as a shift of the velocity vector $\vec{v}$ according to Eq. (12);
(c) the Wigner function can be negative in some regions of phase space, therefore it is not exactly a probability distribution;
(d) the new added parts of the Eq. (11) are Stern and Gerlach force and the torque of the quantum particle spin (respectively).

It should be noted that the quantum mechanical perturbation$\delta\widehat{\vec{v}}$ (according to Eq. 12) disappears in the electrostatic case.

3 Kinetic Simulation of Disturbances and Wakes in Plasma with Quantum Particle Spin, Quantum Image Particle Method

Let's consider induced electric $\overrightarrow{E_{\text{ind}}}$and induced magnetic$\overrightarrow{B_{\text{ind}}}$ fields of arbitrary, unlimited values so that

$$\vec{E} = \vec{E}_0 + \overrightarrow{E_{\text{ind}}} \tag{17}$$

$$\vec{B} = \vec{B}_0 + \overrightarrow{B_{\text{ind}}} \tag{18}$$

Then we have the Wigner equation, Eq. (11), in the following form:

$$\begin{aligned}
&\frac{\partial f_W(\vec{r},\vec{v},\vec{s},t)}{\partial t} + \left(\vec{v} + \delta\widehat{\vec{v}}\right)\cdot\frac{\partial f_W(\vec{r},\vec{v},\vec{s},t)}{\partial \vec{r}} + \frac{e_i}{m_i}\left(\left[\vec{v} + \delta\widehat{\vec{v}}, \widehat{\vec{B}}\right] + \widehat{\vec{E}}\right) \\
&\cdot\frac{\partial f_W(\vec{r},\vec{v},\vec{s},t)}{\partial \vec{v}} + \frac{\mu}{m_i}\frac{\partial}{\partial \vec{r}}\left(\left(\vec{s} + \frac{\partial}{\partial \vec{s}}\right)\cdot\widehat{\vec{B}}\right)\cdot\frac{\partial f_W(\vec{r},\vec{v},\vec{s},t)}{\partial \vec{v}} \\
&\frac{+2\mu}{\hbar}\left(\vec{s}\times\left(\widehat{\vec{B}} + \delta\widehat{\vec{B}}\right)\right)\cdot\frac{\partial f_W(\vec{r},\vec{v},\vec{s},t)}{\partial \vec{s}} = Q_i(\vec{r},\vec{v},\vec{s},t)
\end{aligned} \tag{19}$$

$$\begin{aligned}
\delta\widehat{\vec{v}}(\vec{r},t) = {} & ie_i\frac{\hbar}{m_i^2}\frac{\partial}{\partial \vec{v}} \times \int_{-1/2}^{1/2} \xi d\xi \vec{B}_0\left(\vec{r} + i\frac{\hbar}{m_i}\xi\frac{\partial}{\partial \vec{v}}, t\right) \\
& + ie_i\frac{\hbar}{m_i^2}\frac{\partial}{\partial \vec{v}} \times \int_{-1/2}^{1/2} \xi d\xi \overrightarrow{B_{ind}}\left(\vec{r} + i\frac{\hbar}{m_i}\xi\frac{\partial}{\partial \vec{v}}, t\right)
\end{aligned} \tag{20}$$

$$\begin{aligned}
\widehat{\vec{E}}(\vec{r},t) = {} & \int_{-1/2}^{1/2} d\xi \vec{E}_0\left(\vec{r} + i\frac{\hbar}{m_i}\xi\frac{\partial}{\partial \vec{v}}, t\right) \\
& + \int_{-1/2}^{1/2} d\xi \overrightarrow{E_{\text{ind}}}\left(\vec{r} + i\frac{\hbar}{m_i}\xi\frac{\partial}{\partial \vec{v}}, t\right)
\end{aligned} \tag{21}$$

$$\begin{aligned}
\widehat{\vec{B}}(\vec{r},t) = {} & \int_{-1/2}^{1/2} d\xi \vec{B}_0\left(\vec{r} + i\frac{\hbar}{m_i}\xi\frac{\partial}{\partial \vec{v}}, t\right) \\
& + \int_{-1/2}^{1/2} d\xi \overrightarrow{B_{\text{ind}}}\left(\vec{r} + i\frac{\hbar}{m_i}\xi\frac{\partial}{\partial \vec{v}}, t\right)
\end{aligned} \tag{22}$$

$$\begin{aligned}
\delta\widehat{\vec{B}}(\vec{r},t) = {} & -i\frac{\hbar}{m_i}\int_{-1/2}^{1/2} \xi d\xi \vec{B}\left(\vec{r} + i\frac{\hbar}{m_i}\xi\frac{\partial}{\partial \vec{v}}, t\right)\overleftarrow{\nabla}_{\vec{r}}\cdot\vec{\nabla}_{\vec{v}} \\
& - i\frac{\hbar}{m_i}\int_{-1/2}^{1/2} \xi d\xi \overrightarrow{B_{\text{ind}}}\left(\vec{r} + i\frac{\hbar}{m_i}\xi\frac{\partial}{\partial \vec{v}}, t\right)\overleftarrow{\nabla}_{\vec{r}}\cdot\vec{\nabla}_{\vec{v}}
\end{aligned} \tag{23}$$

wheree_i is electric charge and m_i is mass of the "i" particle.

Let's rewrite the Eq. (19) to see the effect of the perturbation operator $\delta\widehat{\vec{v}}$

$$\begin{aligned}&\frac{\partial f_W(\vec{r},\vec{v},\vec{s},t)}{\partial t}+\vec{v}\cdot\frac{\partial f_W(\vec{r},\vec{v},\vec{s},t)}{\partial\vec{r}}+\frac{e_i}{m_i}\left(\left[\vec{v},\hat{\vec{B}}\right]+\hat{\vec{E}}\right)\cdot\frac{\partial f_W(\vec{r},\vec{v},\vec{s},t)}{\partial\vec{v}}\\&\quad=Q_i(\vec{r},\vec{v},\vec{s},t)\frac{+\mu}{m_i}\frac{\partial}{\partial\vec{r}}\left(\left(\vec{s}+\frac{\partial}{\partial\vec{s}}\right)\cdot\hat{\vec{B}}\right)\cdot\frac{\partial f_W(\vec{r},\vec{v},\vec{s},t)}{\partial\vec{v}}\\&\quad+\frac{2\mu}{\hbar}\left(\vec{s}\times\left(\hat{\vec{B}}+\delta\hat{\vec{B}}\right)\right)\cdot\frac{\partial f_W(\vec{r},\vec{v},\vec{s},t)}{\partial\vec{s}}-\delta\hat{\vec{v}}\cdot\frac{\partial f_W(\vec{r},\vec{v},\vec{s},t)}{\partial\vec{r}}\\&\quad-\frac{e_i}{m_i}\left[\delta\hat{\vec{v}},\hat{\vec{B}}\right]\cdot\frac{\partial f_W(\vec{r},\vec{v},\vec{s},t)}{\partial\vec{v}}\end{aligned} \tag{24}$$

The right part of Eq. (24) contains the image particle sources terms, corresponding simulated collision operators, which are similar to non-quantum case outlined in previous paper Ponomarev (2024a).

It should be noted that we have the excellent correspondence here for the case of extremely small quantum effects. The imaginary particle sources term (in Eq. (7) in the previous paper Ponomarev (2024a)) contains e instead of e_i and m instead of m_i because $e = e_i$ and $m = m_i$ for the cases considered in this previous paper

$$C^{\text{image}}(\vec{r},\vec{v},t)=\frac{-e}{m}\left(\left[\vec{v},\overrightarrow{B_{\text{ind}}}\right]+\overrightarrow{E_{\text{ind}}}\right)\cdot\frac{\partial f_i(\vec{r},\vec{v},t)}{\partial\vec{v}} \tag{25}$$

This Eq. (25) has been derived there as a part of the Boltzmann equation (Eq. 8) in the previous paper Ponomarev (2024a) for the classical (non-quantum) case

$$\begin{aligned}&\frac{\partial f_i(\vec{r},\vec{v},t)}{\partial t}+\vec{v}\cdot\frac{\partial f_i(\vec{r},\vec{v},t)}{\partial\vec{r}}+\frac{e}{m}\left(\left[\vec{v},\overrightarrow{B_0}\right]+\overrightarrow{E_0}\right)\cdot\frac{\partial f_i(\vec{r},\vec{v},t)}{\partial\vec{v}}=Q_i(\vec{r},\vec{v},t)\\&\quad+C^{\text{image}}(\vec{r},\vec{v},t)+\sum_{j=1}^{N}C_{ij}\left(f_i(\vec{r},\vec{v},t),f_j(\vec{r},\vec{v},t)\right)\end{aligned} \tag{26}$$

and the corresponding isodensity contours have been plotted as well.

4 Conclusions

In this article, for the first time, the quantum Image Particle Method is formulated for quantum levels of descriptions to introduce a new method for simulation of charged particle fluxes, which allows the effects of generated electric and magnetic fields to be included in addition to all previous results for simulation of charged particle fluxes in external electric and magnetic fields, taking into considerations spin of quantum particles. The spin is considered using the spin transform or Q-transform.

According to this newly formulated quantum Image Particle Method, the corresponding simulated operators of collisions with these image particles describe the effects of the generated electric and magnetic fields.

The generated electric and magnetic fields can be unlimited and unsteady. The spin transform or Q transform and the Wigner model were applied in this paper for kinetic simulation of plasma wakes formulated in terms of general electromagnetic fields. These spin transformations and Wigner's formulation are natural tools for describing quantum transport and for analysing the decoherence of macroscopic quantum states when interweave with the environment.

This article shows the availability of quantum mechanical models with particle spin in addition to previous classical physics, classical mechanics models. In addition to the quantum generalization, the described models demonstrate an obvious comparison and correspondence with previous both analytical and numerical Boltzmann and Vlasov models.

References

Bogolyubov N.N. (1946) Kinetic equations (in Russian) Z. Eksp, Teor. Fiz. 16, 691. (Kinetic Equations (in English) [JETP 10, 265 (1946)])

Brodin G, Zamanian J (2022) Quantum kinetic theory of plasmas. Rev Mod Plasma Phys 6:4

Carati D, Ponomarjov MG (2003) Statistical description of currents induced by two electron cyclotron counter propagating waves. Electron cyclotron emission and electron cyclotron heating. In: Giruzzi G (ed) Proceedings of the 12th joint workshop. Held 13–16 May 2002 in Aix-en-Provence, France. (Association Euratom-CEA sur la Fusion, France). World Scientific Publishing Co. Pte. Ltd., pp77–82. ISBN #9789812705082

Curtright TL, Fairlie DB, Zachos CK (2018) A concise treatise on quantum mechanics in phase space. World Scientific Publishing. ISBN 978-981-4520-43-0

Gracia Bondá JM, Marshall TW, Santos E (1993) A phase-space description of the Stern-Gerlach phenomenon. Phys Lett A 183:19

Gun'ko YF, Ponomarev MG (1993) The charged particles emission in a magnetic field. Vestnik Sankt-Peterburgskogo Universiteta Ser 1 Matematika Mekhanika Astronomiya 2:89–94

Gun'ko YF, Ponomarev MG (1995a) The charged particle layer expansion in a magnetic field. Vestnik Sankt-Peterburgskogo Universiteta Ser 1 Matematika Mekhanika Astronomiya 4:105–108

Gun'ko YF, Ponomarev MG (1995b) The charged particle distribution due to emission in a magnetic field. Astronomische Nachrichten 316(1):17–21

Iovea M, Neagu M, Stefanescu B, Mateiasi G, Halai H, Amos M, Ponomarev M, Kappatos V, Selcuk C, Gan T-H, Gierl C (2012) Preliminary NDT investigation of sintered powder metallurgy parts by high-resolution TDI based X-ray digital radiography. In: European international powder metallurgy congress and exhibition. Euro PM 2012, vol 1. Basel, Switzerland. Code 105676

Iovea M, Neagu M, Stefanescu B, Mateiasi G, Clarke A, Nicholson I, Ponomarev M, Kappatos V, Selcuk C, Gan T-H (2013) PM parts fast in-line X-ray digital radiography. In: International powder metallurgy congress and exhibition. Euro PM 2013. European PM conference proceedings, vol 1. European Powder Metallurgy Association (EPMA)

May J, Ponomarev M, Kuball S, Gallardo J (2006) A case for new statistical software testing models. In: Annual reliability and maintainability symposium. RAMS'06, article number 1677399. Newport Beach, CA, pp 349–353. Category number CH3744; Code 69734. https://doi.org/10.1109/RAMS.2006.1677399

National Institute of Standards and Technology, USA, accessed 22 February 2025. https://physics.nist.gov/cgi-bin/cuu/Value?gem

Ponomarev MG (2022a) A novel physical model to enhance precision and performance of 3-dimensional force sensors. Process Geomedia 2(32):1589–1600

Ponomarev MG (2022b) Experimental validation of novel physical model for improvement of sensing 3-dimensional fluid flow loads and responses in real sea conditions with South Western Mooring Test Facility (SWMTF). Process Geomedia 2(32):1579–1589

Ponomarev MG (2022c) A novel physical model to enhance precision and performance of multi-dimensional force sensors. In: Physical and mathematical modeling of earth and environment processes—2022, proceedings of 8th international scientific conference-school. Springer, pp 347–364. ISBN 978-3-031-25961-6. https://doi.org/10.1007/978-3-031-25962-3_35

Ponomarev MG (2023a) Exact analytical expression for maximum heat flux from impulsive sources of charged particles in magnetic field. Process GeoMedia VII. https://doi.org/10.1007/978-981-99-6575-5_31

Ponomarev MG (2023b) Physical foundations of mechanics: part 1: Newtonian mechanics. World of Science LLC, pp 1–83

Ponomarev MG (2023c) Physical foundations of mechanics: part 2: relativistic mechanics. World of Science LLC, pp 1–78

Ponomarev MG (2024a) New image particle method for kinetic simulation of wakes in plasma with weak and strong coupling. In: Proceedings of the 9th international conference on physical and mathematical modelling of earth and environmental processes. PMMEEP 2023. Springer proceedings in earth and environmental sciences. Springer, Cham. https://doi.org/10.1007/978-3-031-54589-4_59

Ponomarev MG (2024b) Quantum kinetic theory and the image particle method for kinetic simulation of wakes in plasma. In: Chaplina T.(eds) Processes in GeoMedia—Volume VIII. Springer Geology. Springer, Singapore. https://doi.org/10.1007/978-981-97-6627-7_11

Ponomarev MG, Johanning L, Parish D (2010) Enhancing precision and reliability of tri-axial load cells for mooring load measurements. In: Proceedings of the 3rd international conference on ocean energy (ICOE2010). ISBN 978-84-693-5467-4

Ponomarev MG, Garrison BJ, Vickerman JC, Webb RP (2011) A molecular dynamics study of a 5 keV C60 fullerene impact on a two-component organic molecular sample. Surf Interface Anal 43(1–2):107–111

Ponomarev MG, Kappatos V, Selcuk C, Gan T-H, Amos M, Halai H, Gierl C, Iovea M (2012a) PM dimensional control: development of a digital radiographic inspection technique for production friendly quality assessment of powder metallurgy parts. In: Proceedings of the international euro powder metallurgy congress and exhibition. Euro PM, vol 1. Basel, Switzerland. Code 105676

Ponomarev MG, Selcuk C, Gan T-H (2012b) Construction of pattern recognition system optimized for X-ray inspection of plastic electronics and OLED displays. In: 51st annual conference of the British Institute of Non-Destructive Testing 2012. NDT 2012. BINDT. Northamptonshire, pp 354–360. Code 107048

Ponomarev MG, Kappatos V, Selcuk C, Gan T-H, Amos M, Halai H, Gierl C, Iovea M (2013) Digital radiographic inspection technique for production friendly quality assessment of PM parts. Powder Metall 56(2):92–95

Ponomarev MG, Selcuk C, Gan T-H, Amos M, Nicholson I, Iovea M, Neagu M, Stefanescu B, Mateiasi G (2013b) Non destructive testing: a defect detection and classification system for automatic analysis of digital radiography images of powder metallurgy parts. In: International powder metallurgy congress and exhibition. Euro PM 2013. Gothenburg, Sweden. Code 110976

Ponomarev MG, Selcuk C, Gan T-H, Amos M, Nicholson I, Iovea M, Neagu M, Stefanescu B, Mateiasi G (2014) Defect detection and classification system for automatic analysis of digital radiography images of PM parts. Powder Metall 57(1):17–20

Ponomarev MG, Johanning L, Parish D (2022) Experimental validation of novel physical model for improvement of sensing multidimensional fluid flow loads and responses in real sea conditions with South Western Mooring Test Facility. In: Physical and mathematical modeling of earth and environment processes—2022, proceedings of 8th international scientific conference-school.

Springer, pp 365–376. ISBN 978-3-031-25961-6. https://doi.org/10.1007/978-3-031-25962-3_36
Ponomarjov MG (1995) Disturbances of the ambient magnetoplasma due to its interactions with object surfaces. Imaginary emission method. Far-wake of objects moving through a rarefied plasma at different angles to the ambient magnetic field. Planet Space Sci 43(10–11):1419–1427
Ponomarjov MG (1996) Imaginary-emission method for modeling disturbances of all magnetoplasma species: reflecting and absorbing objects in motion through a rarefied plasma at different angles to the ambient magnetic field. Phys Rev E 54(5):5591–5598
Ponomarjov MG (1997a) Pressure of charged particle flows in ambient magnetic fields. Astron Nachr 318(3):187–192
Ponomarjov MG (1997b) New ways of protecting astronomical equipment and solar batteries of spacecrafts. Bull Am Astron Soc 29:1023
Ponomarjov MG (1998) Outer atmosphere and wake of space objects, kinetic simulation. Disturbances of ambient magnetoplasma due to diffuse reflecting bodies in motion. APS Div Plasma Phys Meet Abstr. Abstract id. U9Q.09
Ponomarjov MG (1999) Kinetic simulation of stratifications and flute structures of charged particle jets and wakes in the ambient magnetic field. Bull Astron Soc 31(4):1157
Ponomarjov MG (2000a) Space flows and disturbances due to bodies in motion through the magnetoplasma. Astrophys Space Sci 274(1):423–429
Ponomarjov MG (2000b) Simulation of oscillations in charged particle systems under the ambient magnetic field control. In: 2nd international conference. Control of oscillations and Chaos (COC 2000), vol 1. St. Petersburg, Russia. Code 57512. Proceedings (Cat. No.00TH8521), pp 167–170. https://doi.org/10.1109/COC.2000.873548
Ponomarjov MG (2000c) 3D collisional kinetic simulation of stratifications and flute structures of plasma flows and wakes in external magnetic fields. Bull Am Astron Soc 32:1022
Ponomarjov MG (2000d) 3D kinetic dynamical models of ionized HII clouds in external magnetic field. In: Ionized Gaseous Nebulae. Mexico City. Meeting abstract id. 44
Ponomarjov MG (2000e) 3D collisional kinetic simulation of stratifications and flute structures of plasma flows and wakes in external magnetic fields. In: 32nd annual meeting of the division for planetary sciences. Session 15. Outer planets IV-aurorae and magnetospheres. Pasadena, CA. Abstract 15.10. 1022
Ponomarjov MG (2000f) 3D time-dependent kinetic simulation of turbulent plasma flows under the effect of external magnetic field. In: American physical society, 42nd annual meeting of the APS division of plasma physics combined with the 10th international congress on plasma physics. Québec City. Canada Meeting ID: DPP00. Abstract id. NO1 007. https://ui.adsabs.harvard.edu/abs/2000APS.DPPNO1007P/abstract
Ponomarjov MG (2001a) Acceleration and transport of particles in collisionless plasmas: wakes due to the interaction with moving bodies. Astrophys Space Sci 277(1):39–44
Ponomarjov MG (2001b) Kinetic simulation of magnetic field effects on stratifications and flute structures of space plasma flows and wakes of bodies. In: 2001 joint assembly American geophysical union spring meeting. Boston, MA. Abstracts SM52B-04
Ponomarjov MG (2001c) Kinetic simulation of magnetic field effects on wakes of meteoroids imaginary emission method, vol 495. European Space Agency (Special Publication) ESA SP, pp 295–300
Ponomarjov MG (2001d) Kinetic modeling magnetic field effect on ion flows, disturbances and wakes in space plasma. In: Büchner J, Dum CT, Scholer M (eds) Space plasma simulation: proceedings of the sixth international school/symposium. ISSS-6. Garching, Germany. Schaltungsdienst Lange o.H.G, Berlin, p 328
Ponomarjov MG (2002) 3D time-dependent kinetic simulation of space plasma disturbances due to moving bodies with the ambient magnetic field effect. Adv Space Res 29(9):1397–1402
Ponomarjov MG, Carati D (2003) Kinetic simulations of relativistic electron flows in time-dependent electromagnetic fields. In: Electron cyclotron emission and electron cyclotron heating, conference proceedings. World Scientific Publishing Co Pte Ltd, pp 137–142

Ponomarjov MG, Carati D (2004a) Search for optimal 3D wave launching configurations for the acceleration of charged particles in a magnetized plasma: resonant moments method. In: American physical society. 46th annual meeting of the division of plasma physics. Savannah, GA. Meeting ID: DPP04. Abstract id. RP1.042

Ponomarjov MG, Carati D (2004b) Acceleration of charged particles by crossed cyclotron waves, resonant moments method. In: 35th COSPAR scientific assembly. Paris, France, p 2851

Ponomarjov MG, Carati D (2004c) Acceleration of electron populations by crossing EC waves in an external magnetic field. In: 13th joint workshop on electron cyclotron emission and electron cyclotron resonance heating (conference paper). Nizhny Novgorod, Russia. https://ec13.iapras.ru/papers/Ponomarjov.pdf

Ponomarjov MG, Carati D (2004d) Enhanced acceleration of charged particles by crossing electromagnetic waves in a magnetized plasma, resonant moments method. https://hal.archives-ouvertes.fr/hal-00001926v1

Ponomarjov MG, Carati D (2006) Enhanced acceleration of electrons populations by crossing electron cyclotron waves in an ambient magnetic field, resonant moments method. Adv Space Res 38(8):1576–1581

Ponomarjov MG, Carati D (2004) Search for optimal 2D and 3D wave launching configurations for the largest acceleration of charged particles in a magnetized plasma. Resonant Moments Method. arXiv 2004. https://arxiv.org/format/physics/0411075. arXiv:physics/0411075 [physics.plasm-ph]

Ponomarjov MG, Gunko YF (1995) Kinetic modeling of charged particle cloud expansion and emission in magnetic and electric fields. Planet Space Sci 43(10–11):1409–1418

Scully MO, Wódkiewicz K (1994) Spin quasi-distribution functions. Found Phys 24:85

Stratonovic RL (1970) Contribution to the quantum nonlinear theory of thermal fluctuations. Z Eksp Teor Fiz 58:1612 (Sov Phys -JETP 31:864)

Webb RP, Ponomarev MG (2007) Molecular dynamics simulation of low energy cluster impacts on carbon nanotubes. Nucl Instrum Methods Phys Res Sect B Beam Interact Mater Atoms 255(1):229–232

Weinbub J, Ferry DK (2018) Recent advances in Wigner function approaches. Appl Phys Rev 5:041104

Zamanian J, Marklund M, Brodin G (2010) Scalar quantum kinetic theory for spin-1/2 particles: mean field theory. New J Phys 12:043019

The Influence of Cosmophysical Factors on Anthropogenesis

A. V. Zhabin, V. G. Gadiyatov, D. A. Dmitriev, and A. A. Zhabina

Abstract The paper considers the influence of cosmic events on the evolutionary transformations of primates in the Cenozoic period. The basis of evolution is the competition between species, genera and higher taxa in acquiring new qualities that contribute to the capture of more resources. The driving forces of anthropogenesis are catastrophes of cosmic origin. The main species of primates are shown, which are the ancestors of not only humans, but also all living monkeys. At the same time, new species appeared quite suddenly, being better adapted to the environment. Correlative connections between cataclysms and the emergence of new species and genera of hominids are given. It is concluded that the recognition of Eternity unites atheism and religion. The Universe as a whole personifies a system that has no beginning and end, that is, it is "immortal". But it, in turn, consists of objects that have, although long in time, but finite boundaries of existence, changing, like in a kaleidoscope, their location in space.

Keywords Anthropogenesis · Evolution · Catastrophes · Crises · Parapithecus · Dryopithecus · Australopithecines · Neanderthals · Cro-magnons · Modern man

1 Introduction

The current concept of the evolution of primates in general and humans in particular is very problematic in terms of defining the age limits of the existence of a particular species. Part of the reason is that some researchers, based on their scientific constructions, often move time boundaries in one direction or another. Even

A. V. Zhabin · D. A. Dmitriev
Voronezh State University, Voronezh, Russia

V. G. Gadiyatov (✉)
Voronezh State Technical University, Voronezh, Russia
e-mail: gadiatovvg@mail.ru

A. A. Zhabina
"Ecological Center" LLC, Voronezh, Russia

T. Chaplina (ed.), *Processes in GeoMedia—Volume IX*, Springer Geology,
https://doi.org/10.1007/978-981-96-9302-3_23

more confusing is the information about disasters related to space events. Despite the presence of many inconsistencies, we tried to find correlations between these seemingly distant manifestations of natural processes. Namely, between the existence of a particular type of hominid in a certain period of time and the catastrophes that occurred on Earth, which, in our opinion, were the key driving forces of the processes of anthropogenesis.

The epistemological problems of general evolutionary processes and specific particular transformations in a number of apelike and human-like individuals should be considered from the point of view of such forms of cognition as philosophical and scientific. The latter makes it possible to operate with specific scientific knowledge currently available in the fields of astronomy and anthropology. We tried to philosophically comprehend all the results of the development of these sciences available to us, reducing them into a single system of dependencies on each other.

Different authors have considered the possibilities of the formation of life on the planets of the Solar system, including from inert matter (Grant 1980). At the beginning of this process, according to this researcher, a chemical evolution should take place, consisting in the accumulation of organic compounds in primary reservoirs. In other words, an organic broth will form. This was favored by two conditions—the absence of living organisms and free oxygen. It is these factors that cause the breakdown of simple organic molecules, preventing the formation of complex organic compounds such as amino acids. Even in the process of chemical evolution, nucleic acid molecules with genetic activity arise. According to the author, they are "naked" genes. However, they already have the ability to replicate, reproduce, and mutate. Replication repeats mutated forms. Reproduction continues up to certain limits set by the amount of environmental resources. Because of this, there will be competition between genes or their carriers, and as a result, those who will be able to adapt to environmental conditions will survive. Therefore, the trend towards evolution is based on the acquisition of new qualities that contribute to the capture of more resources.

2 The Main Stages of Evolutionary Transformations

Based on different criteria, it is possible to consider the evolutionary transformations of all biological forms in general, and hominids in particular. The problem of anthropogenesis is of particular importance in connection with the expansion of biological scientific knowledge in the evolutionary cluster of humanity. This narrative is the main one in the research of leading scientific organizations. Figure 1 shows a diagram of the evolutionary stages of primates from ape-like ancestors to modern humans. It is sufficiently eliminated, that is, all unnecessary forms representing dead-end branches of evolutionary transformations have been eliminated.

The first primates, the hypothetical ancestors of narrow-nosed and anthropoid apes, appeared in the Paleocene-Eocene period. Probably, another catastrophe that occurred at the Cretaceous-Paleogene boundary (about 66 million years ago) finally destroyed the dinosaurs, thereby freeing up ecological niches that were occupied by

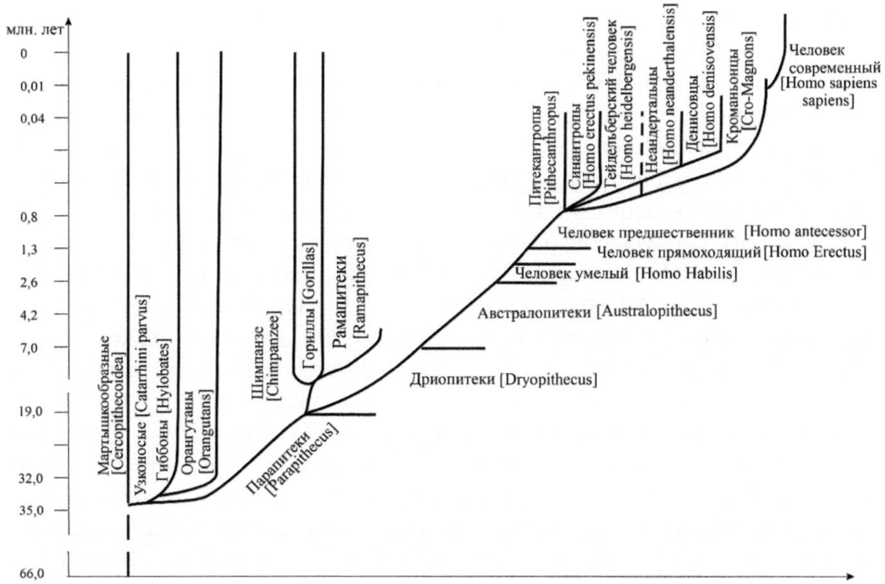

Fig. 1 Stages of hominid evolution (based on materials from Grant (1980), Mors et al. (2023))

mammals, including the ancestors of future apes and humans. The cataclysm that occurred can be considered to be the fall of a space object onto the surface of the earth with the resulting impact processes.

In the Oligocene, 32–35 million years ago, which is the approximate age of the Popigai Impact Event, the descendants of the first mammals, the parapithecus, divided into gibbons, orangutans, and our ancestors, who inhabited the tropical forests of Africa until about 19.0 million years ago. Apparently, during these 32.0–19.0 million years, no physiological or social changes in their appearance and lifestyle occurred.

After 19.0 million years, apparently due to a planetary cataclysm, the dryopithecus appeared—direct descendants of the parapithecus. Within the dryopithecus population, the ramapithecus, the ancestors of chimpanzees and gorillas, emerged. For a long time, from 19.0 to 7.0 million years ago, no remains of hominids, except for dryopithecus and ramapithecus, were found. More precisely, direct descendants of the first hominids who continued the evolutionary transformations, the final result of which should have been Homo sapiens, have not been found.

From 7.0 million years to 2.6 million years ago came the time of the dominance of the Australopithecus. The upper limit (2.6 million) is an important boundary at which the division of primates into monkeys and humans occurred. Homo habilis appeared, which was the ancestor of Homo erectus. About 1.3 million years ago is the time of the appearance of the predecessor of Homo (Homo antecessor).

800 thousand years ago, another division occurred. Homo antecessor divided into three branches: Pithecanthropus, Sinanthropus, and Heidelberg man. Pithecanthropus (Homo pithecanthropus) lived in what is now Indonesia. Remains of Sinanthropus (Homo sinanthropus) were found in China near Beijing. Both of these species were apparently dead ends on the evolutionary tree. Heidelbergers (Homo heidelbergensis) inhabited Europe, being the ancestors of Neanderthals (Homo neanderthalensis) and Denisovans (Homo denisovensis). After that: about 40 thousand years ago, only Cro-Magnons (Homo cromagnon man), who came from Africa, remained on Earth. In the range of 8–13 thousand years ago, Cro-Magnons gave rise to modern man. All other human populations were destroyed as a result of catastrophic events that occurred on Earth (Zhabin and Gadiyatov 2024a).

In the classification of disasters developed by different authors, the main ones can be considered exogenous and endogenous criteria (Armand et al. 1999). These events can be divided by type and time of origin. By type, earthquakes, volcanism, tsunamis, floods and floods, landslides, avalanches, atmospheric and space phenomena are distinguished. By the time of occurrence, catastrophic events are divided into rapid and prolonged (Svitoch 2010). The first of them last from minutes to a day, sometimes longer (it is logical to call them "cataclysms"). They have destructive consequences, which, however, are overcome quite quickly.

Long-term processes last for years and millennia, while the natural environment changes relatively quickly, but without dramatic background. The term "crisis" is more appropriate for defining such disasters. But, apparently, in the vast majority of cases, crises are a continuation of the consequences of cataclysms. Examples include the extinction of Australopithecus and the appearance of the first representatives of Homo. some kind of insert or m.

A review of the main species of hominids, which are the ancestors not only of humans, but also of all monkeys living today, and the time of their appearance and extinction, shows that new species arose quite suddenly, and between them there are long time breaks without any changes in the appearance and anatomical structure of these creatures. The appearance of a new species may indicate the suddenness of a catastrophic event, when the predecessor species has one alternative, either to adapt to the newly appeared natural conditions, or to disappear. At the same time, a large enough group should adapt (to produce healthy offspring to increase the number of individuals in the population), and not individual individuals.

During the studied periods of quiet development, there were no changes in the appearance of individual representatives of the fauna or the biocenosis as a whole. And why should they change if the natural environment fully corresponds to their habitual way of life. Therefore, it seems to us that evolutionary variations depend entirely on catastrophic changes in the environment. They can be caused by both cataclysms and crises.

3 Discussion

Currently, the very understanding of scientific explanation in the field of anthropogenesis has changed significantly. Based on modern, reinterpreted data, the Marxist version of the natural-scientific explanation of this process is being rejected. In connection with this approach, theological creations (within the framework of theology) and philosophical hypotheses, such as "catastrophism" and "jerkiness", receive scientific status. The phenomenological ideas about the metaphysical understanding of human essence have not been forgotten either.

Presumably, the causes that contributed to the disappearance of certain species of organisms and their replacement (extinction) with more competitively capable ones should have affected all spheres and the entire territory of the planet. And only cosmic manifestations can do that. The latter provoked earthly cataclysms, many of which were not even mentioned in the legends of numerous generations of people and reflected in the scriptures. These are earthquakes, fires, tsunamis, and hurricanes. Altogether, it affected the environment, causing its very rapid restructuring. Living organisms simply did not have time to react to such a change in the situation and many species disappeared.

But in some cases, as a result of long-term crises, but caused by cataclysms, the climate changed from humid to arid, and sometimes to a turnaround. The forest area decreased, the food supply changed, and other, not always noticeable, environmental changes occurred. As a result, Darwinian perestroika (evolution) was carried out, which still ended in another cataclysm.

And, here, the time interval during which the dryopithecus was divided into the ancestors of chimpanzees, gorillas and humans, perhaps, can be associated with an impact event. An example is the catastrophe that occurred about 19.0 million years ago (Iosifova 1992; Kholmovoy 2008), the traces of which are noted in the Voronezh region on the northern outskirts of the village of Gorelki. According to the results of the conducted research (Zhabin et al. 2021), it can be assumed that a sufficiently large cosmic body fell into the sea basin. Although this happened thousands of kilometers from the habitat of the parapithecines, it could hardly have affected the change in the natural environment that contributed to their extinction and the emergence of a more complex form of the great ape—dryopithecus. This leads to the conclusion that our planet collided with a comet. As it approached the Earth, it began to disintegrate into individual fragments, which, upon impact on the Earth's surface, could cause drastic changes in the natural environment in significant areas.

Further, for almost 12.0 million years, apparently nothing prevented Dryopithecus and Ramapithecus from existing peacefully, until Australopithecus appeared at the turn of 7.0 million years ago. Of course, it is highly doubtful that the latter contributed to the extinction of the former. It's just that the disappearance of some and the flourishing of others were dictated by certain variations in the natural environment.

Australopithecines are brought closer to humans by the weak development of the jaws, the absence of large protruding canines, a grasping hand with a developed thumb, a supporting foot, and a pelvis adapted for walking upright. The brain is relatively large (530 cm^3), but structurally it differs little from the brain of modern great apes. In terms of volume, it was no more than 35.0% of the average size of the modern human brain. The body size was also small, no more than 120–140 cm in height, and the physique was slender. It is assumed that the size difference between males and females of Australopithecines was greater than that of modern hominids. There is no evidence that Australopithecines used natural shelters or created temporary ground shelters for the night. However, there is no doubt that they could, like modern chimpanzees, spend the night in trees, building temporary nests there. The lifespan of Australopithecines, which were constantly threatened by predators and some large herbivores, was insignificant and averaged 20 years. Almost none of them lived to be 40 years old.

It is believed that climate change contributed to the appearance of Australopithecines. The climate became drier, as a result of which the boundaries of forests began to shift towards the equator. The very first Australopithecines began to explore the shrouds that appeared in place of forests. However, the reasons that led to such drastic changes in the natural environment remain unclear. The impact processes that took place at that time can be attributed to such reasons.

Let us turn to the African continent as an example of impact manifestations. On the territory of Morocco there is a large structure in the form of a regular circle with a diameter of about 40 km, which is called the "Eye of the Sahara". There is a similar formation to the west of it, but its diameter is about 5 km (Fig. 2, indicated by an arrow in the picture). Looking at satellite maps of the deserts of the world, we find similar structures almost everywhere. For example, in the deserts of Iran (Fig. 3), in the European part of Russia. For example, in the vicinity of the Donskoy Sakha tract, on the border of the Voronezh, Rostov, and Volgograd regions, satellite maps show similar, perfectly even circles with a white border around the circumference (Fig. 4). Their diameter ranges from the first meters to several kilometers.

Examining these structures and reconstructing the geological situation, it can be stated that, from the end of the Neogene to the beginning of the Holocene, the territory of the northern hemisphere was repeatedly subjected to collisions of the Earth with space objects. At the same time, we believe that the consequences of these environmental impacts of simultaneously falling masses of relatively small debris may exceed those from falling asteroids of hundreds and thousands of meters in size. Moreover, the collision of even very large fragments (but not of critical size) with the Earth's surface at one point is unlikely to have an impact on the entire territory of the planet. Grandiose eruptions of supervolcans can serve as an analogue of cosmic events of this kind, which sometimes led to the extinction of some local populations of organisms, but, of course, not in all environments of the planet.

Examining the materials testifying to the fall of cosmic bodies to the surface of the earth, it can be assumed with a certain degree of certainty that we have certain ideas about the events that initiated the transformation in the environment and the adaptation of newly emerging representatives of biota to it. Our point of view is supported by other researchers (Raup and Sepkoski 1984), who claim that the main cause of all mass extinctions are impact events. Here it is necessary to take into account the emergence of new species after extinctions, which does not occur during

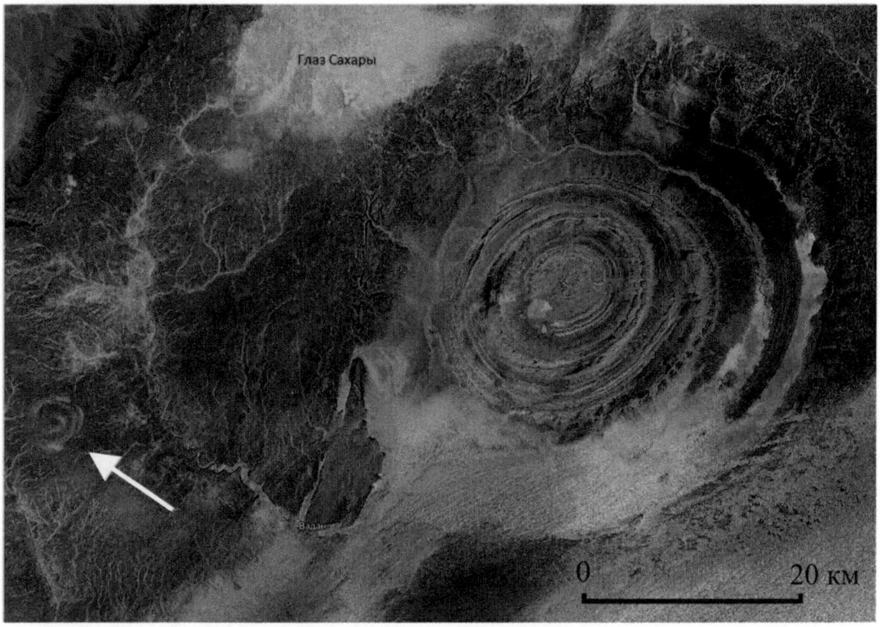

Fig. 2 The impact crater "The Eye of the Sahara". Satellite map fragment

terrestrial catastrophic events. Apparently, space objects contain some components that contribute to the emergence of new forms of biota (which persist both when exposed to low temperatures and ionizing radiation from space, and high temperatures when they collide with the Earth's surface). This interpretation of the current provisions once again confirms the cosmic hypotheses of the origin of life on our planet.

About 2.6 million years ago, an Altan asteroid measuring about 4.0 km across fell on the shelf between South America and Antarctica. The consequences were disastrous. Giant tsunami waves up to several hundred meters high hit the adjacent land where the Australopithecines lived. The waves threw marine fauna and flora deep into the territory, washed away all living and inert objects into the ocean. It should be noted that Australopithecines have never left the territories of eastern and southern Africa. If a very small part of the Australopithecus population survived, it was doomed to extinction due to its small number. In place of the Australopithecines, which had more ape–like features, the first humans appeared—Homo habilis (a skilled man).

Ghana is home to the relatively small Bosuletwa crater (10.5 km in diameter), which is 1.3 million years old. Its formation is associated with the fall of a cosmic body. By analogy with other impact events, it can be assumed that there are craters of the same age on Earth. The age of 1.3 million years is the time of the appearance of the human predecessor (Homo antecessor), the first human in Europe, considered the direct ancestor of the Heidelberg man (Homo heidelbergensis), as well as Neanderthals and Denisovans (Mors et al. 2023).

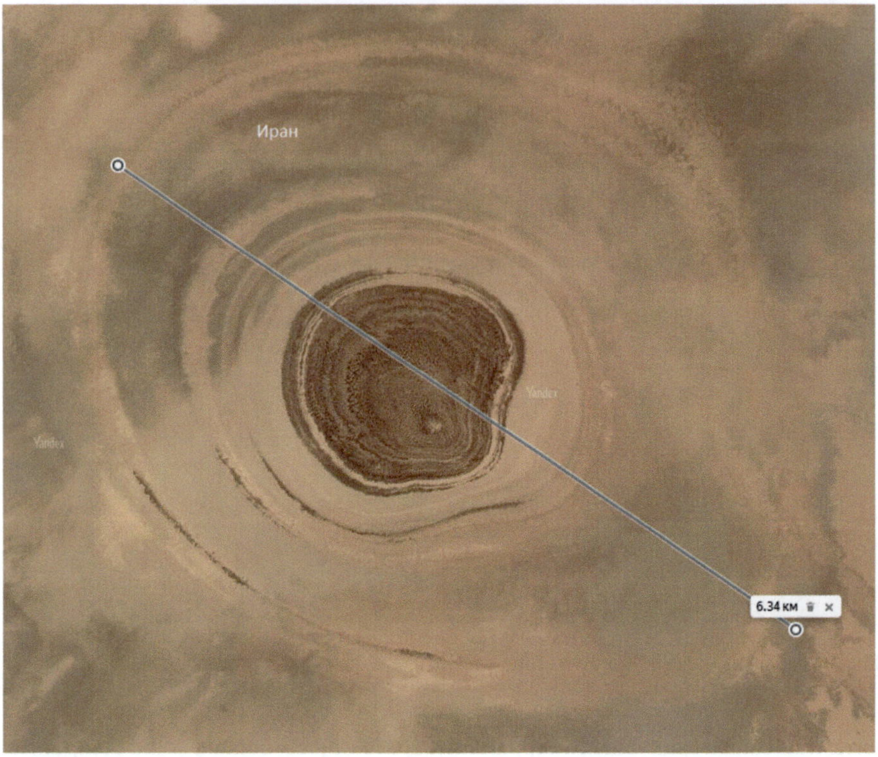

Fig. 3 The desert in the center of Iran. The impact crater on the satellite map

At the time limit of 40 thousand years, a cosmic body measuring at least one kilometer across fell into Lake Ladoga, forming a funnel with a diameter of about 100 km. Such an impact on the earth's surface led to the activation of local volcanism. As a result, about 1,500 m^3 of ash material was released into the atmosphere. Such a significant amount of volcanic dust caused the Paleolithic nuclear winter. Apparently, this space fragment was part of a larger body that fell into the waters of the Arctic Ocean or the North Atlantic. The space alien caused a giant tsunami with a height of 1.0–1.5 km, which washed away all the biota from the plains of Europe, North America and Asia. At the same time, tsunamis could have brought fragments of crystalline rocks from the Baltic and Canadian shields, giving rise to myths about cover glaciations. During this catastrophe, populations of Neanderthals, Denisovans, and large representatives of the mammoth fauna were undermined and doomed to slow extinction (Damuth 1981). Once again, the Cro-Magnons, who came from Africa, gradually settled these territories without encountering any resistance (Zhabin and Gadiyatov 2024b).

At the turn of the late Pleistocene—early Holocene (about 10,000 years ago), the northern hemisphere of the Earth was subjected to intense carpet bombing by space objects. Most of these were relatively small fragments, but there were large ones

Fig. 4 The east of Voronezh region. The impact crater on the satellite map

that fell into the waters of the Arctic Ocean and formed the Amundsen, Nansen, Submariners, and Makarov basins. The giant tsunamis that occurred simultaneously with these events repeated the catastrophe that occurred 30 thousand years before (Firestone et al. 2007). Only this time Cro-Magnons and the last representatives of other human populations and mammoth fauna who survived the previous cataclysm died (Koeberl et al. 2012; Collins et al. 2012). The Cro-Magnons, who came last from Africa and settled the mountainous regions of the aborigines, and therefore the survivors (the same Cro-Magnons, but who came earlier), quickly settled the empty territories (Zhabin and Zhabina 2019). These were already people of the modern type of Homo sapiens.

There are other opinions related to space issues. For example, Salop (1977) believes that all evolutionary changes in biota recorded in the geological history of the earth are associated with harsh ionizing radiation coming from deep space. Or with a decrease in the thickness of the ozone layer. Such phenomena lead to the rapid and massive extinction of large groups of fauna and flora and the subsequent, almost sudden flourishing of new forms belonging to the same groups. As a result, higher-ranking taxonomic units appear more often. That is, almost always, intensively developing forms had relatively primitive predecessors who received a powerful impetus for their progress during these revolutionary periods. At the same time, a number of thriving groups suddenly (by geological standards) completely cease to exist. And here it is appropriate to note that the appearance of the first true ancestors of humans, with a very high degree of probability, was influenced by the harsh radiation from the supernova explosion, which occurred approximately in the same time interval with the fall of the Altan asteroid.

People are often interested in the following problem. If humans evolved from monkeys, then why aren't we seeing these transformations now? Answering this question, it is necessary to indicate, first of all, the duration of such changes, which occur from tens of thousands to millions of years. People don't live that long and can't register such transformations, even when they manifest. It is necessary to take into account the age of the theory of evolution. She's not even two hundred years old. But the most important thing is this. The ancestors of modern apes, despite all the catastrophic events that also affected them, missed their chance to acquire intelligence, and remained at the same primitive level of development.

Without denying Darwin's theory and evolution, which takes place according to the laws of Nature, it should be noted that the ongoing evolutionary transformations have never led to primitive forms in geological history. They were always getting more complicated. Strange as it may seem, the conclusion is that we are dealing with a controlled evolution. What kind of force controls these processes? Science currently has no answer to this profound philosophical question. In any case, in our opinion, the God mentioned in religious literature has nothing to do with this. So far we have no direct evidence, but everything points to the fact that the world is ruled by a civilization of the fourth type (Kardashov 1964), using the resources of the entire universe.

4 Conclusion

The time intervals in which primitive forms of hominids appeared at first were highlighted above, which, as they approached modernity, became more complicated until modern man appeared. One can only assume which or what events caused the appearance of the ancestor of narrow-nosed and great apes, as well as the division of parapithecines into the ancestors of gibbons, orangutans and humans. For tens of millions of years, the processes of sedimentation and erosion have erased all traces of past disasters.

The recognition of Eternity unites atheism and religion. The universe as a whole represents a system that lacks a beginning and an end, that is, it is "immortal." Although, in turn, it consists of objects that have, albeit long-term, but finite limits of existence, changing their location in space, as in a kaleidoscope. When such a concept as time duration is used, it must be understood that even a trillion years, compared with eternity, is a moment. Moreover, the period of evolution from narrow-nosed monkeys to modern humans is about 30 million years, and of this period, the actual existence of Homo sapiens accounts for several thousand years. It's less than a moment. However, man considers himself and his civilization to be the crowning achievement of evolution. Society spends significant amounts of money on the creation of devices with which we listen to outer space, hoping to find their own kind. But the Cosmos is silent, or trying to contact us, but by other methods. That is, we speak different languages. And let's not think much about the fact that our communication methods are not used by other civilizations.

What can we say about other Worlds, when even in the history of our planet there are different ways of writing, as means of communication, which differ sharply from each other. These include the cuneiform script of the Sumerians, the nodular script of the Aztecs and Incas, hieroglyphs of the peoples of Southeast Asia, European alphabets, etc. Some of them have not been deciphered to this day.

References

Armand AD, Lyuri DI, Zherikhin VV et al (1999) Anatomy of crises. Nauka, Moscow, p 238

Collins GS, Melosh HJ, Osinski GR (2012) The impact-cratering process. Elements 8(1):25–30. https://doi.org/10.2113/gselements.8.1.25

Damuth J (1981) Popylation density and body size in mammals. Nature 290:699–700. https://doi.org/10.1038/290699a0

Firestone RB, West A, Kennett JR et al (2007) Evidence for an extraterrestrial impact 12900 years ago that contributed to the megafaunal extinctions and the younger Dryas cooling. PNAS 104:1601616021

Grant V (1980) Evolution of organisms. Translated from English by Fomina NO. Edited by Dr. of Biological Sciences Mednikov BM. MIR, Moscow, 409 p.

Iosifova YI (1992) On the age of the ash-containing strata in the section near the village of Gorelka in the Voronezh region. Stratigr Phaneroz Cent East Eur Platf 36–59

Kardashov NS (1964) Transmission of information by extraterrestrial civilizations. Astron J 41(2):282 287

Kholmovoy GV (2008) Morphoscopic features of Neogene and quaternary ashes of the Voronezh region. Bull State Univ Ser Geol 1:19–22

Koeberl C, Claeys P, Hecht L, McDonald I (2012) Geochemistry of impactites. Elements 8(1):37–42. https://doi.org/10.2113/gselements.8.1.37

Mors M, Papagianni D (2023) Neanderthals. Underestimated competitors of homo sapiens. Discoveries. Research. Facts. Translation from English by Deinichenko PG. Eksmo, Moscow, 272 p.

Raup DM, Sepkoski JJ (1984) Periodicity of extinctions in the geologic past. https://doi.org/10.1073/pnas.81.3.801

Salop LI (1977) On the connection between glaciations and stages of rapid changes in the organic world with cosmic phenomena. Bull Mosc Soc Nat 52(1):5–32. Geological Department, Moscow State University Publishing House

Svitoch AA (2010) Catastrophic events in the paleogeography of the Pleistocene. Bull Mosc Univ Ser 5 Geogr 3:11–19

Zhabin AV, Gadiyatov VG (2024a) Creationism as an integral part of evolutionism. Process Geoenviron 3(41):2576–2583; Armand AD, Lyuri DI, Zherikhin VV et al (1999) Anatomy of crises. Nauka, Moscow, 238 p

Zhabin AV, Gadiyatov VG (2024b) Craters of cosmic origin on the territory of the central black earth regions. Process Geoenviron 1(39):2358–2366

Zhabin AV, Zhabina AA (2019) The influence of impact events on global changes in the natural environment in the late Pleistocene and Holocene. In: Global climate changes: regional effects, models, forecasts. Proceedings of the international scientific and practical conference, vol 1. Voronezh, pp 40–43

Zhabin AV, Ignatenko IM, Donetsky SV (2021) On the origin of volcanic ash deposits in the Voronezh region. Reg Geosyst 45(3):441–454. Publishing House of Belgorod State University, Belgorod

Multidimensional Digital Investigation of the Morphometric Properties of Sand Removed from the Gas Condensate Field Reservoir on the Basis of MCT Analysis

V. V. Khimulia

Abstract This paper presents the results of statistical and geometric digital analysis of the morphological properties of sand removed from the reservoir of a gas condensate field as a result of geomechanical modeling of sand production processes. X-ray computed tomography (CT) and digital core analysis techniques were used to construct three-dimensional models of sand particles derived from tomography data. These models facilitated a detailed quantitative examination of particle shapes and sizes. The analysis was conducted on sand produced during mechanical testing of reservoir rocks using the "hollow cylinder" scheme to investigate sand production and borehole fracture mechanisms. Particle sphericity was evaluated using the Krumbein and Sheppard methods, and size distribution data provided insights into the fractional composition of the reservoir fracture products. A thorough understanding of the sand's morphometric properties is essential for optimizing the design and selection of downhole equipment, as well as for improving geomechanical and hydrodynamic reservoir models. The insights obtained from X-ray CT imaging complement laboratory experiments and in-situ studies, providing a more comprehensive understanding of reservoir behavior.

Keywords Morphometric analysis · Sphericity · Granulometry · Sand production · Digital core analysis · Computed tomography of materials · Reservoir properties · Gas condensate field · Digital granulometric analysis · Rock fracture

1 Introduction

The expansion and development of complex wells in hydrocarbon fields have become a priority, requiring the integration of advanced technologies and innovative approaches. These efforts aim to enhance operational efficiency while minimizing the risks associated with accidents. This need is particularly critical in challenging

V. V. Khimulia (✉)
Ishlinsky Institute for Problems in Mechanics RAS, Moscow, Russia
e-mail: valery.khim@gmail.com

T. Chaplina (ed.), *Processes in GeoMedia—Volume IX*, Springer Geology,
https://doi.org/10.1007/978-981-96-9302-3_24

environments, such as gas and gas condensate reservoirs on Russia's Arctic shelf. These formations are characterized by their high porosity and weak mechanical strength, making them susceptible to stress-induced deformation and fracturing. Such geological features demand careful planning and technological solutions to maintain reservoir stability and ensure sustainable production.

Stress-induced deformation and fracturing in weakly cemented rocks can significantly impact the filtration properties of reservoirs, leading to reduced well productivity (Li et al. 2021). These changes disrupt the flow of fluids, making extraction less efficient and posing challenges to maintaining production rates. Additionally, these fractures can escalate into more severe structural damage, potentially causing environmental hazards and economic losses. For example, uncontrolled sand production from the reservoir can create blockages in wells and damage surface and subsurface equipment, further compounding operational difficulties (Iziumchenko et al. 2018).

One of the main challenges in addressing these issues is the limited ability to observe and analyze well wall failures and sand production directly under field conditions (Popov and Petrakov 2021). The bottomhole zone, where these processes occur, is often inaccessible to monitoring and measurement equipment. This limitation creates a significant knowledge gap, making it difficult to develop real-time solutions to prevent or mitigate these problems (Karev 2018). As a result, researchers and engineers must rely on alternative methods, such as laboratory experiments and modeling, to better understand the physical and mechanical behaviors of reservoir rocks (Karev et al. 2020a).

Comprehensive laboratory studies play a critical role in overcoming the challenges of inaccessible field conditions (Vasiliev et al. 2014). These studies allow researchers to replicate the conditions of real reservoirs, enabling a detailed examination of the processes occurring near the wellbore. Geomechanical modeling is one of the most effective approaches in this regard, as it simulates stress distribution, rock deformation, and fluid flow in rocks. Such models provide valuable insights into the factors that influence well stability and allow engineers to design safer and more efficient operational parameters for well management. This knowledge is instrumental in defining safe operating limits, optimizing downhole equipment design, and minimizing the environmental impact of hydrocarbon extraction (Gaidukov and Orynbaev 2010). Ultimately, these advancements contribute to the sustainable development of hydrocarbon resources, ensuring that operational goals are met while reducing potential hazards.

The morphometric characteristics of sand expelled from reservoir formations during processes such as sand production and rock failure provide critical insights into the mechanical and filtration properties of the reservoir. Understanding particle shape, size, and sphericity helps in identifying the conditions under which sand production occurs and its impact on well productivity. Sand particles influence the permeability of filters and the overall stability of the reservoir. Without this knowledge, it becomes challenging to design effective downhole equipment, prevent blockages, and optimize the extraction process. Studying these characteristics also supports the development of more accurate geomechanical and hydrodynamic models, which are essential for predicting reservoir behavior and minimizing risks associated with sand production.

Digital tomographic methods, such as X-ray computed tomography (CT), enable researchers to study the internal structure of sand particles in a non-destructive and highly precise manner (Menke et al. 2022). CT scans produce detailed three-dimensional images of sand samples, allowing for the quantitative analysis of particle size, shape, and distribution (Lichun and Mian 2014; Diaz et al. 2017). Using advanced image processing techniques, researchers can segment the particles and measure key parameters such as sphericity and roundness. These methods also make it possible to observe internal cracks or voids within the particles, which could affect their mechanical stability. By employing digital tomographic techniques, researchers can generate a comprehensive dataset for further analysis and modeling.

The use of digital tomographic methods offers several advantages over traditional approaches. First, these techniques are non-destructive, allowing repeated analysis of the same sample without altering its structure. Second, the three-dimensional nature of CT data provides a more complete understanding of particle geometry compared to two-dimensional methods like microscopy. Third, high-resolution imaging allows researchers to capture fine details of particle morphology, enabling more accurate and reliable measurements. Furthermore, the integration of CT data into computational models enhances the ability to predict and mitigate challenges associated with sand production. Overall, digital tomographic methods offer a robust and versatile toolset for improving reservoir management and optimizing extraction processes.

2 Research Objects and Methodology

This study focused on the digital granulometric and morphometric analysis of fracture products from a gas condensate reservoir. Sand specimens produced during "hollow cylinder" mechanical tests were analyzed using X-ray computed tomography (CT) to determine characteristic particle sizes, shapes, and distributions.

The core material studied was taken from the gas condensate reservoir of Russia's Arctic shelf. The rocks consisted of highly porous, weakly cemented, low-strength sandstones. For mechanical testing, cube-shaped samples measuring 40 mm per side with a central hole were prepared. These samples were tested on a TILTS facility (Karev et al. 2020b) using the "hollow cylinder" scheme (Klimov et al. 2015) to record the intensity and volume of sand production. The X-ray computed tomography procedure and image acquisition were carried out using a high-resolution ProCon X-Ray CT-MINI microtomograph (https://procon-x-ray.de/en/ct-mini).

The Procon CT-MINI is a compact, high-resolution micro-CT scanner designed for non-destructive imaging of small-scale samples in scientific and industrial applications. It utilizes X-ray computed tomography (CT) to generate detailed three-dimensional reconstructions of internal structures with micrometer-level resolution. The device features an X-ray source with a voltage range of 20–100 kV and a maximum power of 10 W, providing high-contrast imaging of a wide variety of materials. Equipped with a high-sensitivity flat-panel detector, the CT-MINI achieves a spatial resolution down to 5 μm, allowing precise visualization of fine structural details in materials such as geological samples, biological tissues, and engineered components. Despite its compact size, it offers an adjustable field of view up to 40 mm in diameter and 50 mm in height, making it a versatile tool for laboratory-based research and quality control.

The device is widely used in geosciences, materials science, and biomedical research, where high-resolution imaging is essential for analyzing internal microstructures. Its intuitive software enables automated data acquisition, reconstruction, and analysis, generating detailed volumetric datasets for quantitative evaluation. The CT-MINI supports multiple scanning modes, including high-resolution and fast-scan options, to balance image quality and scanning time. With its fully enclosed shielded design, it ensures safe operation without the need for additional radiation protection measures. The CT-MINI's non-destructive approach allows repeated scanning of the same sample under different conditions, making it particularly valuable for studying dynamic processes such as fluid migration in porous materials or deformation under mechanical stress.

3 Image Reconstruction and Data Processing

Image reconstruction was performed using the VGSTUDIO software, and the processing of 3D images was carried out with the Geodict Math2Market GmbH software (https://www.math2market.de/). To evaluate the geometric, statistical, and morphometric characteristics of the sand particles, the MatDict (Rief et al. 2021) and PoroDict modules were utilized. Based on these approaches, the sphericity of sand particles was assessed using multiple methods. Morphometric analysis of grains, as part of granulometry, is widely used in geology to study the dynamics of sediment transport and deposition, as well as in technologies involving different fractions of mechanically generated sedimentary rocks (Samodurov and Kutyrlo 2014).

The primary quantitative parameter of grain morphology is sphericity. The sphericity and roundness of sand grains significantly influence their strength and resistance to stress: the lower these parameters, the higher the likelihood of particle fracture (Magadova et al. 2012). A decrease in sphericity and roundness also reduces the permeability of sand within filters or sand plugs and increases filtration resistance. Sand grains typically exhibit diverse shapes—rarely perfectly round and smooth, but often oval, flat, or triangular.

Numerous methods exist to determine particle sphericity. In this study, approaches based on the methods of Krumbein and Sheppard (Shvanov 1969) were employed. Krumbein's sphericity is calculated as the cubic root of the ratio of the lengths of the three principal axes of an ellipsoid inscribed within the grain. The parameter values range from 0 to 1, where 1 represents a perfect sphere. Sheppard's sphericity is determined as the ratio of the inner radius of the grain to its equivalent radius. The equivalent radius is the radius of a sphere with the same volume as the grain, while the inner radius is the largest sphere that can fit inside the grain. These sphericity values also range from 0 to 1, with 1 representing a perfect sphere.

4 Experimental and Digital Results

In the specimens prepared for testing using the "hollow cylinder" scheme, a 10 mm diameter hole was drilled at the center of the end face along one of the principal axes. During the experiment, air at a pressure of approximately 1 bar was introduced through a channel in the upper active tip of the compression plate, which was aligned with the hole in the sample. Throughout the experiment, uniform and gradually increasing compressive stresses were applied to the cubic sample's end faces. The deformations of the specimen were recorded in three directions, and the mass of sand expelled from the sample by the airflow was measured. Characteristic results of the experiments are presented in Fig. 1. As can be seen from the graphs, sand removal started at 2.6 MPa and specimen failure occurred at 32 MPa. The mass of the transported sand was less than 2 g. The extracted sand samples were then scanned in a computed tomography equipment. After acquiring a dataset, image reconstruction was performed to generate a three-dimensional representation of the sand particles.

Figure 2a presents a section of a tomography projection of the fracture products, where sand granules are shown in gray, and air in black. The acquired images underwent preprocessing and segmentation. To enhance image clarity and simplify the segmentation process, the Non-Local Means (NLM) filter was applied. Figure 2b displays a close-up of a sand projection after reconstruction and application of smoothing filters. Segmentation was conducted using the Global Thresholding and Otsu methods to refine grain boundaries. The results of the segmentation are shown in Figs. 2c and 3a, where the extracted grain boundaries and individual particles are highlighted in red within a 3D model.

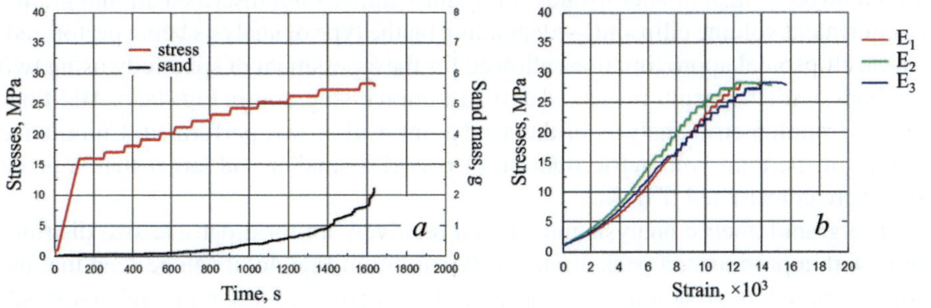

Fig. 1 Typical result of sand production tests: **a** loading program and sand removal curve during the experiment; **b** stress–strain curves of a cubic specimen along three axes

Fig. 2 **a** A section of a tomography projection of the fracture products; **b** close-up of a sand projection after reconstruction and application of smoothing filters; **c** the result of the boundaries segmentation

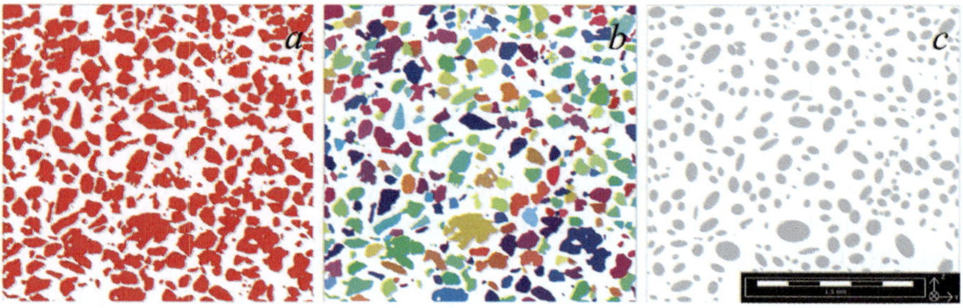

Fig. 3 **a** The result of the image segmentation; **b** the result of grain separation in the model; **c** the output of an algorithm that approximates the particles with ellipsoids

For the statistical analysis of the 3D sand models, the MatDict and PoroDict modules from the GeoDict software package were utilized. Automated grain numbering was performed, with boundary detection for each individual grain. Subsequently, geometric properties of the particles were assessed, and a statistical analysis was carried out. Figure 3c shows the result of one of the algorithms used, which involves matching ellipsoids fitting into grains, circumscribed around grains, or equivalent volume ellipsoids—depending on the type of analysis being performed. These ellipsoidal approximations allowed for the assessment of sphericity using two methods, with the results presented as distribution histograms in Fig. 4a, b. Additionally, a granulometric analysis of the sand composition was performed. Cumulative curves of the total volumetric fraction of particles smaller and larger than a given size were constructed (Fig. 4c).

The granulometric analysis indicated a relatively uniform particle size distribution, with a pronounced peak around 100 μm. From Fig. 4c, it can be seen that the fracture products predominantly consist of fine sand (0.25–0.1 mm) and coarse silt (0.1–0.05 mm) fractions. The characteristic particle size distribution values were as follows: D10: 45 μm, D50: 95 μm, D90: 137 μm. Larger particles ranging from 150 to 354 μm were also present, though their volumetric fraction was relatively

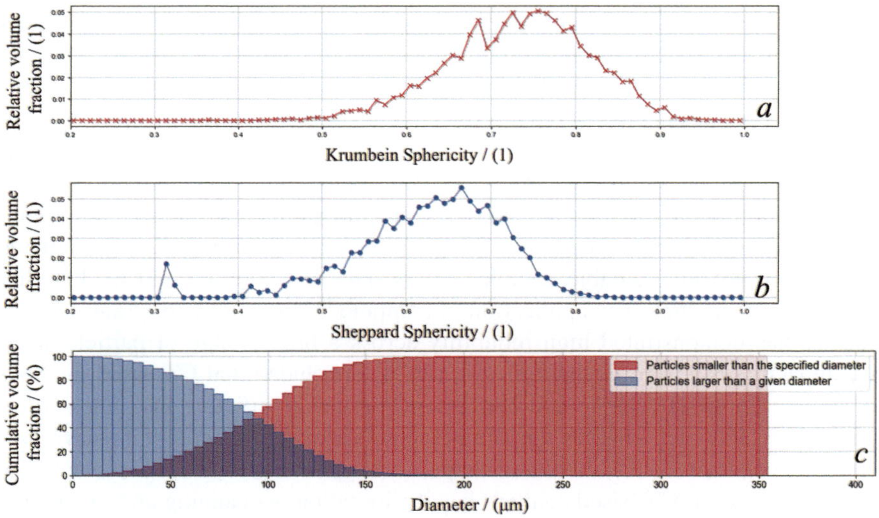

Fig. 4 **a, b** Histograms of sand particle sphericity, computed using two different approaches: the Krumbein method and the Sheppard method; **c** cumulative curves of the total volumetric fraction of particles

low. First stage of digital particle size analysis characterized matrix fracturing at the microscale and helped optimize downhole sand control equipment. The ideal gravel filter should have a median particle size up to 0.6 mm, balancing effective sand retention with permeability.

The MatDict and PoroDict modules were also employed for the morphological characterization of sand particles. Various methods were used to assess particle sphericity. In the constructed 3D models, individual grain boundaries were computationally identified, followed by a detailed pointwise analysis of the granules. The obtained data were then subjected to statistical processing. Figure 4a, b show histograms of sand particle sphericity, computed using two different approaches: the Krumbein method and the Sheppard method, as described in the methodology section.

Both histograms indicate that the majority of sand granules exhibit sphericity values ranging from 0.4 to 0.9, with peak values between 0.7 and 0.8, corresponding to a high degree of sphericity. The distribution appears uniform, without distinct local extrema. A singular peak in the Sheppard histogram may correspond to fractured material from the drilled hole. The high sphericity of the particles suggests a strong ability of the fracture products to withstand compressive stress. Such loads may occur following the formation of a sand plug near the wellbore, accumulation of sand around filters, fluid filtration processes, or fracture closure after pressure changes in the well or hydraulic fracturing operations.

Morphometric analysis provides key parameters such as sphericity, roundness, and aspect ratio, which influence particle transport dynamics, filter clogging potential, and erosion effects on downhole and surface equipment. High-sphericity particles tend to flow more easily through production systems, whereas angular particles are more likely to cause abrasion and blockages. Additionally, assessing particle shape helps refine geomechanical and hydrodynamic models by improving predictions of sand migration, permeability variations, and formation stability under changing reservoir conditions. These insights are essential for optimizing sand control strategies, selecting appropriate filtration technologies, and mitigating operational risks associated with excessive sand production. The results obtained through X-ray computed tomography provide valuable complementary data to laboratory and field studies of reservoir properties. The application of computed tomography for granulometric analysis has demonstrated high reliability across a broad range of particle sizes. A comparative study in Houghton et al. (2024) concluded that CT-based methods are among the most reliable and convenient, provided sufficient instrument resolution. This approach enables the simultaneous extraction of comprehensive data on particle characteristics, including shape, sorting, volume, surface area, and even porosity. However, CT-based analysis has its limitations—scanning and processing image datasets require significantly more time than conventional granulometric methods. Moreover, the accuracy of the final results depends on the ratio between the instrument's resolution and the particle size under investigation.

5 Summary and Conclusions

The paper presents the results of digital analysis of granulometric and morphometric characteristics of fracture products of rocks of weakly cemented reservoir of a gas condensate field. The studies were carried out using X-ray computed tomography and digital analysis methods based on image segmentation, which allowed to create three-dimensional models of sand specimens and quantitatively analyze the shape and size of particles. Specimens of sand removed from the rocks during mechanical testing of the reservoir using the "hollow cylinder" scheme were used to study the process of sand production. First stage of digital particle size analysis characterized matrix fracturing at the microscale and helped optimize downhole sand control equipment. The ideal gravel filter should have a median particle size up to 0.6 mm, balancing effective sand retention with permeability. The morphometric characteristics of the transported sand, including sphericity and particle size distribution, were determined. It is shown that the particles of the transported sand have a high degree of sphericity, which increases their resistance to compressive stresses under operating conditions. Comparative analysis by the Krumbein and Sheppard methods for sphericity assessment confirmed a uniform distribution of sphericity values in the range of 0.4–0.9. The composition of fracture products is dominated by fine sand and coarse silty fractions. The results showed that the applied method of X-ray computed tomography is an effective tool for detailed granulometric analysis. The obtained results allow to supplement and refine geomechanical and hydrodynamic models of the reservoir, as well as to adjust the optimal parameters of downhole equipment to control and minimize sand occurrence. The findings of these studies enable the prediction of bottomhole zone fracturing patterns and provide insights into the optimal specifications for downhole equipment. Selecting appropriate equipment helps mitigate adverse processes near the wellbore, enhances the quality of the extracted raw gas, and lowers refining costs.

Acknowledgement The research was supported by the Russian Science Foundation grant № 23-77-01037, https://rscf.ru/project/23-77-01037/

References

CT-MINI by ProCon X-Ray GmbH. Available at https://procon-x-ray.de/en/ct-mini

Diaz M et al (2017) Surface roughness characterization of open and closed rock joints in deep cores using X-ray computed tomography. Int J Rock Mech Min Sci 98:10–19. https://doi.org/10.1016/j.ijrmms.2017.07.001

Gaidukov LA, Orynbaev BA (2010) Influence of near-wellbore zones on the productivity of gas wells. Proceed MIPT 2(2):86–91 (in Russian)

GeoDict—The digital material laboratory: [digital source]. https://www.math2market.de/ (Date of request: 13.10.2024)

Houghton JE, Behnsen J, Duller RA, Nichols TE, Worden RH (2024) Particle size analysis: a comparison of laboratory-based techniques and their application to geoscience. Sediment Geol 464:106607

Iziumchenko DV, Mandrik EV, Melnikov SA, Ploskov AA, Moiseev VV, Kharitonov AN, Pamuzhak SG (2018) Operation of gas wells in conditions of active water and sand production. Gas Ind News 1(33) (in Russian)

Karev VI (2018) Geomechanical approach to improving the efficiency of the operation of underground gas storages. In: Physical and mathematical modeling of earth and environment processes. Springer proceedings in earth and environmental sciences. Springer International Publishing, Switzerland, pp 150–158

Karev VI, Kovalenko YF, Ustinov KB (2020a) Geomechanical and physical modeling of deformation processes in gas storage reservoirs under cyclic pressure changes. Gas Ind 4(808):46–52 (in Russian)

Karev V, Kovalenko Y, Ustinov K (2020b) Advances in oil and gas exploration and production. Springer International Publishing, Switzerland, 166 p. https://doi.org/10.1007/978-3-030-26608-0

Klimov DM, Karev VI, Kovalenko YF (2015) Geomechanical approach to the development of new oil extraction technologies at great depths. Process Geomed 4:45–51 (in Russian)

Li W, Frash LP, Welch NJ, Carey JW, Meng M, Wigand M (2021) Stress-dependent fracture permeability measurements and implications for shale gas production. Fuel 290:119984. https://doi.org/10.1016/j.fuel.2020.119984

Lichun J, Mian C (2014) 3D imaging of fractures in carbonate rocks using X-ray computed tomography technology. Carbonates Evaporites 29(2):147–153. https://doi.org/10.1007/s13146-013-0179-9

Magadova LA, Silin MA, Glushchenko VN (2012) Oilfield chemistry. Technological aspects and materials for hydraulic fracturing: a textbook for universities. Gubkin Russian State University of Oil and Gas, Moscow, 423 p. (in Russian)

Menke HP et al (2022) Using nano-XRM and high-contrast imaging to inform micro-porosity permeability during stokes–brinkman single and two-phase flow simulations on micro-CT images. Front Water 4:935035. https://doi.org/10.3389/frwa.2022.935035

Merkus HG (2009) Particle size, size distributions and shape. In: Particle size measurements: fundamentals, practice, quality, p 13–42. https://doi.org/10.1007/978-1-4020-9016-5_2
Popov MA, Petrakov DG (2021) Study of gas wells operation regimes in complicated conditions. Perm J Petrol Min Eng 21(1):36–41. https://doi.org/10.15593/2712-8008/2021.1.6
Rief S, Hilden J, Planas B (2021) MATDICT user guide. GeoDict release 2022. Accessed 8 Dec 2021. https://doi.org/10.30423/userguide.geodict2022-matdict
Samodurov VP, Kutyrlo VE (2014) Morphometry of clastogenic rocks of quaternary deposits of Belarus (in Russian)
Sheppard AP, Sok RM, Averdunk H (2006) Analysis of rock microstructure using high-resolution X-ray tomography. In: Proceedings of the international symposium of the society of core analysts
Shvanov VN (1969) Sand rocks and methods of their study: (distribution, structures, textures). Nedra, Leningrad (in Russian)
Vasiliev VA, Gunkina TA, Grishin DV, Golod GS (2014) Diagnosis of the near-wellbore zone using gas dynamic studies. Gas Ind 2:20–23 (in Russian)

Development of a System for Monitoring the Thermal State of the Snow Cover

D. M. Frolov, V. E. Gagarin, A. V. Koshurnikov, and G. A. Rzhanitsyn

Abstract The work is devoted to studies of the thermal state of snow, which are carried out with the help of the developed measuring system. The monitoring system can be useful for assessing the thermal state of melting glaciers, preparation of ski slopes on them, as well as for understanding the mechanical properties of the snow column important for construction and transportation of cargo in polar regions. The paper presents the author's developed methodology and system for monitoring the thermal state of the snow strata.

Keywords Snow · Thermal monitoring · Glaciers · Winters · Ski slopes

According to the Russian Hydrometeorological Center, 2024 became the second warmest year in Russia meteorological history since 1891. 2020 was even warmer. The warmest 2024 was in the Central, Southern, North Caucasian Federal Districts and Crimea. The average temperature for the year became the second highest for the European and Asian territories of Russia, as well as for the Northwestern and Siberian Federal Districts.

In a large part of the country it exceeded the norm by 1 °C, and in the southwest of the central region and new territories, in Siberia and in some places in the Arctic—by 2° and more. The summer average monthly temperatures, as well as November and December, were extremely warm in Russia. In the Northern Hemisphere, the year as a whole became the warmest in the history of observations. It is also record warm in Canada, the USA, China, India, Europe, and North Africa. Almost everywhere in the hemisphere the average annual air temperature exceeds the norm by 1–2 °C and more. In North America—by 1–3 °C and more. The last year was slightly colder than usual only in some water areas of the Pacific Ocean, as well as in Alaska, Iceland and the eastern coast of Greenland (https://meteoinfo.ru/novosti/99-pogoda-v-mire/20778-2024-god-v-rossii-vtoroj-samyj-teplyj).

D. M. Frolov (✉) · V. E. Gagarin · A. V. Koshurnikov · G. A. Rzhanitsyn
Lomonosov Moscow State University, Moscow, Russia
e-mail: denisfrolovm@mail.ru

T. Chaplina (ed.), *Processes in GeoMedia—Volume IX*, Springer Geology,
https://doi.org/10.1007/978-981-96-9302-3_25

The anomaly of surface temperature according to the report of the Copernicus agency (European Union Earth Observation Program) compared to 1991–2020 was +1.60 °C compared to the pre-industrial period (https://climate.copernicus.eu/global-climate-highlights-2024#:~:text=2024%20had%20a%20global%20average,exceed%201.5%20above%20that%20level).

According to a consolidated analysis of six datasets by the World Meteorological Organization, the global mean surface air temperature was 1.55 °C above the 1850–1900 average. This means, most likely, that humanity has just experienced the first calendar year when the global average temperature was more than 1.5 °C above the 1850–1900 average (Hughes et al. 2023).

UNESCO and the World Meteorological Organization (WMO) have declared 2025 the International Year of Glacier Conservation, aimed at raising awareness of the critical role of glaciers in the climate system and the planet's water balance. Special attention is expected to be paid to issues related to the accelerated melting of glaciers, which poses a global environmental and economic threat.

Covering more than 700,000 km^2, glaciers contain about 70% of the world's freshwater resources. However, their area is rapidly shrinking due to climate change. The Intergovernmental Panel on Climate Change (IPCC) predicts that glacier melting will continue in all regions of the world throughout the twenty-first century.

The International Year, led by UNESCO and WMO, will bring together the efforts of more than 75 international organizations and 35 countries. The organizations are calling on governments, the scientific community, business and civil society to actively participate.

The main thrusts of the initiative are:

- **Enhanced monitoring**: Expand global glacier observation systems to improve data collection and analysis.
- **Early warning systems**: developing technologies to prevent glacier-related hazards.
- **Water management**: promoting sustainable water use in glacier-dependent regions.
- **Cultural heritage preservation**: protecting traditional knowledge and cultural values associated with glacial environments.
- **Youth engagement**: engaging the next generation in combating climate change and glacier conservation.

Earlier, cryospheric scientists in the report "State of the Cryosphere 2024" warned of significant consequences of global warming. In particular, they said that if the temperature on the planet exceeds 1.5 °C, it could lead to a large-scale loss of ice in Greenland and West Antarctica, which would cause serious socio-economic costs. The International Year of Glacier Conservation therefore aims to unite the global community in the fight to preserve the unique natural resources on which the sustainable future of the planet depends.

In this regard, monitoring of the thermal state of the snow column and glaciers is important from a scientific and practical point of view, as well as for comprehensive observation of the physical and mechanical properties of the snow column during

Fig. 1 Measuring system: measuring rod with temperature sensors spaced 5 cm apart, portable battery logger and battery charger

the preparation of winter roads, snow runways, ski slopes, etc. in order to determine, for example, mechanical stability and metamorphism processes taking place in the snow strata (Biresselioglu et al. 2020; Ludescher et al. 2018; Stroeve and Notz 2018; Wang et al. 2024; Faiz et al. 2024; Voitkovsky 1999; Abele 1990; Richter 1945; Wang et al. 2025b; Osokin 2017; Wang et al. 2025a; Yue et al. 2025).

We used a measuring system consisting of a measuring rod with temperature sensors spaced 5 cm apart and a portable battery logger and charger (Fig. 1) to study the temperature in the snow column:

The measuring rod was placed vertically in the snow thickness. Measurements were made once an hour. The measurement data were then transferred to a computer and graphs-diagrams of changes in the thermal state of the snow column were plotted in the MATLAB program environment (Fig. 2).

Figure 2 shows both periods of cyclonic weather prevalence, when the difference between day and night temperatures is not so great and the night cold wave does not spread into the snow thickness, and periods of anticyclonic weather, when the night cold wave penetrates deep into the snow thickness.

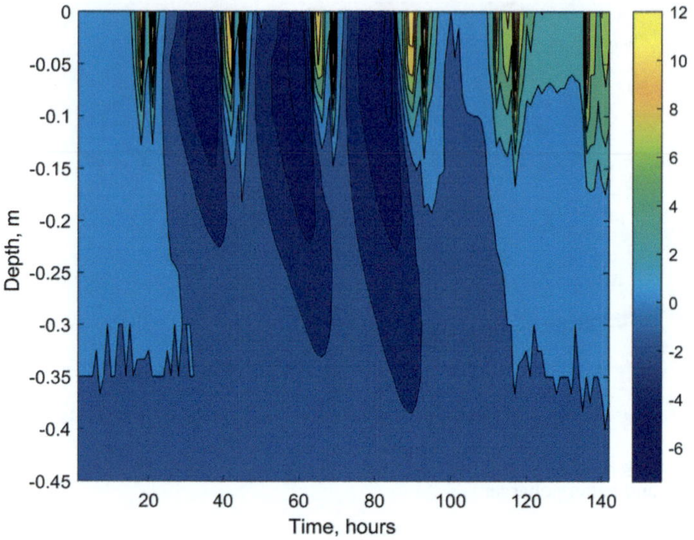

Fig. 2 Diagram-diagram of changes in the thermal state of the snow strata

Acknowledgement The work was carried out in accordance with the state budgetary themes of the Geography Department of MSU "Danger and risk of natural processes and phenomena" (121,051,300,175-4) and "Evolution of the cryosphere under climate change and anthropogenic impact" (121,051,100,164-0).

References

Abele G (1990) Snow roads and runways, vol 90, no 3. US Army Cold Regions Research and Engineering Laboratory

Biresselioglu ME, Demir MH, Solak B et al (2020) Investigating the trends in arctic research: the increasing role of social sciences and humanities. Sci Total Environ 729:139027. https://doi.org/10.1016/j.scitotenv.2020.139027

Faiz H, Alam M, Ng S et al (2024) Carbon dioxide emission evaluation of biochar based vegetation concrete for ecological restoration projects. Case Stud Constr Mater e03874. https://doi.org/10.1016/j.cscm.2024.e03874

https://climate.copernicus.eu/global-climate-highlights-2024#:~:text=2024%20had%20a%20global%20average,exceed%201.5%20above%20that%20level

https://meteoinfo.ru/novosti/99-pogoda-v-mire/20778-2024-god-v-rossii-vtoroj-samyj-teplyj

Hughes KA, Boyle CP, Morley-Hurst K et al (2023) Loss of research and operational equipment in Antarctica: balancing scientific advances with environmental impact. J Environ Manag 348:119200. https://doi.org/10.1016/j.jenvman.2023.119200

Ludescher J, Yuan N, Bunde A (2018) Detecting the statistical significance of the trends in the Antarctic sea ice extent: an indication for a turning point. Clim Dyn 53(1–2):237–244. https://doi.org/10.1007/s00382-0184579-3

Osokin NI, Sosnovsky AV, Chernov RA (2017) Snow thermal conductivity coefficient and its variability. Earth's Criosphere 21(3):60–68

Richter GD Snow cover, its formation and properties Institute of Geography of the USSR 1945, 121c
Stroeve J, Notz D (2018) Changing state of Arctic sea ice across all seasons. Environ Res Lett 13(10):103001. https://doi.org/10.1088/1748-9326/aade56
Voitkovsky KF (1999) Fundamentals of glaciology. Nauka, Moscow, p 255
Wang D, Yang L, Zhang Q et al (2024) Comprehensive evaluation of energy consumption and carbon emissions of asphalt pavement recycling technology. Case Stud Constr Mater 20:e02987. https://doi.org/10.1016/j.cscm.2024.e02987
Wang C, Shirley IA, Wielandt S, Fiolleau S, Lamb JR, Uhlemann S et al (2025a) Advancing the understanding of snow accumulation, melting, and associated thermal insulation using spatially dense snow depth and temperature time series. Geophy Res Lett 52:e2024GL114189. https://doi.org/10.1029/2024GL114189
Wang Z, Zhang Sh, Ze Z, Ch X, Zhou KYuan M (2025b) Study on the hydrothermal effect in ice materials under the polar glacier runway dynamic load. Case Stud Constr Mater 22:e04214. https://doi.org/10.1016/j.cscm.2025.e04214
Yue P, Pei B, Zhang J, Huang N (2025) Dynamic process of dry snow slab avalanche formation: theory, experiment and numerical simulation. Geosci 15:201. https://doi.org/10.3390/geosciences15060201

Creation of a Three-Dimensional Geomechanical Model for the Construction of a Multi-storey Building on a Previously Constructed Pile Foundation

G. V. Alekseev

Abstract This Chapter investigates the impact of multi-story building construction on an existing pile foundation in dense urban settings. Through analytical modeling, the study assesses the foundation's operational safety and determines the need for reconstruction, protective measures, or structural strengthening. The analysis evaluates whether the current pile foundation can be reused and identifies necessary engineering interventions to ensure long-term stability. The results highlight the critical role of an integrated design approach in urban redevelopment, demonstrating how proactive measures can minimize risks to adjacent structures while optimizing project feasibility.

Keywords Capital construction object · Pile foundation · Mathematical modeling · Geomechanical model · Repeated loading · Construction structures · Recommendations

Structures located within the open area designated for new construction must maintain reliable and consistent functionality throughout both the construction phase and subsequent operation. In cases where existing buildings or infrastructure are present within the intended development zone, comprehensive scientific and technical supervision is essential to determine whether continued safe use is feasible, or whether reconstruction or modernization is necessary (Kartozia 2015; Neguritsa 2018).

The objective of this study is to develop a geomechanical model of the foundation system for a 25-storey building being constructed on a site previously prepared with a pile foundation for a residential structure. A segment of the existing pile field intersects with the footprint of the new development.

Engineering survey data indicates that the existing foundation consists of precast driven piles with a solid, square cross-section measuring 30 × 30 cm, joined by

G. V. Alekseev (✉)
National Research Moscow State University of Civil Engineering, Moscow, Russia
e-mail: alekseevg@mgsu.ru

T. Chaplina (ed.), *Processes in GeoMedia—Volume IX*, Springer Geology,
https://doi.org/10.1007/978-981-96-9302-3_26

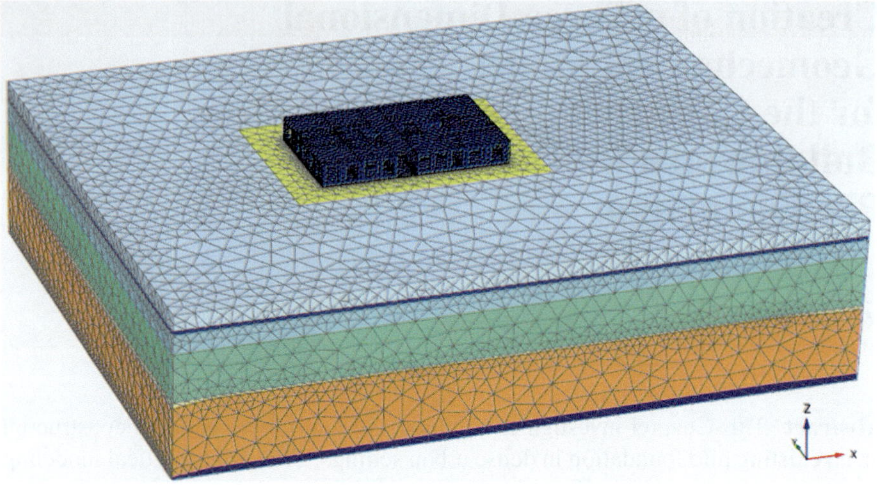

Fig. 1 Calculation model

welded connections and extending to a total length of 14.0–15.0 m. These piles are installed from a concrete base located at the excavation level, arranged in a grid pattern with 1.1–1.2 m spacing in both directions. A monolithic reinforced concrete slab was previously cast over the pile heads, but it had been dismantled at the time of the survey.

To properly formulate the geomechanical problem and accurately account for the interaction between the structure and its foundation, the system is considered as an integrated unit comprising the foundation, subsoil, and underground structural components (Neguritsa et al. 2024; Tsvetkov 2006). The geomechanical simulation is performed using the finite element method implemented in the *PLAXIS 3D* software suite. This method utilizes a volumetric approach, ensuring high accuracy and computational reliability.

The project site is located in an area with a complex geological profile, which requires careful consideration in the computational model. Geological and structural models of the subsurface were developed based on data obtained from detailed engineering and geological investigations conducted on-site. These data are essential for constructing a representative model that closely reflects the actual ground conditions.

The computational model is a soil-rock massif with dimensions of 125 × 100 m, characterised by a heterogeneous composition and varying layer thicknesses according to the geological survey data (Fig. 1). This approach increases the accuracy with which the model's behaviour can be predicted under conditions close to reality (Alekseev et al. 2023; Neguritsa et al. 2022).

When building the model, the following boundary conditions were assigned to the fixings (Fig. 2).

On the surface of the computational domain—free movement of nodes in all directions;

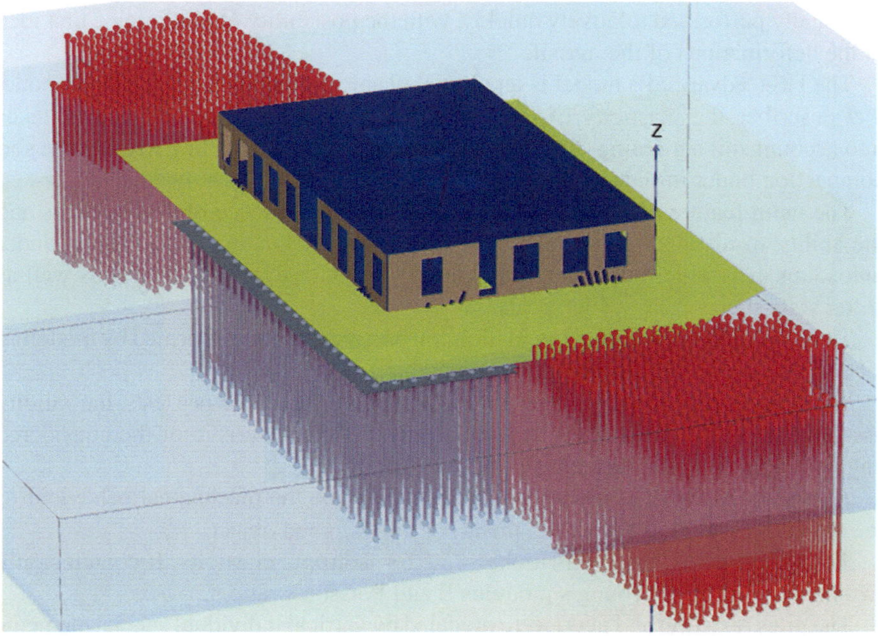

Fig. 2 Finite element model (at the last modeling stage). Geomechanical model in dimensions 120 × 100 × 31 m: 1—existing pile field; 2—bearing structures of the building (basement floor made of monolithic reinforced concrete.); 3—new (projected) piles; 4—sand cushion (between the pile field and the foot of the foundation slab)

On the lateral boundaries of the computational domain—free movement in the vertical direction (along the Z axis) and sliding along the boundary;

On the bottom of the computational domain–prohibition of movements in all directions.

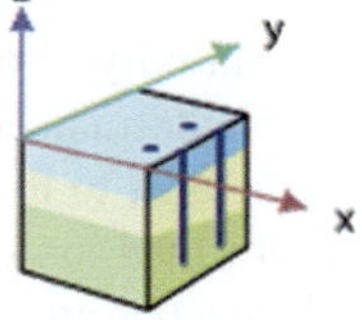

The following mathematical models were used to simulate the mechanical behaviour of engineering geological elements when changing their stress–strain state:

- linear-elastic ideal-plastic Mohr–Coulomb (MC) model;
- elastic–plastic model with hardening (Hardening soil or HS).

Application of MC model is reasonable for general representation of soil behavior during numerical simulation of the considered system in the first approximation. Since each layer of the soil massif has a constant average stiffness, the calculations are usually performed relatively quickly, with the possibility of obtaining a first idea of the deformations of the massif.

The HS ('advanced') model is an elastic–plastic variant of the hyperbolic model that describes the plasticity of soils under shear hardening. In addition, it takes into account soil hardening in compression, which allows modeling irreversible soil compaction under initial compression.

The main feature of the HS model is the stress dependence of soil stiffness and the ability to distinguish between soil stiffness under loading and stiffness under unloading. It is widely used for modeling the behavior of sands, gravels as well as softer soils such as clays, silts, peats.

The validity of the adopted foundation soil parameters was confirmed by modeling laboratory tests in the SoilTest sub-programme.

When assigning the soil parameters and analyzing the laboratory tests, the validity assessment was carried out for the stress intervals characteristic of the considered engineering geological element (EGE).

A Linear Elastic (LE) model was used to describe the mechanical behaviour of the structural materials (concrete, metal) of the designed object.

The LE model is based on Hooke's law for isotropic elasticity. It contains only two main parameters: Young's modulus E and Poisson's ratio υ.

The piles (existing and new) were modeled by vertical individual special elements 'Embedded beam'.

The interaction between the vertical elements and the surrounding soil (soil cluster element) was modeled using special interface elements. An elastic–plastic model is used to describe the interface behaviour.

When selecting materials, there are two types of resistance to consider: Skin resistance, which is the resistance on the lateral surface, and Base resistance, which is the resistance under the end of the beam. Evaluation of lateral surface and base resistance is based on a fracture criterion that takes into account the elastic and plastic behaviour of the model. Relatively small displacements are observed in the interaction region between the pile or anchor bolt and the soil or rock in the case of elastic deformation. Conversely, in the case of plastic deformation, permanent sliding may occur, significantly affecting the stability of the system under consideration.

The calculation was carried out for the main load combination. The values of forces transmitted from the upper structures to the foundations of the Project were taken iteratively (3 iterations were performed) based on the results of the calculation of the load-bearing structures of the Project on an elastic base according to the second group of limit states (Fig. 3).

When performing calculations of the bases of underground parts of structures, it is necessary to take into account the structural nonlinearity associated with changes in the design scheme during construction, technological features of erection and the sequence of construction operations (Neguritsa et al. 2021; Merkin et al. 2023; Alekseev et al. 2023).

Modeling of the stress–strain state (STS) of the system 'Foundation-Foundation-Earth part' and analysis of the results were carried out taking into account the technology of works and the initial modeling of stress–strain state (STS) of the soil mass.

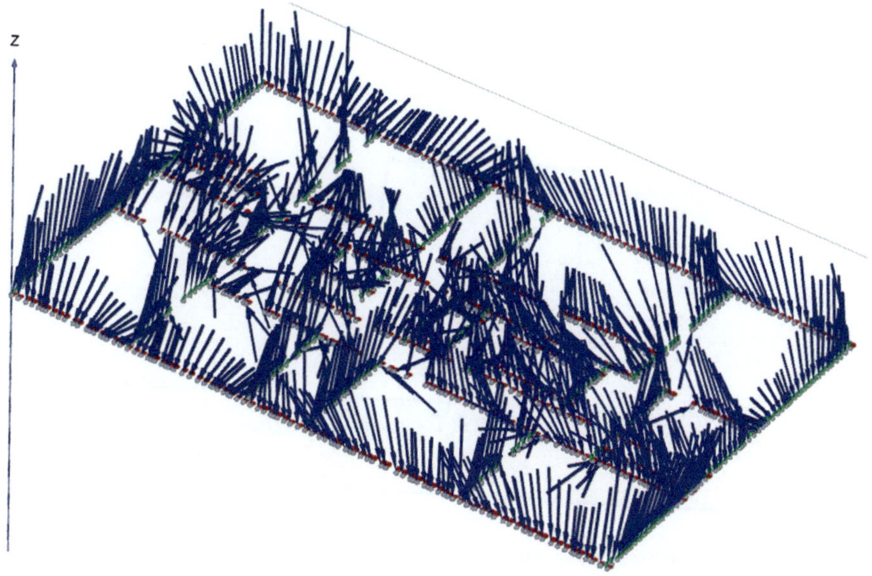

Fig. 3 Vectors of load application on the foundation slab of the Object (surfaces are disabled for clarity)

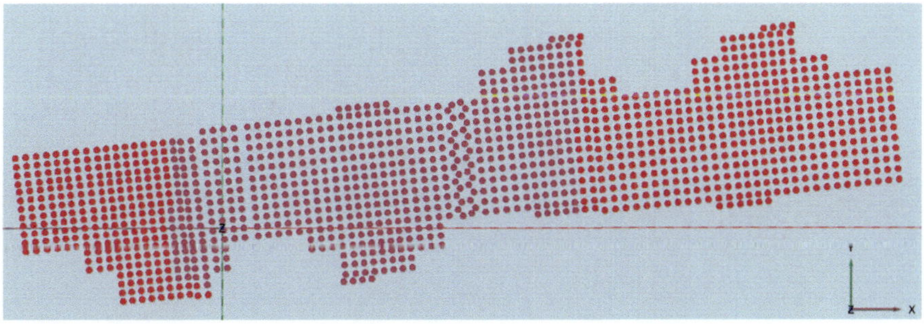

Fig. 4 Loading the computational domain with the dead weight of the soil, modeling the existing piles and determining the initial stress–strain state of the soil mass. Zeroing of displacements

Only static effects are taken into account in the calculations. Dynamic and temperature effects are not taken into account. Seasonal fluctuations of groundwater horizon are also not taken into account (Figs. 4, 5, 6, 7 and 8).

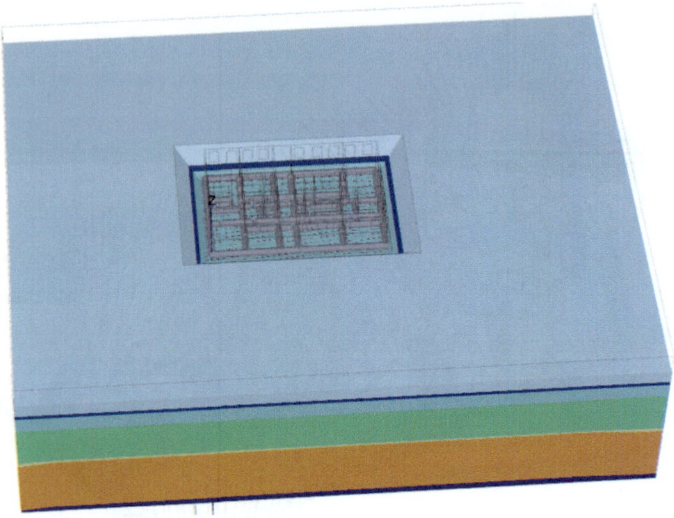

Fig. 5 Modeling of excavation of a pit

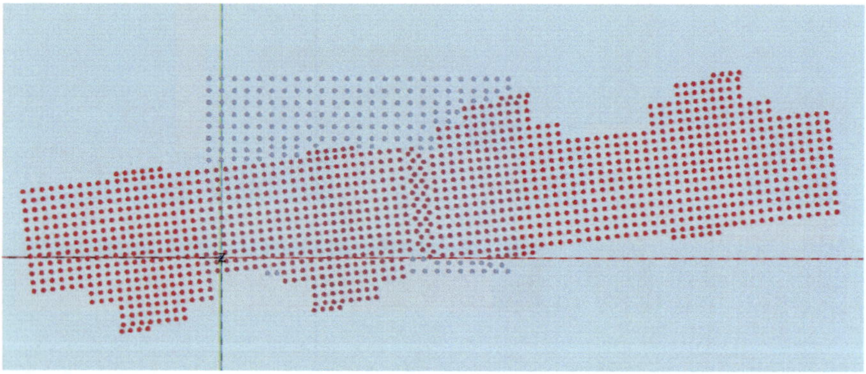

Fig. 6 Installation of additional piles. Zeroing of displacements

1 Results

The geomechanical stability calculation of the considered system was successfully performed using the proposed methodological approach. This indicates that, once all loads from the structures have been applied to the foundation, the soil massif is in a pre-limit state. In this state, the zones of plastic flow beneath the building's foundation remain closed, preventing any loss of the soil's bearing capacity. The calculation results confirm the stability of the model under study.

The bearing capacity of the system was analysed in three iterations at each stage of the calculation. Analysis of the results shows good convergence in the nature

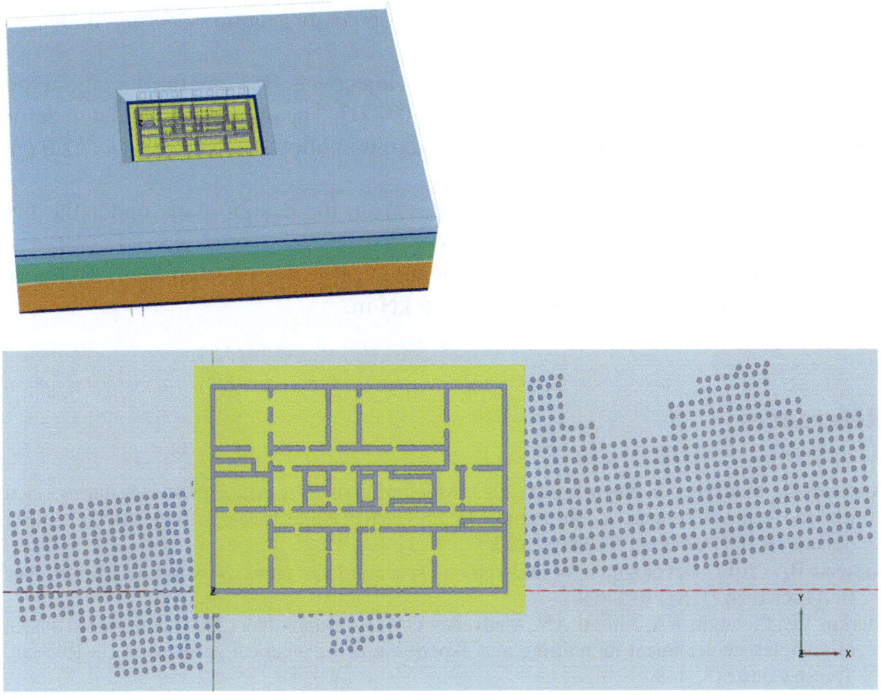

Fig. 7 Arrangement of sand cushion

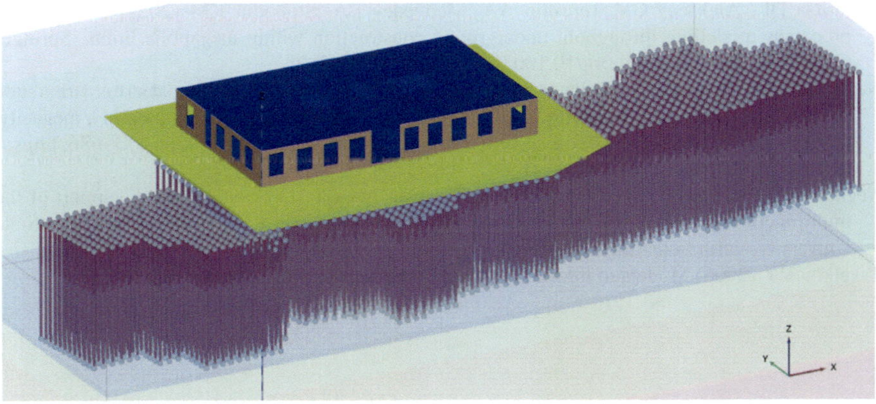

Fig. 8 Modeling of the building structure, backfill and load application

and values of the precipitation and stress isopole distributions acting under the foundations of the structure.

The maximum predicted settlement of the Object was 10.1 cm, the average settlement was 9.7 cm, relative difference $\Delta s/L = 0.0007$. The predicted calculated deformation of the Object does not exceed the maximum allowable values $s_u = 12.0$ cm, $\Delta s/L = 0.0016$.

The maximum vertical force in the piles from the design loads under the first group of limit states was Nmax = 557 kN.

The maximum value of bending moments in piles from design loads under the first group of limit states was Mmax = 32.5 kN·m.

References

Alekseev G., Medvedev E, Neguritsa D (2023) Monitoring of deformation processes during scientific and technical support of construction. In: E3S web of conferences, vol 410, p 02045. https://doi.org/10.1051/e3sconf/202341002045

Kartozia BA (2015) Development of underground space of large cities. New trends. Min Inf Anal Bull (Sci Tech J) (S1):615–630

Merkin VE, Pichugin AA, Simuti AN, Medvedev GM, Kuzyanov NA (2023) Automated system of complex geotechnical monitoring and experience of its practical application (in Russian). Transp Constr (3):4–8

Neguritsa D (2018) The problems of monitoring the deformation processes in the integrated development of the underground space of metropolitan cities. In: E3S web of conferences, vol 56, p 02027. https://doi.org/10.1051/e3sconf/20185602027

Neguritsa DL, Alekseev GV, Tereshin AA, Medvedev EA, Slobodin KM (2022) Deformation processes modelling throughout underground construction within megapolis limits. Springer Geol:215–227. https://doi.org/10.1007/978-3-030-76328-2_23

Neguritsa D, Alekseev G, Medvedev E, Slobodin K, Tereshin I (2024) Modelling of stress-strain state of artificially improved foundation during construction of a residential complex in a megacity. In: Springer proceedings in earth and environmental sciences, part F3114, pp 673–686. https://doi.org/10.1007/978-3-031-54589-4_68

Neguritsa DL, Alekseev GV, Medvedev EA, Tereshin AA (2021) Geomechanical support of the multifunctional centre construction in megapolis conditions (in Russian). Izvestiya Tula State University. Earth Sciences 3:286–296

Tsvetkov VY (2006) Models in information technologies. Max Press, p 104

Behavior and Characteristics of the First Faraday Wave on the Free Surface of Water

E. A. Danilova

Abstract The results of an experimental study on the lowest wave mode of Faraday gravity waves arising on the free surface of a liquid under vertical oscillations of a container are presented. Resonance dependencies, wave profiles, wave steepness, and the formation of a planar jet were investigated. The experiments were conducted in a rectangular container with variable liquid depth, using high-speed imaging and data processing in the ImageJ software. It was shown that an increase in liquid depth changes the slope of the resonance curves, which is associated with nonlinear effects. Transitions from regular waves to jet splashes, accompanied by the formation of a planar jet and intense splashing, were recorded. The experimental results confirmed the theoretical models of Faraday parametric resonance and revealed important characteristics of wave interactions with the sidewalls of the container, which are of practical relevance for engineering applications.

Keywords Faraday waves · Resonance curves · Wave profiles · Wave steepness · Planar jet

1 Introduction

Parametric excitation of gravity standing waves on the free surface of a liquid is an important phenomenon in hydrodynamics, associated with the formation of stable wave structures under periodic external excitation. Standing waves arising under such conditions exhibit unique characteristics depending on the liquid depth, frequency, and amplitude of the oscillations. Studying these waves has both theoretical and practical significance: understanding their behavior is essential for predicting loads on engineering structures, analyzing the stability of liquid transport systems, and solving problems related to wave impacts on coastal structures.

In experiments dedicated to the study of the destruction of standing surface gravity waves, both the lowest wave mode (e.g., liquid oscillations in a cylindrical container

E. A. Danilova (✉)
Ishlinsky Institute for Problems in Mechanics RAS, Moscow, Russia
e-mail: ElizDanilov@yandex.ru

T. Chaplina (ed.), *Processes in GeoMedia—Volume IX*, Springer Geology,
https://doi.org/10.1007/978-981-96-9302-3_27

(Hogrefe et al. 1998; Zeff et al. 2000) and the second mode (e.g., liquid oscillations in a narrow rectangular container (Jiang et al. 1998; Kalinichenko 2009) were considered. A common feature of wave motion in both cases was the presence of antinodes at the center and walls of the container. Previous works primarily focused on the formation of jet ejections observed at the center of the container, while the influence of the container walls was excluded. However, when waves interact with the container walls, a so-called planar jet forms. At shallow liquid depths, the pressure exerted by the wave on the wall can reach significant values.

The aim of this work is to experimentally investigate the lowest wave mode of Faraday waves arising on the surface of an oscillating liquid. In hydrodynamics, Faraday waves are defined as standing waves excited by parametric resonance in a liquid subjected to vertical oscillations. The frequency of such waves is a multiple of half the oscillation frequency of the container ($\omega \sim \Omega/2$).

2 Description of the Experiment

The experiment examined standing gravity waves generated in a vertically oscillating rectangular container filled with liquid. Vertical oscillations were produced using a unique experimental setup at the Ishlinsky Institute for Problems in Mechanics RAS ("Studies of Dynamics and Structure of Oscillatory Flows" Test Bench (DSO) n.d.). Similar to previous studies (Kalinichenko et al. 1995) and (Kalinichenko 2024), two-dimensional wave motions under the primary Faraday resonance conditions were investigated.

A rectangular container with dimensions of 50 × 4 × 50 cm, filled with liquid (tap water at room temperature) to a depth h, underwent harmonic vertical oscillations with amplitude s and frequency Ω (see Fig. 1). The variable parameter of the system was defined as $g^*(t) = -g(1 + \frac{s\Omega^2}{g}\cos\Omega t)$ in the reference frame attached to the oscillating container, where g is the gravitational acceleration. Two-dimensional liquid motion was studied. Experiments were conducted for four liquid depths h: 5, 7.5, 10 and 20 cm.

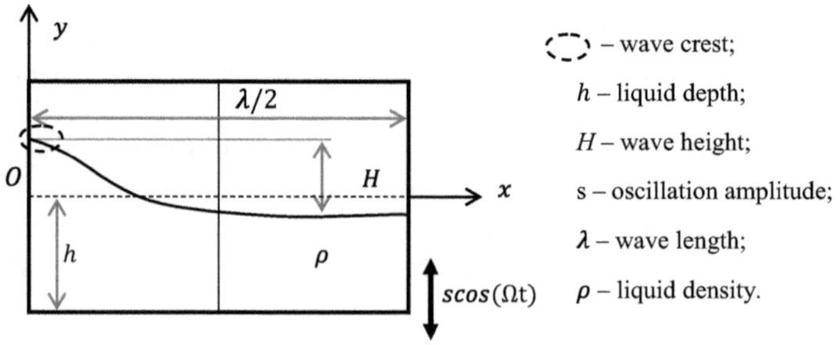

Fig. 1 Schematic diagram of surface wave parameters

Wave motions were recorded using high-speed imaging at resolutions of 1920 × 1080 and frame rates of 120 and 1000 fps. The resulting video material was processed in ImageJ to extract experimental data.

3 Results

3.1 Resonance Curves

The graphs in Fig. 2 show the resonance dependencies $H = H(\Omega)$ for the first oscillation mode at different liquid depths h. The data points represent resonance dependencies obtained from video recordings of wave heights corresponding to specific oscillation frequencies, while the lines depict theoretical curves based on Eq. (1) (Nesterov 1969).

$$\begin{cases} k^2 \frac{H^2}{4} \omega_n \beta = \pm\left(\omega_n - \frac{\Omega}{2}\right) + \varepsilon\omega^2 \frac{1}{2\Omega} \\ \beta = \frac{1}{64} th^{-4}kh(2th^6kh + 3th^4kh + 12th^2kh - 9) \end{cases} \quad (1)$$

The results demonstrate that the theory closely aligns with experimental data. A pronounced change in the slope of the resonance curves was observed with varying liquid depth. The parameter β, characterizing the slope of the resonance curves, is negative for shallow depths (h = 5, 7.5, 10 cm) but becomes positive for greater depths (20 cm), indicating a critical depth effect (Table 1). This is associated with the manifestation of nonlinear effects, whereby at large wave heights, the waves begin to break down due to gravitational or inertial forces, forming jets, spray, or cavitation. Consequently, as the liquid depth increases, the parameter β changes its sign, and the resonance curve reverses its slope.

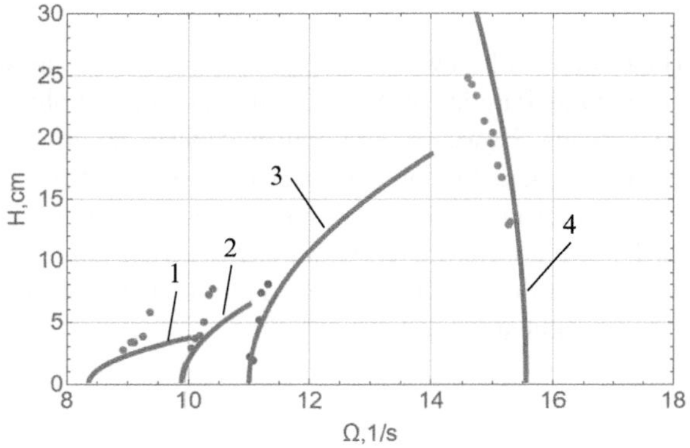

Fig. 2 Resonance dependencies at s = 1.9 cm: 1—h=5 cm, h/λ = 0.05; 2—h=7.5 cm, h/λ = 0.075; 3—h=10 cm, h/λ = 0.1; 4—h=20 cm, h/λ = 0.2

Table 1 Calculated parameter β

Depth h, cm	Wavenumber k	Frequency ω, 1/s	Parameter β
5	0.06	4.33	–14.34
7.5		5.21	–2.75
10		5.86	–0.8
20		7.24	0.05

3.2 Wave Profiles

The main characteristics of the surface waves are the wave height H and the oscillation frequency $\Omega/2$. Equation (2) describes the wave profiles in Lagrangian variables (Nesterov 1969). Figure 3 compares theoretical and experimental profiles of regular gravity waves for a liquid depth h of 10 cm.

$$\begin{cases} x = a - HA_1 sink\alpha cos\psi + H^2 kA_2(\alpha, \psi) \\ y = HA_3 cosk\alpha cos\psi + H^2 kA_4(\alpha, \psi) \\ \psi = \frac{\Omega t}{2} + \theta \end{cases} \quad (2)$$

Three characteristic wave heights H were identified:

- A: Small wave height, corresponding to the onset of regular surface waves.
- B: Wave of ultimate height, where the height and period remain stable over time.
- C: Transition to planar jet, where a planar jet forms upon wave–wall interaction.

For cases A and B, the resonance dependencies confirm theoretical models of Faraday parametric resonance, while case C exhibits weak nonlinearity.

3.3 Wave Steepness

The dimensionless wave steepness $\Gamma = \frac{H}{\lambda}$ serves as an indicator of wave regularity. For regular Faraday waves, Γ remains small at shallow depths and low excitation frequencies. However, as depth and wave height increase, nonlinear effects such as droplet detachment, jet formation, and intense splashing emerge. The system becomes unstable, and its behavior becomes unpredictable. The experimental data for Γ were approximated by a quadratic function: $\Gamma = 5.11x^2 + 0.199x + 0.003$

The form of the approximated curve is shown in Fig. 4, where the experimental data are indicated by dots.

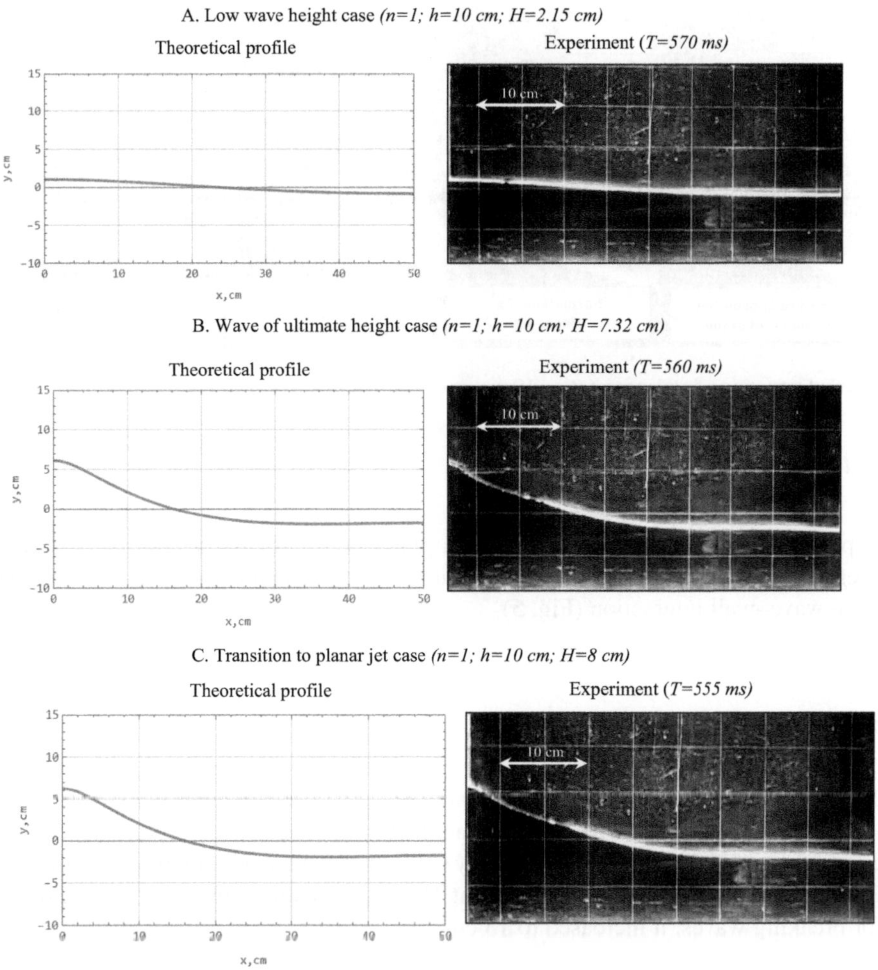

Fig. 3 Wave profiles (n=1, h=10 cm)

Fig. 4 Dependence of limiting wave steepness on dimensionless liquid depth (n = 1)

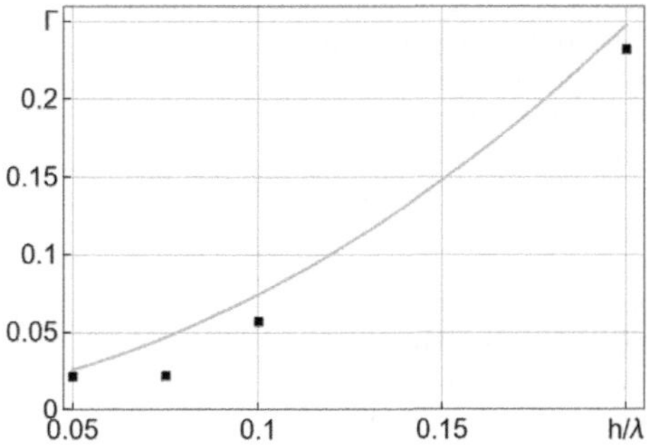

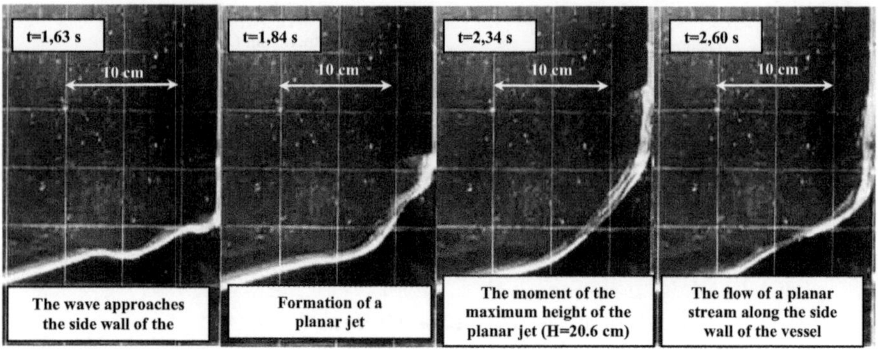

Fig. 5 Formation of a planar jet (h = 10 cm, irregular wave)

3.4 Planar Jet

Experiments revealed the formation of a planar jet, particularly prominent during irregular waves. The planar jet is a thin liquid layer that spreads along the sidewall upon wave–wall interaction (Fig. 5).

3.5 Free Surface Velocity

The velocity of the free surface at the container wall is a key parameter characterizing liquid motion in regions where waves interact with boundaries. Temporal dependencies of the crest height $H_{10}(t)$ were analyzed for regular and breaking waves (Fig. 6). For regular waves, the maximum height H_{10_max} was approximately 3.5 cm, while for breaking waves, it increased to 13 cm.

Fig. 6 Time dependence of height H_10 (t) for the first wave mode (h = 10 cm, s = 1.9 cm): 1.2—regular wave, wave period T = 562 ms, (according to the results of video recording and approximating function, respectively); 3.4—collapsing wave, wave period T = 548 ms (according to the results of the video recording and the approximating function, respectively)

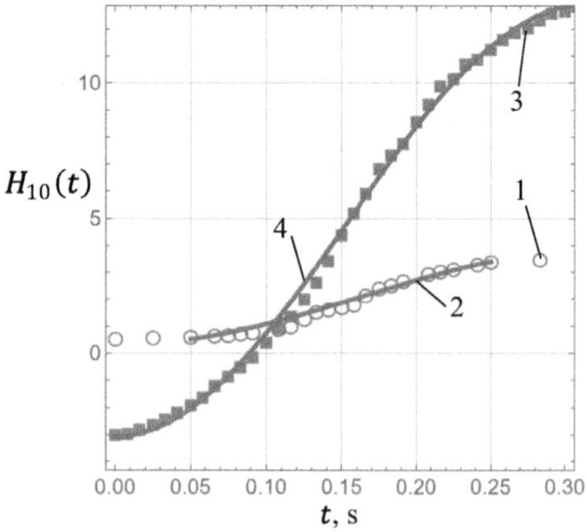

The maximum vertical velocity was 16.28 cm/s for regular waves and 75.3 cm/s for breaking waves, indicating energy concentration near the walls.

The temporal dependencies are approximated by the following functions:

For the regular wave: $H_{10}(t) = 2.82 - 1.69\cos(5.82t) - 0.81\cos(11.64t)$

For the breaking wave: $H_{10}(t) = -5.71\cos(5.82t) - 4.75\cos(11.64t) + 7.4$

4 Conclusion

The article presents an experimental study of the lowest wave mode of Faraday waves arising on the surface of a vertically oscillating liquid. Key parameters, including resonance dependencies, wave profiles, free surface velocity, and planar jet formation, were examined. The experiments provided detailed insights into wave structures, resonance characteristics, nonlinear effects, and wave–boundary interactions.

The results confirmed theoretical models of parametric resonance. Resonance curves exhibited depth-dependent slopes due to nonlinear effects. Wave profiles showed stability at low heights and transition to instability at higher energies, accompanied by planar jet formation and splashing. The velocity of the planar jet at the wall was 4.6 times higher than that of the free surface in regular waves, highlighting energy concentration near boundaries. The dependence of wave steepness on dimensionless liquid depth characterized nonlinearity and served as an indicator of instability transitions.

The work was supported by RSF Grant no. 23–19-00,451. Experiments were conducted using the DSO test bench (a unique facility at the Ishlinsky Institute for Problems in Mechanics RAS).

References

"Studies of Dynamics and Structure of Oscillatory Flows" Test Bench (DSO). Unique Research Facility "GFK IPMech RAS": http://www.ipmnet.ru/uniqequip/gfk/#aboutDSO

Hogrefe JE, Peffley NL, Goodridge CL, Shi WT, Hentschel HGE, Lathrop DP (1998) Power law singularities in gravity capillary waves. Physica D 123(1):183–205. https://doi.org/10.1016/S0167-2789(98)00120-1

Jiang L, Perlin M, Schultz WW (1998) Period tripling and energy dissipation of breaking standing waves. J Fluid Mech 369:273–299. https://doi.org/10.1017/S0022112098001785

Kalinichenko VA, Nesterov SV, Sekerzh-Zen'kovich SY, Chaikovskii AA (1995) Experimental study of surface waves with Faraday resonance excitation. Fluid Dyn 30(1):101–107. https://doi.org/10.1007/BF02029933

Kalinichenko VA (2009) Breaking of Faraday waves and jet launch formation. Fluid Dyn 44:577–586. https://doi.org/10.1134/S0015462809040115

Kalinichenko VA (2024) Kinematics of the first wave Faraday mode on the side wall of a rectangular vessel. Fluid Dyn 59(5):1028–1037. https://doi.org/10.1134/S0015462824603802

Nesterov SV (1969) Parametric excitation of waves on the surface of a heavy liquid. Morskie Gidrofiz Issled 3(45):87–97. [in Russian]
Zeff BW, Kleber B, Fineberg J, Lathrop DP (2000) Singularity dynamics in curvature collapse and jet eruption on a fluid surface. Nature 403(6768):401–404. https://doi.org/10.1038/35000151

A Simple Model of Ice and Snow Thermodynamics Applied to the Rybinsk Reservoir Within the Framework of Three-Dimensional Numerical Simulation

D. S. Gladskikh, E. V. Mortikov, R. A. Ahtamyanov, and V. A. Lomov

Abstract A simple description of the thermodynamics of ice and snow is implemented in a three-dimensional numerical model of an inland water body. Numerical experiments aimed at studying the periods of freeze-up and ice break-up on the Rybinsk Reservoir are conducted. The results of the modelling were compared with both field measurement data and satellite images obtained using remote sensing of the Earth. It is confirmed that the model correctly describes the ice regime of the water object under study.

Keywords Numerical simulation · Inland water bodies · The Rybinsk reservoir · Ice regime · Thermodynamics

D. S. Gladskikh (✉)
A.V. Gaponov-Grekhov Institute of Applied Physics RAS, Nizhny Novgorod, Russia
e-mail: daria.gladskikh@gmail.com

D. S. Gladskikh · E. V. Mortikov · R. A. Ahtamyanov · V. A. Lomov
M.V. Lomonosov Moscow State University, Research Computing Center, Moscow, Russia

D. S. Gladskikh · E. V. Mortikov · R. A. Ahtamyanov
Moscow Center for Fundamental and Applied Mathematics, Russian Federation, Moscow, Russia

E. V. Mortikov
G.I. Marchuk Institute of Numerical Mathematics RAS, Moscow, Russia

R. A. Ahtamyanov · V. A. Lomov
Faculty of Geography, M.V. Lomonosov Moscow State University, Moscow, Russia

R. A. Ahtamyanov
Hydrometcenter, Moscow, Russian Federation

V. A. Lomov
A.M. Obukhov Institute of Atmosphere Physics RAS, Moscow, Russia

T. Chaplina (ed.), *Processes in GeoMedia—Volume IX*, Springer Geology,
https://doi.org/10.1007/978-981-96-9302-3_28

1 Introduction

The processes of ice formation and breaking up are an important part of the hydrological cycle of many inland water objects. The conditions of the ice regime have a strong influence on the ecological system of lakes and are an indicator of changes in climatic factors (Adrian 2009; Karetnikov and Naumenko 2008). Ice formation and breaking up directly affects the aquatic ecosystem of lakes. Ice cover significantly modifies lake hydrodynamics, inhibiting wind stress, so that vertical mixing is sustained only by natural convection (Oveisy et al. 2012; Farmer 1975). Therefore, proper modelling of the lake during the ice-covered season requires accurate modelling of the ice cover and its effects on heat and momentum transfer. Thus, to correctly describe the inland water objects during the ice season, accurate modelling of the ice and its effects on heat and momentum transfer is necessary.

Several thermodynamic models of ice formation have been developed in recent decades: mostly simplified three-dimensional models of ice formation and transport, or more complex one-dimensional (vertical) models that do not allow for water density dynamics and biogeochemistry. Most of these models use simplified transformations for the energy fluxes during ice formation (Maykut and Untersteiner 1971). Patterson and Hamblin (Patterson and Hamblin 1988) proposed a one-dimensional (vertical) model (DYRESM-I) that included a thermodynamic model of lake mixing in the water column beneath the ice and accounted for the two-dimensional effects of partial ice cover. The mixed lake ice (MLI) model by Rogers et al. (Rogers et al. 1995) extended the DYRESM-I model to include new processes such as white ice formation and snow albedo variability to solve specific problems for mid-latitude water bodies.

In this study, a model of ice and snow thermodynamics given in Rogers et al. (1995) taking into account the modifications proposed in Oveisy et al. (2012) and (Hipsey et al. 2019) is implemented in a three-dimensional inland water body model proposed by the authors. The ice formation algorithm is a quasi-stable solution of the heat transfer equation for three layers (blue ice, white ice and snow). The governing equations of the model are the Reynolds-averaged Navier–Stokes equations in the hydrostatic approximation and the Boussinesq approximation.

2 The Three-Dimensional Model of an Inland Water Objects

Main Equations of the Model

The main research tool is a three-dimensional numerical model of an inland water body, based on a unified hydrodynamic code that combines DNS, LES and RANS approaches for calculating geophysical turbulent currents at high spatial and temporal resolution (Mortikov 2016; Mortikov et al. 2019; Kadantsev et al. 2021; Gladskikh et al. 2021, 2023; Gladskikh and Mortikov 2024).The numerical model includes equations of hydrodynamics in a stratified turbulent rotating layer of liquid in the shallow water approximation, as well as equations for heat and salinity transfer taking into account horizontal and vertical diffusion. The general equations of the model are as follows:

$$\frac{\partial u}{\partial t} = -A(u) + D_H(u, \lambda_m) + D_z(u, K_m + \nu) - g\frac{\partial \eta}{\partial x} - \frac{g}{\rho_0}\frac{\partial}{\partial x}\int_z^{\eta} \rho dz' + fv, \quad (1)$$

$$\frac{\partial v}{\partial t} = -A(v) + D_H(v, \lambda_m) + D_z(v, K_m + \nu) - g\frac{\partial \eta}{\partial y} - \frac{g}{\rho_0}\frac{\partial}{\partial y}\int_z^{\eta} \rho dz' - fu, \quad (2)$$

$$\nabla \cdot u = \frac{\partial u}{\partial x} + \frac{\partial v}{\partial y} + \frac{\partial w}{\partial z} = 0, \quad (3)$$

$$\frac{\partial T}{\partial t} = -A(T) + D_H(T, \lambda_h) + D_z(T, K_h + \chi), \quad (4)$$

$$\frac{\partial S}{\partial t} = -A(S) + D_H(T, \lambda_h) + D_z(S, K_h + \chi), \quad (5)$$

$$\rho = \rho(T, S), \quad (6)$$

$$\frac{\partial \eta}{\partial t} + (u_h \cdot \nabla_h)\eta = w. \quad (7)$$

Here, $\mathbf{u} = (u, v, w)$ is the velocity vector, η is the free surface deviation from the equilibrium state, f is the Coriolis parameter, T is the temperature, S is the salinity, ρ is the density, K_m (λ_m) and K_h (λ_h) are coefficients of vertical (horizontal) turbulent viscosity and temperature conductivity, respectively, ν and χ are coefficients of molecular viscosity and temperature conductivity, respectively, and z is the vertical coordinate going from the water bottom $z = -H(x, y)$ to the surface. In addition, $A(q)$ is the advection operator, and $D_H(q, \lambda)$ and $D_z(q, K)$ are the operators defining horizontal and vertical diffusion with the coefficients λ and K, respectively.

The numerical method for solving the system of equations is based on conservative finite-difference discretization methods on rectangular grids and the use of a semi-implicit method for time approximation, in which advective transport and horizontal diffusion are described by explicit schemes (Mortikov et al. 2019).

As for the description of turbulent transport, the model uses two-parameter $k - \varepsilon$ scheme (Burchard 2007) based on the gradient hypothesis and the turbulence kinetic energy (TKE) balance equation. The additional equation in this scheme is for the dissipation rate (ε) of the TKE. Thus, the scheme has the following form:

$$\frac{\partial k}{\partial t} = \frac{\partial}{\partial z}\frac{K_m}{Sc_k}\frac{\partial k}{\partial z} + P + B - \varepsilon, \quad (8)$$

$$\frac{\partial \varepsilon}{\partial t} = \frac{\partial}{\partial z}\frac{K_m}{Sc_\varepsilon}\frac{\partial \varepsilon}{\partial z} + \frac{\varepsilon}{k}(C_{1\varepsilon}P - C_{2\varepsilon}\varepsilon + C_{3\varepsilon}B). \quad (9)$$

Here, $P = K_m\left[\left(\frac{\partial u}{\partial z}\right)^2 + \left(\frac{\partial v}{\partial z}\right)^2\right]$ corresponds to the production of TKE due to velocity shear and $B = -\frac{g}{\rho}K_h\frac{\partial \rho}{\partial z}$ describes the production or consumption of energy due to the action of buoyancy forces. Also in the given expressions Sc_k, Sc_ε are the turbulent Schmidt numbers for the TKE and the dissipation rate, respectively, $C_{1\varepsilon}$, $C_{2\varepsilon}$, $C_{3\varepsilon}$ are empirical constants (see, for example, (Burchard and Deleersnijder 2001)). In the inland water body model, the following values are used, in accordance with (Mortikov et al. 2019): $C_{1\varepsilon} = 1.44$, $C_{2\varepsilon} = 1.92$, $C_{3\varepsilon}$ is assumed to be equal to -0.4 for stable stratification and 1.14 for unstable. The exchange coefficients are given as

$$K_m = C_\varepsilon\frac{k^2}{\varepsilon}, \quad (10)$$

$$K_h = C_{\varepsilon,T}\frac{k^2}{\varepsilon} = \frac{C_{\varepsilon,T}}{C_\varepsilon}K_m = Pr_T^{-1}K_m. \quad (11)$$

The stability functions C_ε, $C_{\varepsilon,T}$ are assumed to be constant in the ''standard'' $k-\varepsilon$ scheme, the values $C_\varepsilon = 0.09$ and $Pr_T = 1.25$ are used. The model also implements stability functions according to Gladskikh et al. (2023); Gladskikh and Mortikov 2024; Baas et al. 2008; Schumann and Gerz 1995).

On the free surface, the kinematic condition is satisfied and the momentum and heat fluxes are assumed to be given. The logarithmic layer approximation is used to calculate the bottom friction. On the solid boundaries, the conditions of impermeability and the equality of the heat flux to zero are assumed. Dirichlet boundary conditions are used for the kinetic energy of turbulence and the dissipation rate.

The Model of Ice and Snow Thermodynamics

To account for the ice regime, the description given in Rogers et al. (1995) was implemented in the model, taking into account the modifications proposed in Oveisy et al. (2012) and (Hipsey et al. 2019). For each cell on the surface of the reservoir, the stationary equation of heat transfer between the atmosphere and the water column (Semtner 1976) is solved for three conjugate layers of blue ice (crystalline ice), white ice ("snow" ice) and snow:

$$K\frac{\partial^2 T}{\partial^2 z} = \frac{1}{\rho_n c_n}\frac{\partial Q}{\partial z}, \quad (12)$$

where K is the heat diffusivity, ρ_n is the snow or ice cover density, c_n is the specific heat of snow or ice cover, Q is the heat flux. The penetrating radiation is $I = I_{sw}\exp[-\lambda(h-z)]$, where the incoming solar radiation I_{sw} has been divided into visible (hereinafter—subscript 1) and near infrared (hereinafter—subscript 2) parts of the spectra according to Kirk (1994); Boër 1980), and λ is a distinct attenuation coefficient for each component and ice and snow layer. The system of equations of type (12) for three distinct layers (i = blue ice, e = white ice and s = snow) is supplemented by boundary conditions:

$$T_i = T_f \quad z = 0 \tag{13}$$

$$\left.\begin{aligned} T_i &= T_e \\ K_i \frac{\partial T_i}{\partial z} &= K_e \frac{\partial T_e}{\partial z} \end{aligned}\right\} \quad z = h_i \tag{14}$$

$$\left.\begin{aligned} T_e &= T_s \\ K_s \frac{\partial T_s}{\partial z} &= K_i \frac{\partial T_i}{\partial z} \end{aligned}\right\} \quad z = h_i + h_e \tag{15}$$

$$T_s = T_a \quad z = h_i + h_e + h_s \tag{16}$$

Here, subscripts are f = freezing, a = air. So, we can get the solution:

$$Q_0 + Q_{lw} + Q_s + Q_e + Q_r = 0, T_a < T_f = 0\,°\mathrm{C} \tag{17}$$

$$Q_0 + Q_{lw} + Q_s + Q_e + Q_r = \rho_n L_w \frac{dh_n}{dt}, T_a = T_f = 0\,°\mathrm{C} \tag{18}$$

$$Q_f = Q_O - A_1 I_{sw}\{1 - \exp[-(\lambda_{s1}h_s + \lambda_{e1}h_e + +\lambda_{i1}h_i)]\} -$$

$$-A_2 I_{sw}\{1 - \exp[-(\lambda_{s2}h_s + \lambda_{e2}h_e + +\lambda_{i2}h_i)]\} - Q_{si}h_s \tag{19}$$

$$Q_w = -K_w \frac{dT_w}{dz}, \tag{20}$$

where Q_0 is the incident solar radiation, Q_{lw} is the net long-wave radiation, Q_s is the sensible heat flux, Q_e is the latent heat flux and Q_r is the heat flux due to rainfall. L_w is the latent heat of fusion of water, $A_1 = 0.7$, $A_2 = 0.3$ (Kirk 1994; Boër 1980). Q_{si} is the heat flux per volume released from white ice formation resulting from flooding of snow. Q_f is the flux from ice to water and Q_w is the flux from water to the ice, K_w is the water conductivity. When heat fluxes create a freezing temperature on the free surface, ice begins to form. The imbalance of heat fluxes between ice and the water column leads to a change in the thickness of blue ice:

$$\frac{dh_i}{dt} = \frac{Q_f - Q_w}{\rho_i L_W}. \tag{21}$$

Snow accumulates on the surface of blue ice; snow can be flaky or watery, so its density can vary significantly. Snow can also be significantly compacted over time. Snow density variance due to compaction (Koren et al. 1999) is used in the model:

$$\rho_{s,c} = \rho_s \frac{exp(Bw_s) - 1}{Bw_s}, \tag{22}$$

where $\rho_{s,c}$ is the compacted snow density, ρ_s is the average snow density, w_s is the water equivalent in the snow and B is calculated from Oveisy et al. (2012). Unlike Rogers et al. (Rogers et al. 1995), who assumed constant density of the fallen snow as 80 kg m^{-3}, the fallen snow density is assumed to be a function of air temperature (Gottlieb 1980), and the average snow density is parameterized as given in Oveisy et al. (2012); Hipsey et al. 2019).

When calculating the flux of incoming solar radiation, differences in the albedo of ice and snow are taken into account. Albedo has a wide range of values (see (Henderson-Sellers 1984)), which depend mainly on the temperature and thickness of ice and snow. Note that the surface albedo (depending on the surface temperature) is calculated based on the state at the previous time step. Albedo variation is particularly important when modelling mid-latitude water bodies, where meteorological conditions can change rapidly, and observations show that the albedo of ice and snow decreases with increasing temperature. We used variable albedo values calculated according to Vavrus et al. (1996).

If the weight of the snow exceeds the buoyancy of ice, then white ice is formed from the snow as a result of flooding. The thickness of the flooded snow layer may be calculated by comparing the thickness of snow (h_s), to the maximum that can be supported by the existing blue ice ($h_{s,max}$).

$$\Delta h_e = h_s - h_{s,max}, \tag{23}$$

$$h_{s,max} = \frac{h_i(\rho_w - \rho_i) + h_e(\rho_w - \rho_e)}{\rho_s}. \tag{24}$$

A similar formulation as (Rogers et al. 1995) was used to calculate the heat flux per volume released from white ice formation resulting from flooding of snow:

$$Q_{si} = \frac{(T_w c_a + L_f)\rho_w \frac{\Delta h_e}{h_s}\left(1 - \frac{\rho_s}{\rho_w}\right)}{\Delta t}. \tag{25}$$

where L_f is a latent heat of fusion.

3 Object of the Study

The Rybinsk Reservoir is an artificial reservoir, the water area of which is located in the Yaroslavl, Tver and Vologda regions of the Russian Federation. The reservoir has a valley-type structure of its bed. The water exchange coefficient is about 1.4 year^{-1}, which classifies the Rybinsk reservoir as an object with slow water exchange. The

Rybinsk reservoir is very large in area—4550 km^2 at the full supply level of 102 m in Baltic System. The amplitude of the level during the year can reach 4.9 m, however, this happens quite rarely (Litvinov and Roshchupko 2007). The average depth is 5.6 m and the maximum is 30 m. Ice formation on the reservoir usually begins from mid-November and ice cover disappeared in mid-April. To determine specific dates of freezing and ice breaking up, data from in situ measurements and satellite images are used.

Ice cover measurements were carried out on the Rybinsk Reservoir in January and April 2022. During measurements at stations located throughout the reservoir area, ice cores were taken to determine the thickness and structure of the ice cover and the height of the snow cover was determined using a snow measuring rail. In January, the ice cover thickness at most measurement stations was about 35–40 cm. It can be seen that on the tributaries of the reservoir, as well as in areas where the influence of the tributaries is noticeable, the ice is somewhat thinner (about 30 cm) due to the influence of the water current. The thickest ice cover is observed in the central part of the water area. By April, this trend continues, and an increase in ice thickness is also clearly visible (at some stations more than 60 cm). The ice cover in April is more porous in structure, which indicates an imminent weakening of its thickness. It is worth noting that snow surveys showed that the snow on the ice had melted almost completely by April, and at rare stations the thickness of the snow cover reaches 10 cm. Thus, the field surveys cover two periods of ice formation on the Rybinsk Reservoir—the active development of the ice cover in January and the state before melting in April.

Remote sensing of the Earth allows us to determine not only the date of ice conditions change, but also ice cover fields for large bodies of water if it is established and destroyed unevenly. Photographs of the Earth's surface in the visible spectrum range were obtained from the open Internet resource "Worldview" (see NASA Worldview (n.d.)). As an example, photographs are given before and after the ice breaking up of the Rybinsk Reservoir in 2022 (Fig. 1).

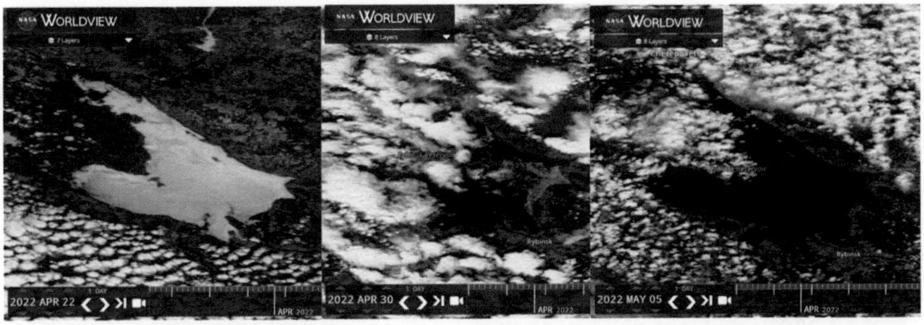

Fig. 1 Terra/MODIS satellite images for 22.04., 30.04. and 05.05.2022 in the visible range on the territory of the Rybinsk Reservoir

4 Numerical Experiments

As part of the study of the ice and show regime using the implemented description, a three-dimensional model configuration corresponding to the conditions of the Rybinsk Reservoir was used. The data of the ERA5Land reanalysis (Hersbach et al. 2020) were used as the meteorological forcing, namely: air temperature and humidity, atmospheric pressure, zonal and meridional components of wind speed, downward fluxes of long-wave and short-wave (solar) radiation, precipitation rate. The numerical experiment covered the period from May 1, 2021 to January 1, 2023. The duration of freeze-up, dates of ice freezing and breaking up, the thickness of the layers of blue ice, white ice and snow, as well as the temperature regime of the reservoir in the period under consideration were studied. The arrays of measurements and satellite data described above were used for comparison with the numerical results.

Figure 2 shows the results of ice and snow thickness calculations compared with the measurement data. Figure 3 shows the measurement points. Figure 4 shows the time-depth dependence of the average (vertical) temperature and the thickness of the layers of blue ice, white ice and snow in 2021–2023.

Above in Fig. 1, it is shown that, according to satellite images, the ice cover began to disintegrate at the end of April—on April 22, most of the reservoir was covered with ice, minor banks were visible, as well as open areas near the hydroelectric power station and at the confluence of the Mologa and Volga. On April 30, only traces of drifting ice remained in the southern part of the water area, and by May 5, the reservoir was completely free of ice cover. The calculations demonstrated agreement with the results of satellite data.

Thus, the model is adapted to take into account the ice regime, and the results of the calculations demonstrate sufficient agreement with the reference data. Note the fact that the measurements were carried out at different points and have a spread in values. It can be concluded that under these conditions, the model with the implemented block for calculating the thermodynamics of ice and snow satisfactorily describes the ice regime in inland water bodies using the Rybinsk Reservoir as an example.

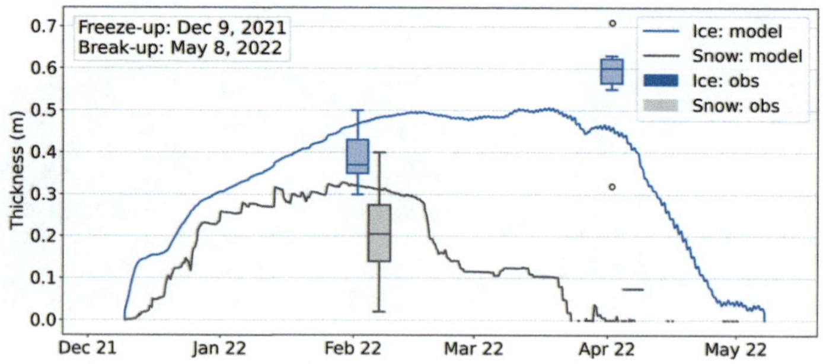

Fig. 2 Results of calculations of ice and snow thickness on the Rybinsk Reservoir in 2021–2022, compared with measurement data

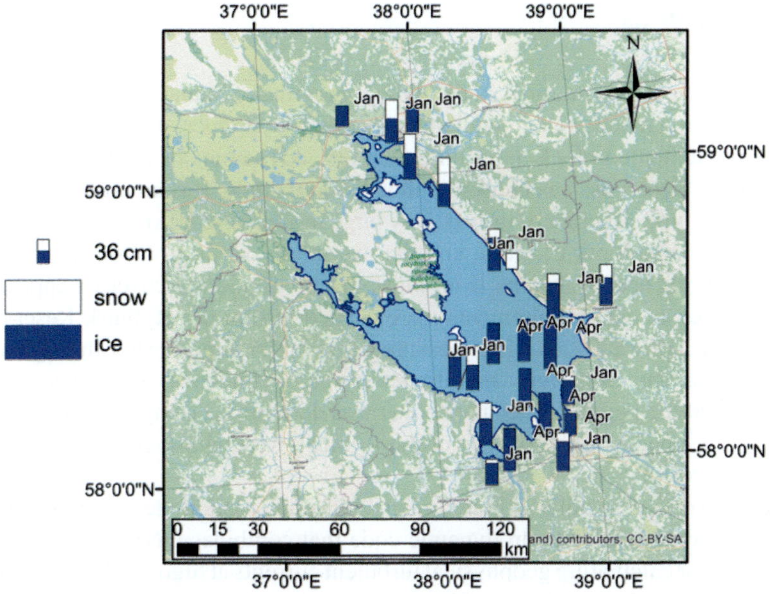

Fig. 3 Map of ice and snow measurement points on the Rybinsk Reservoir from January to March 2022

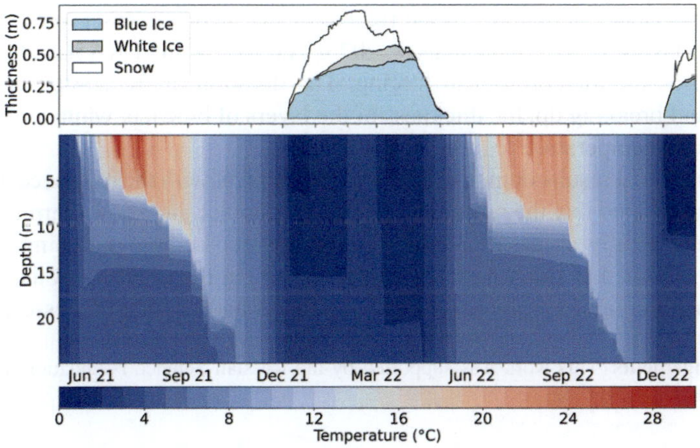

Fig. 4 The time-depth dependence of the average (vertical) temperature and the thickness of the layers of blue ice, white ice and snow in 2021–2023

5 Conclusions

The paper presents a numerical study of the ice regime of inland water bodies using a three-dimensional numerical model of thermohydrodynamics of lakes and reservoirs with an implemented block describing the thermodynamics of ice and snow cover. The Rybinsk Reservoir was chosen as the object of study. It is an artificial reservoir with a valley-type structure of its bed, very large in area of 4550 km^2 at the full supply level, the average depth of 5.6 m and the maximum depth of 30 m. Ice formation on the reservoir usually begins from mid-November and ice cover disappeared in mid-April. Ice cover measurements were carried out on the Rybinsk Reservoir in January and April 2022. During measurements at stations located throughout the reservoir area, ice cores were taken to determine the thickness and structure of the ice cover, and the height of the snow cover was determined using a snow measuring rail. Satellite images of the surface of the Rybinsk Reservoir were also used in this work as reference data.

The main research tool is a three-dimensional numerical model of an inland water body, based on a unified hydrodynamic code that combines DNS, LES and RANS approaches for calculating geophysical turbulent currents at high spatial and temporal resolution. A simple description of ice and snow thermodynamics was implemented in the model. For each cell on the surface of the reservoir, the stationary equation of heat transfer between the atmosphere and the water column is solved for three conjugate layers of blue ice (crystalline ice), white ice ("snow" ice) and snow.

The numerical calculations of ice regime of the Rybinsk Reservoir covered the period from May 1, 2021 to January 1, 2023. The data of the ERA5Land reanalysis were used as the meteorological forcing. The duration of freeze-up, dates of ice freezing and breaking up, the thickness of the layers of blue ice, white ice and snow, as well as the temperature regime were studied.

The numerical results demonstrated good agreement with the reference data, thus, it can be concluded that the proposed simple model of ice and snow thermodynamics can be successfully applied to describe inland water bodies, including long-term calculations aimed at studying lakes and reservoirs in the interests of the problems of the contribution of such water bodies to weather and climate processes.

Acknowledgements This work was supported by the Russian Science Foundation, project No. 23-77-01032.

References

Adrian R et al (2009) Lakes as sentinels of climate change. Limnol Ocean 54(6), Part 2:2283–2297

Baas P, de Roode SR, Lenderink G (2008) The scaling behaviour of a turbulent kinetic energy closure model for stably stratified conditions. Bound-Layer Meteorol 127:17–36

Boër KW (1980) The terrestrial solar spectrum. Solar energy technology handbook, part A. Dekker, pp 65–87

Burchard H (2007) Applied turbulence modelling in marine waters. Springer
Burchard H, Deleersnijder E (2001) Stability of algebraic non-equilibrium second-order closure models. Ocean Model 3(1–2):33–50
Farmer DM (1975) Potential temperatures in deep freshwater lakes. Limnol Ocean 20(4):634–635
Gladskikh D et al (2023) Turbulent transport in a stratified shear flow. J Mar Sci Engineering 11(1):136
Gladskikh DS, Mortikov EV (2024) On parameterization of dissipative processes in turbulent transport models for description of thermohydrodynamics and biogeochemistry of stratified inland water bodies. Izv Atmos Ocean Phys 60(3):286–296
Gladskikh DS, Stepanenko VM, Mortikov EV (2021) The effect of the horizontal dimensions of inland water bodies on the thickness of the upper mixed layer. Water Resour 48:226–234
Gottlieb L (1980) Development and applications of a runoff model for snowcovered and glacierized basins. Hydrol Res 11(5):255–272
Henderson-Sellers B (1984) Engineering limnology, Pitman Advanced Pub. Program, Boston
Hersbach H et al (2020) The ERA5 global reanalysis. Q J R Meteorol Soc 146(730):1999–2049
Hipsey MR et al (2019) A General Lake Model (GLM 3.0) for linking with high-frequency sensor data from the Global Lake Ecological Observatory Network (GLEON). Geosci Model Development 12(1):473–523
Kadantsev E, Mortikov E, Zilitinkevich S (2021) The resistance law for stably stratified atmospheric planetary boundary layers. Q J R Meteorol Soc 147(737):2233–2243
Karetnikov SG, Naumenko MA (2008) Recent trends in lake Ladoga ice cover. Hydrobiology 599(1):41–48
Kirk JTO (1994) Light and photosynthesis in aquatic ecosystems. Cambridge University Press
Koren V et al (1999) A parameterization of snowpack and frozen ground intended for NCEP weather and climate models. J Geophys Res: Atmospheres 104(D16):19569–19585
Litvinov AS, Roshchupko VF (2007) Long-term and seasonal water level fluctuations of the Rybinsk Reservoir and their role in the functioning of its ecosystem. Water Resour 34:27–34
Maykut GA, Untersteiner N (1971) Some results from a time-dependent thermodynamic model of sea ice. J Geophys Res 76(6):1550–1575
Mortikov EV (2016) Numerical simulation of the motion of an ice keel in a stratified flow. Izv Atmos Ocean Phys 52:108–115
Mortikov EV, Glazunov AV, Lykosov VN (2019) Numerical study of plane Couette flow: turbulence statistics and the structure of pressure–strain correlations. Russ J Numer Anal Math Model 34(2):119–132
NASA Worldview. https://worldview.earthdata.nasa.gov
Oveisy A, Boegman L, Imberger J (2012) Three-dimensional simulation of lake and ice dynamics during winter. Limnol Oceanogr 57(1):43–57
Patterson JC, Hamblin PF (1988) Thermal simulation of a lake with winter ice cover. Limnol Oceanogr 33(3):323–338
Rogers CK, Lawrence GA, Hamblin PF (1995) Observations and numerical simulation of a shallow ice-covered midlatitude lake. Limnol Oceanogr 40(2):374–385
Schumann U, Gerz T (1995) Turbulent mixing in stably stratified shear flows. J Appl Meteorol (1988–2005):33–48
Semtner AJ (1976) A model for thermodynamic growth of sea ice in numerical investigations of climate. J Phys Oceanogr 6:379–389
Vavrus SJ, Wynne RH, Foley JA (1996) Measuring the sensitivity of southern Wisconsin lake ice to climate variations and lake depth using a numerical model. Limnol Oceanogr 41(5):822–831

On Fluvial Paleostructures of the Gelendzhik Bay and the Mezyb River Near Divnomorskoe on the Black Sea Shelf According to Seismic–acoustic Profiling Data

M. S. Klyuev, A. A. Schreider, A. S. Zverev, V. G. Gainanov, A. E. Sazhneva, A. L. Brekhovskikh, and I. Ya. Rakitin

Abstract The surrounding topography, bathymetry and structure of the bottom strata of the Gelendzhik Bay and the mouth of the Mezyb River in Divnomorskoe on the Black Sea shelf are considered. As a result of continuous seismic–acoustic profiling using the Geont-Shelf complex, fluvial paleostructures of the bottom strata were discovered. Seismic-stratigraphic analysis of the bottom strata of the Gelendzhik Bay and the mouth of the Mezyb River in Divnomorskoe on the Black Sea shelf was carried out based on the original methods of the authors. The main parameters of the discovered intra-bottom fluvial paleostructures are presented and their mapping is performed. Assumptions are made about possible mechanisms of formation of the discovered structures.

Keywords Gelendzhik Bay · Mouth of the Mezyb River in Divnomorskoe · Relief and topography · Regression and transgression of sea level · Fluvial paleostructures · Seismo-acoustic sections · Seismo-stratigraphic analysis · Mapping

Gelendzhik Bay (GB, Fig. 1) is located on the north-eastern coast of the Black Sea (coordinates 44°34.1′ N 38°03.0′ E). In the north-west–south-east direction, it is about 4.8 km long, about 3.5 km wide, the strait between the north-western Thin

M. S. Klyuev (✉) · A. A. Schreider · A. E. Sazhneva · A. L. Brekhovskikh · I. Ya. Rakitin
Shirshov Institute of Oceanology, Russian Academy of Sciences, Moscow, Russia
e-mail: mklyuev@mail.ru

A. S. Zverev
Vernadsky Institute of Geochemistry and Analytical Chemistry, Russian Academy of Sciences, Moscow, Russia
e-mail: zverev@geokhi.ru

V. G. Gainanov
Geological Faculty, Lomonosov Moscow State University, Moscow, Russia

T. Chaplina (ed.), *Processes in GeoMedia—Volume IX*, Springer Geology,
https://doi.org/10.1007/978-981-96-9302-3_29

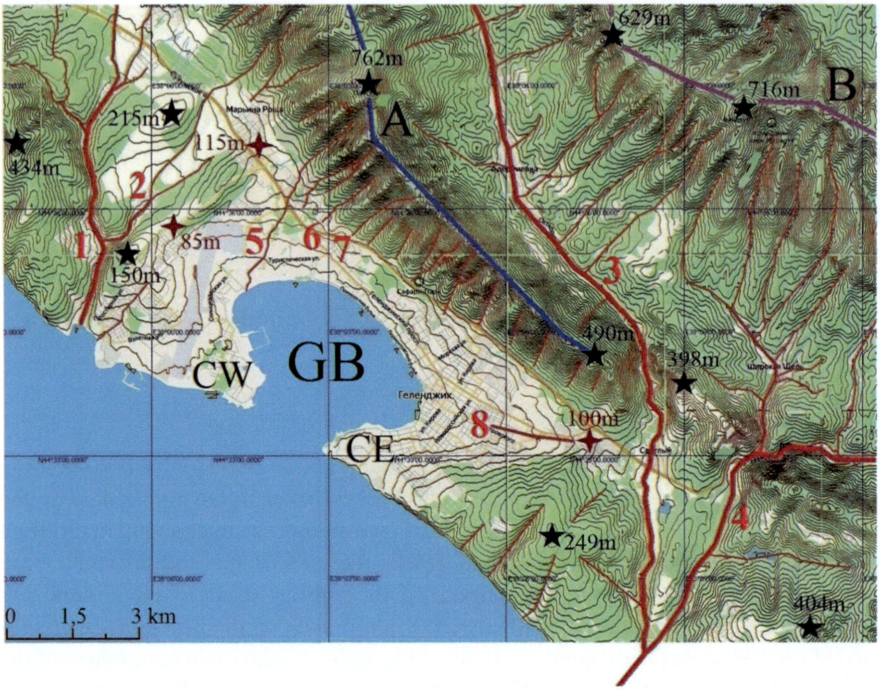

Fig. 1 Topographic map of the adjacent land of the Gelendzhik bay

(CW) and south-eastern Thick (CE) capes about 1.9 km and an area of about 12 km^2, where the space of the modern town of Gelendzhik is formed around it (Fig. 1).

Gelendzhik Bay is enclosed to the north-east by the Markotkh Ridge with elevations up to 762 m near the town (see Fig. 1, A, blue line) and the Kotsekhur Ridge (see Fig. 1, B, purple line) with elevations up to 716 m near the town. It is sheltered to the north-west by the Doob Mountain with an elevation of about 434 m and to the south-east by an elevation of about 249 m. Here, black asterisks denote localised altitudes and red asterisks denote localised watersheds.

To the west and north-west of Gelendzhik Bay flows the Yashamba (Ashamba) River (1 in Fig. 1, red line) with its tributary the Dry Valley Creek (2 in Fig. 1, red line), separated from it by watersheds of 85 and 115 m (see Fig. 1).), and to the east and south-east flow the rivers Aderba (3 in Fig. 1, red line) and Mezyb (4 in Fig. 1, red line), separated by the 249 m high Markotkh Ridge and 100 m high watershed (see Fig. 1).

The bay is drained by streams from the south-western slope of the Markotkh ridge (5—Kupriyanova Gap, 6—Bobrukovaya Gap, 7—Gorbunova Gap, etc., thin red lines in Fig. 1), Kuplizi Creek (8 in Fig. 1, thin red line) as well as streams from the south-eastern slope of Mount Doob and the north-western slope of 249 m height.

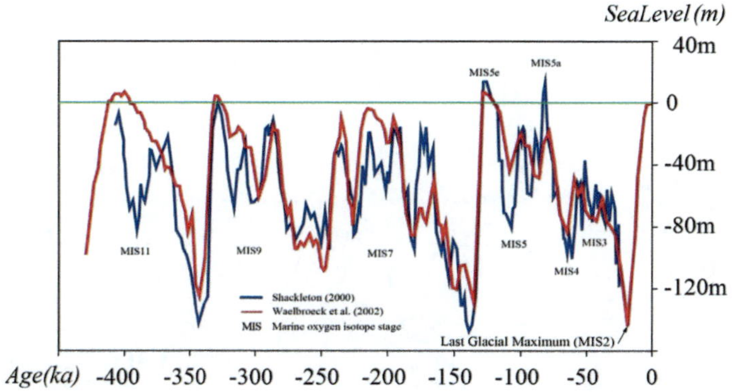

Fig. 2 Change in the Pleistocene ocean level over the last 400 thousand years: red-(Waelbroeck et al. 2002), blue-(Shackleton 2000). Modified

Thus, the total catchment area of the Gelendzhik Bay, bounded by Mount Doob from the north-west, the Markotkh Ridge from the north-east and the 249 m height from the south-east, is approximately 40 km^2.

According to geological drilling data, the upper part of marine sediments of the bottom strata (up to 15 m) of Gelendzhik Bay is represented mainly by sands, pelites and coquina with inclusions of silt, pebbles, clay and peat bog (Shcheglov 1994).

As is known, over the last hundreds of thousands of years, sea level has experienced cycles of regression (lowering) and transgression (rising), associated in particular with glaciation (glaciation) and deglaciation (melting of glaciers) or tectonic movements of the Earth's surface (Fig. 2). During regression periods, the sea level dropped by tens of metres (to about 120 m about 18,000 years ago), so that a river network was formed on the drained areas of the former seabed under aerial conditions, which was then flooded by the sea during transgression periods and buried in marine sediments, forming fluvial palaeostructures of the shelf strata.

Fluvial palaeostructures of the bottom strata are widespread practically in all areas of the World Shelf (Klyuev et al. 2023), including the Black Sea shelf, one example of which is the fluvial palaeostructures of the Yashamba River and the Sukhoi Valley Creek in the bottom strata of the Goluboy (Rybatskaya) Bay near Gelendzhik (Klyuev et al. 2022). It is natural to assume that fluvial palaeostructures may be present in the bed strata of Gelendzhik Bay and the adjacent water area.

To verify this assumption, continuous seismo-acoustic profiling (CSP) of the bottom strata of Gelendzhik Bay was carried out, which is the most effective means of detecting and identifying shelf fluvial palaeostructures according to the original technology developed by the authors (Klyuev et al. 2023, 2022; Brekhovskikh et al. 2022). A seismo-acoustic complex of continuous profiling 'Geont-Shelf' with a boomer-type transmitter with a central frequency of 1500 Hz, electrical energy of the pulse of about 300 J and a parcel emission rate of 0.5 with a spatial resolution of the order of several metres (without convolutional processing) was used (Klyuev et al. 2022, 2023; Brekhovskikh et al. 2022).

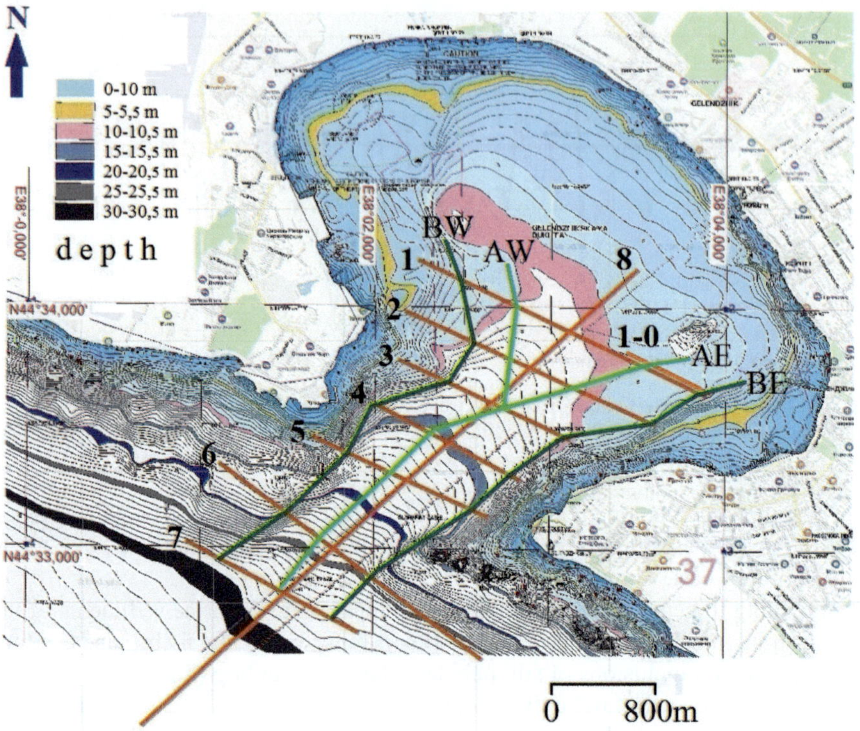

Fig. 3 Bathymetric map of the Gelendzhik Bay and adjacent water area with location of CSP transects and boundaries of intra-bottom fluvial palaeostructures. Numbers of transects 1–8 (left and above)

The research was conducted from the R/V 'Ashamba' (length 15 m, width 4 m, draft 1.5 m, displacement 27 tonnes) of the Southern Branch of the Shirshov Institute of Oceanology of the Russian Academy of Sciences. Seven longitudinal (along the main coastline) and one transverse transects of the Continuous seismic profiling CSP were made, the spatial location of which is shown by orange segments 1–8 on the bathymetric map of the Gelendzhik Bay and adjacent water area (Fig. 3). A set of seismo-acoustic features developed in Klyuev et al. (2023); Klyuev et al. 2022; Brekhovskikh et al. 2022) was used to detect, reveal and identify fluvial paleostructures of the bottom strata.

The obtained longitudinal profiles CSP 1–7 (Fig. 4) clearly show the bottom surface and bottom strata structures of the Gelendzhik Bay (the conditional depth is given assuming a sound velocity in sediments of 1500 m/s). Seismo-stratigraphic analysis of this set of consecutive profiles (according to Klyuev et al. (2023); Klyuev et al. 2022; Brekhovskikh et al. 2022)) shows that the following reflecting intra-bottom boundaries of the acoustic impedance jump can be distinguished: R0—bottom surface (yellow), R1—first reflection multiple (green), R2—palaeo-channel boundary (red, R2A—eastern branch, R2B and R2C—western branches, R2D—palaeo-channel bottom), R3—boundary in the sedimentary fill of the palaeo-channel (blue).

On this basis and taking into account geological drilling data (Shcheglov 1994), it was determined that the main intra-bottom structure of the Gelendzhik Bay, first discovered by the authors in 2023, is a bifurcated paleo-channel of U-shaped transverse form (profiles 1–2), transitioning to V-shaped (profiles 3–7), cut in the rocky marl base of the bay and filled with sedimentary fill. The spatial arrangement of this palaeo-channel is shown in Fig. 3, where the dark green lines BW and BE indicate the western and eastern boundaries of the palaeo-channel respectively, and the light green lines AW and AE indicate the longitudinal axes of the palaeo-channel. The width of the channel from profiles 3–7 is about 900 m, the depth to the bottom varies from 12 to 18 m and the length of the surveyed section of the palaeo-channel is about 4 km. The palaeo-channel axis bifurcates at its origin to the north-west AW (towards the Yashamba River) and south-east AE (towards the Aderba and Mezyb rivers) from profile 3 towards profile 1.

The edges of the palaeo-channel (reflecting boundary R2) show traces of abrasion in the form of cuts in the layers of rocky marl bedrock, possibly caused by the impact of mudflow-like flows during flood periods (Fig. 5). The transverse profile of the sedimentary fill of the palaeo-channel shows one main reflecting boundary R3, indicating a jump in acoustic impedance, and several additional ones (Figs. 4, 5, 6 and 7), which represent different stages of sedimentation processes.

The size and shape of the palaeo-channel under consideration indicate high fluvial activity during the drainage of the Gelendzhik Bay during periods of regression under aerial conditions in the past. This may have been due to the strong concentration of atmospheric runoff from a catchment area of about 40 km^2 into the bowl of the Gelendzhik Bay and then into the channel of the palaeo-channel penetrating to the sea. It can also be assumed that there was an overflow of water from the Yashamba and Aderba rivers (as well as the Mezybi) across watersheds of 85 m, 115 m and 100 m respectively during catastrophic floods and inundations in the past. It is also possible that in the past the Yashamba and Aderba (and also the Mezyb) Rivers flowed into the bowl of the Gelendzhik Bay, and only later changed their mouths to their present state as a result of tectonic activity in the region.

In addition, the mouth of the Mezyb River is located 10 km south-east of Gelendzhik Bay in the village of Divnomorskoye, Krasnodar Krai of the Russian Federation (Fig. 1.6). The Mezyb River originates on the northern slope of Mount Mikhailovka (height 788 m) near the Mikhailovsky Pass in the Krasnodar Territory of the Russian Federation, has a length of about 17 km, merges with the Aderba River (length of about 20 km) 2.5 km from the mouth and flows into the Black Sea in Divnomorskoye village at the coordinates 44°30.03′ N 38°07.65′ E (Fig. 6).

The bottom marine sedimentary strata near the mouth of the Mezyb River is mainly sand, pelite and coquina with inclusions of silt, pebble, clay and peat bog

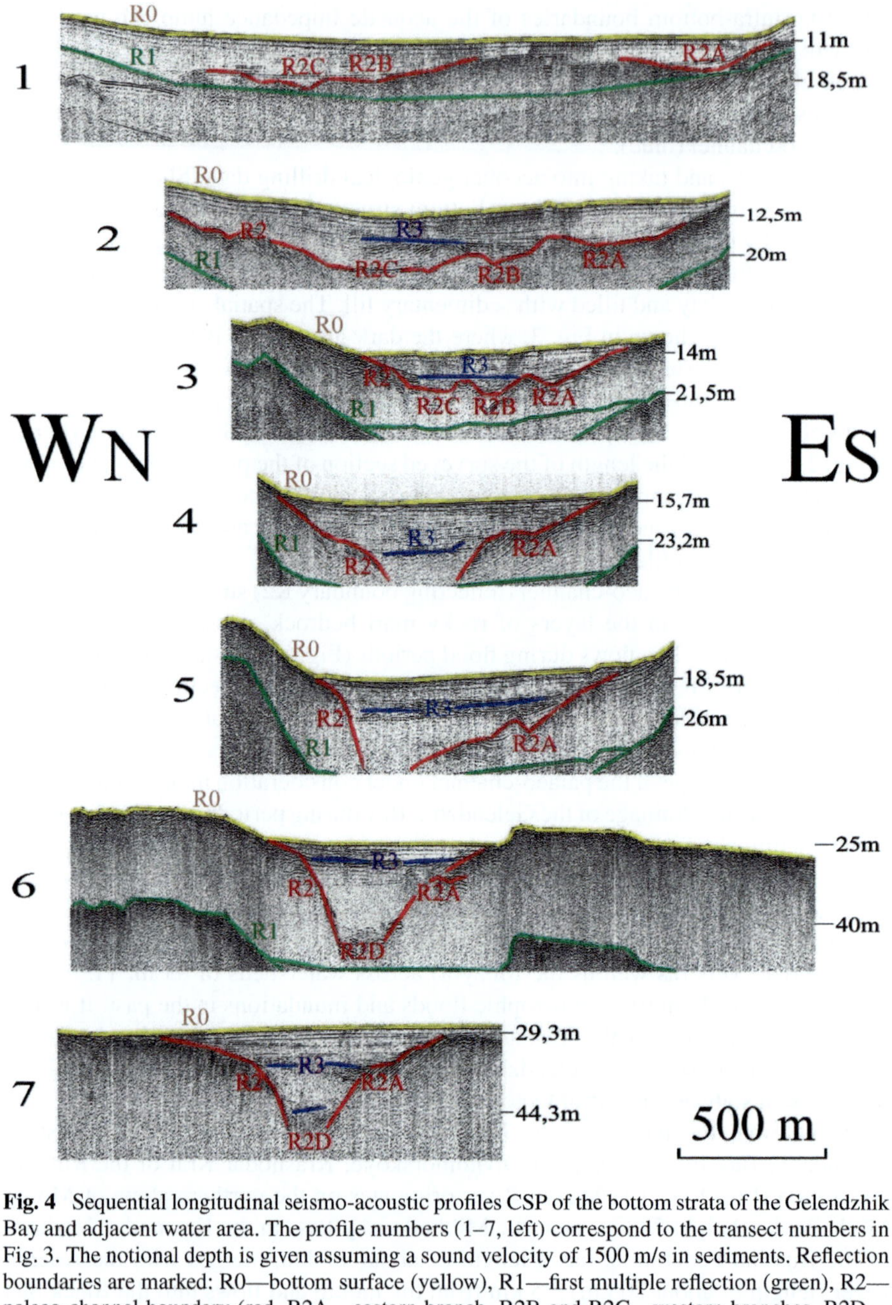

Fig. 4 Sequential longitudinal seismo-acoustic profiles CSP of the bottom strata of the Gelendzhik Bay and adjacent water area. The profile numbers (1–7, left) correspond to the transect numbers in Fig. 3. The notional depth is given assuming a sound velocity of 1500 m/s in sediments. Reflection boundaries are marked: R0—bottom surface (yellow), R1—first multiple reflection (green), R2—palaeo-channel boundary (red, R2A—eastern branch, R2B and R2C—western branches, R2D—palaeo-channel bottom), R3—boundary in the sedimentary fill of the palaeo-channel (blue)

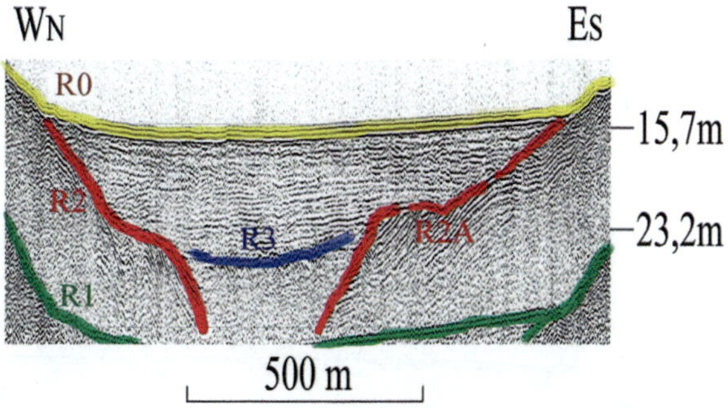

Fig. 5 Seismo-acoustic profile of the CSP on the tack no.4. The designations are the same as in Fig. 4

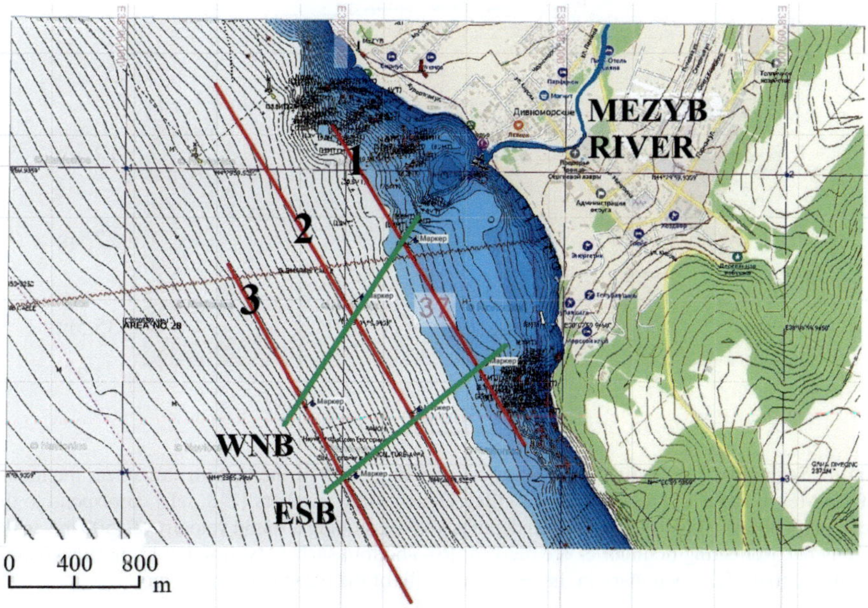

Fig. 6 Topographic and bathymetric map of the area of the Mezyb River mouth in Divnomorskoye village. The isobath spacing is 0.5 m. Red sections 1, 2, 3 show the positions of CSP transects. Green sections WNB and ESB show the positions of the west-north and east-south boundaries of the bottom palaeo-channel of the Mezyb River, respectively

on a marl base with localised outcrops of marl on the bottom surface (Kuklev et al. 2014).

It is also natural to assume that fluvial palaeostructures may also be present in the bottom marine sedimentary strata near the mouth of the Mezyb River in Divnomorskoye, especially since signs of such structures were detected as early as 2019 (Klyuev et al. 2021).

To test this assumption, continuous seismo-acoustic profiling CSP of the bottom marine sedimentary strata near the mouth of the Mezyb River in Divnomorskoye

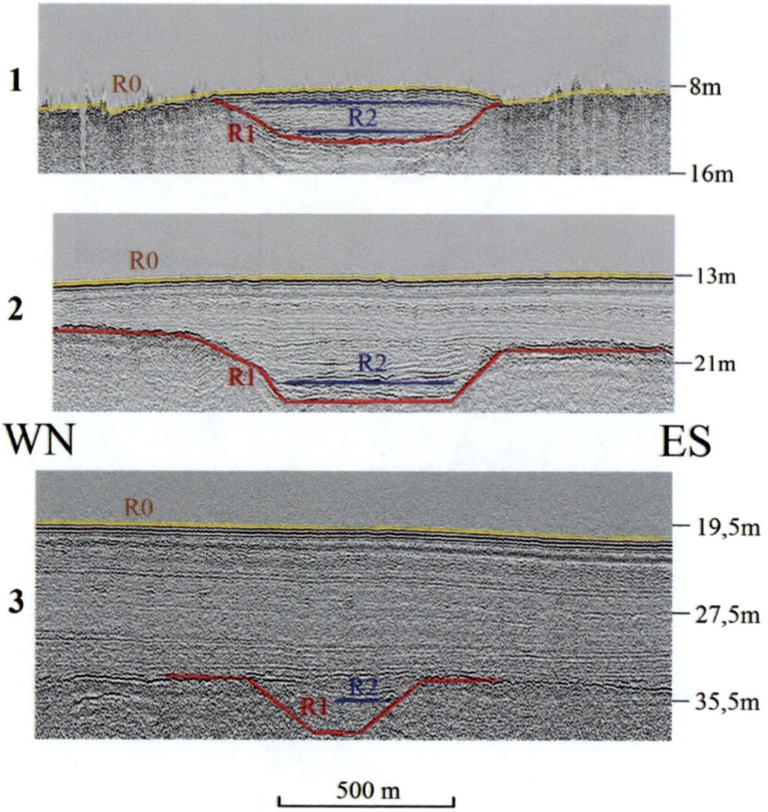

Fig. 7 Sequential longitudinal seismo-acoustic sections CSP of the bottom strata near the mouth of the Mezyb River in Divnomorskoye village. The profile numbers (1–3, left) correspond to the section numbers in Fig. 6. The notional depth is given assuming a sound velocity of 1500 m/s in the sediments. Reflecting boundaries are marked: R0—bottom surface (yellow), R1—palaeo-channel boundary (red), R2 - boundary in the sedimentary fill of the palaeo-channel (blue)

was carried out, which is the most effective means of detecting and identifying shelf fluvial palaeostructures according to the original technology developed by the authors (Klyuev et al. 2022,2023; Brekhovskikh et al. 2022). The seismic–acoustic complex of continuous profiling 'Geont-Shelf' with a boomer-type transmitter with a central frequency of 1500 Hz, electrical energy of the pulse of about 300 J and a parcel emission rate of 0.5 with a spatial resolution of the order of several metres (without convolutional processing) was used (Klyuev et al. 2022, 2023; Brekhovskikh et al. 2022).

The research was also conducted from the R/V 'Ashamba' (length 15 m, width 4 m, draft 1.5 m, displacement 27 tonnes) of the Southern Department of the Shirshov Institute of Oceanology of the Russian Academy of Sciences. In the process of research 3 longitudinal (along the shore) transects of the CSP were made, the spatial location of which is shown by red segments 1–3 on the topographic and bathymetric map of the mouth of the Mezyb River (Fig. 6). A set of seismo-acoustic features developed in Klyuev et al. (2023); Klyuev et al. 2022; Brekhovskikh et al. 2022) was also used to detect, reveal and identify fluvial paleostructures of the bottom strata.

The obtained longitudinal CSP P 1–3 cross sections (Fig. 7) clearly show the bottom surface and marine bottom strata structures near the mouth of the Mezyb River (the conditional depth is given assuming a sound velocity in the sediments of 1500 m/s). Seismo-stratigraphic analysis of this set of consecutive profiles (according to Klyuev et al. (2023); Klyuev et al. 2022; Brekhovskikh et al. 2022; Kuklev et al. 2014; Klyuev et al. 2021)) shows that the following reflecting intra-bottom boundaries of the acoustic impedance jump can be distinguished: R0—bottom surface (yellow), R1—palaeo-channel boundary of the Mezyb River (red), R2 - boundary in the sedimentary fill of the palaeo-channel (blue).

On this basis, the main intra-bottom structure near the mouth of the Mezyb River, first discovered by the authors in 2023, was found to be a U-shaped transverse palaeo-channel of the Mezyb River (sections 1–3), cut into the rocky marl bedrock near the shore (section 1) and further into the surrounding sediments (sections 2, 3), which is filled with sedimentary fill. The spatial location of this palaeo-channel is shown in Fig. 6, where the green lines WNB and ESB indicate the west-north and east-south boundaries of the palaeo-channel, respectively.

In section 1, the channel is about 850 m wide, with a depression of about 4 m. Here the palaeo-channel appears to be cut in a marl environment, as evidenced by its surface indentation, with the sedimentary layered fill of the palaeo-channel clearly visible. In section 2, the channel is about 750 m wide, with a deepening of about 4 m at a depth of 12 m to the seabed. Here the palaeo-channel is deepened into the sedimentary strata. In section 3, the channel is about 500 m wide, with a deepening of about 4 and 16 m deep into the seabed. Here, the palaeo-channel is even deeper into the sedimentary stratum, which results in a significant deterioration of its contrast due to the limited sensitivity of the equipment. The length of the surveyed section of the palaeo-channel is about 1.5 km, taking into account the capabilities of the equipment used.

In conclusion, we note that the shelf fluvial paleo-channels can be potential reservoirs of valuable rare natural mineral resources (Ainemer and Konshin 1982), eroded and deposited by ancient water flows in their channels, which determines the relevance of detection and mapping of these structures.

The work was carried out within the framework of the state assignment of the Shirshov Institute of Oceanology FMWE-2024–0018 and state orders of the Vernadsky Institute of Geochemistry and Analytical Chemistry of the Russian Academy of Sciences.

References

Ainemer AI, Konshin GI (1982) Placers of the World Ocean shelf zones. Nedra, Leningrad, Izd, p 263

Brekhovskikh AL, Klyuev MS, Sazhneva AE, Schreider AA, Zverev AS (2022) Technology of seismic acoustic detection and research of river Paleostructures of the sea bottom of the Coastal Zone and Its approbation in Blue Bay. In: Chaplina T. (eds) Processes in GeoMedia, vol. 4, pp. 345–356. Springer Geology. Springer, Cham. https://doi.org/10.1007/978-3-030-76328-2_36

Klyuev MS, Schreider AA, Brekhovskikh AL, Sazhneva AE, Zverev AS (2021) On some aspects of detection and study of river palaeostructures of the coastal zone of the seafloor by means of broadband seismoacoustic profiling and GPS-positioning. Geophysics 3:56–65

Klyuev MS, Schreider AA, Zverev AS, Brekhovskikh AL, Rakitin IY, Sazhneva AE (2022) On the boundaries of Paleostructures of the Ashamba River in the Blue Bay Near Gelendzhik according to broadband continuous seismic and acoustic profiling and GPS positioning. In: Karev VI (eds) Physical and mathematical modeling of earth and environment processes, pp 11–19. Springer Proceedings in Earth and Environmental Sciences. Springer, Cham. https://doi.org/10.1007/978-3-030-99504-1_2

Klyuev M, Schreider A, Zverev A (2023) Shelf Fluvial Paleo structures: Seabed seismic acoustic view. Earth and Environmental Sciences Library. Springer, Cham, p 79, 70 ill. ISBN: 978–3–031–27519–7 https://doi.org/10.1007/978-3-031-27520-3

Kuklev SB, Evsyukov YD, Rudnev VI (2014) Catastrophic flooding in the Gelendzhik region. Transformations of land and seabed relief. LAP LAMBERT Academic Publishing, Saarbrucken, Deutschland, p 80

Shackleton NJ (2000) The 100,000-year ice-age cycle identified and found to lag temperature, carbon dioxide, and orbital eccentricity. Science 289:1897–1902

Shcheglov AP (1994) History of marine and continental sedimentation in the Gelendzhik Bay for the last 20 thousand years. Bulletin of the Commission for the study of the Quaternary period no.61. Nauka, Moscow, pp 47–50

Waelbroeck C, Labeyrie L, Michel E, Duplessy JC, McManus JF et al (2002) Sea-level and deep water temperature changes derived from benthic foraminifera isotopic records. Quat Sci Rev 21(1–3):295–305. https://doi.org/10.1007/978-3-031-54798-0

Advancements in Geomechanical and Digital Core Analysis for Reservoir Characterization

V. V. Khimulia

Abstract This article explores recent advancements in geomechanical testing and digital core analysis for reservoir characterization. It highlights the influence of stress conditions on rock deformation, filtration properties, and sand production processes. The study reviews various laboratory techniques, including uniaxial and triaxial compression tests, as well as true triaxial loading apparatus (TTLA), which enable a more comprehensive understanding of rock behavior under complex stress conditions. Additionally, the integration of digital technologies, such as computed X-ray tomography, machine learning for core analysis, and numerical simulations, is discussed in relation to improving the accuracy of reservoir modeling and wellbore stability assessments. Special attention is given to Russian and global research efforts, emphasizing the need for further development of integrated geomechanical and digital approaches for optimizing hydrocarbon recovery. The results of recent research conducted in the geomechanics laboratory of IPMech RAS are presented. The presented results of comprehensive studies allowed to determine optimal parameters for safe and efficient operation of specific fields, including determination of permissible drawdowns in wells and characteristics of downhole equipment. The conducted research forms the basis for an integrated approach to studying reservoir properties to determine well operation parameters based on computer tomography, digital core analysis, laboratory experiments, and unique mechanical tests. The findings suggest that a combined use of experimental and digital methods provides deeper insights into reservoir heterogeneity, stress-dependent permeability changes, and long-term rock behavior, which are critical for sustainable resource extraction.

Keywords Rock mechanics · True triaxial testing · Digital core analysis · Computed X-ray tomography · Reservoir characterization · Sand production · Geomechanical modeling · Micro-CT imaging · Stress–strain behavior

V. V. Khimulia (✉)
Ishlinsky Institute for Problems in Mechanics RAS, Moscow, Russia
e-mail: valery.khim@gmail.com

T. Chaplina (ed.), *Processes in GeoMedia—Volume IX*, Springer Geology,
https://doi.org/10.1007/978-981-96-9302-3_30

1 Introduction

In recent years, extensive laboratory studies have been conducted on core samples obtained from reservoirs of various oil and gas fields, along with hydrodynamic and geophysical well studies aimed at identifying patterns in how stress conditions affect rock permeability (Guo et al. 2024). It has been established that permeability can undergo significant and often irreversible changes under stress within the rock mass, depending on factors such as rock structure, lithological composition, mechanical and deformation properties, depth of occurrence, reservoir pressure, as well as wellbore construction and operating conditions (Huo et al. 2024; Sun et al. 2024; Zhu et al. 2024).

The construction of underground facilities and the extraction of hydrocarbons and solid minerals alter the three-dimensional stress state of rock formations. Studying the deformation processes in response to stress changes helps to determine fracture mechanisms and track the evolution of mechanical and filtration properties over time. This knowledge is critical for preventing failures, improving the reliability of engineering structures, and ensuring the safe and efficient operation of wells and mining excavations.

Most of these studies rely on laboratory experiments, as rock properties are influenced by natural geological conditions and cannot be precisely determined by purely theoretical calculations. Experimental approaches allow for a more detailed examination of how stress variations impact rock structure and provide scientific justification for managing their mechanical and filtration characteristics.

2 Modern Laboratory Methods for Studying Rock Strength and Deformation

Advanced laboratory methods for investigating the strength and deformation properties of rock formations are diverse. Currently, a wide range of standardized methodologies and specialized equipment are available for assessing mechanical properties. These include uniaxial compression tests, conventional triaxial compression, compression in oedometers that prevent lateral expansion under vertical loading, and suffosion compression, which evaluates the ability of saline rocks to reduce volume due to salt leaching under prolonged water filtration and sustained compressive loads. The study of the physical and mechanical characteristics of rocks under stress is most often carried out using triaxial testing equipment based on the Karman principle (Karman 1911).

The Karman method is widely recognized and employed in numerous laboratories worldwide. These industrially produced installations allow for tests under traditional pseudo-triaxial loading, where the load is distributed across the entire specimen surface but can be adjusted only along two directions: along the axis and radially. Test specimens typically have a cylindrical shape. This method is extensively used for determining rock strength properties (Zong et al. 2020a), analyzing the influence of initial fracture networks on their mechanical behavior (Zhang and Li 2019; Yang et al. 2018), studying creep properties (Zong et al. 2020b; Wang et al. 2018; Hamza and Stace 2018), and examining filtration and rheological parameters under pseudo-triaxial compression conditions (Nguyen et al. 2018; Yang and Hu 2020; Martyushev et al. 2019).

However, one of the key limitations of such systems is their inability to fully replicate the actual geomechanical conditions of rock formations, where the stress state is inherently three-dimensional. Additionally, testing anisotropic rocks using these setups presents significant challenges, as these materials exhibit different mechanical properties along various directions, requiring a more adaptable approach to experimental modeling.

3 Global Research in True Triaxial Testing and Its Applications in Geomechanics

Research using true triaxial loading apparatus (TTLA) provides the most comprehensive data on the mechanical characteristics of rock formations. Despite their significant advantages over conventional Karman-type triaxial systems, TTLA devices remain relatively uncommon worldwide. The primary obstacles to their widespread adoption are the complexity of their design, the high cost of equipment, and the lack of mass production until recent years (GCTS Testing Systems n.d.; GDS Instruments n.d.). Unlike conventional triaxial tests, where the specimen is cylindrical, TTLA systems use parallelepiped-shaped specimens, which are subjected to independent loading along three mutually perpendicular directions.

Similar to conventional triaxial testing, TTLA systems are utilized for analyzing the mechanical behavior of rocks of various lithotypes under different loading conditions. These systems enable the study of how initial rock structure and defects influence deformation and strength properties, as well as the role of poroelastic and viscoplastic deformation mechanisms. Additionally, TTLA devices facilitate investigations into how pore pressure affects mechanical responses and the impact of stress variations on filtration processes. The ability to model realistic stress conditions in rock formations and to analyze the behavior of geomaterials under anisotropic loading makes TTLA systems particularly valuable for accurately reproducing natural subsurface conditions.

Several leading research institutions specializing in true triaxial rock testing include the University of Mons, Belgium (J.-P. Tshibangu & F. Descamps) (Kwasnievski et al. 2013; Georgieva et al. 2020), University of Toronto, Canada (R.P. Young & M.H.B. Nasseri) (Young et al. 2013), Curtin University, Australia (M. Sarmadivaleh) (Minaeian et al. 2020; Rasouli 2013; Rezagholilou and Sarmadivaleh 2019; Nabipour 2013; Minaeian 2014), China University of Mining and Technology (Miao Xiexing) (Li et al. 2019), Colorado School of Mines, USA (Azra Nur Tutuncu,

Ali I. Mese) (Baizhanov et al. 2019), and Northeastern University, China (Xia-Ting Feng) (Feng et al. 2024).

At the National Institute of Advanced Industrial Science and Technology in Japan, studies focus on the reactivation of fracture surfaces within rock samples and their impact on mechanical and filtration properties (Asahina et al. 2019a, b). Research on the strength characteristics of sandstones (Feng et al. 2019) and ultra-strong materials under independent loading conditions is actively conducted at Northeastern University, China (Feng et al. 2018; Zhao et al. 2018a, b, 2019), with a particular emphasis on anisotropy in rock strength (Feng et al. 2020). Additional fundamental studies on the strength and filtration properties of rock formations are being carried out in France (Couture and Bésuelle 2019), Saudi Arabia (Garcia et al. 2018), Chongqing University, China (Duan et al. 2020), and the University of Oklahoma (Vachaparampil and Ghassemi 2017).

Several research teams in Russia focus on studying the mechanical behavior of geomaterials under complex loading conditions. At Lomonosov Moscow State University (MSU), researchers investigate the properties of soils under low loads, using a dedicated test stand from Geotek. Meanwhile, at the Institute of Mining, Siberian Branch of the Russian Academy of Sciences, research is being conducted on hydraulic fracturing modeling under true triaxial compression conditions, using large-scale synthetic rock samples (Rubtsova and Skulkin 2017).

The Schmidt Institute of Physics of the Earth RAS also conducts comprehensive studies on rock deformation under non-uniform loading conditions, using the GCTS RTR-4500 system (Veselovsky et al. 2022). However, this system presents several technical limitations for geomechanical modeling, such as restrictions on applied loads along different axes due to sample-fluid contact in pressure conditions and challenges in measuring deformations along one of the principal axes.

Recent studies and practical experience in developing complex hydrocarbon fields have demonstrated that heterogeneity and anisotropy of reservoirs significantly influence mechanical and mass transfer processes in formations. This highlights the need to consider these factors when developing and improving technological solutions.

4 Integration of Digital Technologies in Rock Mechanics and Oil and Gas Research

To investigate rock failure mechanisms in laboratory conditions, methods such as acoustic emission analysis (Huang et al. 2022; Wu et al. 2019) and computed X-ray tomography (Cao et al. 2021; Lu et al. 2020; Botao et al. 2020; Lichun and Mian 2014) are widely applied. Computed tomography of heterogeneous media is based on reconstructing the spatial distribution of the X-ray attenuation coefficient using scanned projection data. This process generates a three-dimensional (3D) model, providing insight into the internal structure of rock samples (Lichun and Mian 2014). Specialized software enables the determination of element sizes, distribution, sphericity, and anisotropy (Diaz et al. 2017). The 3D models are further utilized for numerical modeling of filtration and conductivity processes, as well as for statistical analysis of porosity, fracture networks, and grain structures. Segmentation techniques are required to convert X-ray absorption data into spatial representations of rock components, allowing for high-resolution quantitative analysis (Gerke et al. 2021; Ar et al. 2020).

Computed X-ray tomography is among the most effective non-destructive methods for analyzing rock internal structures, including the formation and propagation of fractures during failure. However, due to the high cost and technical complexity of this equipment, its accessibility remains limited. Despite this, leading oil and gas companies worldwide have increasingly integrated digital technologies, including core digitalization, into their research and development strategies (Dinariev et al. 2021). In recent years, computed tomography and digital core analysis have become essential tools for evaluating reservoir properties, significantly improving the accuracy and scope of geological data. In this regard, digital core structure and properties research technologies are considered as one of the key trends in the industry. Large foreign oil and gas companies and scientific teams are already actively using and improving digital core and computed tomography methods (Tian et al. 2025; Chi et al. 2024; Wang et al. 2025; Chen et al. 2024), significantly outperforming Russian research groups in terms of implementation level and accumulated experience.

In Russia, several research teams are actively involved in the development and application of digital analysis technologies. These include Moscow Institute of Physics and Technology, Skolkovo Institute of Science and Technology (Skoltech), Tyumen Industrial University, Kazan Federal University, Don State Technical University, and other scientific institutions. However, the focus areas and application scopes vary significantly across these research groups.

For instance, at Kazan Federal University, recent research has focused on the effect of porous structure heterogeneity on hydrocarbon displacement processes. Studies have examined interfacial contact areas between injected and displaced fluids, as well as the overall efficiency of displacement mechanisms (Zakirov and Khramchenkov 2022). Meanwhile, at Tyumen Industrial University, researchers have investigated the depth of core clogging by drilling mud, assessed mineral composition using micro-CT, and explored experimental methods for simulating thermal effects on Bazhenov formation source rocks (Ponomarev et al. 2021).

A collaborative effort between Schmidt Institute of Physics of the Earth RAS, Moscow State University (MSU), and Skoltech has focused on applying machine learning to digital core model analysis, as well as evaluating various approaches to core modeling and multiphase filtration simulation at the pore scale (Gerke et al. 2021). K. Gerke's team is actively developing new domestic software for digital analysis and numerical modeling across multiple scales.

Additionally, a significant volume of commercial non-destructive core studies is conducted by "Microscopy and Analysis Systems," led by Shklover et al. (2021a, b). One of the company's key development areas involves integrating artificial intelligence technologies and creating a comprehensive database of core material properties. In the oil and gas sector, digital core research in Russia is being carried out by research centers affiliated with major developers, including Gazprom Neft's R&D center (Morozov et al. 2022), SurgutNIPIneft (Gasanova and Kayukov 2022), and Rosneft (Stepanov et al. 2024). However, the combination of digital analysis with geomechanical testing for addressing key challenges in oil and gas production is still pursued by only a few select research teams.

Notably, external research groups have yet to conduct large-scale studies involving the simultaneous application of pore-scale and macro-fracture digital analysis, numerical modeling of filtration processes, and true triaxial testing in wellbore conditions. This is primarily due to limited access to specialized equipment and the high level of expertise required to carry out such studies.

5 Geomechanical Challenges of Sand Production and Wellbore Integrity

Gas field reservoirs are often weakly consolidated, highly permeable sandstones with low strength, making them prone to creep deformation. This factor is critical in predicting long-term well stability, particularly in open-hole completions. If a highly plastic rock surrounds an uncased wellbore, it may undergo progressive deformation, leading to sand production, clogging of the filter, mechanical damage, and even wellbore collapse.

The investigation of sand production and physical modeling of geomechanical processes is often conducted using artificial rock samples, which may not fully replicate natural reservoir behavior (Lu et al. 2020; Kozhagulova et al. 2020; Arora et al. 2020). In some cases, research is performed on coal samples (Duan et al. 2020; Cao et al. 2021; Liu et al. 2018) or non-reservoir rocks, with most studies focusing on hydraulic fracturing mechanics (Wu et al. 2019; Heng et al. 2019).

The geomechanical aspects of sand production in wells are currently being studied by collaborative teams from Fulbright University and Nazarbayev University (Kozhagulova et al. 2020), using synthetic sandstone samples. Additionally, University of Tulsa (Zhong et al. 2019) and Waseda University (Papamichos et al. 2021a) are conducting research on wellbore wall failure and sand production mechanisms in natural rock samples, employing various geomechanical testing techniques (Papamichos et al. 2021b; Walle et al. 2019).

Earlier studies (Papamichos et al. 2010) combined pseudo-triaxial wellbore stability tests with computed tomography analysis, providing detailed insights into the mechanisms of sand production and wellbore collapse.

In Russia, both academic research groups (Leontyev et al. 2017; Poroshin et al. 2020) and specialized oil and gas research centers (Pavlov et al. 2021; Melikov et al. 2018; Pyatakhin 2012) are engaged in solving sand production and wellbore failure problems. The primary areas of research include developing and optimizing sand control equipment and well completion techniques (Leontyev et al. 2017; Betekhtin and Kostin 2017; Gilaev et al. 2021), strengthening the near-wellbore zone, well trajectory optimization, perforation techniques (Pyatakhin 2009), flow rate regulation (Iziumchenko et al. 2018), uniaxial compression testing, and investigating the effects of reservoir pressure on sand migration (Bliznyukov et al. 2010).

However, most studies rely on field data analysis, model material experiments, mathematical simulations, and conventional uniaxial or pseudo-triaxial testing methods. A significant portion of research on stress–strain state effects on reservoir rock filtration, deformation, and strength properties is carried out at the Geomechanics Laboratory of the Institute for Problems in Mechanics RAS (IPMech RAS).

6 Innovative Approaches to Reservoir Characterization and Well Stability at IPMech RAS

At IPMech RAS, researchers investigate rocks from oil and gas reservoirs across different regions of Russia, with lithologies varying from shallow formations (~100 m) to deep reservoirs (~7 km), exhibiting different pressure anomaly coefficients (Karev et al. 2020). Studies focus on rock deformation and failure processes at the wellbore wall, including sand production under compressive loads. A comprehensive set of experimental and theoretical studies at IPMech RAS has helped refine geomechanical-filtration models. Current research aims to develop an advanced creep model for geomaterials, differing from conventional metal creep models (Karev et al. 2019).

Using the unique TTLA system (TILTS Triaxial Independent Load Test System), the research team has published over 200 scientific papers and obtained 11 patents, including international ones. Since 2021, the laboratory has also been utilizing computed X-ray tomography and digital core analysis, enabling quantitative and qualitative assessment of reservoir rock structures, as well as statistical analysis and numerical simulations based on 3D images (Khimulia and Barkov 2022; Khimulia 2024; Khimulia et al. 2024).

The integration of digital core analysis with geomechanical testing has significantly enhanced our understanding of rock failure processes, structural changes, and filtration-reservoir properties across various rock types, including low-permeability reservoirs.

As part of the successfully completed RSF Project No. 23-77-01,037, a comprehensive study of a northern shelf reservoir was conducted, aiming to define well operation safety parameters and mitigate sand production risks. Geomechanical tests and digital analysis of the pore space of the gas field reservoir were carried out within the framework of the project. In particular, pore space characteristics such as porosity, pore size distribution, and geodesic tortuosity of percolation paths were studied. For this purpose, high-resolution scans were performed with subsequent creation of digital rock twins before and after geomechanical impact. Figure 1 shows the created models, as well as the results of the performed analysis of pore size distribution and

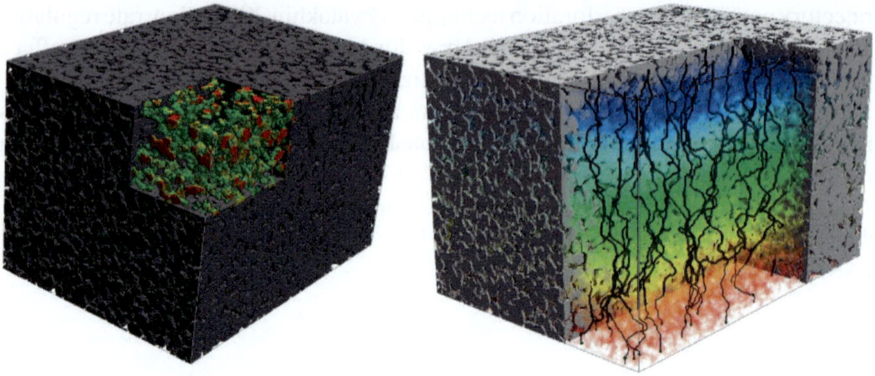

Fig. 1 The created models of pore space, as well as the results of the performed analysis of pore size distribution and pore space tortuosity estimation

pore space tortuosity estimation. In the left picture, individual pores are shown in the order of increasing size from green to red, and in the right picture, black lines visualize percolation paths, and the color shows the average length of the paths from one face of the specimen to the opposite face. For example, the average geodetic tortuosity before testing was estimated to average 1.05–1.06. It changed slightly after testing, with an increase for some specimens to 1.07–1.09, indicating a slight decrease in permeability after geomechanical impact.

One of the types of tests being conducted were geomechanical modeling tests of processes in the vicinity of the wells. In this case, cubic specimens were subjected to a complex, true triaxial stress state simulating stresses on the well wall during well pressure changes. More detailed schemes of experiments are presented in Karev et al. (2020). The three components of the applied stresses corresponded to the three principal stresses acting in the vicinity of the boreholes: radial, circular, and axial (Khimulia et al. 2024). The permeability of the specimen is measured during the experiment. These experiments allow, among other things, to establish critical stress values at which the rock fracture occurs at the borehole wall. Based on these data, allowable drawdowns can be calculated to maintain wellbore stability and prevent the development of negative effects in the reservoir.

Figure 2 shows characteristic results of true triaxial experiments for rocks of the Arctic gas condensate field. Figure 2a shows the loading program during the experiment and the permeability change curve, and Fig. 2b shows the stress–strain curves of the specimen tested in the first part of the experiments. As can be seen, the permeability during the experiment changed insignificantly, but decreased during the growth of shear stresses. The specimen fractured at a stress of 30 MPa, which corresponds to a well drawdown less than 0.12 MPa.

One type of geomechanical experiments were experiments on the "hollow cylinder" scheme to study the processes of fracture and sand production. For this purpose, cubic specimens with a 40 mm face and a hole in the center were compressed by uniform stresses, during which the process of sand removal from the hole was

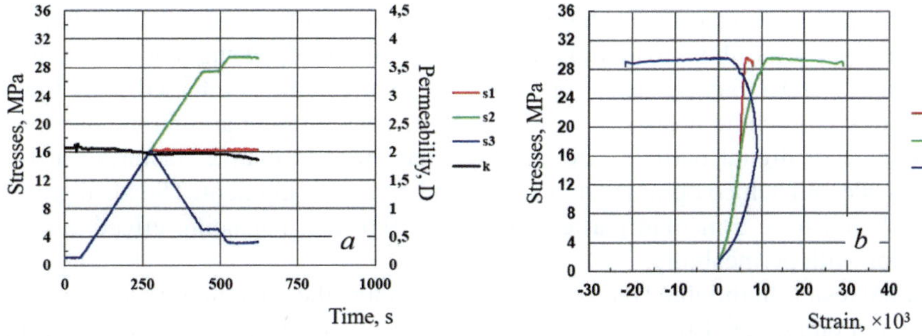

Fig. 2 The loading program during the experiment with permeability change curve and stress–strain curves for characteristic specimen

recorded. This made it possible to assess the intensity and conditions of the beginning of sand removal. Also, to study the failure processes of the holes, the specimens were scanned after testing and the cracks and blowouts formed were analyzed. Figure 3 shows some cross sections of the created 3D models of specimens after testing. Gray represents rock, black—cracks and air.

Test results revealed that, despite the reservoir's isotropic elastic properties, its strength characteristics exhibit significant anisotropy. In a horizontal well, pressure drawdowns at the bottomhole can cause rock failure at the upper and lower contour points, leading to specific breakouts. In the case of a vertical borehole, directionally oriented dropouts will not form because the shear stresses act in the same plane of isotropy. In this case, cracks will form along the entire perimeter of the borehole, similar to the example of specimens presented in Fig. 3. Additionally, the high deformability of the reservoir rock suggests that even minor pressure drops may induce creep, progressively pushing the formation toward the sand screen. This

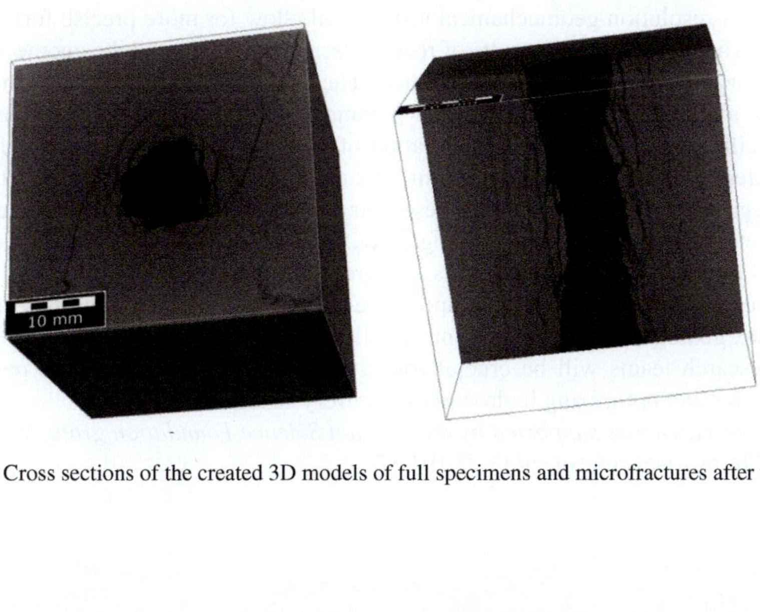

Fig. 3 Cross sections of the created 3D models of full specimens and microfractures after testing

uneven pressure can result in deformation or collapse of the screen. The hollow cylinder-type tests allows to identify the stress conditions triggering sand production, its intensity under varying stress states, and the total volume of released sand. Fracturing of the borehole walls led to radial macrocracks forming around the opening. First stage of digital particle size analysis characterized matrix fracturing at the microscale and helped optimize downhole sand control equipment. The ideal gravel filter should have a median particle size up to 0.6 mm, balancing effective sand retention with permeability.

Initial results highlighted the necessity of expanding these methods to other Arctic fields and emphasized the importance of anisotropy considerations, creep effects, advanced digital analysis techniques, and numerical modeling for more precise reservoir forecasting and optimization of well parameters.

7 Conclusion

The analysis presented in this article underscores the significant influence of stress conditions on the deformation and filtration properties of reservoir rocks. The use of laboratory methods such as triaxial compression and true triaxial loading has provided valuable insights into rock behavior under various loading conditions. However, traditional experimental techniques have limitations in fully replicating in situ subsurface conditions, making it essential to integrate digital technologies for enhanced rock characterization. Computed X-ray tomography, machine learning, and advanced numerical simulations are proving to be indispensable tools for improving the accuracy of reservoir models and predicting well performance.

Future research should focus on refining geomechanical models that account for rock anisotropy, stress-dependent permeability changes, and long-term deformation effects such as creep. Expanding the use of digital core analysis and integrating it with high-resolution geomechanical testing will allow for more precise forecasting of reservoir behavior. The results of recent research conducted in the geomechanics laboratory of IPMech RAS are presented. The presented results of comprehensive studies allowed to determine optimal parameters for safe and efficient operation of specific fields, including determination of permissible drawdowns in wells and characteristics of downhole equipment. The conducted research forms the basis for an integrated approach to studying reservoir properties to determine well operation parameters based on computer tomography, digital core analysis, laboratory experiments and unique mechanical tests. The results obtained from studies on Arctic shelf reservoirs demonstrate the importance of extending these approaches to other complex geological settings. Continued collaboration between academic and industrial research teams will be crucial for advancing the understanding of reservoir mechanics and optimizing hydrocarbon recovery strategies.

*The research was supported by the Russian Science Foundation grant № 23–77-01037,*https://rscf.ru/project/23-77-01037/.

References

Ar Rushood I et al (2020) Segmentation of X-ray images of rocks using deep learning. In: SPE annual technical conference and exhibition: Virtual. Paper SPE-201282-MS, p 13. https://doi.org/10.2118/201282-MS

Arora K, Gutierrez M, Hedayat A (2020) Physical modeling of lined tunnel in squeezing ground conditions. In: Geo-Congress 2020: Engineering, monitoring, and management of geotechnical infrastructure. American Society of Civil Engineers, Reston, VA, pp 335–344

Asahina D, Sato M, Takahashi M (2019a) Laboratory observations of fracture plane reactivation induced by pore pressure in Kimachi sandstone. In: Proceedings of rock dynamics summit, pp 101–105. https://doi.org/10.1201/9780429327933-14

Asahina D, Pan P, Sato M et al (2019b) Hydraulic and mechanical responses of porous sandstone during pore pressure-induced reactivation of fracture planes: an experimental study. Rock Mech Rock Engineering 52:1645–1656. https://doi.org/10.1007/s00603-018-1706-8

Baizhanov B, Katsuki D, Tutuncu AN et al (2019) Experimental investigation of coupled geomechanical, acoustic, and permeability characterization of berea sandstone using a novel true triaxial assembly. Rock Mech Rock Engineering 52:2491–2503. https://doi.org/10.1007/s00603-019-01816-6

Betekhtin AN, Kostin DK (2017) Laboratory studies of sand production control systems. In: SPE Russian petroleum technology conference

Bliznyukov VY et al (2010) Issues of sand-producing formation operation. The Influence of Reservoir Pressure on Sand Production from the Reservoir during production well operation. Engineer-Oilman. 1:11–22 (In Russian)

Botao LIN et al (2020) Study on fracture propagation behavior in ultra-heavy oil reservoirs based on true triaxial experiments. Pet Explor Dev 47(3):651–660

Cao W et al (2021) Fracture behavior and seismic response of naturally fractured coal subjected to true triaxial stresses and hydraulic fracturing. Fuel 288(119618)

Chen M et al (2024) A novel reservoir classification method for sandstone reservoir evaluation using multi-scale digital rock method. J Pet Explor Prod Technology:1–14

Chi P et al (2024) From digital rock to digital wellbore: multiscale reconstruction and simulation. Adv Geo-Energy Res 13(1):1–6

Couture C, Bésuelle P (2019) Diffuse and localized deformation of a porous Vosges sandstone in true triaxial conditions. In: E3S web of conferences, vol 92, no 06007. https://doi.org/10.1051/e3sconf/20199206007

Diaz M et al (2017) Surface roughness characterization of open and closed rock joints in deep cores using x-ray computed tomography. Int J Rock Mech Min Sci 98:10–19

Dinariev OY et al (2021) Digital core analysis in oil and gas field development design. Bus J NeftegazRU (5):50–58. (In Russian)

Duan M, Jiang C, Gan Q et al (2020) Study on permeability anisotropy of bedded coal under true triaxial stress and its application. Transp Porous Media 131:1007–1035. https://doi.org/10.1007/s11242-019-01375-y

Feng X-T et al (2020) Evolution of the mechanical and strength parameters of hard rocks in the true triaxial cyclic loading and unloading tests. Int J Rock Mech Min Sci 131(104349)

Feng XT et al (2024) A test system for microwave-assisted dual-mode mechanical cutting of hard rock under true triaxial compression. Rock Mech Rock Engineering:1–20

Feng X-T, Kong R, Zhang X, Yang C (2019) Experimental study of failure differences in hard rock under true triaxial compression. Rock Mech Rock Engineering 52:2109–2122. https://doi.org/10.1007/s00603-018-1700-1

Feng X, Zhao J, Zhang X et al (2018) A novel true triaxial apparatus for studying the time-dependent behaviour of hard rocks under high stress. Rock Mech Rock Engineering 51:2653–2667. https://doi.org/10.1007/s00603-018-1516-z

Garcia AV, Rached RM, Santamarina JC (2018) Large-scale true triaxial apparatus for geophysical studies in fractured rock. Geotech Test Journal 41(4):821–829

Gasanova AV, Kayukov DA (2022) On the possibilities of application and prospects for the development of the Digital Core technology using X-ray tomography. In: Current problems of Arctic hydrocarbon field development: Arkhangelsk: Northern (Arctic) Federal University named after M. V. Lomonosov, pp 3–11. (In Russian)
GCTS Testing Systems. https://www.gcts.com/?s=prod_ver&p=products&ID=199#tab-1 (accessed: 17 Jan 2025)
Georgieva T, Descamps F, Gonze N et al (2020) Stability assessment of a shallow abandoned chalk mine of Malogne (Belgium). Eur J Environ Civ Engineering:1–15. https://doi.org/10.1080/19648189.2020.1762752
Gerke KM et al (2021) Modern approaches to pore space scale digital modeling of core structure and multiphase flow. Georesources 23(2):197–213. https://doi.org/10.18599/grs.2021.2.20
Gilaev GG et al (2021) Methods for combating sand production in production wells. Oil Gas Bus 19(5):119–127. (In Russian)
Guo J et al (2024) The influence of mining stress paths on rock damage and permeability. Environ Earth Sciences 83(10):1–20
Hamza O, Stace R (2018) Creep properties of intact and fractured muddy siltstone. Int J Rock Mech Min Sciences 106:109–116
Heng S et al (2019) Experimental and numerical study on the non-planar propagation of hydraulic fractures in shale. J Pet Sci Eng 179:410–426
Huang J et al (2022) Pre-peak acoustic emission characteristics of tight sandstone failure under true triaxial stress. J Nat Gas Sci Engineering 102(104576)
Huo J, Selvadurai APS, Meguid M (2024) Permeability evolution in a fracture in granite during isotropic compression. Rock Mech Rock Engineering:1–15
GDS Instruments. https://www.gdsinstruments.com/gds-products/gds-true-triaxial-apparatus (accessed: 17 Jan 2025)
Iziumchenko DV et al (2018) Operation of gas wells under conditions of active water and sand production. Gas Sci Bulletin 1(33):235–242. (In Russian)
Karev VI, Klimov DM, Kovalenko YF, Ustinov KB (2019) Experimental study of rock creep under true triaxial loading. In: Proceedings of the Russian Academy of Sciences. Mechanics of solids, no 6, pp 30–37. (In Russian)
Karev VI, Kovalenko YF, Ustinov KB (2020) Geomechanics of oil and gas wells. In: Advances in oil and gas exploration and production. Springer International Publishing, Cham, p 166
Karman T (1911) Festigkeitsversuche unter allseitigem Druck. Z Ver Deutsch Ing
Khimulia VV, Barkov SO (2022) Analysis of internal structural changes in low-permeability reservoir rocks using computed tomography under controlled stress unloading. Curr Issues Oil Gas 4(39):27–42. (In Russian)
Khimulia VV (2024) Application of digital core analysis technology to study the filtration and reservoir properties and structure of high-permeability formations in underground gas storage. Russ J Earth Sci 24(5):ES5012. https://doi.org/10.2205/2024es000928
Khimulia V, Karev V, Kovalenko Y, Barkov S (2024) Changes in filtration and capacitance properties of highly porous reservoir in underground gas storage: CT-based and geomechanical modeling. J Rock Mech Geotech Eng 16(8):2982–2995. https://doi.org/10.1016/j.jrmge.2023.12.015
Kozhagulova A et al (2020) Experimental and analytical investigation of sand production in weak formations for multiple well shut-ins. J Pet Sci Engineering 195(107628)
Kwasnievski M et al (eds) (2013) True triaxial testing of rocks. CRC Press/Balkema, Leiden, p 365
Leontyev DS, Kleshchenko II, Tsidrik NS (2017) Technology for reducing sand production in oil wells. News of Higher Educational Institutions. Oil and Gas 5:72–74 (In Russian)
Li Z, Wang L, Lu Y et al (2019) Experimental investigation on true triaxial deformation and progressive damage behaviour of sandstone. Scientific Rep 9(3386). https://doi.org/10.1038/s41598-019-39816-9
Lichun J, Mian C (2014) 3D imaging of fractures in carbonate rocks using X-ray computed tomography technology. Carbonates Evaporites 29(2):147–153

Liu Y et al (2018) Permeability evolution of anthracite coal considering true triaxial stress conditions and structural anisotropy. J Nat Gas Sci Engineering 52:492–506
Lu Y et al (2020) Experimental and theoretical analysis of hydraulic fracturing and gas fracturing of rock under true triaxial compressions. Eng Fract Mechanics 234(107100)
Lu M et al (2020) Verification of fracture reorientation and analysis of influence factors in multiple fracturing treatment. Geofluids 2020:1–13
Martyushev DA, Galkin SV, Shelepov VV (2019) Influence of the stress state of rocks on matrix and fracture permeability in various lithological-facies zones of the Tournaisian-Famennian oil deposits of the Upper Kama region. Vestnik of Moscow University. Geology (5):44–52. (In Russian)
Melikov RF et al (2018) Optimization of well construction and operating parameters for horizontal wells in the PK1 formation of the Kharampur NGC. In: Moscow. Russian Oil and Gas Technical Conference SPE. No. 191635, pp 15–17
Minaeian V (2014) True triaxial testing of sandstones and shales. Doctoral dissertation. Curtin University. Australia, p 220
Minaeian V, Dewhurst DN, Rasouli V (2020) An investigation on failure behaviour of a porous sandstone using single-stage and multi-stage true triaxial stress tests. Rock Mech Rock Engineering 53:3543–3562. https://doi.org/10.1007/s00603-020-02134-y
Morozov VV et al (2022) Experience in applying the digital core analysis technology at the sarkała field. PRONEFT Prof Oil Journal (3):65–70. (In Russian)
Nabipour A (2013) Experimental and numerical study of ultrasonic monitoring of hydraulic fracture propagation. Doctoral dissertation. Curtin University, Australia, p 222
Nguyen TS, Li Z, Su G, Nasseri MHB, Young RP (2018) Hydro-mechanical behavior of an argillaceous limestone considered as a potential host formation for radioactive waste disposal. J Rock Mech Geotech Eng 10(6):1063–1081. https://doi.org/10.1016/j.jrmge.2018.03.010
Papamichos E et al (2021a) Sand rate from laboratory tests to field predictions. In: 55th US rock mechanics/Geomechanics symposium. OnePetro
Papamichos E et al (2021b) Sand mass production in anisotropic stresses from lab to field predictions. In: SPE annual technical conference and exhibition. OnePetro
Papamichos E et al (2010) Hole stability of red wildmoor sandstone under anisotropic stresses and sand production criterion. J Pet Sci Engineering 72(1–2):78–92
Pavlov VA et al (2021) Justification of well operation modes for the Cenomanian gas reservoir of the Kharampur field based on geomechanical modeling results. Expo Oil Gas 2(81):41–46. (In Russian)
Ponomarev AA et al (2021) Application of X-ray microtomography of core in oilfield geology. Georesources 23(4):34–43. (In Russian)
Poroshin MA, Tananykhin DS, Grigoriev MB (2020) Analysis of laboratory methods for studying the sand production process in oil field development. Bull Eurasian Sci 12(2):22. (In Russian)
Pyatakhin MV (2009) Critical flow rate of rock failure in the near-wellbore zone of a horizontal well. Gas Industry 7:36–39. (In Russian)
Pyatakhin MV (2012) Geomechanical problems in well operation. Moscow: Gazprom VNIIGAZ, p 266. (In Russian)
Rasouli VA (2013) True triaxial stress cell (TTSC) used for simulations of real field operations in the lab. In: Kwasniewski M et al (eds) True triaxial testing of rocks, vol 4, pp 311–318. CRC Press/Balkema, Leiden
Rezagholilou A, Sarmadivaleh M (2019) Design and application of a high-performance true triaxial stress cell. Insights Min Sci Technology 1(5):555573. https://doi.org/10.19080/IMST.2019.01.555573
Rubtsova EV, Skulkin AA (2017) On physical modeling of the measurement hydraulic fracturing process in model samples under non-uniform loading. Probl Subsoil Use 2(13):42–46. (In Russian)
Shklover VY et al (2021a) Study of morphology and crystalline structure of natural silicas using electron microscopy and electron diffraction methods. Crystallogr 66(4):626–635. (In Russian)

Shklover VY et al (2021b) Experience in creating a database for studying micro- and nanometer void space of reservoirs in the Digital Core technology. Inf Syst Technol Geol Oil Gas Production:28–39. (In Russian)

Stepanov SV et al (2024) The concept of applying digital core research results in reservoir simulation practice. PRONEFT Prof Oil Journal 9(4):141–151. (In Russian)

Sun K et al (2024) Impact of effective stress on permeability for carbonate fractured-vuggy rocks. J Rock Mech Geotech Eng 16(3):942–960

Tian Y et al (2025) Digital rock modeling of deformed multi-scale media in deep hydrocarbon reservoirs based on in-situ stress-loading CT imaging and U-Net deep learning. Mar Pet Geology 171(107177)

Vachaparampil A, Ghassemi A (2017) Failure characteristics of three shales under true-triaxial compression. Int J Rock Mech Min Sciences 100:151–159

Veselovsky RV et al (2022) Collective use center of the schmidt institute of physics of the Earth RAS "Petrophysics, Geomechanics, and Paleomagnetism". Geodyn Tectonophysics (2). (In Russian)

Walle LE, Berntsen AN, Papamichos E (2019) Novel true triaxial apparatus applied to the study of sand mass production under anisotropic stress conditions. In: 14th ISRM congress. OnePetro

Wang H et al (2025) Constructions of multi-scale 3D digital rocks by associated image segmentation method. Front Earth Science 12(1518561)

Wang Z, Shen M, Gu L, Zhang F (2018) Creep behavior and long-term strength characteristics of greenschist under different confining pressures. Geotech Test Journal 41(1):55–71. https://doi.org/10.1520/GTJ20170143

Wu S et al (2019) Shear-tensile fractures in hydraulic fracturing network of layered shale. J Pet Sci Engineering 183(106428)

Yang S, Hu B (2020) Creep and permeability evolution behavior of red sandstone containing a single fissure under a confining pressure of 30 MPa. Sci Reports 10(1900). https://doi.org/10.1038/s41598-020-58595-2

Yang S, Huang Y-H, Ranjith PG (2018) Failure mechanical and acoustic behavior of brine saturated-sandstone containing two pre-existing flaws under different confining pressures. Eng Fract Mechanics 193:108–121. https://doi.org/10.1016/j.engfracmech.2018.02.021

Young RP, Nasseri MHB, Lombos L (2013) Imaging the effect of the intermediate principal stress on strength, deformation and transport properties of rocks using seismic methods. In: Kwasniewski M et al (eds) True triaxial testing of rocks, pp 311–318. CRC Press/Balkema, Leiden

Zakirov T, Khramchenkov M (2022) Identifying the representative elementary volume for various two-phase flow patterns in porous media using a statistical approach: Lattice Boltzmann Study. Available at SSRN. 2022. https://doi.org/10.2139/ssrn.4286295

Zhang H, Li CC (2019) Effects of confining stress on the post-peak behaviour and fracture angle of fauske marble and iddefjord granite. Rock Mech Rock Engineering 52:1377–1385. https://doi.org/10.1007/s00603-018-1695-7

Zhao J, Feng X-T, Zhang X et al (2018a) Time-dependent behaviour and modeling of Jinping marble under true triaxial compression. Int J Rock Mech Min Sciences 110:218–230

Zhao J, Feng X-T, Zhang X et al (2018b) Brittle-Ductile transition and failure mechanism of jinping marble under true triaxial compression. Eng Geology. 232:160–170

Zhao J, Feng X-T, Zhang X, Yang C (2019) Brittle and ductile creep behavior of Jinping marble under true triaxial stress. Eng Geology 258(105157). https://doi.org/10.1016/j.enggeo.2019.105157

Zhong R et al (2019) Experimental investigation of fracture-based wellbore strengthening using a large-scale true triaxial cell. J Pet Sci Eng 178:691–699

Zhu N, Cheng P, Yu Q (2024) Experimental study on the time-dependent gas permeability of fractures in shales. Rock Mech Rock Engineering 57(1):695–718

Zong Y, Han L, Meng Q, Wang Y (2020a) Strength properties and evolution laws of cracked sandstone samples in re-loading tests. Int J Min Sci Technology 30(2):251–258. https://doi.org/10.1016/j.ijmst.2019.03.004

Zong Y, Han L, Jin Y, Zhao W, Meng L (2020b) Experimental investigation on the post-peak short-term and creep behavior of fractured sandstone. Energies 13:598. https://doi.org/10.3390/en13030598

Axisymmetric Distribution of Oil Products in Arctic Conditions

V. P. Pakhnenko

Abstract The article considers the process of axisymmetric and one-dimensional spreading (e.g., spill is in a river mouth or canal) of oil products in Arctic conditions, in particular on the ice surface. The dynamics of crude oil and motor oil slick spreading on the ice surface is analyzed under different experimental parameters, such as oil product temperature, initial oil product volume, ice surface, and environment temperature. It is shown that low temperatures and rough ice structure significantly affect the viscosity of oil products, their adhesion to surfaces and the rate of spreading. The paper discusses the key factors determining the behavior of oil slicks (temperature, roughness, spreading process, etc.) and their influence on the efficiency of traditional cleaning methods.

Keywords Experimental work · Oil products · Oil spills · Arctic conditions · Ice · Axisymmetric spreading

1 Introduction

Intensive human activity to develop the natural resources of the World Ocean includes exploration and extraction of minerals, transportation of goods, use of energy and biological resources, as well as many other aspects. One of the inevitable consequences of such activity is the emergence and spread of pollution of different kinds. Today oil is one of the most widespread substances polluting natural waters: The volume of its transportation by sea reaches about 2.2 billion tons per year (www.itopf.org). With such large transportation volumes, accidental spills occur periodically, as a result of which oil or oil products are released into the aquatic environment. It is believed that about 0.03% of oil transported by tankers is lost due to various reasons. In Russia, annual losses of oil and oil products reach 4.8 million tons, and about 12% of them are caused by tanker and oil platform accidents.

V. P. Pakhnenko (✉)
A.Yu. Ishlinsky Institute for Problems in Mechanics, Moscow, Russia
e-mail: terkin95@mail.ru

T. Chaplina (ed.), *Processes in GeoMedia—Volume IX*, Springer Geology,
https://doi.org/10.1007/978-981-96-9302-3_31

Oil spills are one of the most serious environmental problems, having a negative impact on ecosystems, economy and human health at the same time. When oil hits the water surface, it spreads in a thin film and also partially adheres to objects in the spill area. The resulting layer of adhered oil is a complex multi-component system whose properties and behavior largely determine the scale and consequences of an accidental spill. History knows many examples of major environmental disasters caused by large tanker accidents. Such accidents include the recent major spill in the Kerch Strait, when a storm caused the hulls of two Russian oil tankers to break apart. Such accidents have extremely severe consequences for marine and coastal ecosystems, as well as significant economic costs and lengthy cleanup programs.

Often spills occur on the free water surface in the ocean or sea, which corresponds to the case of an axisymmetric spill, and each such spill has a different degree of severity and impact on the environment. However, even more destructive impact on ecosystems is caused by oil spills occurring in water areas located near the coastline, as their consequences affect not only aquatic flora and fauna, but also land ecosystems of coastal zones. Elimination of consequences in such areas can take decades, requiring enormous efforts and financial expenditures.

In many cases, an oil spill is considered to be a two-dimensional phenomenon (e.g., spreading over large areas of oceans, and seas or bays). However, there are specific conditions in which oil products spread predominantly one-dimensionally, such as along the walls of canals, rivers, or narrow water bodies (Hoult and Suchon 1970). This geometric formulation of the problem requires the use of specialized theoretical and experimental models, since the mechanisms of oil buildup and the methods of cleaning such objects differ from those used in more open water spaces.

At the same time, in a number of situations (e.g., a point release of oil into a calm or conditionally enclosed water area), spreading can become axisymmetric. In such a case, oil spreads from the central point (exit or leakage zone) to the periphery, forming annular boundaries of the spill. Such geometry is characteristic, in particular, of oil product releases from ice wells or pipeline ruptures under ice in Arctic conditions, where the free surface of water is limited in radius and side walls may be absent or distant.

The Arctic and subarctic areas have recently attracted particular interest, as these regions are considered promising in terms of discovering and developing new oil fields, and there are plans to significantly increase production there in the future. Low temperatures and prolonged ice cover complicate both the spreading process and subsequent spill cleanup. According to Gong et al. (2014), low temperatures increase the viscosity of oil and increase its adhesion to hard surfaces, making cleanup much more difficult. Oil aging processes, including vaporization of light fractions and oxidation of heavy components, are exacerbated at low temperatures and lead to the formation of denser water–oil emulsions. Such emulsions are more difficult to remove by standard methods. It should also be taken into account that surfaces with micropores and roughnesses are able to retain significant volumes of oil contaminants, as oil droplets are easily " locked" in depressions and ice has such properties. In addition, at low temperatures oil becomes more viscous and less mobile, which repeatedly enhances its adhesion (Dickins 2011).

Based on comprehensive studies (Fingas 2017), it has been established that the most effective methods of cleaning of adhered oil are approaches combining physical, chemical and biological methods. However, the peculiarities of the Arctic regions (first of all, low temperature and a significant proportion of the area covered by ice) may require specific technologies that take into account not only the properties of oil but also the dynamics of interaction between water, ice and oil products.

Studying the characteristics of the adhered oil layer remains an important task, since its physical and chemical parameters directly determine the effectiveness of cleaning measures. At the same time, the processes of transformation of this layer under the influence of external factors (temperature, wind, solar radiation, etc.) can significantly change the structure and toxicity of contaminants, significantly complicating their removal. The study of spreading of not only crude oil but also motor oil on the ice surface was carried out in this work.

2 Axisymmetric Spreading on the Ice Surface

In Chaplina et al. (2024), a simplified mathematical model is presented in which the chemical composition and all thermodynamic parameters of the media are assumed to be constant values: All media are homogeneous and isotropic, and the mass of the spreading slick is unchanged. Since the transport velocities of water and oil (or motor oil) are much lower than the speed of sound in them, all media are considered incompressible in any propagation model.

In the process of oil products spreading on the ice surface, several main factors should be taken into account. First, in the gravitational field there is a transition of potential energy of the system "oil-product-ice-air" into kinetic energy. This is due to constant changes in the geometric characteristics of the spill over time and the fact that oil (or motor oil) tends to take a position with a lower potential energy.

Secondly, when oil products come into contact with ice and air, the sum of surface energies changes, because the areas of contact between "oil product-air," "oil product-ice," and "air-ice" change. The spreading of oil products is accompanied by rearrangement of interfacial boundaries, which significantly affects the overall energy balance of the system.

The third important factor is the inhibition of oil product spreading by the rough ice surface and possible interaction with the water mass (especially when the spill occurs directly under the ice). The oil product, filling irregularities and micropores, forms so-called stagnant zones, which leads to a decrease in the mass directly involved in further movement. This additional retention of oil products in ice depressions and cracks can significantly reduce the velocity and change the character of spreading.

In Kistovich and Chaplina (2023), it was shown that in a simplified model it is possible to neglect viscous heat losses in oil and air, but it is important to take into account the influence of rough ice surface on the dynamics of spreading. The model of rough ice, as well as the process of oil spreading on it, is described by means of statistical parameters that allow us to consider the phenomenon on average.

Numerous laboratory experiments (Konno et al. 2002) and field observations of oil spills on ice (Mackay et al. 1975) show that in most cases the oil slick has irregular outlines. Since analytical description of constantly changing irregular slick shapes is an extremely difficult task, in this paper we study the spreading approximated to the axisymmetric case.

The theory of this spreading is described in Chaplina et al. (2024). Let at time τ the slick boundary be described by the function $L(\sigma, \tau)$ in the cylindrical coordinate system (l, σ, z) with the origin at the point O located on the ice surface (the z—axis is directed against the gravity vector). The concept of the average radius of the spot is introduced

$$L(\tau) = \overline{L(\sigma, \tau)} = \frac{1}{2\pi} \int_0^{2\pi} L(\sigma, \tau) d\sigma$$

and its standard deviation

$$\delta_L(t) = \sqrt{\frac{1}{2\pi} \int_0^{2\pi} (L(\sigma, \tau) - L(\tau))^2 d\sigma}$$
$$= \sqrt{\left\langle \left(L(\sigma, \tau) - \overline{L(\sigma, \tau)}\right)^2 \right\rangle} = \sqrt{\overline{L^2(\sigma, t)} - L^2(\tau)},$$

from which follows the relation

$$\overline{L^2(\sigma, \tau)} = L^2(\tau) + \delta_L^2(\tau).$$

These relations are important in the context of determining which spreading can be considered to approximate the axisymmetric case. If the condition $\frac{\delta_L(\tau)}{L(\tau)} \ll 1$ is satisfied for all time instants, then the spreading slicks are considered to approximate the axisymmetric case. In order to evaluate whether the spreading of an oil slick in an experiment or in a real situation is axisymmetric or not, the following procedure is carried out. First, the values of $L(\tau)$ and $\delta_L(\tau)$ are calculated from the slick shape according to the relations given earlier, and then the validity of the condition $\frac{\delta_L(\tau)}{L(\tau)} \ll 1$ is checked.

Let us consider the features of the experimental work in the case of axisymmetric spreading of oil products on the ice surface. Figure 1 shows photos from the experiment with spreading of crude oil of Mamontovskoye field $\rho_0 = 878$ kg/m^3, $\nu_0 = 2.15 \cdot 10^{-5} \text{m}^2/s$ and Volga M8B-SAE 20 API motor oil $\rho_0 = 883.7$ kg/m^3, $\nu_0 = 3.34 \cdot 10^{-4} \text{m}^2/s$.

It is worth noting that Arctic conditions affect many parameters of oil product spreading. Low temperatures significantly increase the viscosity of oil products and reduce their mobility. As the viscosity increases, the resistance to movement increases and the rate of spreading decreases. In some cases, at sufficiently low temperatures, a part of light fractions of petroleum products may vaporize, and the residue of heavier fractions may partially solidify or sharply thicken on the cold ice surface. This leads

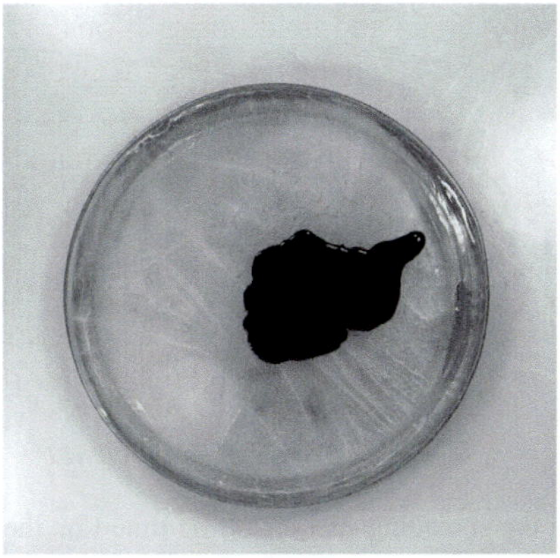

a) crude oil from the Mamontovskoye field

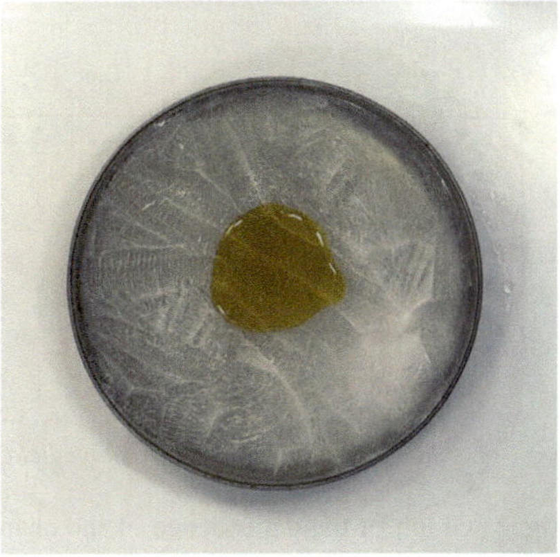

b) motor oil Volga M8B-SAE 20 API

Fig. 1 Images of the process of axisymmetric spreading on the ice surface

to the formation of "viscous fronts" and significantly slows further spreading. If the temperature of oil products at the spill is higher than the ambient temperature, the effect of local melting of ice under the thin film of oil products is possible (especially in the first moments). However, most often this process is insignificant and quickly slows down after they cool down.

The structure and microrelief of the ice surface also has a strong influence. Ice is seldom perfectly smooth, but often has cracks, roughness, a near-surface layer of snow–ice mixture, etc. All of these irregularities and pores can trap fluid and cause some oil to accumulate in depressions, cracks or under snow deposits. As a consequence, roughness and microporosity factors affect the adhesion of oil to ice. A rough surface can hold an order of magnitude more oil than a smooth surface.

Due to such a variety of factors affecting the spreading process, a large number of experiments have been conducted with different initial parameters: Ice temperature, ice surface size, shape and roughness, oil product temperature, volume and spill method, the variety of the spreading oil product itself, and ambient temperature. In this paper, the influence of the type of oil product, its volume and spill method on the spreading process on the ice surface will be discussed in more detail.

The first series of experiments was conducted with a uniform supply of oil product to the ice surface. The tank with oil product was placed almost closely above the ice surface so that the height from which it spreads did not affect the spillage process. Experiments were conducted with the volume of oil products $V = 2.5$ ml, $V = 5$ ml, and $V = 10$ ml. According to the results of a series of experiments, the change in the surface area of the spreading slick S from the time t of spreading was calculated in Fig. 2.

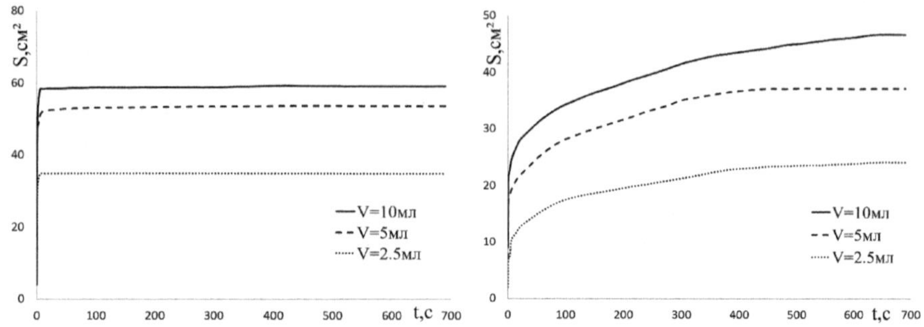

a) crude oil from the Mamontovskoye field *b) motor oil Volga M8B-SAE 20 API*

Fig. 2 Graph of the dependence of the change in the spreading spot area S on time t on the ice surface

These results clearly demonstrate the differences in the spreading processes of crude oil and motor oil. Crude oil slick with volume $V = 2.5$ ml reached its maximum spreading area in time $t = 6$ s, motor oil in the same volume required $t = 360$ s for this. That is, the parameters of the petroleum product in Arctic conditions very strongly affect the process of its spreading, the temperature of the environment and ice had a much greater effect on crude oil. And naturally the initial volume of the spreading oil product influenced the process of spreading, the more it was, the more the area of contamination was larger and the time of spreading increased.

A series of experiments were also conducted with an alternative method of oil product delivery. The petroleum product was placed in a narrow vertical flask to add height and increase the rate at which the oil product reaches the ice surface. Because in natural conditions, the causes of the accidental spill and the spreading process can be very different from each other, which in turn can affect the spreading patterns (Fig. 3).

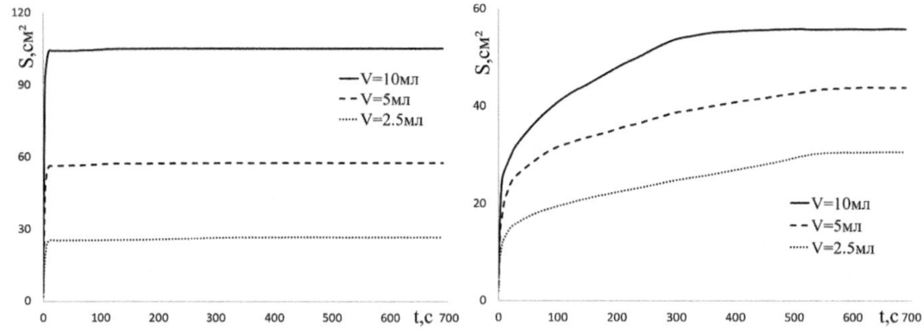

a) crude oil from the Mamontovskoye field *b) motor oil Volga M8B-SAE 20 API*

Fig. 3 Graph of dependence of change of spreading spot area S on time t on the ice surface

The change in the method of delivering petroleum products had a significant impact on the results. Taking a closer look at the same volume of $V = 10$ ml, for crude oil the maximum spill area reached $S = 105.2\ \text{cm}^2$, whereas under the previous spreading method it was $S = 59.3\ \text{cm}^2$. (For smaller oil volumes, the differences were less noticeable, since the oil spill very quickly reaches its maximum spread area.) For motor oil with a volume of $V = 10$ ml, the maximum spill area reached $S = 55.8\ \text{cm}^2$, while under the previous method it was $S = 46.6\ \text{cm}^2$.

In addition, to evaluate how the ambient, ice, and oil product temperatures influenced the spreading process, these parameters were monitored in each experiment using a thermal imager. We will illustrate this with an example from the motor oil spreading experiments (Fig. 4).

It should be noted that the larger the initial volume of the motor oil, the slower it cooled, and this occurred at a later point in time. Consequently, the oil spread faster and covered a larger area of the ice surface. A similar pattern was observed in the case of the oil spill.

In the early stages of the spill, the spreading front moved significantly faster (while the oil products were still quite fluid and warmer), but as they cooled and

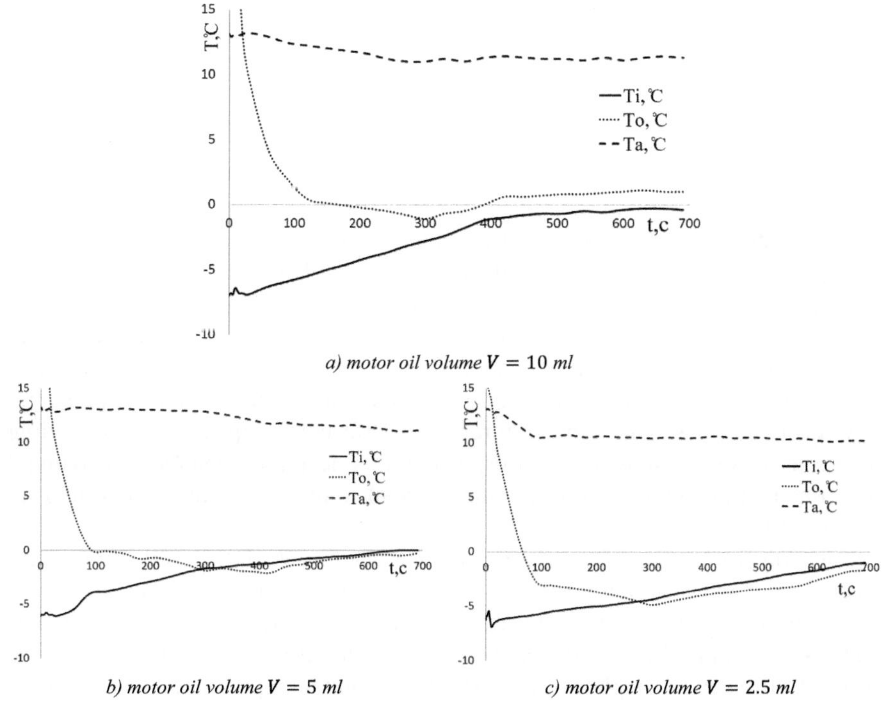

a) motor oil volume **V** = 10 *ml*

b) motor oil volume **V** = 5 *ml*

c) motor oil volume **V** = 2.5 *ml*

Fig. 4 Graph showing the changes in T_o °C—the motor oil temperature, T_i °C—the ice surface temperature, and T_a °C—the ambient temperature over time t their viscosity increased, the spreading rate decreased. The dependence of the spill radius on time differs substantially from the classical formula for liquid spreading on a rigid substrate at room temperature.

In this work, the presence of a snow layer on the ice surface was not taken into account. When petroleum products come into contact with snow, they are quickly absorbed, forming a heterogeneous suspension (an oil–snow mixture). This fundamentally changes the spreading mechanisms, as the saturated snow impedes free flow of the liquid. Additionally, if there is a thin layer of meltwater or if spring thawing has begun, petroleum products can mix with the water and form emulsions (complex "water-in-oil" or "oil-in-water" structures), which affects both spreading rate and area. However, these processes will be addressed in future research.

Thus, the axisymmetric spreading of oil products on ice in Arctic conditions is a complex, multifaceted process. Its characteristic features include increased viscosity and decreased mobility of oil at low temperatures; a high degree of influence from the temperatures of the ice surface, ambient environment, and oil product on the spreading process; and the effect of the immediate release parameters (initial height and velocity of the petroleum product), which significantly impact how it behaves on the ice. The possibility of partial infiltration or freezing and the substantial influence of microrelief and cracks in the ice further complicate this phenomenon.

3 Conclusion

A series of experiments was carried out under various initial conditions, yielding a substantial body of experimental data on the spreading dynamics of crude oil and motor oil over ice in both axisymmetric and one-dimensional cases. These two types of oil product spreading differ significantly in their characteristics and spreading mechanisms. In the first case, oil disperses radially from the source, while in the second, it moves along bounding walls or through narrow passages, which greatly affect the dynamics of the contamination slick. Accounting for these specific cases and their differences is essential to correctly predict the scale and rate of spreading.

Arctic conditions (low temperatures, ice surfaces) complicate the spreading process, as the oil's viscosity increases, its mobility decreases, and there is potential for partial melting of the ice (due to initial thermal contrast) and subsequent freezing of the oil film.

Low temperatures and the nature of the surface on which the spill occurs have a decisive influence on the behavior of crude oil and motor oil. As they cool, they become more viscous, adhere more strongly to the ice surface (and walls), and maintain the shape of the slick for longer. Additionally, the structure and properties of the surface itself (such as roughness, hydrophobicity, and pore presence) directly affect the spreading rate and the amount of oil that adheres. Smooth, hydrophilic materials retain relatively small quantities of oil products, whereas rough ice surfaces, especially those with numerous cracks and pores, can hold considerable volumes, forming stagnation zones. This reduces the "active" mass involved in spreading and complicates both the theoretical description of the process and practical cleanup efforts.

The research was carried out within the framework of State Assignment 124,012,500,441-6 (FFGN-2024-0007).

References

Chaplina TO, Kistovich AV, Pachnenko VP (2024) Axisymmetric spreading of petroleum products on the ice cover. Doklady Earth Sci, издательство Maik Nauka/Interperiodica Publishing (Russian Federation) 516(1):851–856. https://doi.org/10.1134/s1028334x24601007

Dickins D (2011) Behavior of oil spills in ice and implications for arctic spill. Response. https://doi.org/10.4043/22126-MS

Documents & Guides. The international tanker owners pollution federation limited, ITOPF. www.itopf.org

Fingas M (2017). Oil Spill Science And technology. Elsevier. https://doi.org/10.1016/C2015-0-04851-1

Gong Y, Zhao X, Cai Z, O'Reilly SE, Hao X, Zhao D (2014) Mar Pollut Bull 79(1–2):16–33. https://doi.org/10.1016/j.marpolbul.2013.12.024

Hoult DP, Suchon W (1970) The spread of the oil in a channel. Fluid Mech Lab, Department of Mechanics Engineering, Massachusetts Institute of Technology, p 25

Kistovich AV, Chaplina TO (2023) Energy approach to the description of the dynamics of hydrocarbon spills. Phys Fluids 35:072109. https://doi.org/10.1063/5.0159629

Konno A, Izumiyama K, Sakai S (2002) Experimental study on spreading of oil under ice cover. In: Proceedings of the twelfth international offshore and polar engineering conference, Kitakyushu, Japan, pp 821–826

Mackay D, Leinonen PJ, Overall JCK, Wood BR (1975) The behavior of crude oil spilled on snow. Geology Arctic:10–20

Synthesis and Magnetic Characterization of Dispersed Titanomagnetite Microcrystals Prepared by the Modified Glass–Ceramic Method

Natalia S. Bezaeva, Dmitriy A. Chareev, and Myriam Kars

Abstract Titanomagnetites ($Fe_{3-x}Ti_xO_4$) are common magnetic minerals in igneous and metamorphic rocks and also occur in meteorites. This paper presents a new synthesis method of model rocks—submicron grains of titanomagnetite, quasi-uniformly distributed in a glass–ceramic matrix (much needed for rock magnetic experiments in the laboratory). We used as a basis the glass–ceramic method (Worm and Markert, Phys Earth Planet Inter 46:263–269, 1987) and introduced there some modifications, which allowed synthesizing samples without using a high-temperature gas-mixing furnace and corresponding gas equipment unavailable in most laboratories. We carried our four independent synthesis runs with different synthesis parameters and described the obtained samples: its geometry, density and distribution of titanomagnetite grains in a non-magnetic matrix, samples' magnetic properties such as Curie temperature, magnetic susceptibility, and magnetic hardness. Titanomagnetite inclusions demonstrate a predominantly single-domain behavior and are characterized by low titanium content ($x < 0.1$). The described synthesis method is promising for obtaining artificial (titano)magnetite-bearing rock samples with controlled properties such as titanium content in titanomagnetites, size, concentration, and distribution of magnetic grains in a glass–ceramic matrix and can be used for investigating rock magnetic effects in laboratory conditions.

Keywords Titanomagnetite · Synthesis · Glass–ceramic method · Magnetic properties

N. S. Bezaeva (✉) · D. A. Chareev
Vernadsky Institute of Geochemistry and Analytical Chemistry, Russian Academy of Sciences, Moscow, Russia
e-mail: bezaeva@geokhi.ru

M. Kars
Center for Advanced Marine Core Research (Now: Marine Core Research Institute), Kochi University, Nankoku, Japan

International Ocean Drilling Programme, University of Plymouth, Plymouth, UK

T. Chaplina (ed.), *Processes in GeoMedia—Volume IX*, Springer Geology,
https://doi.org/10.1007/978-981-96-9302-3_32

1 Introduction

Magnetic properties of rocks are carried out by ferrimagnetic minerals therein. The most common ferrimagnetic minerals in terrestrial rocks are Ti-free (Fe_3O_4) and Ti-bearing magnetite ($Fe_{3-x}Ti_xO_4$), hematite (Fe_2O_3), pyrrhotite ($Fe_{1-x}S$), goethite (α-FeO(OH)), and greigite (Fe_3S_4) (Dunlop and Ozdemir 1997). These minerals often occur in natural rocks in form of complicated assemblages, and magnetic properties of such rocks may be hard to interpret. In addition, magnetic properties of rocks are highly sensitive to the chemical composition of constituent magnetic grains, its size, and shape as well as its concentration and distribution within the rock matrix. The later causes interaction or non-interaction between magnetic grains.

To deepen our understanding of rock magnetic effects it is necessary to use artificial samples with well-known or predefined magnetic properties. While choosing or creating a synthesis method of model rock, i.e., sample with magnetic grains quasi-uniformly distributed in a "non-magnetic" matrix it is important to control not only grain chemical composition, but also its size, shape, and uniformity of distribution.

Worm and Markert (1987) suggested a so-called glass–ceramic method of synthesis. This method is widely in use in rock magnetism for laboratory investigations of different physical phenomena and processes (e.g., Khakhalova and Moskowitz 2019). It allows preparing Ti-free, Ti-poor, or Ti–rich magnetite particles quasi-uniformly dispersed in a glass–ceramic matrix using a gas-mixing furnace. Chemical composition of grains is controlled by variation of initial reagents and their proportions as well as application of different gas mixtures in the furnace. Grain size is controlled via variation of annealing time and other synthesis parameters (Worm and Markert 1987). One considerable disadvantage of this method is that it requires a gas-mixing furnace, which is unavailable in most laboratories.

Titanomagnetite is known to be one of the most common magnetic minerals, which occurs in a variety of geological environments. In this paper we synthesized titanomagnetite grains with low titanium content dispersed in a glass–ceramic matrix. The purpose of this paper was to introduce a new method of synthesis, modified after (Worm and Markert 1987), which does not require a gas-mixing furnace and to characterize in details synthesis products and their magnetic properties.

2 Materials and Methods

2.1 Description of Synthesis

We describe below the method of synthesis of model rocks in form of glass–ceramic matrix with dispersed titanomagnetite particles therein. It also applies for preparation of dispersed Ti-free magnetite particles.

In Worm and Markert (1987), microcrystalline titanomagnetite was produced in two steps. At the first step, a mixture of $CaCO_3$, $NaCO_3$, K2CO_3, $FeC_2O_4{\cdot}2H_2O$, and TiO_2 was pressed into a pellet and annealed in a platinum crucible in a furnace at a temperature of 1400 °C for 3–7 h in a flow of $CO+CO_2$ gas mixture to buffer oxygen fugacity, and, as a consequence, the control of the degree of iron oxidation. At the second step, the resulting glass was left at a temperature of 600–1000 °C for 5–60 h, also in a flow of $CO+CO_2$ mixture. In this case, titanomagnetite precipitated in the glass as a separate phase, and the fixed fugacity of oxygen did not allow the formation of ferric iron. After that, the samples were quenched in liquid nitrogen.

In this paper in order to produce dispersed microcrystalline titanomagnetite particles in a "non-magnetic" matrix (i.e., artificial rock samples) we applied the above-described glass–ceramic method (Worm and Markert 1987) with some modifications, which allowed avoiding the usage of high-temperature gas-mixing furnace. The starting materials were SiO_2 obtained from silica gel annealed at 85 °C, as well as $CaCO_3$, $NaCO_3$, K_2CO_3, $FeC_2O_4{\cdot}2H_2O$, and TiO_2. The resulting substances were mixed in a molar ratio of 27:28:3:2.4:4:1 and thoroughly ground. This mixture was placed in a platinum ampoule, and the platinum ampoule was placed in a quartz tube of a suitable inner diameter sealed at one end. The other end of the tube was connected to a vacuum pump. The opposite end of the tube was heated by the flame of an oxygen burner for several minutes with the formation of a stable white glow. After that, the ampoule with the contents were quenched in cold water. The content of the platinum ampoule was in form of black sinter.

The resulting sinter together with a platinum ampoule was annealed in a quartz ampoule together with a mixture of FeO/Fe_3O_4 or Fe_3O_4/F_2O_3 (Fig. 1). The mixture of iron oxides was spatially separated from the contents of the platinum ampoule. Oxide mixtures buffer the oxygen fugacity and, accordingly, play a role similar to that of the gas mixture in (Worm and Markert 1987). In order to vary titanomagnetite grain size and composition, we ran four independent rounds of synthesis with variable synthesis parameters and synthesized four samples. Synthesis parameters are specified in Table 1.

Thus, we synthesized four different samples of model rocks with titanomagnetite inclusions dispersed in a glass–ceramic matrix. Each sample was annealed together with a specific buffer for a given time and temperature (see Table 1). Full characterization of synthesized samples, including their magnetic properties is presented below in the *Results and discussion* section (see below).

2.2 Experimental Methods and Protocols

Secondary electron (SE) and backscattered electron (BSE) images as well as elemental analyses for samples are acquired at the Vernadsky Institute (Moscow, Russia) using Tescan MIRA 3 scanning electron microscope (SEM) with an energy-dispersive X-ray spectrometer (EDS).

Magnetic properties of samples were studied at Kochi University (Japan) using a magnetic balance, a vibrating sample magnetometer (VSM) and a magnetic susceptibility meter (except for low-temperature measurements of magnetic susceptibility,

Fig. 1 Quartz tube (1) assembled for annealing the sample at a given oxygen fugacity (2nd synthesis step, see text). The platinum ampoule with sample (4) and a small open quartz glass ampoule (2) with an oxygen buffer (3) are inserted into the quartz tube (1). The total length of the large quartz tube (1) is 110 mm and its internal diameter is 10 mm

Table 1 Synthesis parameters for all produced samples

Sample id	Buffer	Annealing time (h)	Annealing temperature (°C)
CD-778	Fe_3O_4/F_2O_3	15	850
CD-779	FeO/Fe_3O_4	15	850
CD-780	FeO/Fe_3O_4	19	750
CD-836	FeO/Fe_3O_4	45	650

see below). A part of VSM measurements were also carried out at the Institute for Rock Magnetism (IRM, Minneapolis, USA).

Thermomagnetic curves were obtained by heating samples up to 600 °C in vacuum using NMB-89 magnetic balance manufactured by Natsuhara-Giken at a heating rate of 10 °C per minute in a magnetic field of 300 mT. The hysteresis loops (with parameters such as saturation magnetization M_s, saturation remanent magnetization M_{rs}, and coercivity B_c) and backfield remanence demagnetization curves (with parameters such as M_{rs} and remanent coercivity B_{cr}), as well as first-order reversal curve (FORC) diagrams were acquired using a Princeton Measurements Corporation vibrating sample magnetometer (VSM) with a maximum magnetic field of 1 T. Low-field magnetic susceptibility (χ_0) was measured using Agico MFK1 kappabridge. The magnetic susceptibility at low temperatures (down to −196 °C) was measured using Agico MFK1 kappabridge at CEREGE (Aix-en-Provence, France).

Conventional first-order reversal curves (FORCs; Pike et al. 1999) were measured for five samples. Sixty-four FORCs were measured for samples CD-778 and CD-779 with a 1 T saturating field (100 ms averaging time; 4 mT field increment, B_u: −50 to +50 mT, B_c: 0–120 mT). Seventy FORCs were measured for samples CD-780, CD-836 (2 subsamples) with a 1.2 T saturating field (100 ms averaging time; 8.2 mT field increment, B_u: −100 to +1000 mT, B_c: 0–300 mT). Regular FORC diagrams with evenly spaced field steps were processed using the FORCinel software (Harrison and Feinberg 2008), with constant smoothing factor (SF).

3 Results and Discussion

3.1 Description of Synthesized Samples

As described above, we conducted four independent synthesis runs (see Table 1 for synthesis parameters) and each time obtained a single sample of irregular shape and several millimeters in size. Each of the four samples was divided into subsamples and used for both magnetic measurements (see below) and preparation of polished sections for scanning electron microscopy. Figure 2 shows the combined SE-BSE images for samples CD-778 and CD-779 (see Table 1).

The lightest phase in the images represent titanomagnetite inclusions (also indicated by arrows, see Fig. 2). Sample CD-778 contains two generations of titanomagnetite grains. The first generation is titanomagnetite inclusions in pores (see "*1*" in Fig. 2a). The second generation most likely formed later and is composed of titanomagnetites that precipitated as a rim along the edges of individual fragments of the glass–ceramic matrix (see "*2*" in Fig. 2b). Titanomagnetite inclusions in all four synthesized samples are of submicron size, which corresponds to single-domain/pseudo-single-domain titanomagnetite grains (Dunlop and Özdemir 1997). Elemental analyses (EDS) confirmed titanomagnetites, however, it was not possible to determine their chemical composition and the proportion of titanium because of their submicron size. Due to the micron size of the electron beam itself and the excitation zone, the composition of titanomagnetite grains is averaged with the composition of the adjacent glass–ceramic matrix during elemental analysis, which leads to deviation of the final analyses. Therefore, such analyses are not presented in this paper. The glass–ceramic matrix in all samples is quite porous (Fig. 2) and in some cases appears heterogeneous. This does not affect magnetic analyses, since the glass–ceramic matrix does not contribute to magnetic measurements. In terms of chemical composition, the matrix does not correspond to any of the known natural rocks, however, it is convenient for creating model rocks, since it is resistant to heating up to hundreds of degrees.

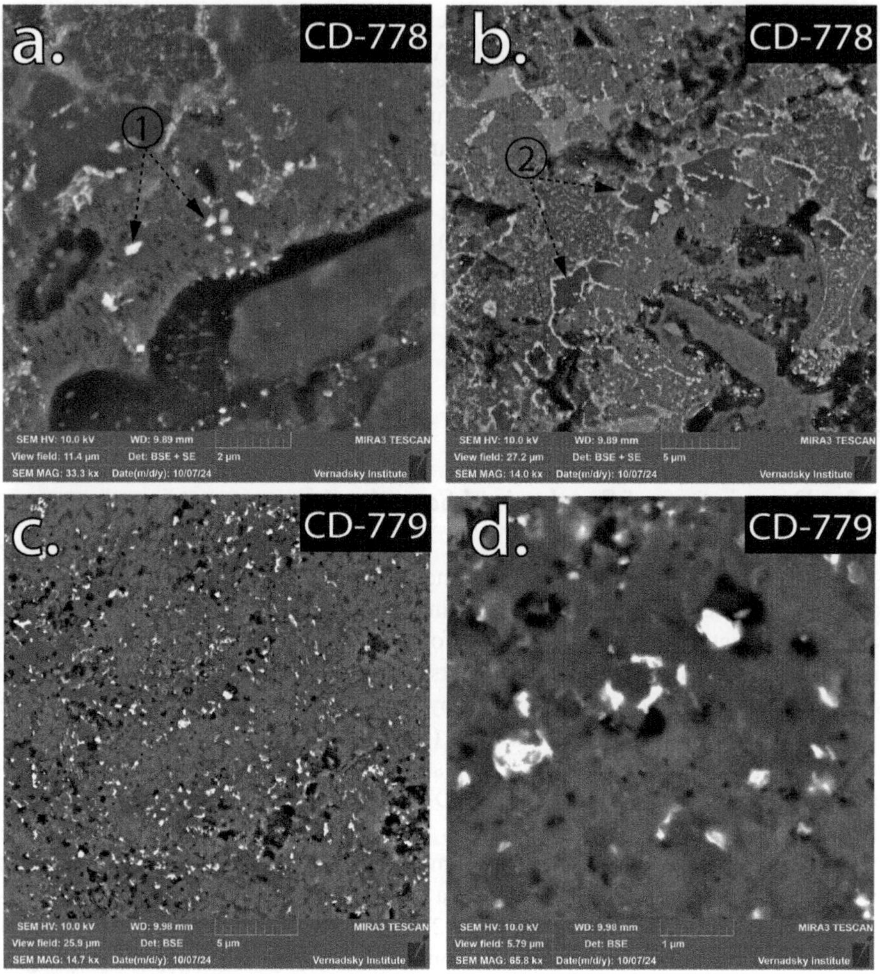

Fig. 2 Combined SE-BSE images of CD-778 и CD-779 samples (see Table 1). Light inclusions are titanomagnetite grains. *1*—titanomagnetite inclusions of first generation, *2*—titanomagnetite inclusions of second generation (see text for details)

3.2 Magnetic Properties of Synthesized Samples

Thermomagnetic curves $M_s(T)$ were measured for all samples up to 600 °C. In order to avoid oxidation of the samples during heating, the samples were heated under vacuum. Figure 3 shows the thermomagnetic curves (heating–cooling cycle) for the sample CD-778 (Table 1).

The Curie temperature determined from the heating curve is <570 °C, which is very close to the Curie point of pure magnetite (580 °C) and indicates a very low titanium content in titanomagnetites $x < 0.1$ (Lattard et al. 2006). Titanium content

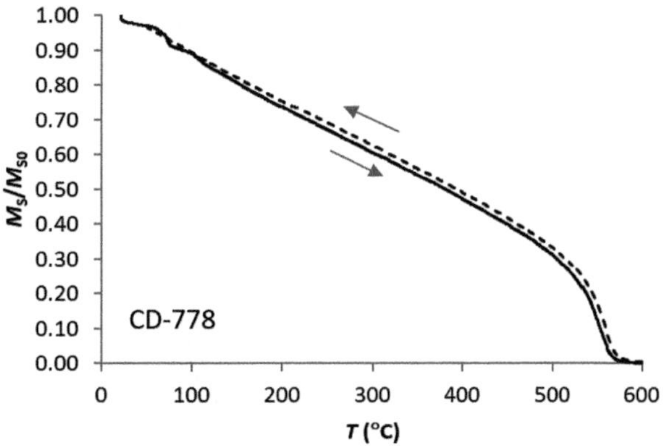

Fig. 3 Normalized thermomagnetic curves for the sample CD-778 (Table 1). The experiment was conducted in vacuum in the presence of a magnetic field of 0.3 T. The heating curve is shown as a solid line and is indicated by an arrow. The cooling curve is shown as a dotted line and is also indicated by an arrow

in titanomagnetite inclusions of the other samples is also low, as indicated by the corresponding Curie points of the samples. At the same time, measurements of the magnetic susceptibility of samples in the temperature range from room temperature to liquid nitrogen temperature (−196 °C) did not reveal a low-temperature phase transition in magnetite also known as Verwey transition (Verwey 1939), which is observed in magnetite at temperatures of ~120 K (−153°C). It is known that the Verwey transition is suppressed in magnetite while titanium content $x > 0.1$ (Dunlop and Özdemir 1997).

Figure 4 displays a characteristic hysteresis loop (Fig. 4a) and backfield remanence demagnetization curve for the subsample of the sample CD-778 (Table 1). Table 2 displays the magnetic characteristics of this sample and the other three synthesized samples. The hysteresis loops and backfield remanence demagnetization curves for other three samples are similar (see Table 2 for details).

The parameters in Table 2, namely the B_{cr}/B_c and M_{rs}/M_s ratios, indicate the presence of pseudo-single-domain grains in the synthesized samples (Dunlop and Özdemir 1997). However, FORC diagrams (Fig. 5) give another insight.

FORC diagrams are displayed in Fig. 5. Samples CD-778 and CD-779 show measurement noise at low B_c values. Despite this analytical artifact, it is possible to describe sample CD-778 with a single-domain (SD) signature showing closed contour lines and a negative region at the bottom-left corner of the FORC diagram. Its coercivity peak is between 30 and 40 mT. The contour lines are widespread and divergent at low coercivities suggesting the presence of coarser magnetic minerals (PSD-size). The measurement noise is more dominant for sample CD-779, which makes its description and interpretation more challenging. Samples CD-780 and both subsamples of CD-836 clearly exhibit a SD signature (closed contour lines, and bottom-left corner negative region) with a coercivity peak around 60–70 mT. The contour lines are slightly divergent toward the very low coercivities, which may suggest the presence of coarser magnetic minerals.

The presence of SD signature is in line with single-domain threshold for titanomagnetite: 200–600 nm depending on the titanium content (Dunlop and Özdemir 1997) and the results of studying the samples using scanning electron microscopy

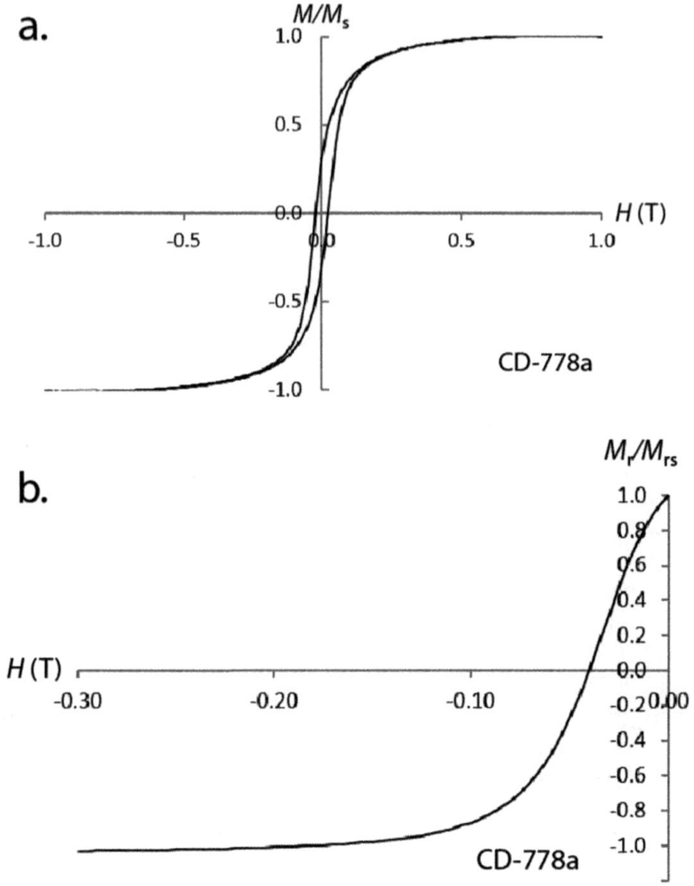

Fig. 4 **a** Hysteresis loop measured on sample CD-778a (fragment of sample CD-778, see Table 1), M is magnetization, M_s is saturation magnetization; **b** Backfield remanence demagnetization curve measured on sample CD-778a. M_r is remanent magnetization, M_{rs} is saturation remanent magnetization, H is magnetic field strength

Table 2 Magnetic properties of synthesized samples. m is mass in mg, χ_0 is magnetic susceptibility in m^3/kg, B_c is coercivity in mT, B_{cr} is remanent coercivity in mT, M_s is saturation magnetization in Am^2/kg

Образец	m	χ_0	B_c	B_{cr}	M_s	B_{cr}/B_c	M_{rs}/M_s
CD-778a	57.8	22.4×10^{-6}	22.5	40.4	2.73	1.8	0.306
CD-779a	114.2	2.71×10^{-6}	28.6	56.3	0.46	2.0	0.246
CD-780a	102.7	11.6×10^{-6}	39.8	72.6	2.83	1.8	0.288
CD-836a	93.4	5.2×10^{-6}	41.2	74.9	0.99	1.8	0.315

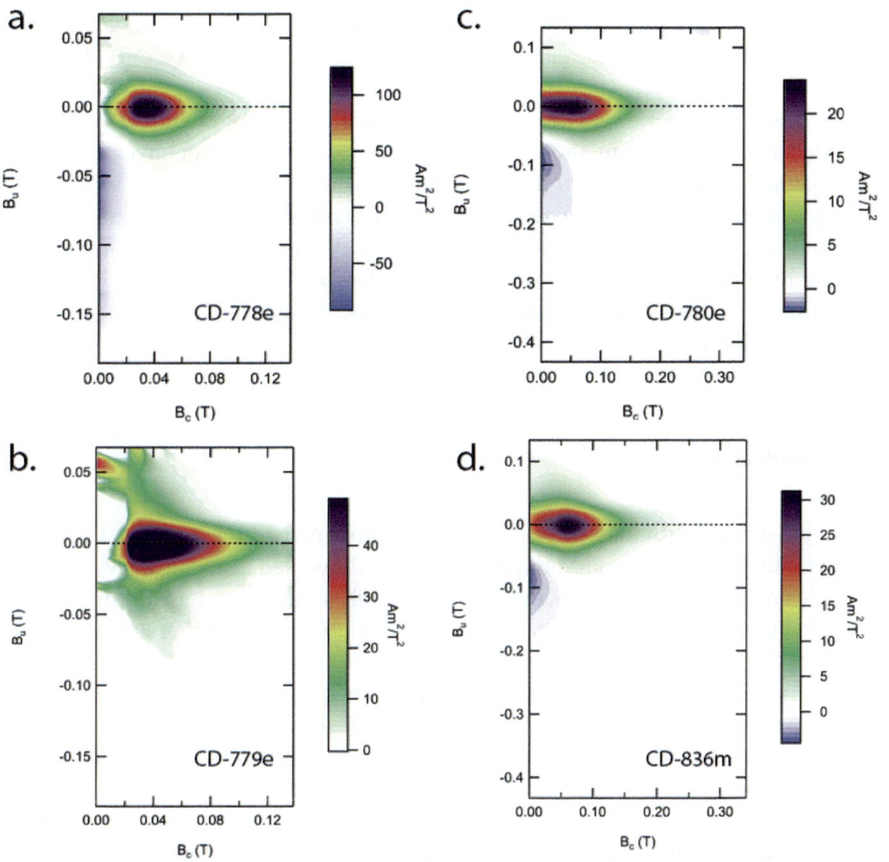

Fig. 5 First-order reversal curve (FORC) diagrams for synthesized samples **a** CD-778e, **b** CD-779e, **c** CD-780e and **d** CD-836 m (see Tables 1 and 2 for synthesis parameters and rock magnetic parameters, respectively). FORC diagrams were processed using the FORCinel software (Harrison and Feinberg 2008), see text for details

(see the previous section *Description of synthesized samples*). Overall, the analyzed samples show a dominant SD signature, with a probable mixture of SD and PSD grains.

In this paper, it is shown that the glass–ceramic synthesis method (Worm and Markert 1987) can be adapted for the synthesis of model rocks without the use of high-temperature gas-mixing furnaces with a controlled gas atmosphere. In such case, the magnetic properties of the synthesized samples, as well as the size and distribution of titanomagnetite grains in the glass–ceramic matrix, vary depending on the synthesis parameters (see Table 1). Thus, the titanium content in titanomagnetites can be influenced by varying the quantity and quality of the starting materials, and in particular titanium oxide (if magnetite without titanium impurities is required, titanium oxide is not used). The size, quantity, and distribution of magnetic grains in the glass–ceramic matrix, as well as the density/porosity of the matrix, are directly affected by the temperature and annealing time of the samples (see Table 1 and Fig. 2).

Thus, in this paper, we presented an accessible to many chemical and material science laboratories method of synthesis of model rocks for investigation of rock magnetic effects, when a model rock represents titanomagnetite grains with given parameters (size, shape, concentration), uniformly dispersed in a glass–ceramic matrix. Such samples are suitable for any laboratory experiments including experiments at high pressures and temperatures as well as particle radiation for laboratory simulation of various physical effects on rocks and extraterrestrial matter on Earth and in the Solar system.

4 Conclusions

1. In this work, a previously known glass–ceramic method for synthesizing model rock samples (Worm and Markert 1987) is modified and tested in a new capacity, which made it possible to exclude the use of high-temperature gas-mixing furnace with a controlled atmosphere, which is unavailable in most laboratories.
2. Using the modified glass–ceramic synthesis method, we carried out four independent runs of synthesis of titanomagnetite grains quasi-uniformly distributed in a glass–ceramic matrix (model rock). Each synthesis run had different parameters, which allowed variation in the size and shape, concentration, and distribution, as well as the chemical composition of the magnetic grains in the sample.
3. Titanomagnetite grains in the synthesized samples have a submicron size, low titanium content ($x < 0.1$), and demonstrate predominantly single-domain behavior, which is consistent with the results of observations on a scanning electron microscope. At the same time no Verwey transition is observed in the samples.
4. The modified method of synthesis of model rocks with inclusions of magnetite or/and titanomagnetites with different titanium content is promising for synthesizing model rocks with given magnetic properties for investigating rock magnetic effects in laboratory conditions.

Acknowledgements This work was conducted under the GEOKHI RAS state assignment.

Conflict of Interest All authors declare that they have no conflicts of interest.

References

Dunlop DJ, Özdemir Ö (1997) Rock magnetism. In: Fundamentals and frontiers. Cambridge University Press, Cambridge, p 573. https://doi.org/10.1017/CBO9780511612794

Harrison RJ, Feinberg JM (2008) FORCInel: an improved algorithm for calculating first order reversal curve distributions using locally weighted regression smoothing. Geochem Geophys Geosyst 9:Q05016. https://doi.org/10.1029/2008GC001987

Khakhalova E, Moskowitz BM (2019) High-temperature observation of transdomain transitions in vortex states in intermediate titanomagnetite. J Geophys Res Solid Earth 124:7604–7615. https://doi.org/10.1029/2019JB017857

Lattard D, Engelmann R, Kontny A, and Sauerzapf U (2006) Curie temperatures of synthetic titanomagnetites in the Fe–Ti–O system: Effects of composition, crystal chemistry, and thermomagnetic methods. J Geophys Res Solid Earth 111:B12S28. https://doi.org/10.1029/2006JB004591

Pike CR, Roberts AP, Verosub KL (1999) Characterizing interactions in fine magnetic particle systems using first order reversal curves. J Appl Phys 85(9):6660–6667. https://doi.org/10.1063/1.370176

Verwey EJW (1939) Electronic conduction of magnetite (Fe_3O_4) and its transition point at low temperature. Nature 144:327–328

Worm H-U, Markert H (1987) The preparation of dispersed titanomagnetite particles by the glass-ceramic method. Phys Earth Planet Inter 46(1–3):263–269. https://doi.org/10.1016/0031-9201(87)90189-0

References

Changes in Total Ozone Content, Ultraviolet Radiation Fluxes, and Cosmic Rays Over Northern Europe

A. V. Kholoptsev, R. G. Shubkin, and G. F. Batrakov

Abstract Among the factors that affect the oncological morbidity of the population in the Northern European regions, there are changes in the fluxes of incoming ultraviolet radiation. One of the reasons for these changes is changing distribution of the total ozone content above them. It is hypothesized that changes in the total ozone content and ultraviolet radiation fluxes may be caused by variations in the cosmic ray flux. The purpose of this work is to verify the hypothesis, as well as to identify areas of Northern Europe for which the influence of these factors is significant. It was established that the hypothesis is valid for many parts of Northern Europe. The sites, for which the change in their total areas and the corresponding signs of correlation between the studied processes allows them to be associated with the movement of the North Geomagnetic Pole occurring in the same years, were identified.

Keywords The territory of Northern Europe · Total ozone content · Ultraviolet radiation flux · Cosmic ray flux · Relations · Trends · Spring months

1 Introduction.

Changes in the total ozone content (hereinafter TOC) over any area of the Earth's SRFface largely determine variations in the flux of biologically active, ultraviolet component of solar radiation (hereinafter SRF), which can have a significant impact on public health and is involved in the formation of ground-level ozone. A decrease in TOC by only 1% leads to a 2% increase in cancer incidence in the population (McMichael et al. 2010). Therefore, the decrease in the average value of the

A. V. Kholoptsev
Sevastopol Branch of FSBI "N. N. Zubov State Oceanographic Institute", Sevastopol, Russia

R. G. Shubkin
FSBEI HE "Siberian Fire and Rescue Academy", Zheleznovodsk, Krasnoyarsk Krai, Russia

G. F. Batrakov (✉)
FSBI FRC "Marine Hydrophysical Institute of RAS", Sevastopol, Russia
e-mail: batrakov@gmail.com

T. Chaplina (ed.), *Processes in GeoMedia—Volume IX*, Springer Geology,
https://doi.org/10.1007/978-981-96-9302-3_33

TOC for the entire Earth's atmosphere, revealed in the second half of the twentieth century, caused concern of the world community and led to a number of international agreements aimed at stopping this process.

At the beginning of the twenty-first century, the ozone layer over many regions of the world began to recover. This phenomenon could be caused both by the measure's taken to limit emissions of ozone-depleting substances into the atmosphere and by the action of some natural factors. The question of how significant was the influence of such factors on the TOC changes that took place is not sufficiently studied. Therefore, the development of existing ideas about features of their action over these or those regions and identification of probable trends of further changes in their characteristics is an actual problem of physical geography, meteorology, as well as ecology.

The solution of this problem is of the greatest interest for regions that are densely populated and economically developed, in which the rates of ozone layer recovery occurring in the twenty-first century have been reduced. One of them is Northern Europe, which includes a significant part of the European territory of Russia, located north of Moscow, as well as the territories of the Republic of Belarus, the Baltic States, Scandinavia, and Denmark.

While over the world regions located in the latitudinal belt (60°S–60°N), over the period from 2000 to 2024, the mentioned ozone layer recovery led to an increase in the average TOC to values only 2% inferior to its level in the 1980s, for the countries of Northern Europe these values are now 3.5% below the specified level (https://ozone.unep.org/sites/default/files/2023-09/Ozone_and_UV_Bulletin_1_ru.pdf; https://www.wmo.int/pages/prog/arep/WMOArcticOzoneBulletins2016.html; https://www.unep.org/ru/novosti-i-istorii/press-reliz/ozonovyy-sloy-nakhoditsya-na-puti-k-vosstanovleniyu-chto-pomozhet). As a result, the fluxes of ultraviolet radiation dangerous for public health are also higher here.

For many years, domestic (Alexandrov et al. 1982; Larin 2000; Nikiforova et al. 2016; Zvyagintsev et al. 2013; Khrgian 1973; Kholoptsev et al. 2014; Polyakov et al. 2024) and foreign scientists (Muller 2012; UNEP 2010; Hoffman 2005; Mohanakumar 2011; Lang et al. 2013; Ali et al. 2012; Zepp et al. 2003; Dessler 2000) have been studying the causes of changes in the TOC over various parts of the European territory of Russia and other parts of the Territory of Northern Europe (hereinafter TNE).

It has been established that these processes are significantly affected by changes in the intensity of air exchange between different stratospheric layers caused by changes in the characteristics of atmospheric circulation and other processes (Kholoptsev et al. 2014; Mohanakumar 2011).

The decrease of TOC over some territory is usually observed at warming of the troposphere above it and cooling of the stratosphere (Kholoptsev et al. 2014).

In the months from October to May, a Polar Vortex is formed in the stratosphere above the TNE, in which the air moves at a speed of up to 250 km/h and, as a rule, from west to east. Sometimes the movement of this air temporarily changes its direction, which is associated with a sudden warming of the stratosphere (https://new-science.ru/arkticheskij-polyarnyj-vihr-polnostju-izmenil-napravlenie-vyzvav-rekordnyj-pik-ozona-v-regione/).

The mentioned stratospheric warming occurs at heights of 10–50 km. It is caused by Rossby waves, which arise in the atmospheric air flows as a result of the planet rotation and other processes (Kolstad et al. 2010; Kishore et al. 2009). In the stratosphere above the TNE, these waves interact with the Polar Vortex existing in the indicated months and are capable of causing the mentioned changes in the direction of its air movement.

Such changes are one of the mechanisms generating the vorticity of the considered air current, which affects the delivery to the corresponding stratospheric layers of both ozone-depleting substances and ozone itself (Mohanakumar 2011).

Above the areas of the Earth's SRF face where the flux of the mentioned substances coming from the troposphere is sufficiently large, activation of air exchange in the stratosphere leads to a decrease in TOC. Above the areas where such a flux is small, the result of this process may be opposite (Kholoptsev et al. 2014).

Some influence on the formation of the Polar Vortex vorticity can be exerted by changes in the ionization of the air transported by it. The greater the concentration of ions in this air, the stronger the Lorentz force influences its motion, the greater the vorticity and the more intensive the exchange of substances between stratospheric layers.

The TOC changes over some area of the Earth's SRFface are caused by changes of ozone concentrations in all layers of the stratospheric air column located above it. In each such layer, the ozone concentration increases at the time moments when its real balance is positive. The latter takes place if the increase of the ozone amount in the layer under consideration (hereinafter referred to as the inflow) recorded per time unit exceeds the total loss of this substance (hereinafter referred to as the outflow) (https://new-science.ru/arkticheskij-polyarnyj-vihr-polnostju izmenil-napravlenie-vyzvav-rekordnyj-pik-ozona-v-regione/).

The inflow consists of the amount of ozone formed during the direct reaction of the Chapman cycle and its amount entering the layer under consideration during air exchange with the stratospheric layers located above. The consumption is formed from ozone destroyed in the layer under study and ozone that migrated during air exchange to the stratospheric layers located below (Khrgian 1973).

Ozone destruction in each such layer occurs during the reverse reaction of the Chapman cycle, as well as reactions of the catalytic cycles of the hydrogen, nitrogen, and halogen cycles, which involve substances entering it from below and formed either in the troposphere (nitrogen oxides), on the Earth's SRFface (halogens), or in the subSRFface (hydrogen) (Dessler 2000). Consequently, as the ionization intensity of stratospheric air increases over different regions of the Earth's SRFface, the TOC above them can both increase and decrease.

The main mechanism of stratospheric air ionization is the impact on it of showers of secondary particles formed at the entrance to the Earth's atmosphere of cosmic rays (Maurchev et al. 2019), which are mainly protons with energies up to 10^{16} eV (Stozhkov et al. 2022). Obviously, changes in the fluxes of such cosmic rays (hereinafter referred to as FCR) can have some influence on the dynamics of both the TOC and the SDP. Nevertheless, the areas of the Earth's SRFface where this influence is significant have not been identified.

The energies sufficient to overcome the protective shield created by the geomagnetic field (Jones 2011) are usually possessed by protons arising from various astrophysical processes occurring outside the solar system. Therefore, their flux penetrating into the heliosphere and reaching the Earth's orbit is regulated by solar activity due to the Forbush effect (Erlykin and Wolfendale 2011; https://sites.lebede3v.ru/DNS_FIAN; Masrsh and Svenmark 2000; Todd and Kniveton 2001; Veretenenko and Pudovkin 1996; Raspopov and Veretenenko 2009).

The minimum proton energy sufficient for their entry into the Earth's atmosphere over some TNE site depends on the distance between it and the North Geomagnetic Pole (hereafter NGP) (Maurchev et al. 2019; Stozhkov et al. 2022; Erlykin and Wolfendale 2011; https://sites.lebede3v.ru/DNS_FIAN; Golenkov et al. 1980).

The location of the SGP was first established by James Ross in 1831 (1834). At that time, the pole was located at Cape Adelaide, located on the west coast of the Butia Peninsula (Canadian Arctic Archipelago), which belongs to the Territory of Nunavut (Canada). It was further established that the geomagnetic poles of our planet move. Thus, the SGP since 1900 has been moving toward the Taimyr Peninsula (Russia) along a trajectory that is practically rectilinear (https://travelask.ru/blog/posts/13546-kuda-dvizhetsya-severnyy-magnitnyy-polyus-i-kogda-polyusa-sn; https://web.archive.org/web/20220427002623; https://www.ngdc.noaa.gov/geomag/data/poles/NP.xy). In the period after 2000, the speed of this movement increased to 55 km/year.

At the same time, until 2017, the NGP was also approaching the North Geographic Pole, moving away from the Polar Vortex zone. In 2017, the NGP crossed the Date Change Line and began approaching the Polar Vortex area over Russia.

According to the estimates of the British Geological SRFvey (BGC), in recent years, the speed of the SGP motion has slightly decreased, but its direction has not changed (https://www.rbc.ru/rbcfreenews/67394e8b9a794716d3a3cb81).

Since the process that generates the geomagnetic field is very inertial (Reshetnyak 2018), it is very likely that the motion of the NGP in the same direction will continue in the coming decades. As a result, the distance between the NGP and the section of the Polar Vortex located over the territory of Russia will further decrease, and the FCR entering the atmosphere above the TNE will increase.

The above allows us to assume that there are areas of the TNE where the variations of FCR variations had a significant influence on the changes of mean monthly TOC values for spring months, and the strength of the links between these processes changed in accordance with the displacement of the NGP.

The put forward hypothesis is not trivial, since the changes of TOC in the atmosphere over different TNE sites are also significantly influenced by changes in their local climate and processes causing sudden stratospheric warming. Therefore, its verification is not only of theoretical but also of practical interest.

The state of the ozone layer above TNE has been monitored since 1978. Initially, the TOMS (Total Ozone Mapping Spectrometer) instruments deployed on the Nimbus satellite (NASA) were used (Ionov et al. 2002). From 2005 to the present day, monitoring is continued by the Aura (NASA) satellite, which is equipped with a more advanced OMI (Ozone Meter Instrument) spectrometer.

The World Data Bank of TOC observations covers the entire SRFface of the Earth. It contains daily mean TOC values for grid nodes with a step of 1° × 1° (Muller 2012). This information is also taken into account in the ERA 5 reanalysis, which presents information on the "ozone stock"—the total mass (M, in kg) of its molecules in the atmospheric column over a 1 m^2 area of the Earth's SRFface. This indicator is related to TOC by the relation TOC $= = 100000 * M / \rho$, where ρ is the density of ozone at normal conditions equal to 2.1445 g/dm^3.

For many decades, hydrometeorological stations of Roshydromet have been monitoring changes in PSD with wavelengths of 0.2–0.4 μm in many parts of the European territory of Russia. Similar monitoring is carried out on the territory of other Northern European countries. The results obtained are taken into account in the development of a number of global atmospheric reanalyses, one of the most advanced of which is ERA 5 (Hólm et al. 2020) supported by COPERNICUS.

Monitoring of the variations of cosmic ray fluxes (FCR) entering the atmosphere over the European territory of Russia at the Dolgoprudny Research Station of the P.N. Lebedev Physical Institute of the Russian Academy of Sciences has been carried out since 1957, and its results obtained for the period up to and including 2022 are presented in https://sites.lebede3v.ru/DNS_FIAN.

Nevertheless, the hypothesis has not been verified earlier. The latter significantly limits the possibilities of applying the results of the mentioned monitoring for long-term forecasting of changes in the TOC and SRF, as well as for developing meaSRFes to protect the population from their consequences.

Verification of the put forward hypothesis is the purpose of this paper. In order to achieve it, the following tasks were accomplished:

1. Determination of the trends of TOC and SRF variations over different TNE sites occurring in spring months over the period from 1998 to 2019;
2. Identification of TNE sites for which interannual TOC variations corresponding to spring months of 1998–2022 were significantly related to the variations of FCR entering the atmosphere over Dolgoprudny, Moscow region.

2 Factual Material and Methodology of the Study

In order to identify the peculiarities inherent in the tendencies of interannual changes in TOC and SRF over the TSN, we considered changes in the states of these processes over all its sections corresponding to the nodes of the ERA 5 reanalysis grid, so the information of the ERA 5 reanalysis (https://cds.climate.copernicus.eu/cdsapp!/dataset/reanalysis-era5-presSRFe-levels?tab=form) was used as actual material on the changes in TOC and SRF for the spring months. As a source of information on the FCR changes in the period 1998–2022, the materials of the Dolgoprudnenskaya research station (https://sites.lebede3v.ru/DNS_FIAN) were used.

When solving the first problem, it was taken into account that the solar activity controls the PCL changes. Therefore, as a characteristic of the tendencies of changes in TOC and SRF for this or that spring month occurring over this or that point, we considered the value of the difference between the mean values of these indices, which were calculated for cycles 24 (2009–2019) and 23 (1997–2008) of solar activity.

When solving the second problem, the value of the correlation coefficient of their time series was considered as a characteristic of the relationship between the studied processes, which was evaluated for the time intervals of 1998–2012 and 2008–2022 with a length of 15 years each.

Therefore, in order to identify significant relationships between the studied processes, a technique based on the method of correlation analysis was applied. Linear trends present in the time series under consideration are compensated.

The decision on the significance of the relationship between the FCR series, as well as the TOC and SRF series was made if its reliability was at least 0.9. The latter occurred if the value of the correlation coefficient of the compared series exceeded modulo the level of 90% significance threshold, which was determined by Student's criterion taking into account the number of degrees of freedom. The plots for which the reliability of conclusions about the significance of the relationship under consideration exceeded 0.95 and 0.99 were also taken into account.

The values of the indicators under consideration were determined for all spring months and for each TNE site that corresponds to one or another node of the ERA 5 reanalysis grid.

As follows from the above, the method under consideration makes it possible to test the hypothesis and, for the time periods in which the mean distances from the SGP to the territory of Russia differed significantly, to compare the locations of the TSE sites where the variations of the TOC and SRF were significantly related to the synchronous variations of the FCR with respect to them. Thus, it allows us to verify the presence of a significant influence of the location of NGP on the studied processes and to make an assumption about the trends of changes in their characteristics at least in the next few years.

3 The Results of the Study and Their Analyses

In accordance with the described methodology for solving the first problem, the mean TOC and SRF values were determined for cycles 23 and 24 of solar activity and all TNE sites, and the values of their difference corresponding to the months from March to May were calculated.

The areas of water areas and territories of Russia, where the difference of mean TOC values for solar activity cycle 24 with respect to cycle 23 for the months of March–May took these or those values, are shown in Fig. 1.

As follows from Fig. 1a, for March, the TNE is dominated by the areas over which the TOC have significantly increased over 1998–2019. The most significant increase occurred over the Svalbard archipelago (by 14.6 Dobson units).

In the mainland part of the TNE, the increase of TOC values for the period under consideration reached the largest values (not less than 9 Dobson units) over the territories of the Murmansk and Arkhangelsk regions. A significant decrease in TOC

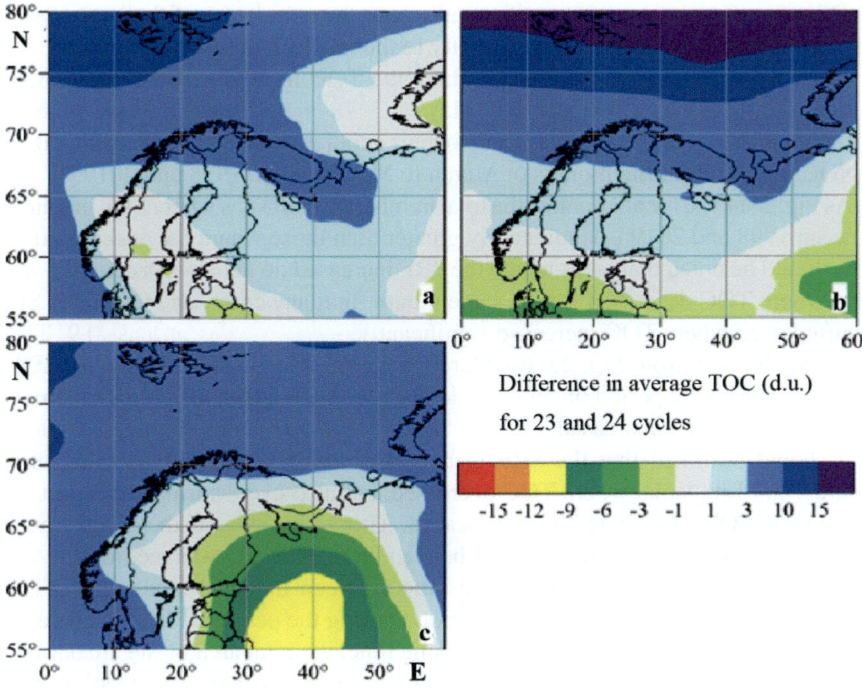

Fig. 1 Areas of the Northern European territory where certain trends of interannual changes in TOC prevailed in the period 1998–2019 for the months: **a** March; **b** April; **c** May

was revealed only in the territories of the Chelyabinsk Oblast, Belarus, Lithuania and Sweden.

Figure 1b shows that for April the distribution of TOC changes by TNE has a zonal character. The largest increase in CCA (by at least 20 Dobson units) was detected for Pt. Barentsburg (Spitsbergen), which is located practically at the northern boundary of the study region.

The largest decrease of this indicator corresponds to the sites of the Chelyabinsk Oblast territory, which is located near its southern border (8.57 Dobson units).

It follows from Fig. 1c that for May in the study region a significant increase of TOC also prevails. This is typical of the Russian regions belonging to the Ural Federal District, the northern part of the territory of the Murmansk Oblast, and also for the coast of Norway and the territories of Denmark, Norway and Sweden in the area of the Danish Straits (Skagerrak, Kattegat, etc.).

It should be noted that the total area of the sites, where the TOC decreased by at least 10 Dobson units, for May is maximum. Dobson units is maximum for May. Such sites mainly belong to the territory of Russia (Central and to a lesser extent the North-Western Federal District) and also Belarus. One of the mentioned sites corresponds to the city of Moscow.

This phenomenon poses a significant threat to the population of these territories, as it has probably contributed significantly to the increase in cancer morbidity.

Thus, the total area of TNE sites, where the TOC in 1998–2019 is decreasing, increased significantly from March to May.

Figure 2 shows the TNE sites where one or another trend of interannual changes in NOC prevailed for the months of March to May between 1998 and 2019.

As Fig. 2 shows, for all months, the total number of TNE sites where SRF increased between 1998 and 2019 is significantly greater than those where they decreased.

It should be noted that areas where the SRF increased do not always coincide with areas where TOC decreased for the same month. In many cases, an increase in SRF occurred even where TOC increased significantly.

As can be seen from Fig. 2a, for March, the total area of SRF plots where TOC increased during the period under consideration is much larger than the total area of its plots where TOC decreased.

It should be noted that the SRF increased not only over Chelyabinsk Oblast, Belarus, and Lithuania, where TOC decreased, but also over Murmansk Oblast and Arkhangelsk Oblast, where CCA increased. This is possible if the cloud optical density decreased over the territories where an increase in TO was recorded.

Figure 2b shows that for April, the ratio between the total areas of the TNE sites where the TOC increased and decreased during the period under consideration remained the same as for March. At the same time, the total area of the areas where the TNE decreased significantly increased in comparison with the same indicator for March.

A decrease in the SRF was recorded over the territories of the regions of the Northwestern and Volga Federal District of Russia, as well as Finland, which satisfactorily correspond to the areas of the TNE where there was an increase in TOC.

Over the southern part of the TNE, where the TOC decreased (Fig. 1b), the SRF, as one could expect, increased. However, over many areas of the Arctic where there was an increase in TOC for April, the SRF for that month also increased.

Figure 2c shows that for May, a significant increase in TOC was recorded over the entire European territory of Russia, the Baltic States, Finland, and many parts of the Scandinavian countries and Denmark.

As can be seen from the comparison of Figs. 2c and 1c, this phenomenon was recorded over all TNE areas, where in the same years there was a decrease in TOC. At the same time, as for the previous months, for May, the increase in the SST was detected over many areas and water areas where the TOC increased.

Thus, it has been established that in spring months, an increase in SRF was recorded at the predominant part of the TNE for 1998–2019. This phenomenon occurred at all TNE sites, where a decrease in TOC was detected for the corresponding month.

In addition, an increase in the TOC increase was also detected at many sites where the SRF increased, mainly in the water areas of certain seas or their coastal territories.

When solving the second problem, it was found that in the periods 1998–2012 and 2008–2022, the TNE sites where the correlation of interannual changes in TOC

Fig. 2 TNE plots, where in the period 1998–2019 one or another tendency of interannual changes in SRF (monthly UV fluxes) prevailed for the months: **a** March; **b**April; **c** May

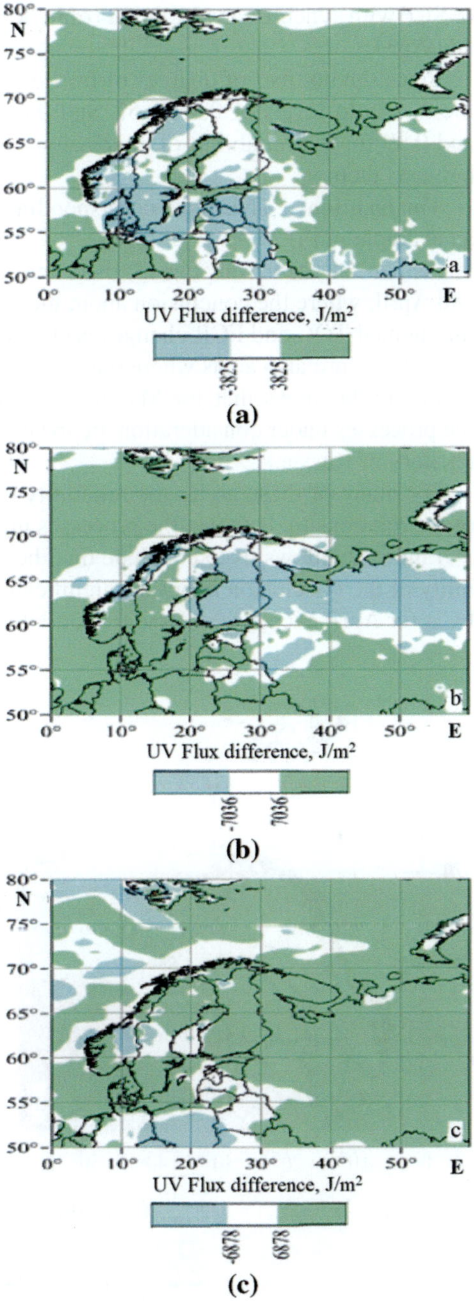

or SRF with synchronous variations of FCR is significant were found only for March and April.

Since the number of degrees of freedom of the studied processes is equal to 15, the threshold levels of 0.45, 0.52, and 0.63 correspond to the values of 0.9, 0.95, and 0.99 for the reliability of conclusions about the significance of the correlation between them.

The mentioned TNE plots corresponding to the period 2008–2022 are shown as an example in Fig. 3.

As can be seen from Fig. 3, there are numerous areas in the TNE for March and April, where the conclusion about the significance of the positive correlation of interannual TOC and FCR changes is characterized by the reliability not lower than 0.95. There are also areas where the reliability of this conclusion exceeds 0.99.

Figure 3a shows that for March the areas of significant positive correlation of the processes under consideration are located over the northern part of the European territory of Russia.

For many of these areas, the reliability of the conclusion about the significance of the relationship between the processes under consideration exceeds 0.99.

Figure 3b allows us to conclude that the required TNE sites, for which the reliability of the conclusion about the significance of the connection between NOC and FCR for April exceeds 0.99 and 0.95, are located only over the North Sea.

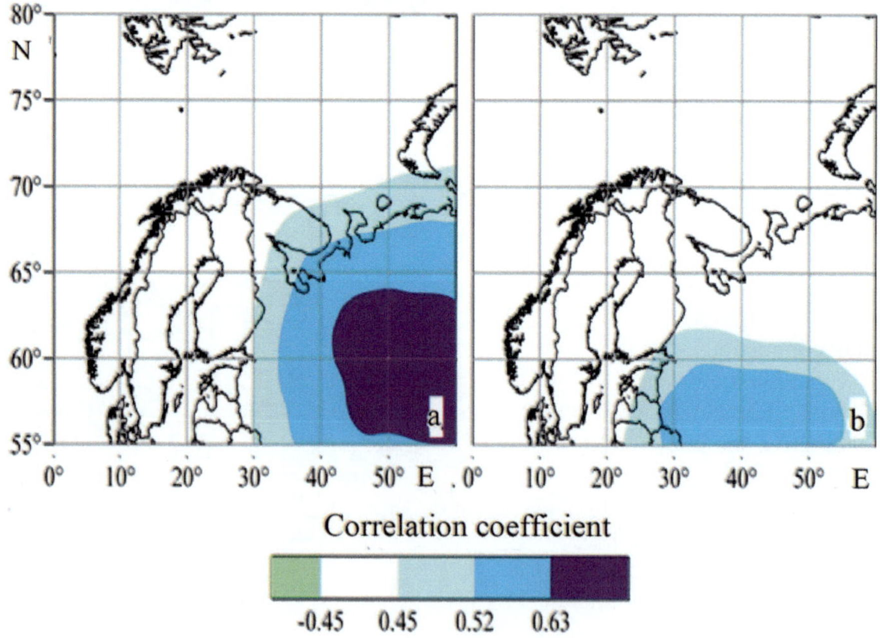

Fig. 3 Areas of Northern Europe, where in the period 2008–2022 the relations of interannual TOC changes with synchronous to them variations of FCR measured at Dolgoprudnenskaya station (Moscow Region) were significant **a** February; **b** March

Directly on land there are numerous TNE sites where the reliability of such a conclusion exceeds 0.9 and 0.95. These sites are mainly located in the territories of the regions of Russia belonging to the Central Federal District, as well as in the territories of Belarus, Estonia, Lithuania, and Denmark.

There are also quite a few TNE sites where the influence of FCR on the process under study in March and April is not significant. This is quite natural, since the TOC changes in these months can be also caused by sudden stratospheric warming and other reasons.

The TNE sites, where the TOC and FCR variations for the period 1998–2012 corresponding to March and April are significantly and positively correlated with each other, are located in many respects similarly. Most of them are also located over the European territory of Russia, the White Sea and the coastal areas of the Barents Sea. In addition, the searched areas also occupy the territory of Finland, the Baltic States and Belarus.

As mentioned above, in the period 1998–2022, the NGP approached the territory of Russia (https://travelask.ru/blog/posts/13546-kuda-dvizhetsya-severnyy-magnitnyy-polyus-i-kogda-polyusa-sn; https://web.archive.org/web/20220427002623; https://www.ngdc.noaa.gov/geomag/data/poles/NP.xy). The NGP approached the territory of Russia (https://travelask.ru/blog/posts/13546-kuda-dvizhetsya-severnyy-magnitnyy-polyus-i-kogda-polyusa-sn; https://web.archive.org/web/20220427002623; https://www.ngdc.noaa.gov/geomag/data/poles/NP.xy). At the same time, in 1998–2012, it was approaching the North Geographic Pole, and its distance to the Polar Vortex area corresponding to the Territory of Nunavut (Canada) was increasing. Between 2008 and 2022. The NGP was at its minimum distance from the North Geographic Pole and its average distance to the Polar Vortex was greater than in 1998–2012. For this reason, the mean values of FCR entering the atmosphere above the Polar Vortex zone were larger in 1998–2012 than in 2008–2022. The average ionization intensity of the stratospheric air forming this vortex was also smaller in the last period.

Therefore, it is acceptable to assume that the change in the location of the NGP that occurred in the period 1998–2022 can be the cause of the detected peculiarity. If this assumption is valid, something similar should be manifested in the FCR and SDP connections.

Figure 4 shows the TNE plots where the correlation of the interannual changes in both SRF and FCR for the months of March–May, estimated for the periods 1998–2012 and 2008–2022, was found to be significant.

From Fig. 4a and b shows that there were few TNE plots for March where the correlation of interannual changes in both SRF and FCR was significant for both 1998–2012 and 2008–2022.

In the first period (Fig. 4b), areas of significant positive correlation of these processes were revealed only in the Murmansk Oblast, and areas of significant negative correlation—in the waters of the Barents Sea near the shores of the islands of the archipelagos of Novaya Zemlya and Spitsbergen.

In the second period (Fig. 4a), the areas of their significant positive correlation were revealed on the Kola Peninsula, in the territories of the Sverdlovsk Oblast and

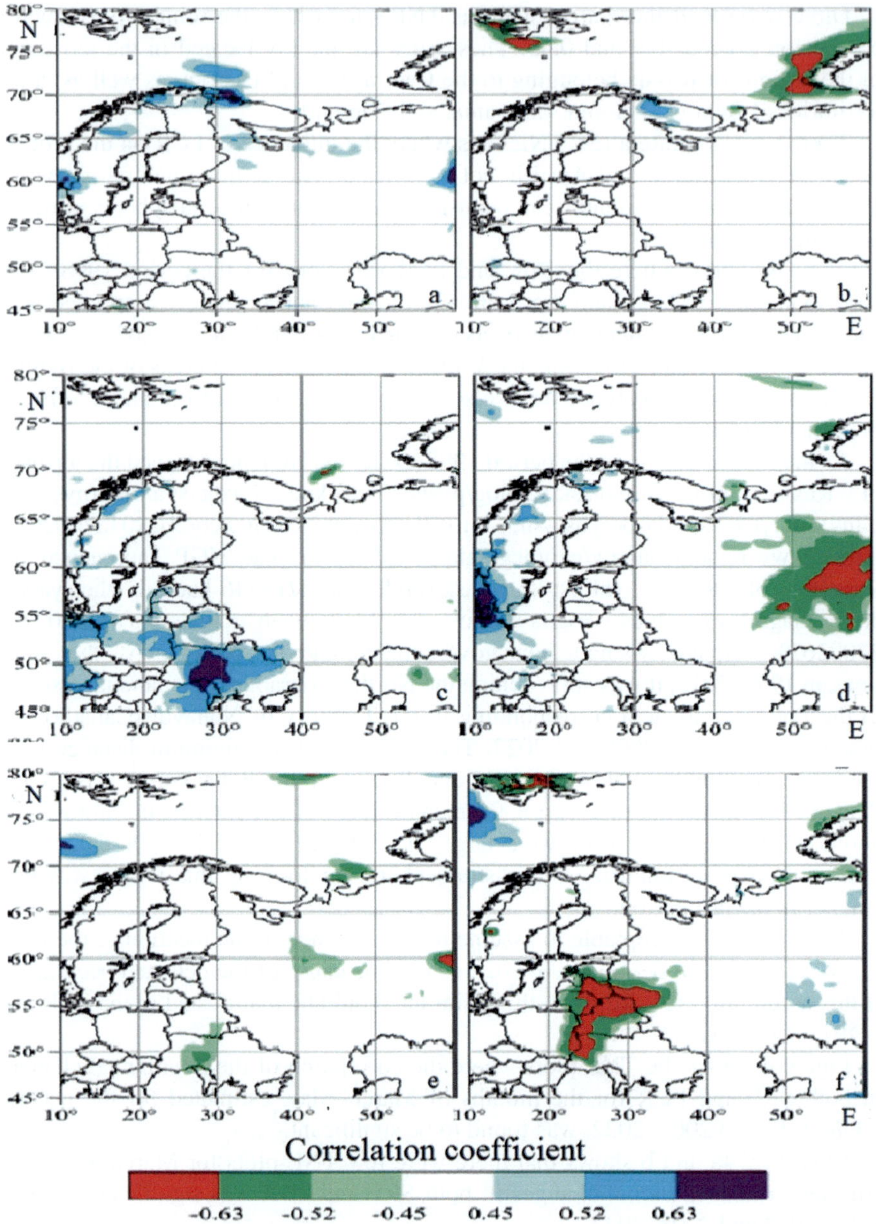

Fig. 4 TNE plots where the SRF and FCR relationships were significant for the following time periods: **a** March, 2008–2022; **b** March, 1998–2012; **c** April, 2008–2022; **d** April, 1998–2012; **e** May, 2008–2022; **f** May, 1998–2012

Karelia, Norway and Sweden, as well as in the Barents Sea water area to the north of Cape Nordkaap. At the same time, the total areas of the identified sites for the period 1998–2012 were slightly larger than in 2008–2022, although they did not differ significantly.

Figure 4c and d show that for April, the total area of TNE sites where the correlation of interannual changes in both SRF and FCR estimated for the period 1998–2012 was significant was significantly larger than for the period 2008–2022. Moreover, it was mainly negative in the first period and positive in the second period.

As can be seen from Fig. 4c, in the period 2008–2022, the detected TNE areas were located mainly over the territories of Ukraine, Poland, and Germany (they were not detected in the European territory of Russia).

Figure 4f shows that in 1998–2012 the TNE sites, where the correlation of the same processes was significant and negative, were located over many regions of the Ural and Volga Federal Districts, as well as the Arkhangelsk Oblast.

The TNE areas where the correlation of the processes under consideration was significant and positive were also identified. These areas belong to the territories of Norway, Sweden, Denmark and Germany.

From Fig. 4d and e shows that for May, the total area of TNE plots where the SRF and FCR correlations were significant was also larger in 1998–2012 than in 2008–2022. In both periods, areas where the significant correlation of these processes was both positive and negative were identified on the TNE.

In 2008–2022, a significant negative correlation of the studied processes was revealed in many areas of the territories of the Vologda and Sverdlovsk regions of Russia, territories of Ukraine, Belarus, as well as in the area of the Barents Sea between Kolguyev Island and Kanin Peninsula.

The area where their correlation was significantly positive was found only over the Norwegian Sea (see Fig. 4d).

The vast area of significant negative correlation of the same processes in 1998–2012 partially included the territories of the Pskov, Tver, and Smolensk oblasts, as well as the Baltic States, Belarus, Ukraine, and Poland. For many of its areas, the reliability of the conclusion about the significance of the relationship between the time series of SRF and FCR exceeds 0.99.

Significantly smaller by area areas of significant positive correlation of the processes under consideration were revealed over the Norwegian Sea and also over the European territory of Russia.

Thus, Fig. 4 indicates that in the period 1998–2012, the total number of TNE sites for which the relationships between the interannual changes of SRF and FCR variations synchronous to them were significant for April and May was significantly larger than in 2008–2022.

Consequently, the assumption that there is a significant influence of the location of NGP on the changes in the TOC and SRF over the TNE seems to be quite reasonable, and the existence of a statistical relationship between them is confirmed.

It follows from this that the significant causes of the detected in 1998–2022 partial recovery of the ozone layer over the Northern Hemisphere, along with the measures aimed at limiting anthropogenic and anthropogenic emissions of freons into the atmosphere, could also include the increase in the distance between the NGP and the Polar Vortex. Since at further movement of the NGP along the same trajectory the distance from it to the Polar Vortex (over the territory of the TNE) will decrease, the revealed feature allows us to assume that the average FCR entering the atmosphere above it will increase during the 25th solar activity cycle (and possibly further), and the intensity of ionization of stratospheric air over the corresponding part of the Polar Vortex zone will increase. The latter will entail the intensification of air exchange between the stratospheric layers located above the TNE and also an increase of the total area of its regions, where the FCR variations significantly influence the TOC and SRF changes.

In the changes of the ozone layer state over many of such areas, as the NGP approaches the territory of Russia, the unfavorable trends that took place in the late twentieth–early twenty-first century will be more and more clearly manifested.

Taking into account the results obtained, the above scenario of further changes in the state of the ozone layer over Europe should be considered as one of the probable ones, and therefore activation of measures aimed at preventing its destruction is very relevant and justified.

The approach of the NGP may lead to consequences that are dangerous for the population and economy of all countries of the studied region. Therefore, the problem of preventing such consequences is of interstate nature, and for its successful solution it is necessary to develop a program of joint actions of the relevant states and the public.

Thus, the aim of the study has been achieved and all the tasks set in it have been solved.

4 Discussion of the Findings

As follows from the above, the obtained results are consistent with the existing ideas about the mechanisms of changes in the ozone layer state (Alexandrov et al. 1982; Larin 2000; Nikiforova et al. 2016; Zvyagintsev et al. 2013; Khrgian 1973; Kholoptsev et al. 2014; Polyakov et al. 2024; Muller 2012; UNEP 2010; Hoffman 2005; Mohanakumar 2011; Lang et al. 2013; Ali et al. 2012; Zepp et al. 2003; Dessler 2000) and the trends manifested in this process over Europe in the late twentieth–early twenty-first century (https://ozone.unep.org/sites/default/files/2023-09/Ozone_and_UV_Bulletin_1_ru.pdf; https://www.wmo.int/pages/prog/arep/WMOArcticOzoneBulletins2016.html; https://www.unep.org/ru/novosti-i-istorii/press-reliz/ozonovyy-sloy-nakhoditsya-na-puti-k-vosstanovleniyu-chto-pomozhet; Nikiforova et al. 2016; Zvyagintsev et al. 2013). They also correspond to the ideas (https://new-science.ru/arkticheskij-polyarnyj-vihr-polnostju-izmenil-napravlenie-vyzvav-rekordnyj-pik-ozona-v-regione/) about the role of FCR variations in the changes in the ionization intensity of different layers of the Earth's atmosphere.

However, some of the facts established have substantial elements of novelty. These include the following facts:

1. TNE areas where the trends in the changes in the TOC and SRF for different spring months that occurred during the period 1998–2019 prevailed were identified. Among them, those areas where there is a correspondence between the trends of these processes, indicating the existence of causal relationships between them, were identified.
2. We have identified the areas of the TNE, where statistical relationships of interannual changes of the TOC and SRF for the spring months with synchronous variations of PCL in the periods 1998–2012 and 2008–2022 were significant. The identified features of these relationships suggest that they are causal in nature.
3. The total areas of the identified sites corresponding to the period 1998–2012 are larger than for 2008–2022. The sign of the SRF and FCR correlation prevailing at such sites is different. A possible reason for this may be the displacement of the NGP and changes in its distance from the Polar Vortex core during the same years.

5 Conclusions

Thus, it has been established:

1. On the territory of Northern Europe, the main part of which belongs to Russia, there are numerous areas for which the tendencies of interannual changes of mean monthly values of total ozone content and monthly fluxes of ultraviolet solar radiation for spring months in the period 1998–2024, qualitatively corresponded to the stated ideas about the consequences of the approach of the North Geomagnetic Pole to it.
2. At many of them there is a correspondence between the trends of the processes under consideration, indicating the significance of the cause-and-effect relations between them. At the same time, there are areas where such correspondence is absent. The latter indicates the significance of the influence of changes in the state of tropospheric cloudiness on changes in the ultraviolet solar radiation fluxes.
3. The areas of the study area for which statistical relations of interannual changes in the mean monthly values of the indicators of the studied processes with synchronous variations of the cosmic ray flux entering the atmosphere above the city of Dolgoprudny, Moscow Region, are significant have been identified. There are also areas for which the considered relations are not significant, which is apparently due to a stronger influence of other factors on the studied processes.

The work was carried out within the framework of the theme of the state assignment of FGBUN FIC MGI on the topic FNNN-2024-0001.

References

Alexandrov EL, Karol IL, Rakipova LR, Sedunov YS, Khrgian AH (1982) Atmospheric ozone and changes in global climate. Gidrometeoizdat L:165

Ali SM, Dash NK, Pradhan A et al (2012) Effects of stratospheric ozone depletion on the environment and agriculture. Int J Adv Res Technol 1(4):31–36

Dessler A (2000) The chemistry and physics of stratospheric ozone. Academic Press, p 520

ERA5 hourly data on pressure levels from 1979 to present. https://cds.climate.copernicus.eu/cdsapp!/dataset/reanalysis-era5-presSRFe-levels?tab=form. https://doi.org/10.24381/cds.bd0915c6

Erlykin AD, Wolfendale AW (2011) Cosmic ray effects on cloud cover and their relevance to climate change. J Atmos Sol Terr Phys 77:1681–1686

Golenkov AE, Okhlopkov VP, Svirzhevskaya AK, Svirzhevsky NS, Stozhkov YI (1980) Planetary distribution of cosmic ray intensity from measurement's in the stratosphere. In: Proceedings of FIAN, vol 122. Nauka, Moscow, pp 3–14

Hoffman MJ (2005) Ozone depletion and climate change. State University press, New York, p 276

Hólm E, Janisková M, Keeley S et al (2020) The ERA5 global reanalysis. Q J R Meteorol Soc 146:1999–2049

Ionov DV, Timofeev YM, Shalamiansky AM (2002) Comparison of satellite (GOME, TOMS) and ground meaSRFements of the total ozone content. Earth Explor Space (3):10–19

Jones CA (2011) Planetary magnetic fields and fluid dynamos. Ann Rev 43:583–614

Kholoptsev AV, Nikiforova MP, Bolshikh AV (2014) The world ocean and the ozonosphere. LAP Lambert Academic Publishing, Saarbrucken, Deutschland, p 522

Khrgian AH (1973) Physics of atmospheric ozone. Gidrometeoizdat L:292

Kishore P, Velicogna I, Venkat Raman M, Jiang JH, Madhavi GN (2009) Planetary waves in the upper stratosphere and lower mesosphere during, arctic major stratospheric warming. Annales Geophysicae 30(10):1529–1538

Kolstad EW, Charlton-Perez AJ (2010) Observed and simulated precursors of stratospheric polar vortex anomalies in the northern hemisphere. Clim Dyn 37(7–8):1443–1456

Lang C, Waugh DW, Liang Q et al (2013) Impact of UV changes from stratospheric ozone depletion on tropospheric ozone. J Hopkins Univ Catalog 3:2–42

Larin IK (2000) Chemistry of the ozone layer and life on Earth. Chem Life XXI Century (7):10–15

Masrsh ND, Svenmark H (2000) Cosmic rays, clouds and climate. Space Sci Rev 94:215–230

Maurchev EA, Balabin YV, Germanenko AV, Mikhalko EA, Gvozdevsky BB (2019) Calculating the ionization rate induced by GCR and SCR protons in earth atmosphere. Solar-Terr Phys 5(3):68–74

McMichael AJ, Lucas R, Ponsonby AL, Edwards SJ (2010) Stratospheric ozone depletion, ultraviolet radiation and health. In: Climate change and human health, pp 159–180

Mohanakumar K (2011) Interaction of stratosphere and troposphere. Fizmatlit, Moscow, p 451

Movement of the north magnetic pole to the territory of Russia slowed down. https://www.rbc.ru/rbcfreenews/67394e8b9a794716d3a3cb81

Muller R (2012) Stratospheric ozone depletion and climate change. RSC Publishing, p 346

New-science.ru. https://new-science.ru/arkticheskij-polyarnyj-vihr-polnostju-izmenil-napravlenie-vyzvav-rekordnyj-pik-ozona-v-regione/

Nikiforova MP, Zvyagintsev AM, Vargin PN et al (2016) Anomalously low levels of total ozone over Siberia at the end of January 2016. Atmos Ocean Opt 30(1):12–19

Polyakov A, Kriukovskikh E, Virolainen Y, Nerobelov G, Kozlov D, Timofeyev Y (2024) Determining the total ozone column from spectral measurement of IKFS-2 in 2015–2022. Izv Atmos Oceanic Phys 60(6):689–698

Raspopov OM, Veretenenko SV (2009) Solar activity and cosmic rays: influence on cloudiness and processes in the lower atmosphere (in memory and on the 75th anniversary of M. I. Pudovkin). Geomagn Aeron 49(2):147–155

Reshetnyak MY (2018) Geodynamo models. Izv High Schools. Radiophysics 61(8–9):605–613

Ross JC (1834) On the position of the north magnetic pole. Philos Trans R Soc Lond J 124:47–52
Scheme of movement of the north magnetic pole towards Russia. https://travelask.ru/blog/posts/13546-kuda-dvizhetsya-severnyy-magnitnyy-polyus-i-kogda-polyusa-sn
Stozhkov YI, Makhmutov VS, Svirzhevsky NS (2022) Studies of comic beams on balloons at the P.N. Lebedev Physical Institute of the Russian Academy of Sciences. Uspekhi Physicheskikh Nauki 192(9):1054–1063
The ozone layer is on the way to recovery, which would avoid global warming by 0.5°. https://www.unep.org/ru/novosti-i-istorii/press-reliz/ozonovyy-sloy-nakhoditsya-na-puti-k-vosstanovleniyu-chto-pomozhet
Todd MC, Kniveton DR (2001) Changes in cloud cover associated with Forbush decreases of galactic cosmic rays. J Geophys Res 106(D 23):32031–32041
Track of the north magnetic pole. https://web.archive.org/web/20220427002623; https://www.ngdc.noaa.gov/geomag/data/poles/NP.xy
UNEP (2010) Environmental effects of ozone depletion and its interactions with climate change: 2010 assessment. UNEP, Nairobi, p 328
Vernov SN. Dolgoprudny Research Station. Laboratory of solar and cosmic ray physics. P. N. Lebedev Physical Institute of the Russian Academy of Sciences. https://sites.lebede3v.ru/DNS_FIAN
Veretenenko SV, Pudovkin MI (1996) Forbush-decay effects of galactic cosmic rays in variations of the total cloudiness. Geomagn Aeron 36(1):153–156
WMO Arctic Ozone Bulletin (2016) (1). https://www.wmo.int/pages/prog/arep/WMOArcticOzoneBulletins2016.html
WMO Ozone Bulletin. [Electronic resource]. https://ozone.unep.org/sites/default/files/2023-09/Ozone_and_UV_Bulletin_1_ru.pdf
Zepp RG, Callaghan TV, Erickson DJ (2003) Interactive effects of ozone depletion and climate change on biogeochemical cycles. Photochem Photobiol Sci 2:51–61
Zvyagintsev AM, Kuznetsov GI, Kuznetsova IN (2013) Spring anomalies of the ozone layer over Russia. Meteorol Hydrol (5):5–13

Impact of Clay Liner Integrity Breaches on Leachate Migration in Municipal Solid Waste Landfills

Ya. N. Parshakova and R. I. Kataev

Abstract This study aims to investigate how the dimensions of breaches in a landfill's clay liner (protective screen) influence the migration of leachate at the base of a municipal solid waste (MSW) landfill. Numerical modeling of pollutant transport was conducted in homogeneous porous media, with the liner represented as a loam layer modeled as a homogeneous porous medium. The breaches are simulated as horizontal discontinuities within the loam layer, with a fixed length of approximately 10 m and variable widths. These discontinuities are entirely filled with a technogenic (engineered) material in the model. The time-dependent advection–dispersion equations governing pollutant transport were solved numerically, as no analytical solution exists for this class of problems.

Keywords Municipal solid waste (MSW) landfills · Contaminant filtration dynamics · Rayleigh–Taylor instability · Computational hydrogeological modeling

1 Introduction

Modern landfill containment systems for municipal solid waste (MSW) comprise complex multilayer barriers incorporating both mineral and synthetic components, whose performance is significantly compromised by mechanical defects of varying origins. Extensive research over recent decades has established that the parameters of such defects—particularly their geometric characteristics and spatial distribution—govern leachate migration patterns (Ding et al. 2021; Giroud and Bonaparte 1989; Li et al. 2024a; Mendes et al. 2010; Rowe and Abdelatty 2012). Controlled laboratory and field experiments reveal a highly nonlinear relationship between defect

Ya. N. Parshakova (✉) · R. I. Kataev
Institute of Continuous Media Mechanics UB RAS, Perm, Russian Federation
e-mail: parshakova@icmm.ru

R. I. Kataev
Perm State University, Perm, Russian Federation

T. Chaplina (ed.), *Processes in GeoMedia—Volume IX*, Springer Geology,
https://doi.org/10.1007/978-981-96-9302-3_34

dimensions and leachate flux intensity, necessitating the development of specialized mathematical models (Giroud et al. 1997).

For clay liners—a conventional component of containment systems—the critical parameter is desiccation crack width induced by wetting–drying cycles. Laboratory studies on compacted clay specimens demonstrate that when crack apertures reach 5 mm, the hydraulic conductivity increases by 2–3 orders of magnitude compared to intact material (Albrecht and Benson 2001). Field monitoring at operational landfills confirms this process is exacerbated by seasonal climatic fluctuations, leading to progressive crack network development over time (Benson et al. 2007). Modern 3D modeling emphasizes the necessity to account for crack orientation relative to flow direction, as this factor may alter predicted contaminant migration rates by several folds (Snow 1968).

For geomembrane barriers, a key component of modern containment systems, leakage rates are governed primarily by defect area rather than linear dimensions. A statistical analysis of defects across 20+ operational landfills in North America and Europe revealed a predominance of 1–5 mm puncture-type defects, with areal densities ranging from 2 to 5 per 1000 m^2 (Rowe and Sangam 2002). Laboratory studies of flow through artificial defects demonstrated strong agreement with modified Giroud–Bonaparte equations for small defects ($\leq$3 mm) (Giroud and Bonaparte 1989). However, larger defects require additional consideration of: turbulent flow effects and local hydraulic gradient variations (Giroud 1997). Of particular significance are composite defect scenarios, where coincident damage in both geomembranes and underlying clay liners creates direct leachate migration pathways (Take et al. 2015).

Experimental verification of mathematical models of filtration through damaged barriers is carried out using various methods, each of which has its own advantages and limitations. In laboratory conditions, the most accurate results are achieved using precision mass flow meters, which allow recording flows with an accuracy of up to 0.5% of the measured value in a wide range of flow rates (Benson et al. 1999). Field studies are based on a set of geophysical methods, including ground-penetrating radar scanning and electrical resistivity tomography, which ensures spatial localization of defects with a resolution of up to 1 mm at a research depth of up to 5 m (Kong et al. 2018). Of particular value to the scientific community are long-term monitoring programs demonstrating the time dynamics of damage development under the influence of operational loads and natural factors (Touze-Foltz et al. 2019).

A promising direction for the development of filtration models is the integration of traditional hydrogeological approaches with machine learning and big data analysis methods. This allows for a significant increase in forecasting accuracy based on extensive arrays of monitoring data collected using modern automated control systems (Kong et al. 2021). Of particular interest are developments in the field of digital twins of protective systems, which allow real-time monitoring of the state of barriers and forecasting the development of potentially dangerous damage (US Epa 2022). These technologies are based on the integrated use of remote sensing data, built-in sensors, and physical and mathematical models, which opens up new opportunities for managing filtration risks at solid municipal waste landfills.

Modern approaches to modeling filtration through damaged barriers take into account not only the static parameters of defects, but also their evolution over time under the influence of a set of factors. In particular, probabilistic models have been developed that describe the process of damage accumulation as a function of the service life, mechanical loads, and environmental conditions (Chaudhary et al. 2024). These models are based on extensive statistical data collected during long-term observations of various types of protective systems and allow one to predict changes in filtration characteristics over time. Particular attention is paid to taking into account synergistic effects, when a combination of various factors (e.g., chemical aging of geomembranes and mechanical loads) leads to accelerated development of damage (Shu et al. 2019).

An important aspect of modern research is the development of methods for assessing the residual life of damaged protective systems. Experimental data indicate that even in the presence of significant damage, barriers can retain partial efficiency due to self-compacting and chemical sorption mechanisms (Reddy et al. 2021). To quantitatively assess this effect, complex models are being developed that take into account not only the hydraulic parameters of damage but also the processes of interaction of the filtrate with the barrier material (Rowe and Yu 2022). These models allow a more informed approach to making decisions about the need to repair or replace protective systems, which is important from an economic and environmental point of view.

Modeling filtration under conditions of combined damage, when defects occur simultaneously in several layers of the protective system, is particularly difficult. Experimental studies show that in this case, the filtration flow can increase nonlinearly due to the formation of direct migration channels (Li et al. 2024b). To take this effect into account, specialized three-dimensional models are being developed that allow the analysis of the spatial configuration of damage and their mutual influence (Chai et al. 2020). These models require significant computing resources, but provide a more accurate forecast of contaminant migration compared to traditional two-dimensional approaches.

Considerable attention in modern research is paid to the development of standardized methods for assessing the state of protective systems. In particular, protocols for classifying damage by the degree of its impact on filtration characteristics have been proposed (US Epa 2022). These protocols are based on the generalization of extensive experimental material and allow for the unification of approaches to assessing filtration risks at various landfills. Of particular importance is the development of non-destructive testing methods that allow for assessing the state of protective systems without violating their integrity (Kong et al. 2023). These methods, including geophysical and thermal imaging technologies, are becoming increasingly popular in the practice of operating solid municipal waste landfills.

Prospects for the development of models of filtration through damaged barriers are associated with the further integration of various approaches and methods. Of particular interest is the combination of physical modeling in laboratory conditions with numerical methods and field observations (Feng et al. 2022). This allows for the creation of more adequate models that take into account a set of factors influencing the filtration process. An important area is also the development of methods for predicting the long-term behavior of damaged barriers, taking into account the aging processes of materials and changes in operating conditions (Touze-Foltz et al. 2023). These developments are of fundamental importance for ensuring the long-term safety of solid municipal waste landfills and preventing environmental pollution. To assess leachate migration rates in this study, we performed numerical modeling of leachate dynamics for various defect sizes in the impermeable barrier of a landfill that has been operational for over 40 years.

2 Characteristics of the Studied Municipal Solid Waste (MSW) Landfill

The landfill is located on the slope of the floodplain of a small river, the relief of the site is uneven and with a general slope in the northeast direction toward the river flowing 300 m to the northeast of the site. Due to the infiltration of atmospheric precipitation through the body of the landfill and biological processes of MSW destruction, contaminated runoff is formed. The filtrate from the landfill spreads along the anti-seepage screen and accumulates in its depressions and settling ponds. The landfill is surrounded by a dam up to 6 m high along the perimeter. In the work (Parshakova et al. 2024), the filtering capacity of the dam protecting the natural environment from the ingress of MSW filtrate is assessed for this landfill. The need to isolate the foot of the dam is shown to prevent filtrate from entering the base of the landfill. The design of the studied object was carried out in the eighties of the last century, a natural layer of loam more than 1 m thick, distributed almost uniformly over the site, was used as an isolating screen. The vertical section in the southeast direction is shown in Fig. 1. The geological structure of the landfill base up to the first impermeable horizon consists of three layers of rocks. On top of them is poured a layer of heterogeneous, uncompacted and compacted, uncompacted and loose soils and waste, which is saturated with landfill filtrate—the body of the landfill (technogenic layer).

Since the landfill is located on a hilly terrain and has been in operation for over 40 years, in addition to the natural ones, unevenness has formed due to deformation of the geological layers. Depressions up to 4–5 m deep have formed in the landfill structure (Fig. 1 at elevation 170), where filtrate accumulates and is unable to flow through the drainage system. In places where filtrate accumulates, the insulating layer may be destroyed. This creates a risk of filtrate diffusion into the soil under the landfill and penetration into groundwater. The rate of groundwater movement varies from 3 to 30 m per day, depending on the slope and permeability of the rocks. Thus, if filtrate penetrates the soil, it can spread through groundwater for tens of kilometers within a year.

The technogenic layer includes polyethylene (20–25%), construction waste (15–30%), textiles (10–20%), waste paper (8–20%), plastic (9–15%), wood (8–20%),

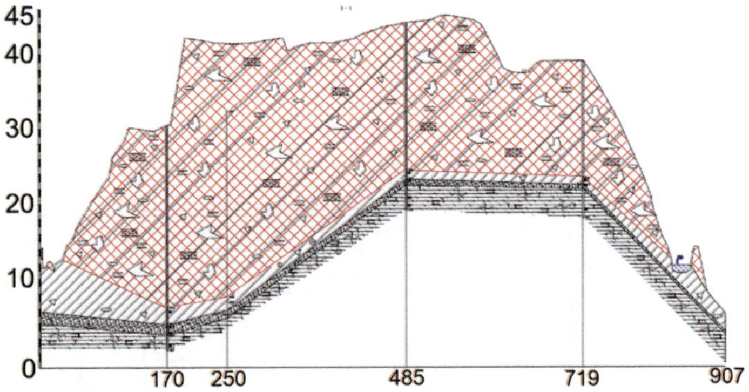

Fig. 1 View of the vertical section of the polygon in the middle of the depressions. Natural data from 2017

glass (7–15%), metal in the form of cables, wire, pieces of reinforcement (7–15%), mineral soil (10%), and others (6–15%). Directly under the technogenic layer is a loam layer. As stated above, the loam acts as a screen preventing the filtrate from penetrating under the landfill body. In-kind measurements revealed inclusions of construction waste, crushed stone, and sand in the loam, which is a violation of its insulating capacity. The layer following the loam with a maximum thickness of up to 1.5 m consists of clayey siltstones. Below it is a layer of argillite, which reaches the aquiclude; its thickness varies within 4 m. The physical characteristics of the rock layers of the base of the landfill are presented in Table 1.

Table 1 Physical characteristics of the landfill base rocks (archived data from field measurements in 2017)

Material Physical characteristics	Technogenic layer—a waste deposit layer in an MSW landfill	Clay loam	Siltstone	Mudstone
ρ_s, density, kg/m^3	$1.25 \cdot 10^3$	$1.55 \cdot 10^3$	$1.74 \cdot 10^3$	$1.67 \cdot 10^3$
C_{ps}, specific heat, J/(kg K)	1180	1973	894	756
m, porosity	0.4	0.702	0.481	0.542
K, permeability, m^2	$1.06{*}10^{-12}$	$1.15{*}10^{-14}$	$1.42{*}10^{-14}$	$4.25{*}10^{-14}$
Filtration coefficient, m/day	0.6	0.0065	0.008	0.024
κ_s, , thermal conductivity coefficient, W/(m K)	1.06	1.36	2.1	1.69
Resistance of rocks, 1/m^2	$9.42{*}10^{11}$	$8.69{*}10^{13}$	$7.06{*}10^{13}$	$2.35{*}10^{13}$
Initial concentration, wt%	0.0047	0.000202	0.000202	0.000202

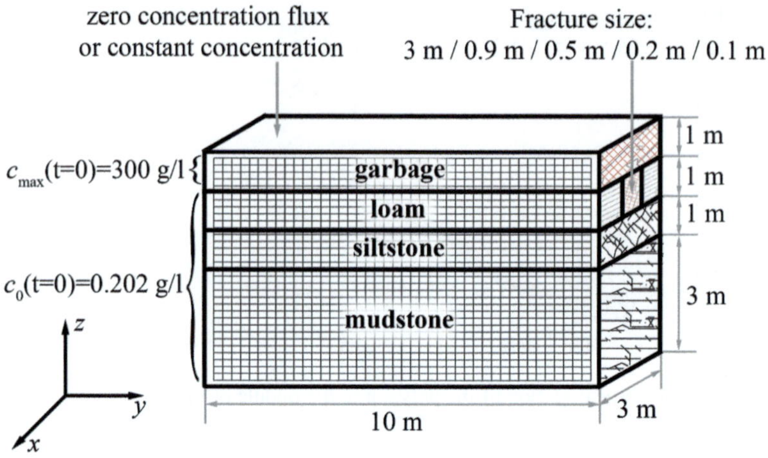

Fig. 2 View of the landfill site in the middle of the local depression of the studied solid municipal waste landfill, used as a calculation scheme for the filtration problem

Since depressions can have different sizes, one general problem of filtration through a parallelepiped consisting of a technogenic layer and rock layers forming the base of the polygon was solved (Fig. 2). The argillite layer contains unconfined groundwater, its thickness to the aquiclude for calculations was taken equal to a thickness of 3 m. The thicknesses of the remaining layers were 1 m each. A section measuring 10 by 3 m in the horizontal plane and 6 m vertically was considered. In the initial state, the filtrate filled the technogenic layer. The development of Rayleigh–Taylor instability in the isothermal case was studied for different sizes of the insulation layer failure.

3 Mathematical Model of Landfill Leachate Propagation

A three-dimensional numerical simulation of leachate propagation from an undrained landfill depression was performed using a laminar flow model in a porous medium. The problem was solved using transient isothermal and non-isothermal approaches. For the numerical implementation, a uniform spatial grid composed of quadrilateral four-node finite elements was generated.

The governing equations for the layer included.

Tensor-form equations of liquid leachate motion

$$\frac{\partial (m\rho)}{\partial t} + \frac{\partial}{\partial x_i}(\rho \mathrm{v}_i) = 0, \tag{1}$$

$$\mathrm{v}_i = \frac{K}{\mu}\left(-\frac{\partial p}{\partial x_i} + \rho g_i\right). \tag{2}$$

where m, and K—porosity and permeability of a porous medium, respectively; ρ—density of liquid, v_i—components of the velocity vector (i = 1, 2, 3), and μ—dynamic viscosity of a liquid. The quadratic dependence of density on concentration was considered. $\rho = \rho_0 + A \cdot c + B \cdot c^2$ ($\rho_0 = 999.993$ kg/m^3, $A = 667.8$, $B = -0.1229$). Heavy impurity does not create unstable density stratification when accumulated near the bottom.

The equation for impurity transport is

$$\frac{\partial}{\partial t}(m\rho c) + \nabla \cdot (\rho \vec{v} c) = -\nabla \cdot \vec{J}. \tag{3}$$

Here $\vec{J}$—diffusion flux of impurity, determined by the expression

$$\vec{J} = -\rho D_m \nabla c, \tag{4}$$

where D_m—given coefficient of molecular diffusion.

The finite volume method (FVM) was employed for all numerical calculations. Spatial discretization of the governing equations was implemented using a second-order accurate scheme. For temporal discretization, an explicit second-order time integration scheme was applied. In the simulations, impurity concentrations were expressed in weight percent (wt%), with values constrained to the range of 0 to 1 wt%.

Boundary conditions

The boundary conditions for Eqs. (1)–(4) are given below for different boundaries of the system.

At the lower boundary of the computational domain (the first aquifer), the no-slip condition and the zero mass flow condition are used:

$$v_x = v_y = v_z = 0, \quad \frac{\partial c}{\partial \mathbf{n}} = 0. \tag{5}$$

At the entrance to the computational domain, the velocity in each rock layer was set, corresponding to the filtration coefficient given for the layers in Table 1 (the velocity vector of the ambient flow is perpendicular to the inlet boundary $\boldsymbol{V} = \{V,\ 0,\ 0\}$); the concentration was set equal to the background concentration of the pollutant C_0 in the water:

$$v_x = V,\ v_y = 0,\ v_z = 0,\ c = C_0 \tag{6}$$

The upper boundary of the region corresponding to the free surface of the liquid was assumed to be non-deformable; the conditions of absence of the normal velocity component, tangential stresses, and impurity flow were considered to be satisfied on it:

$$(\mathbf{v}\,\mathbf{n}) = 0, \ \frac{\partial v_x}{\partial x_z} + \frac{\partial v_z}{\partial x_x} = 0, \ \frac{\partial v_y}{\partial x_z} + \frac{\partial v_z}{\partial x_y} = 0, \ \frac{\partial c}{\partial \mathbf{n}} = 0. \tag{7}$$

Similar conditions were set on the lateral surface of the computational domain.

The conditions at the output of the computational domain consisted in fulfilling the mass balance condition:

$$\oint\limits_{S_{in}} \rho v_{in} dS - \oint\limits_{S_{out}} \rho v_{out} dS = 0. \tag{8}$$

The dynamic viscosity of the leachate was taken to be equal to $\mu = 10^{-3}$pa/s, molecular diffusion coefficient $D = 1.0 \cdot 10^{-9}\,\text{m}^2/\text{c}$. These values correspond to typical values for these quantities characteristic of weakly concentrated aqueous solutions.

The initial data used were the background concentration of the pollutant C_0 in the entire volume and the velocity of the main flow, equal to the velocity at the inlet of the calculation region.

4 Numerical Simulation Results

The computational grid consisted of rectangular elements. Test calculations were performed on four grids with a uniform spatial step, m: 0.075, 0.05, 0.02, and 0.01. For calculations, a 0.075 grid with a thickening near the place of violation of the insulating layer was selected. The process of filtrate spreading from the technogenic layer was considered when the integrity of the anti-filtration screen was violated such that the loam layer at the place of violation is replaced by the technogenic layer and is described by its characteristics. The dynamics of contaminant spreading was studied for a model case of intensive filtrate formation with a concentration of $C_0 = 300\,\text{g/l}$.

The impurity concentration fields at different moments in time are shown in Fig. 3.

The section is constructed vertically in the middle of the computational domain for x = 1.5 m. At the initial moment in time, the filtrate fills the upper, man-made layer. The process is considered when there is no filtrate flow from the landfill body. The filtrate density depends on the impurity concentration in it, so the heavier liquid is on top. Since the denser liquid is higher in the gravity field than the less dense one, the Rayleigh–Taylor instability develops. In concentration convection, the change of modes occurs with time, and not with a change in the process parameters (Rapaka et al. 2008; Slim et al. 2013). The development of instability in a porous medium occurs in such a way that initially, about twenty days later, there is a diffusion mode of linear growth of the concentration front thickness. In the linear growth mode, weak disturbances appear on the diffusion boundary layer. With a more detailed grid, the growth of disturbances is observed earlier, but the spatial appearance of the structures is the same. Next, macroscopic fingers appear in the filtrate flow, which

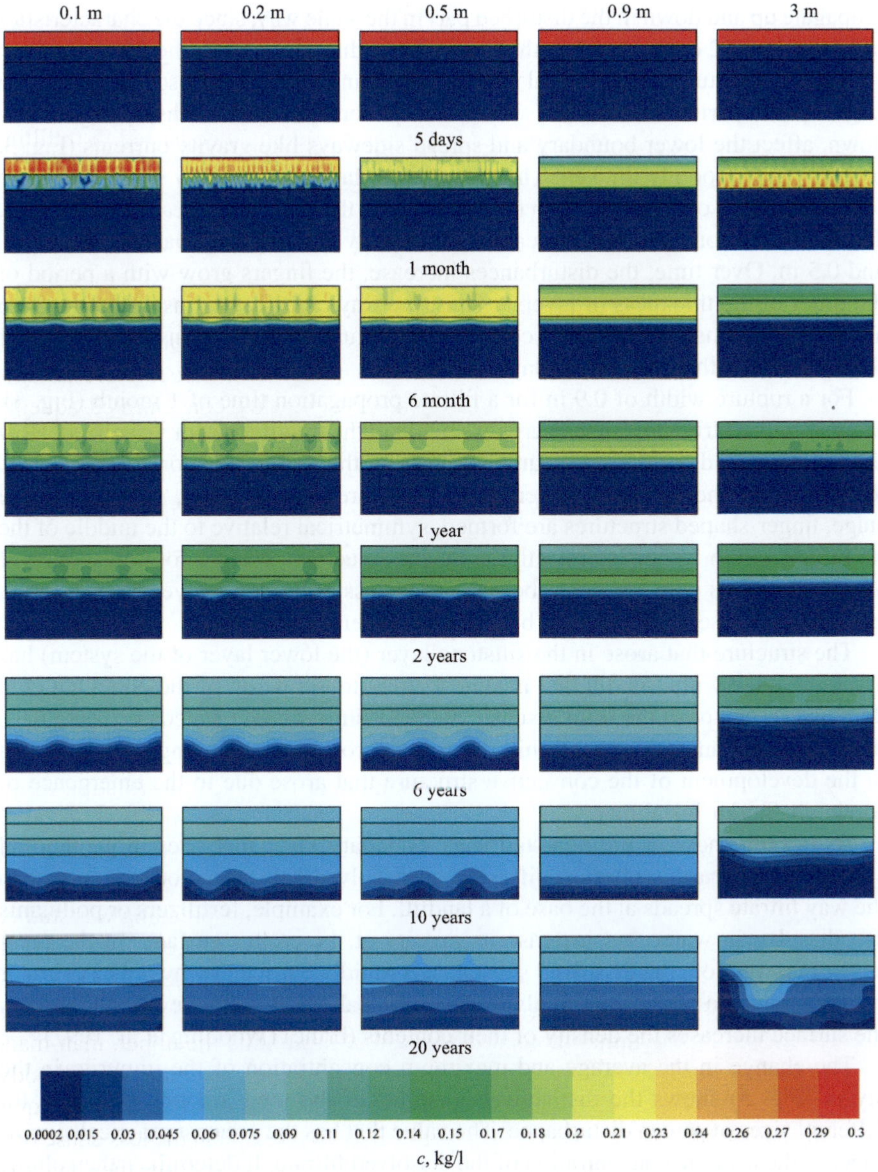

Fig. 3 Concentration fields of impurities for different sizes of damage to the insulating layer at different moments in time in a vertical section along the insulation break

propagate up and down in the disturbed part in the same way, since the characteristics of layers 1 and 2 coincide when the insulation is disturbed. Then the fingers merge: convective structures begin lateral migration and unite, complex descending jets of a larger spatial period are formed (see the field patterns after 6 months). The jets flow down, affect the lower boundary and spread sideways like gravity currents (Fig. 3, 6 years) and smoothly thicken. The entire lower layer is gradually saturated, while for disturbances of 0.9 and 3 m in the lower layer, the impurity spreads by diffusion. Therefore, the concentration increases more slowly than for disturbances of 0.1, 0.2, and 0.5 m. Over time, the disturbances increase, the fingers grow with a period of the order of the thickness of layer 3. Thus, the Rayleigh–Taylor instability leads to an increase in small disturbances compared to disturbances of the insulation layer of the order of the thickness of the layer itself.

For a rupture width of 0.9 m for a filtrate propagation time of 1 month (Fig. 3). Finger-shaped structures in the vertical section in the middle are not visualized. How the filtrate spreads in this case can be seen from the vertical section located across the rupture of the insulation layer, these fields are shown in Fig. 4. At the initial stage, finger-shaped structures are formed, symmetrical relative to the middle of the rupture in width. After lateral migration, the structures merge, forming a pair of vortex structures with a size of about the thickness of the loam layer. So that in the middle vertical section there is a boundary between these vortices.

The structure that arose in the siltstone layer (the lower layer of the system) has a leading edge with a width that is almost equal to the width of the computational domain. The shape of the front is such that the leading edge is located horizontally so that diffusion transfer plays a significant role in propagation, leading to a slowdown in the development of the convective structure that arose due to the emergence of Rayleigh–Taylor instability.

As the thickness of the non-stationary boundary layer increases in geothermal systems in the earth's interior, diffusion of dissolved substances occurs similar to the way filtrate spreads at the base of a landfill. For example, fertilizers or pollutants can dissolve in water and increase its density at the earth's surface. In the presence of convection, the dissolved substance is rapidly carried downwards. A similar phenomenon can occur in groundwater beneath salt lakes, where evaporation from the surface increases the density of their contents (brine) (Wooding et al. 1997).

The change in the average and maximum concentration of the impurity in the layers (Fig. 5) shows the distinctive dynamics of the total amount of filtrate for different sizes of screen disturbance. The value that has the greatest practical significance is the average concentration of the dissolved filtrate. It determines the volume of contaminants that have leaked into the base of the landfill. The average concentration is shown in Fig. 4a as a function of time. For a disturbance size of 3 m, the concentration reaches the same value in 19 years as for smaller values of the width of the disturbed layer. However, the growth of concentration tends to accelerate.

Thus, the average concentration for the following years turns out to be higher. For violations up to 1 m, the maximum value is observed 3 years after the start of the filtrate flow. And it lasts for 2 years. In this case, the aquifers will experience the

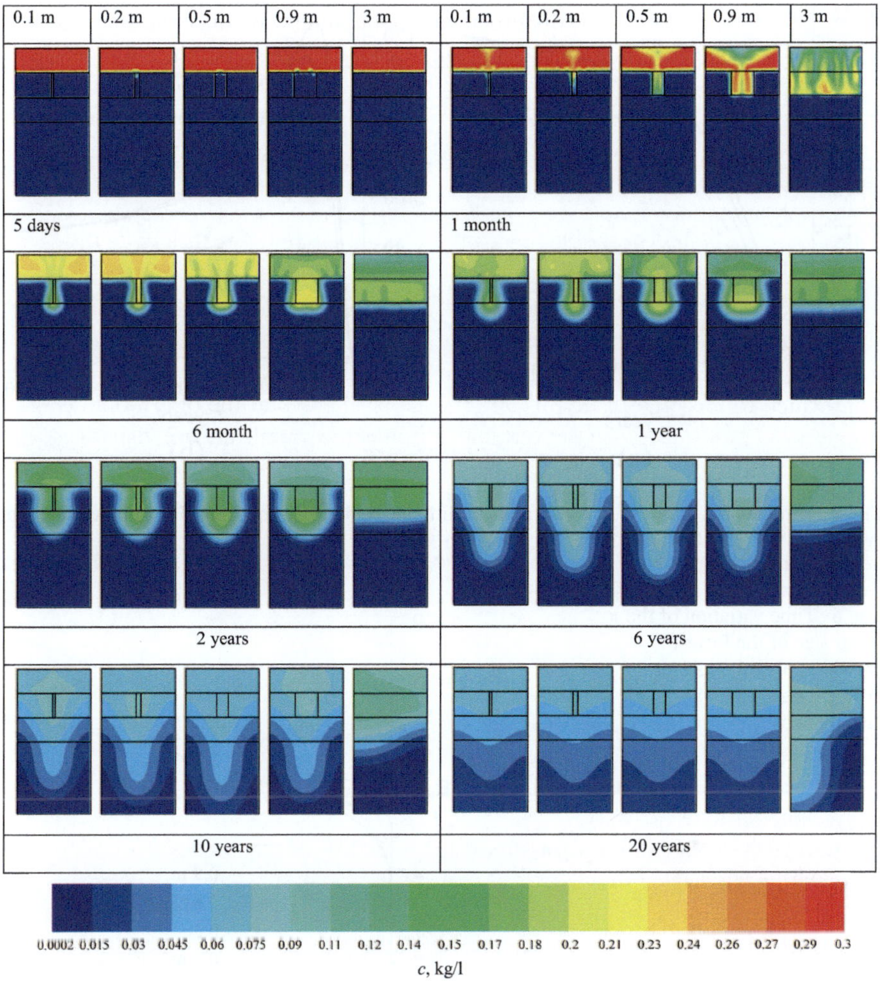

Fig. 4 Concentration fields of impurities for different sizes of damage to the insulating layer at different moments in time in a vertical section across the insulation break

maximum concentration load with an excess of the permissible value by two orders of magnitude.

The change in the linear size of the fastest growing disturbance in the direction of the vertical axis is shown in Fig. 6. At the initial stage of the computational experiment, the process has a purely diffusion character, the resulting concentration profile is described by an analytical solution (Slim et al. 2013) and is determined by the effective diffusion coefficient, closely related to the dispersion of the impurity in a saturated porous medium. For small sizes of the insulation layer disturbance, the effective dispersion of the system is higher than for disturbances of the order of the layer thickness due to the fact that the Rayleigh–Taylor instability occurs, leading to

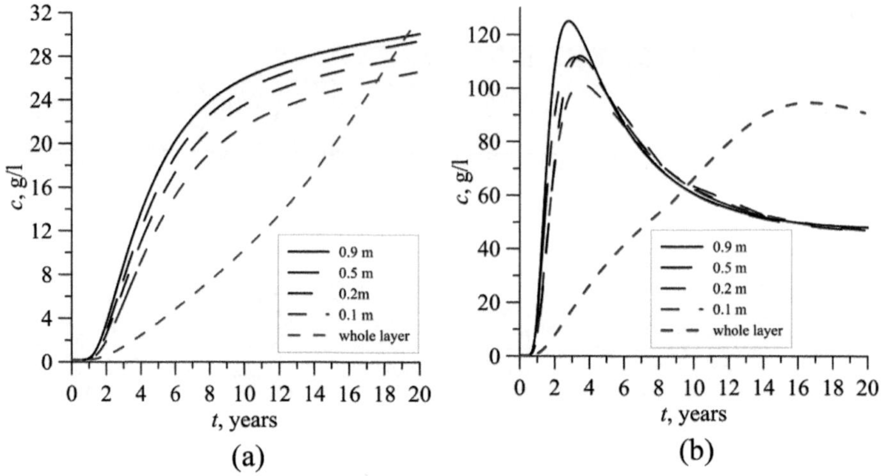

Fig. 5 Time graphs of the average (**a**) and maximum (**b**) impurity concentrations in the mudstone layer

Fig. 6 Time variation of the linear size of the fastest growing polygon base for different sizes of disturbances

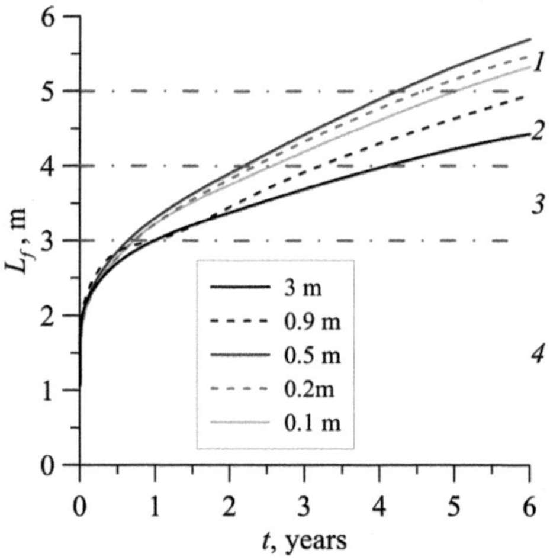

the influx of pure water to the interface. For disturbances wider than the thickness of the insulation layer, the effective dispersion at the initial stage is higher, and as a consequence of this, the Rayleigh–Taylor instability at the base of the polygon develops more slowly.

5 Conclusions

During long-term operation of a solid municipal waste landfill, relief changes of its site and foundation inevitably occur. As a result of such changes, depressions may form under the body of the landfill, in which filtrate accumulates without the possibility of flowing down the drainage system. In case of violation of the integrity of the isolating layer, groundwater pollution occurs due to filtrate seepage into the layers of the landfill base. Numerical modeling of filtrate spreading from an undrained depression of the landfill to the underlying aquifer in the presence of a violation of the isolating layer was carried out. The violations are ruptures in the loam layer, elongated in the horizontal direction. The length of the ruptures is assumed to be about ten meters, while their width varies. The place of the rupture is completely replaced by a technogenic layer.

The results of the numerical modeling showed that at the initial stage, the filtrate spreads by diffusion. Then, due to the development of Rayleigh–Taylor instability, a convective flow arises in the form of vortex structures that ensure the supply of cleaner water closer to the boundary of the base layers and significantly increase the rate of filtrate penetration into its depth. Concentration fingers are formed, the spatial size of which increases over time and upon reaching the lower boundary of the system reaches the order of the thickness of the lower layer. In the filtrate formation layer, intense concentration convection quickly develops in it and penetration of the convective flow into the lower layers of the MSW landfill base is observed. As a result, the development of Rayleigh–Taylor instability is significantly accelerated.

The rate of filtrate propagation due to convection increases significantly and leads to a linear average concentration. With defect sizes significantly smaller than the thickness of the insulating layer, an increase in the rate of propagation of filtration waters by 15–20% is recorded relative to violations comparable to the layer thickness. This effect is explained by the development of Rayleigh–Taylor instability.

Acknowledgements The research was supported by a grant from the Russian Science Foundation Project No. 23-11-00214.

References

Albrecht BA, Benson CH (2001) Effect of desiccation on compacted natural clays. J Geotechn Geoenviron Eng 127(1):67–75. https://doi.org/10.1061/(ASCE)1090-0241(2001)127:1(67)

Benson CH, Daniel DE, Boutwell GP (1999) Field performance of compacted clay liners. J Geotechn Geoenviron Eng 125(5):390–403. https://doi.org/10.1061/(ASCE)1090-0241(1999)125:5(390)

Benson CH, Thorstad PA, Jo HY, Rock SA (2007) Hydraulic performance of geosynthetic clay liners in a landfill final cover. J Geotechn Geoenviron Eng 133(7):814–827. https://doi.org/10.1061/(ASCE)1090-0241(2007)133:7(814)

Chai JC, Nguyen VT, Hino T (2020) Numerical investigation of leakage through defective geomembrane overlain by GCL. Geosynth Int 27(2):198–210. https://doi.org/10.1680/jgein.19.00038

Chaudhary E, Swami D, Joshi N, Reddy KR (2024) Sustainable landfill liner using local soils and wastes amended with bentonite: hydraulic conductivity and stochastic leachate transport modelling. J Geotechn Geoenviron Eng 150(1). https://doi.org/10.1061/JGGEFK.GTENG-11470

Ding XH, Feng SJ, Zheng QT (2021) A two-dimensional analytical model for contaminant transport in a finite domain subjected to multiple arbitrary time-dependent point injection sources. J Hydrol 597. Art. 126318. https://doi.org/10.1016/j.jhydrol.2021.126318

Feng SJ, Chen Z, Chen HX, Li L (2022) Experimental and numerical investigation of landfill composite liner failure characteristics. Waste Manag 144:69–78. https://doi.org/10.1016/j.wasman.2022.03.012

Giroud JP (1997) Equations for calculating the rate of liquid migration through composite liners due to geomembrane defects. Geosynth Int 4(3–4):335–348. https://doi.org/10.1680/gein.4.0094

Giroud JP, Bonaparte R (1989) Leakage through liners constructed with geomembranes—part II. Geotext Geomembr 8(2):71–111. https://doi.org/10.1016/0266-1144(89)90022-8

Giroud J, Soderman KL, Badu-Tweneboah K (1997) Comparison of leachate flow through compacted clay liners in landfill liner systems. Geosynth Int 4(3–4):391–431

Kong L, Chen Y, Li J (2018) Non-destructive testing of geomembrane defects using infrared thermography and electrical leak location. Waste Manag 75:356–365. https://doi.org/10.1016/j.wasman.2018.02.013

Kong L, Xue Q, Liu J, Li J (2021) Machine learning for leakage risk assessment of landfill liners. Environ Sci Technol 55(3):1888–1896. https://doi.org/10.1021/acs.est.0c05996

Kong L, Liu J, Li J (2023) AI-based detection system for geomembrane defects using infrared thermography. J Environ Eng 149(1). Art. 04022045. https://doi.org/10.1061/JOEEDU.EENG-7301

Li T, Sun D, Chen Z, Chen Y, Fu X, Wang L (2024a) Two-dimensional migration characteristics of contaminant through composite liner with arbitrarily distributed defects. Comput Geotechn 174. Art 106652. https://doi.org/10.1016/j.compgeo.2024.106652

Li T, Sun D, Chen Z et al (2024b) Two-dimensional migration characteristics of contaminant through composite liner with arbitrarily distributed defects. Comput Geotechn 174. Art 106652. https://doi.org/10.1016/j.compgeo.2024.106652

Mendes MJA, Touze-Foltz N, Palmeira EM, Pierson P (2010) Influence of structural and material properties of GCLs on interface flow in composite liners due to geomembrane defects. Geosynth Int 17(1):34–47. https://doi.org/10.1680/gein.2010.17.1.34

Parshakova YN, Viskov MV, Kataev RI, Kartavykh NN (2024) Numerical simulation of migration of filtration water of solid municipal waste disposal objects through ground protective structures. Comput Continuum Mech 17(2):24–32

Rapaka S, Chen S, Pawar RJ, Stauffer PH, Zhang D (2008) Non-modal growth of perturbations in density-driven convection in porous media. J Fluid Mech 609:285–303

Reddy KR, Giri RK, Kulkarni HS (2021) Service life assessment of damaged geomembrane liners using machine learning techniques. Geotext Geomembr 49(3):712–726. https://doi.org/10.1016/j.geotexmem.2021.01.005

Rowe RK, Abdelatty K (2012) Modeling contaminant transport through composite liner with a hole in the geomembrane. Can Geotechn J 49(7):773–781. https://doi.org/10.1139/t2012-038

Rowe RK, Sangam HP (2002) Durability of HDPE geomembranes. Geotext Geomembr 20(2):77–95. https://doi.org/10.1016/S0266-1144(02)00005-5

Rowe RK, Yu Y (2022) Modeling of contaminant transport through GMB/GCL composite liners with defects. Can Geotechn J 59(4):487–502. https://doi.org/10.1139/cgj-2021-0234

Shu S, Zhu W, Shi J (2019) A new simplified method to calculate breakthrough time of municipal solid waste landfill liners. J Clean Prod 219:649–654. https://doi.org/10.1016/j.jclepro.2019.02.050

Slim AC, Bandi MM, Miller JC, Mahadevan L (2013) Dissolution-driven convection in a hele-shaw cell. Phys Fluids 25:024101. https://doi.org/10.1063/1.4790511

Snow DT (1968) Rock fracture spacings, openings, and porosities. J Soil Mech Found Div 94(1):73–91

Take WA, Bolton MD, Wong PCP, Yeung FJ (2015) Evaluation of landslide triggering mechanisms in model fill slopes. Landslides 12(1):51–62. https://doi.org/10.1007/s10346-014-0477-z

Touze-Foltz N, Bannour H, Barral C, Stoltz G (2019) A review of the performance of geosynthetics for environmental protection. Geotext Geomembr 47(6):803–813. https://doi.org/10.1016/j.geotexmem.2019.103498

Touze-Foltz N, Stoltz G, Duquennoi C (2023) Long-term performance of geosynthetic barriers: recent advances. Geotext Geomembr 51(1):1–18. https://doi.org/10.1016/j.geotexmem.2022.11.003

US EPA (2022) Technical guidance for evaluating landfill gas emissions from closed or malfunctioning landfill cells. EPA-600/R-22/001. https://www.epa.gov/landfills

Wooding RA, Tyler SW, White I (1997) Convection in groundwater below an evaporating salt lake: I onset of instability. Water Resour Res 33(6):1199–1217. https://doi.org/10.1029/96WR03533

Experiment on Studying Thermophysical Properties of Snow Cover

D. M. Frolov, A. V. Koshurnikov, V. E. Gagarin, and G. A. Rzhanitsyn

Abstract This work is devoted to the study of the thermal properties of snow cover, in particular its thermal conductivity, which plays a key role in understanding the heat exchange between the atmosphere, snow cover and soil. Snow cover represents an important part of the Earth's cryosphere and is subject to significant variations depending on climatic conditions and other factors such as wind, humidity and solar radiation. These variations affect the physical properties and structure of snow, which requires a thorough understanding of the heat and mass transfer processes within the snow thickness. A series of experiments analysed the temperature distribution in snow using modern measurement techniques. The results confirmed the complexity of the heat transfer process through snow and the need for additional research to more accurately determine the effective thermal conductivity. Understanding these processes is critical for predicting avalanche behaviour, assessing the effect of snow on soil freezing and developing strategies for natural resource management in winter conditions.

Keywords Snow column · Laboratory experiment · Thermal conductivity · Thermal sensors · Loggers · Temperature measurement

1 Introduction

The study of thermophysical properties of snow cover, in particular, snow thermal conductivity, which is one of the key characteristics determining the heat exchange between the atmosphere, snow cover and soil, is an important task. Snow cover, being an important component of the Earth's cryosphere, shows significant variability depending on climatic conditions, seasons and other factors. These variations affect its physical properties and structure, which makes a thorough understanding of heat and mass transfer processes in the snow column necessary (Lykov 1967; Kudryavtsev 1974; Samarskiy and Gulin 1989; Tikhonov and Samarskiy 1999; Pavlov 2008;

D. M. Frolov (✉) · A. V. Koshurnikov · V. E. Gagarin · G. A. Rzhanitsyn
Lomonosov Moscow State University Geography Department, Moscow, Russia
e-mail: denisfrolovm@mail.ru

T. Chaplina (ed.), *Processes in GeoMedia—Volume IX*, Springer Geology,
https://doi.org/10.1007/978-981-96-9302-3_35

Voytkovsky 1977; Osokin et al. 2000, 2013; Osokin and Sosnovskiy 2014; Kotlyakov et al. 2018; Chernov 2016; Calonne et al. 2011; Schneebeli and Sokratov 2004; Sturm et al. 1997). The study included laboratory experiments using specially designed snow sample boxes.

Snow cover is characterised by high seasonality of existence and significant variability of its properties and structure. One of the key thermophysical characteristics of the snow cover is thermal conductivity, which determines the heat exchange between the atmosphere, snow cover and soil. This work is aimed at studying heat and mass transfer in the snow column.

Snow consists of ice crystals, air and sometimes liquid water. Its physical properties are subject to significant changes due to external factors such as temperature, humidity and pressure. One of the main reasons for the low thermal conductivity of snow, especially in loose layers of freshly fallen snow, is its porous structure, which hinders heat transfer. This makes it necessary to study the influence of snow density and temperature on its thermal conductivity.

Changes in the structure of snow cover occur as a result of natural compaction and sublimation rounding, which affects its thermal properties. It was previously found that snow density and temperature directly affect its thermal conductivity. Given the complex interactions between snow components, effective thermal conductivity proves to be a difficult parameter to estimate.

The study of snow thermal conductivity has a history of more than a century. The first experiments demonstrated the dependence of the thermal conductivity coefficient on the snow cover density. However, recent studies show a considerable variation in the values of this coefficient among different works. This is due to the variety of measurement methods used in the experiments, as well as different experimental conditions.

Snow metamorphism is a set of processes leading to changes in snow structure under the influence of physical factors. The main processes of metamorphism include ice sublimation, water vapour migration and redistribution of ice matter within the snow column. These processes affect the efficiency of heat and mass transfer in the snow cover, which requires detailed analysis.

2 Research Methodology and Results

Experimental studies were conducted under controlled laboratory conditions, using specially designed boxes for snow samples. Thermal insulation material for snow sample boxes was polyurethane foam plates. In the course of measurements, high-precision temperature sensors were used, which were connected to a data logger for automatic collection of temperature information (logger), including those of our own design. A total of 16 temperature sensors were placed in the 17 cm high snow column at a distance of 1 cm in height from each other. Figure 1 illustrates the external view of the installation for measuring temperature in the snow column in the cold laboratory of the Faculty of Geography.

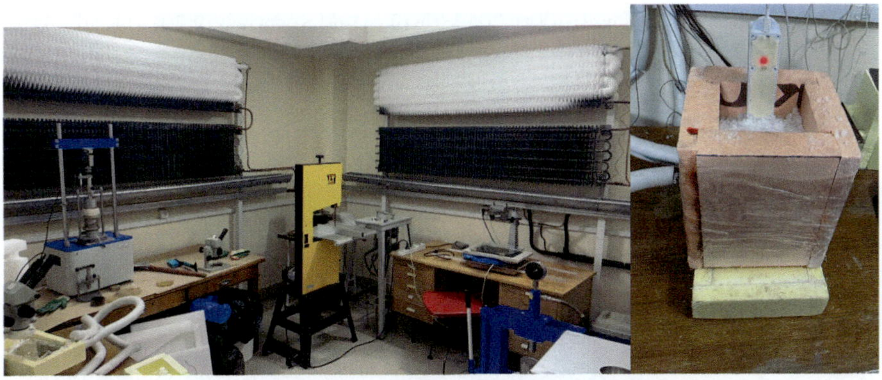

Fig. 1 Cold laboratory of the geography department. Installation for measuring temperature in the snow column

Temperature experiments with snow samples included two series: 'cooling from below' and 'heating from below', which was achieved using a thermostat. This allowed us to analyse the temperature distribution in the samples. Thus, in the process of stopping the cooling of the snow column, a slow redistribution of temperature along the height of the snow sample was observed. The temperature data of the snow column obtained during the experiments were taken from loggers and processed and visualised on a personal computer in the MATLAB software environment. The results of the experiments revealed different temperature gradients depending on the state of the snow cover and its density and also showed a general difference in the temperature distribution in the snow column. An example of the results of graphical processing of the obtained temperature data is presented in Fig. 2.

For example, Fig. 2 shows that because at the beginning of the experiment the temperature at the bottom of the snow column was maintained at −15 degrees Celsius, while at the top the ambient temperature in the cooling chamber was 5 degrees Celsius, there was a strong temperature gradient in the snow column. However, after the snow sample stopped cooling from below, there was a slow redistribution of temperature throughout the snow column, its warming up, as a result of which almost the entire snow column acquired the ambient temperature of −5 degrees Celsius in the cooling chamber.

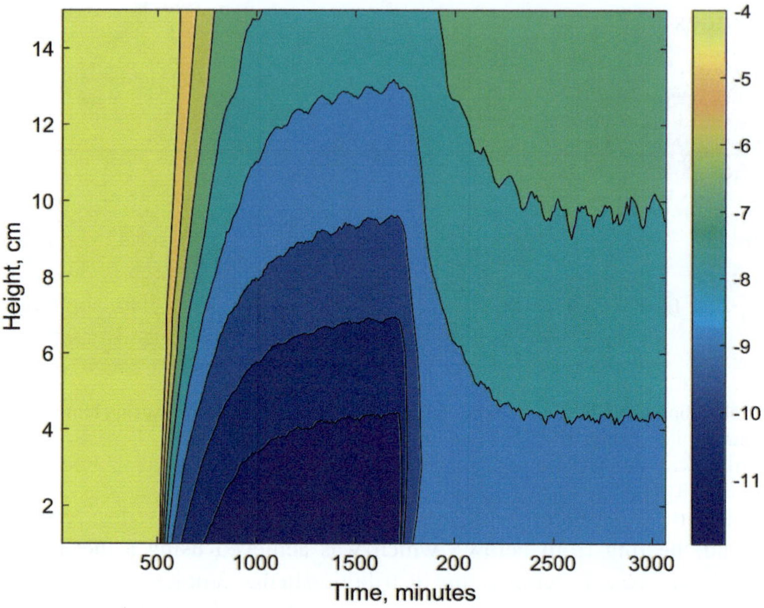

Fig. 2 Temperature distribution in the snow thickness over time

3 Conclusion

The present work confirms the complexity of the mechanisms of heat transfer through snow cover and emphasises the need for further research to more accurately determine the coefficient of effective thermal conductivity of snow. Understanding these processes is critical for correctly predicting avalanche behaviour, assessing the impact of snow on ground freezing and developing methods for natural resource management in winter conditions.

The present work confirms the complexity of the mechanisms of heat transfer through snow cover and emphasises the need for further research to determine more accurately the coefficient of effective Despite extensive research in this area, the problems of heat conductivity determination remain relevant. The obtained data are of practical importance for climate modelling and can be useful for planning measures to prevent negative consequences of natural phenomena.

The work was carried out in accordance with the state budgetary themes of the Geography Faculty of MSU 'Danger and risk of natural processes and phenomena' (121051300175-4) and 'Evolution of the cryosphere under climate change and anthropogenic impact' (121051100164-0).

References

Calonne N, Flin F, Morin S, Lesaffre B, du Roscoat SR, Geindreau C (2011) Numerical and experimental investigations of the effective thermal conductivity of snow. Geophys Res Lett 38:L23501. https://doi.org/10.1029/2011GL049234

Chernov RA (2016) Metamorphism and thermal properties off freshly fallen snow (based on studies in the Moscow region). Ice Snow 56(2):199–206. https://doi.org/10.15356/2076-6734-2016-2-199-206

Kotlyakov VM, Sosnovsky AV, Osokin NI (2018) Estimation of the heat conductivity coefficient of snow by its density and hardness in Western Svalbard. Ice Snow 58(3):343–352. https://doi.org/10.15356/2076-6734-2018-3-34335

Kudryavtsev VA (ed) (1974) Basics of permafrost forecasting in engineering and geological research. Moscow State University Publishing House, Moscow, p 431

Lykov AV (1967) Theory of heat conduction. Higher School, Moscow, p 599

Osokin NI, Samoilov RS, Sosnovskiy AV, Sokratov SA (2000) About herole of some natural factorings oil freezing. MGI 88:41–45

Osokin NI, Sosnovsky AV, Chernov RA (2013) Influence of snow cover stratigraphy on its thermal resistance. Ice Snow 3(123):63–70

Osokin NI, Sosnovskiy AV (2014) Experimental studies of the effective heat conductivity coefficient of the snow cover in West Spitsbergen. Ice Snow54(3):50–58. https://doi.org/10.15356/2076-6734-2014-3-50-58

Pavlov AV (2008) Cryolithozone monitoring. GEO, Novosibirsk, p 230

Samarskiy AA, Gulin AV (1989) Numerical methods. Nauka, Moscow, p 432

Schneebeli M, Sokratov SA (2004) Tomography of temperature gradient metamorphism of snow and associated changes in heat conductivity. Hydrol Process 18:3655–3665

Sturm M, Holmgren J, Konig M, Morris K (1997) The thermal conductivity of seasonal snow. J Glaciol 43(143):26–41

Tikhonov AN, Samarskiy AA (1999) Equations of mathematical physics, 6th edn. Moscow State University, p 800

Voytkovsky KF (1977) Mechanical properties of snow. Nauka, Moscow, p 126

Geodesic Control of Technological Object Condition by Laser Scanning Method

Dmitriy Neguritsa and Ivan Tereshin

Abstract The article presents the results of the control geodetic survey of monolithic reinforced concrete structures of the tank, carried out with the use of 3D laser scanning technology, carried out within the framework of the survey, it is established that the permissible deviations of geometric parameters are exceeded. The study shows the importance of an integrated approach to geodetic surveys in the reconstruction of capital construction of technological facilities, which allows to create an accurate model of the object with minimal time and labor costs.

Keywords Capital construction object · Pile foundation · Mathematical modeling · Geomechanical model · Repeated loading · Construction structures · Recommendations

In the modern world, where technologies are rapidly developing, geodetic monitoring becomes one of the key tasks to ensure their reliability and safety (Neguritsa 2018; Alekseev et al. 2023a). One of the most effective methods for monitoring the condition of technological objects is laser scanning, which allows obtaining accurate and detailed three-dimensional models of objects. This method combines high speed of data acquisition and exceptional accuracy, which makes it an indispensable tool in geodesy and construction. Laser scanning opens new horizons for monitoring the condition of objects, allowing not only to capture the current state but also to analyze changes over time. In the context of rapidly changing infrastructure and the need to comply with strict norms and standards, the use of laser scanning becomes especially relevant (Merkin et al. 2023; Alekseev et al. 2023b).

The object of the study is a storage tank with a pumping station for rainwater runoff. During the survey of the object it is necessary to establish the possibility of its intended use, or in case of its unsatisfactory condition to prepare an engineering model for its reconstruction or renovation of the object, when the existing object is to be demolished, and instead of it on the same territory a new object is built,

D. Neguritsa (✉) · I. Tereshin
University of Science and Technology «MISIS», Moscow, Russia
e-mail: neguritsadl@misis.ru

T. Chaplina (ed.), *Processes in GeoMedia—Volume IX*, Springer Geology,
https://doi.org/10.1007/978-981-96-9302-3_36

Fig. 1 Point cloud from the results of laser scanning view 1

performing not only the functions of the demolished object but also additional ones in accordance with modern requirements (Neguritsa et al. 2022). The condition of the construction object should guarantee their reliable and uninterrupted operation both during and after the completion of construction works (Neguritsa 2018; Alekseev et al. 2023a).

The purpose of this study is to create a spatial model of the structure. The design of the storage tank is partially buried and consists of two independent tanks (compartments). The walls of the tank are inclined and the bottom is horizontal. Horizontal and inclined structures are made on the sub-concrete preparation and extruded polystyrene foam 100 mm thick. The overall stability is ensured by the joint work of the monolithic reinforced concrete frame, spread by the ground of the base. The tank has dimensions in plan 55 × 49 m, depth—3.4 m.

The geodetic survey of the storage tank with rainwater pumping station is performed by laser scanning.

To perform laser scanning, a typical for the whole object section of structures available for measurements at the time of engineering-geodetic works is selected. In the selected area laser scanning of structures is performed, if possible from two (or more) stations, in order to obtain the maximum possible laser scanning point cloud in this area.

A Leica ScanStation P50 laser scanner was used for engineering and geodetic works.

Computer data processing was performed using specialized software, Leica Cyclone, Leica Infiniti, AutoCAD.

The geodetic survey results were used to determine the deviations of the actual geometric parameters of the tank structures from the design dimensions (Figs. 1 and 2):

- dimensions along the longitudinal axis,
- dimensions along the transverse axis,
- vertical dimensions.

The point cloud can be used to measure the actual position of structural elements of the structure (Fig. 3). By comparing the dimensions of the model elements and

Fig. 2 Point cloud from the results of laser scanning view 2

design or execution drawings of the object, it is possible to determine the deviations of the position of individual elements of the structure in relation to the design, or deformations of the object in the process of its operation (Fig. 4).

According to the results of the geodetic survey, the maximum values of deviations of the actual dimensions of the tank structures from the design dimensions were as follows:

- along the longitudinal axis of the structure—up to 466 mm,
- along the transverse axis of the structure—up to 589 mm,
- vertically—up to 218 mm.

Deviations of geometric parameters of structures were determined for the section of structures available at the given object for measurement.

According to the results of geodetic control, uneven elevation of the tank enclosing structures was recorded both in the plane of one face and in the plane of the whole tank.

The height difference on one wall is up to 140 mm, the difference of opposite wall marks is up to 153 mm.

The maximum deviations of the actual dimensions of the tank structures from the design values were:

- along the longitudinal axis of the structure—up to 466 mm;
- along the transverse axis of the structure—up to 589 mm;
- vertically—up to 218 mm.

Thus, geodetic condition monitoring of process facilities using laser scanning is an effective tool that significantly improves the accuracy and efficiency of monitoring. Due to its ability to collect data quickly and accurately, laser scanning not only captures the current state of objects but also tracks changes over time, which is critical to ensure the safety and reliability of infrastructure. The application of this method facilitates deeper analysis and informed decision-making in the management of technological facilities. It is important to note that the development of laser scanning technologies and their integration with modern information systems opens up new opportunities for optimizing design, construction, and operation processes.

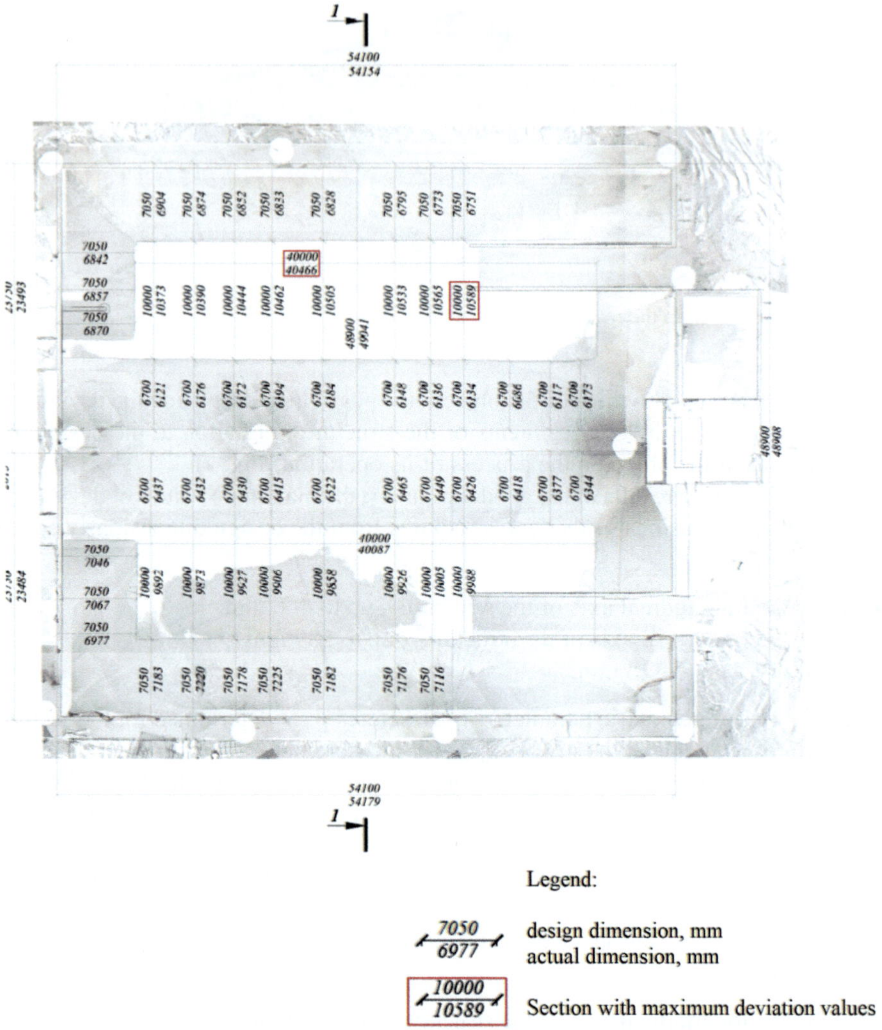

Fig. 3 Determination of deviation of the actual dimensions of the tank structures from the design dimensions along the longitudinal and transverse axes

In the context of constantly growing quality and safety requirements, laser scanning becomes an integral part of effective management and control of technological facilities, ensuring their long-term sustainability and reliability.

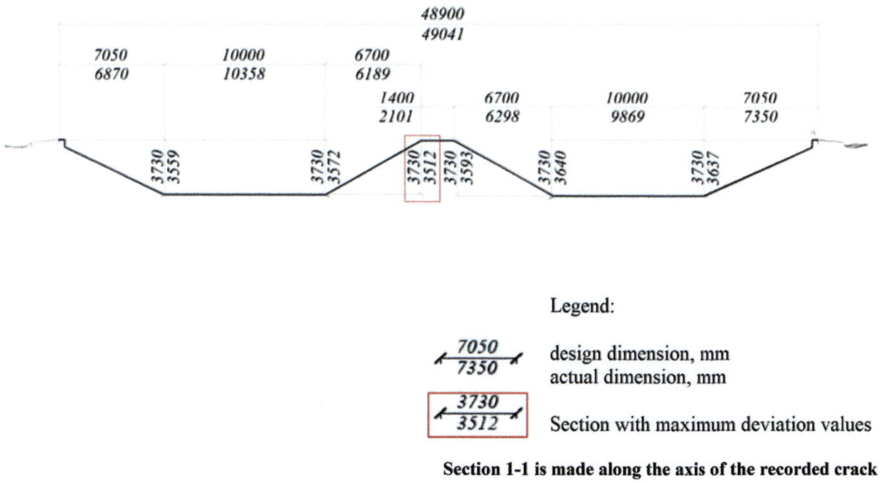

Fig. 4 Determination of vertical deviation of the actual dimensions of the tank structures from the design dimensions

References

Alekseev G, Medvedev E, Neguritsa D (2023a) Monitoring of deformation processes during scientific and technical support of construction. In: E3S web of conferences, vol 410, p 02045. https://doi.org/10.1051/e3sconf/202341002045

Alekseev G, Medvedev E, Neguritsa D (2023b) Monitoring of deformation processes during scientific and technical support of construction. In: E3S web of conferences, vol 410, p 02045. https://doi.org/10.1051/e3sconf/202341002045

Merkin VE, Pichugin AA, Simutin AN, Medvedev GM, Kuzyanov NA (2023) Automated system of complex geotechnical monitoring and the experience of its practical application (in Russian). Transp Constr (3):4–8

Neguritsa D (2018) The problems of monitoring the deformation processes in the integrated development of the underground space of metropolitan cities. In: E3S web of conferences, vol 56, p 02027. https://doi.org/10.1051/e3sconf/20185602027

Neguritsa D, Alekseev G, Tereshin A, Medvedev E, Slobodin K (2022) Deformation processes modelling throughout underground construction within megapolis limits. https://doi.org/10.1007/978-3-030-76328-2_23

Numerical Experiments on Modeling the Black Sea Circulation in 2016 Using Various Degrees of Polynomial Dependence of Density on Temperature and Salinity

Natalia Evstigneeva

Abstract Three numerical experiments were carried out to simulate the Black Sea circulation in 2016 for different types of state equations, which had the form of a polynomial depending on the degree of temperature and salinity. The linear state equation (experiment 1), the equation in the form of O. I. Mamaev (experiment 2), and the EOS-80 equation of state (experiment 3) were considered. The obtained model values were compared with measurement data of temperature and salinity, taken during cruises 87, 89, 91 of the R/V Professor Vodyanitsky in 2016. The worst estimates were obtained in the experiment 1, using the linear state equation, the results of the experiments 2 and 3 were close to each other. The area of relatively cold salt waters in the northwest, noted by observations in the summer season, was more accurately reproduced in experiment 2. The area of warm, poorly desalinated waters along the east coast, noted according to measurement data in the autumn–winter season, was more accurately reconstructed in experiment 3.

Keywords Black Sea · Numerical modeling · Equation of state · Hydrophysical fields · Observational data · Standard deviation

1 Introduction

The density of seawater is not measured directly, but is calculated using formulas for its dependence on temperature, salinity, and pressure, i.e., the equation of state. The dependence of density on these characteristics is quite complex, so only approximate equations of state are currently used (Knudsen-Ekman, Mamaev, International Equation of State 1980 (EOS-80), linear, etc.). The simplest is the linearized equation, in which the expansion coefficients are constant values. O. I. Mamaev proposed a simplified equation of state of seawater (Mamaev 1964, 1987), obtained by the

N. Evstigneeva (✉)
Marine Hydrophysical Institute RAS, Russian Federation, Sevastopol, Russia
e-mail: naevstigneeva@yandex.ru

T. Chaplina (ed.), *Processes in GeoMedia—Volume IX*, Springer Geology,
https://doi.org/10.1007/978-981-96-9302-3_37

least squares method and distinguished by good accuracy compared to other simplified equations. The density was calculated in Bryan and Cox (1972), using simple polynomial formulas, provided that the constant coefficients in those formulas were a function of pressure and were defined in accordance with the classical Knudsen formula for a realistic range of temperatures and salinities in the World Ocean. Approximate formulas for in situ density as a function of potential temperature, salinity, and pressure were defined in Wright (1997) for use in numerical models. An approximation for the UNESCO state equation of seawater was obtained in Brydon et al. (1999) in the form of a polynomial, cubic in potential temperature, quadratic in pressure and linear in salinity. An equation of state of seawater containing 25 terms was presented in McDougall et al. (2003), written for potential temperature, and a method for converting potential temperature to in situ temperature was proposed using specific entropy, based on the Gibbs function. It was found in Roquet et al. (2015) that a simple polynomial equation of state with a quadratic term in temperature, the product of temperature and pressure, and a linear term in salinity, i.e., with only four tuning parameters, was sufficient to model a realistic global circulation. A new type of equation was presented in Grekov et al. (2019) that expressed the density of seawater through independent and in situ measured parameters: temperature, hydrostatic pressure, and sound velocity in explicit form. The presence of an equation of that type allowed in some cases to exclude indirect (through conductivity) measurements of salinity, replacing them with direct measurements of the sound velocity.

The aim of the investigation was to assess the accuracy of calculating the prognostic parameters (temperature and salinity) based on the results of numerical experiments on modeling the Black Sea circulation in 2016 with the help of the model of the Marine Hydrophysical Institute (MHI) (Demyshev 2012), using different degrees of polynomial dependence of density on temperature and salinity in the equation of state.

In accordance with the works (Mamaev 1964, 1987) that showed that it was possible to ignore the compressibility of water when solving problems on ocean dynamics, the calculation of the density of sea water was carried out using the most frequently used equations of state that considered the dependence of the density of sea water only on temperature and salinity. Those were the linear equations of state, the equation in the form of O.I. Mamaev and the International Equation of State of Sea Water, 1980 (EOS-80). To select the state equation that ensured optimal accuracy, we analyzed tables of standard deviations of calculated values of hydrological parameters from observational data in 2016, vertical profiles and features of the spatio-temporal variability of fields of temperature and salinity, obtained by using the mentioned equations of state.

2 Parameters of the Numerical Model and Description of Numerical Experiments

We used a z-coordinate three-dimensional nonlinear model of the Black Sea dynamics (Demyshev 2012). It was based on the full equations of ocean thermohydrodynamics in the Boussinesq approximation, hydrostatics, and incompressibility of seawater. The boundary conditions on the surface, on the bottom, and on solid side walls were described in detail in Demyshev (2012). New approximation schemes for nonlinear terms in the equations of motion, possessing various discrete quadratic invariants, were used. The time step was 96 s. The calculations were performed for one year from January 1 to December 31, 2016 with a horizontal resolution of 1.64 km and 27 vertical horizons. A satellite sea surface temperature was assimilated in the model once a day [https://data.marine.copernicus.eu/product/SST_BS_SST_L3S_NRT_OBSERVATIONS_010_013]. The sea surface fields of tangential wind friction stress (τ_x, τ_y), precipitation *Pr* and evaporation *Ev* fields, total heat fluxes *Hf* and shortwave radiation fluxes on the surface *SWR* were specified every day, obtained from ERA5 atmospheric reanalysis data with a resolution of (1/4)° (Copernicus Climate Change Service, Climate Data Store (2023): ERA5 hourly data on single levels from 1940 to present. https://doi.org/10.24381/cds.adbb2d47). The climatic runoff of rivers and water exchange through straits were taken into account. The vertical turbulent exchange was parameterized by the Mellor-Yamada 2.5 closure model (Mellor and Yamada 1982), and the horizontal diffusion and viscosity were described by the Laplace biharmonic operator with constant coefficients equal to 10^{16} cm^4/s.

In experiment 1 the linear dependence of density on temperature and salinity was considered: $\rho = \rho_0 + \alpha_1^T T + \alpha_1^S S$, where $\rho_0 = 1.000082$, $\alpha_1^T = -35\cdot10^{-7}$, $\alpha_1^S = 802\cdot10^{-6}$ (CGS system).

In experiment 2 the state equation had the form $\rho = \rho_0 + \alpha_1^T T + \alpha_1^S S + \alpha_2^T T^2 + \alpha^{ST} ST$, where $\rho_0 = 1.000082$, $\alpha_1^T = -35\cdot10^{-7}$, $\alpha_1^S = 802\cdot10^{-6}$, $\alpha_2^T = -469\cdot10^{-8}$, $\alpha^{ST} = -2\cdot10^{-6}$ (CGS system).

In experiment 3 the following state equation of 1980 (EOS-80) was considered: $\rho_0 + \left(b_0 + b_1 \cdot T + b_2 \cdot T^2 + b_3 \cdot T^3 + b_4 \cdot T^4\right) \cdot S + \left(c_0 + c_1 \cdot T + c_2 \cdot T^2\right) \cdot S^{3/2} + d_0 \cdot S^2$, where $\rho_0 = 1000.082$, $b_0 = 8.24493\cdot10^{-1}$; $b_1 = -4.0899\cdot10^{-3}$; $b_2 = 7.6438\cdot10^{-5}$; $b_3 = -8.247\cdot10^{-7}$; $b_4 = 5.3875\cdot10^{-9}$; $c_0 = -5.72466\cdot10^{-3}$; $c_1 = 1.0227\cdot10^{-4}$; $c_2 = -1.6546\cdot10^{-6}$; $d_0 = 4.8314\cdot10^{-4}$ (International System of Units).

When comparing the obtained model values with in situ data, we used the data from hydrological measurements of the 87th, 89th, and 91st cruises of the R/V Professor Vodyanitsky (summer, autumn, and autumn–winter seasons of 2016), available in the MHI Oceanographic Data Bank (http://bod-mhi.ru/ru/index.shtml). Figure 1 showed the stations, where data on the temperature and salinity of the Black Sea was obtained.

The temperature and salinity of seawater to a depth of ~1000 m were measured using the CTD SBE 911plus complex. The error for temperature was up to 0.001 °C, for salinity—up to 0.003 PSU (SBE 911 plus measuring complex by Sea-Bird Electronics. URL: http://www.seabird.com/sbe911plus-ctd).

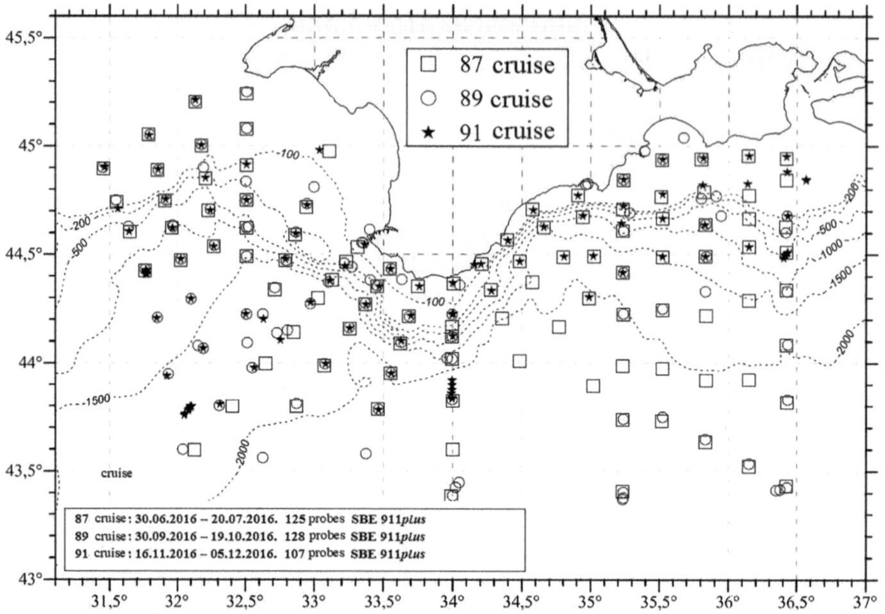

Fig. 1 Map of measurement stations, carried out on the R/V Professor Vodyanitsky in 2016

Time periods under consideration in 2016 were from June 28 to July 18 (125 stations), from September 29 to October 20 (109 stations), and from November 15 to December 5 (80 stations).

3 Tables of Mean-Square Deviations of Calculated Values of *T*, *S* from Observational Data in 2016

The root-mean-square deviations (RMSE) of the model temperature and salinity from those measured for each season for each of the three experiments were calculated. A detailed analysis of the obtained RMSEs was performed for the most significant layers (0–5 m, 5–30 m, 30–100 m) depending on the type of the state equation and the season under consideration. In the Table 1 the minimum RMSE among the three experiments were shown in blue, and the maximum RMSE values among the three experiments for each season and water layer were shown in red.

In the summer season the results of experiment 1 contained large errors in temperature in the 5–30 m layer, the smallest deviations from the observational data for temperature in the 30–100 m layer were observed in experiment 3. In the autumn season the results of experiment 1 contained large errors in temperature in the 30–100 m layer, the smallest deviations from the observational data for temperature for all layers were observed in experiment 2. In the autumn–winter season, the results

Table 1 Standard deviation of temperature and salinity from data, obtained during the cruises of the R/V Professor Vodyanitsky in 2016 for three experiments

TEMPERATURE

	Experiment 1					
	0-5m	5-30m	30-100m	100-300m	300-800m	800-1500m
PV 87	0.40967	7.79207	1.41318	0.22557	0.03037	0.00443
PV 89	0.56553	2.67825	3.67199	0.24532	0.04456	0.01327
PV 91	0.78608	0.64987	1.83224	0.23827	0.05	0.02008
mean	0.58709	3.70673	2.305803	0.236386	0.04164	0.012593

	Experiment 2					
	0-5m	5-30m	30-100m	100-300m	300-800m	800-1500m
PV 87	1.42282	2.09427	0.73739	0.25133	0.0307	0.00468
PV 89	0.61212	2.03318	0.95893	0.31084	0.04426	0.01345
PV 91	0.61045	0.58463	0.75289	0.33422	0.04855	0.0209
mean	0.88179	1.57069	0.81640	0.29879	0.04117	0.01301

	Experiment 3					
	0-5m	5-30m	30-100m	100-300m	300-800m	800-1500m
PV 87	0.93242	5.71566	0.48107	0.27757	0.03581	0.00641
PV 89	1.08276	3.10102	2.49286	0.27272	0.04959	0.0153
PV 91	1.19814	1.27487	2.81991	0.76624	0.06204	0.0152
mean	1.07110	3.36385	1.93128	0.77217	0.23528	0.01230

SALINITY

	Experiment 1					
	0-5m	5-30m	30-100m	100-300m	300-800m	800-1500m
PV 87	0.13304	0.2045	0.49083	0.53973	0.12818	0.0787
PV 89	0.2128	0.20698	0.49869	0.62336	0.15291	0.08578
PV 91	0.10998	0.10213	0.56861	0.76754	0.16163	0.11179
mean	0.15194	0.17120	0.519376	0.643543	0.147573	0.09209

	Experiment 2					
	0-5m	5-30m	30-100m	100-300m	300-800m	800-1500m
PV 87	0.12899	0.09189	0.4978	0.60172	0.13085	0.07901
PV 89	0.30136	0.28189	0.4678	0.70333	0.15353	0.07901
PV 91	0.11792	0.11416	0.55771	0.82158	0.15851	0.11246
mean	0.18275	0.16264	0.507786	0.70887	0.14763	0.09016

	Experiment 3					
	0-5m	5-30m	30-100m	100-300m	300-800m	800-1500m
PV 87	0.11448	0.13866	0.53548	0.63586	0.14522	0.08034
PV 89	0.18011	0.17154	0.50895	0.63586	0.16224	0.08074
PV 91	0.08063	0.08542	0.62574	1.01808	0.52488	0.15696
mean	0.12507	0.13187	0.55672	0.763266	0.277446	0.106013

of experiment 3 contained errors in temperature in the 30–100 m layer, the smallest errors in temperature in all layers were achieved in experiment 2. From the analysis of the averaged standard deviation of temperature for all seasons, it was found that the smallest deviations from the observational data for temperature in the 0–5 m layer were achieved in experiment 1, in the 5–30 m and 30–100 m layers—in experiment 2.

From the analysis of the calculated values of the standard deviation of salinity from the data, obtained during the cruises of the R/V Professor Vodyanitsky in 2016 for three experiments, we noted the following. In the summer season, the salinity error in experiment 1 in the 5–30 m layer increased by 2 times, compared to experiment 2. In the autumn season the salinity error in experiment 3 in the upper layer and 5–30 m layer decreased by 40%, compared to experiment 2. In the autumn–winter season the salinity error in experiment 3 in the upper layer and 5–30 m layer decreased by ~30%, compared to experiment 2. In all the seasons, the smallest deviations from the observational data for salinity in the upper layer and the 5–30 m layer were obtained in experiment 3, and in the 30–100 m layer—in experiment 2.

4 Maps and Features of Spatio-Temporal Variability of the Temperature and Salinity Fields for Different Types of State Equation

In order to compare model fields of temperature and salinity with the distributions of thermohaline characteristics based on the data from the performed cruises (related to one moment in time), the model fields were averaged over the calculated periods. Figure 2 and 3 showed the model and experimental data in the upper water layer for the summer and autumn–winter seasons, respectively.

According to measurements in July (87th cruise), the temperature of upper quasi-homogeneous layer varied from 24 to 27 °C, “tongues” of relatively cold waters were observed, extended in the southern direction, in the northwest and east of the survey; a

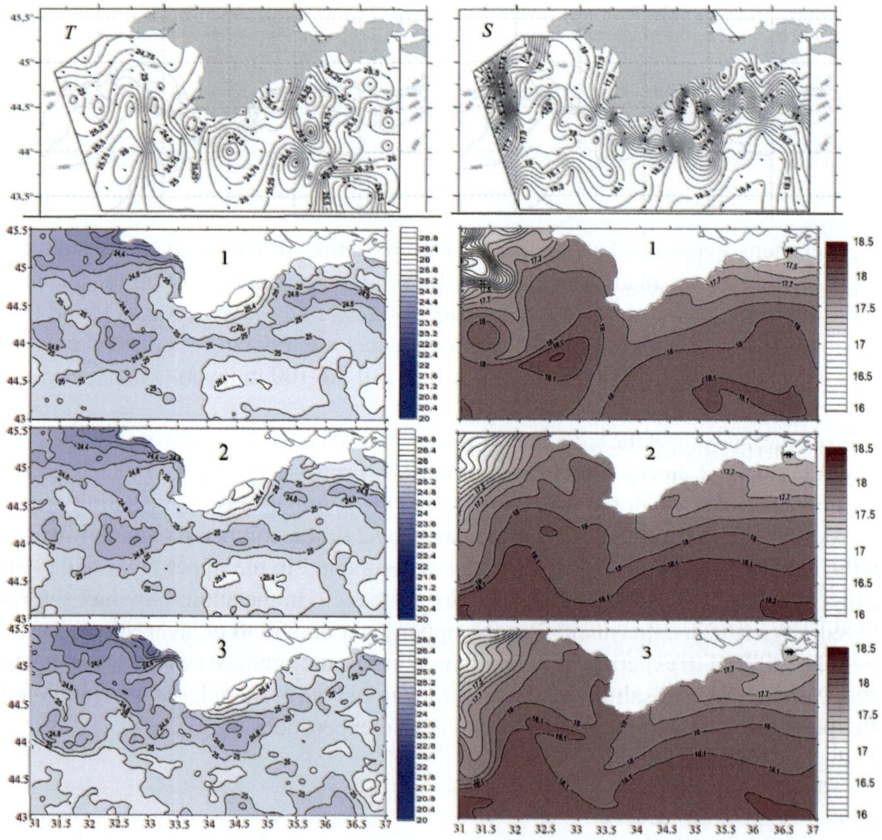

Fig. 2 Distribution of temperature and salinity for the surface layer according to data from cruise No 87 and model fields of temperature and salinity at the upper horizon for three experiments, averaged over the summer season (isotherms were drawn through 0.2 °C, isohalines through 0.1 PSU)

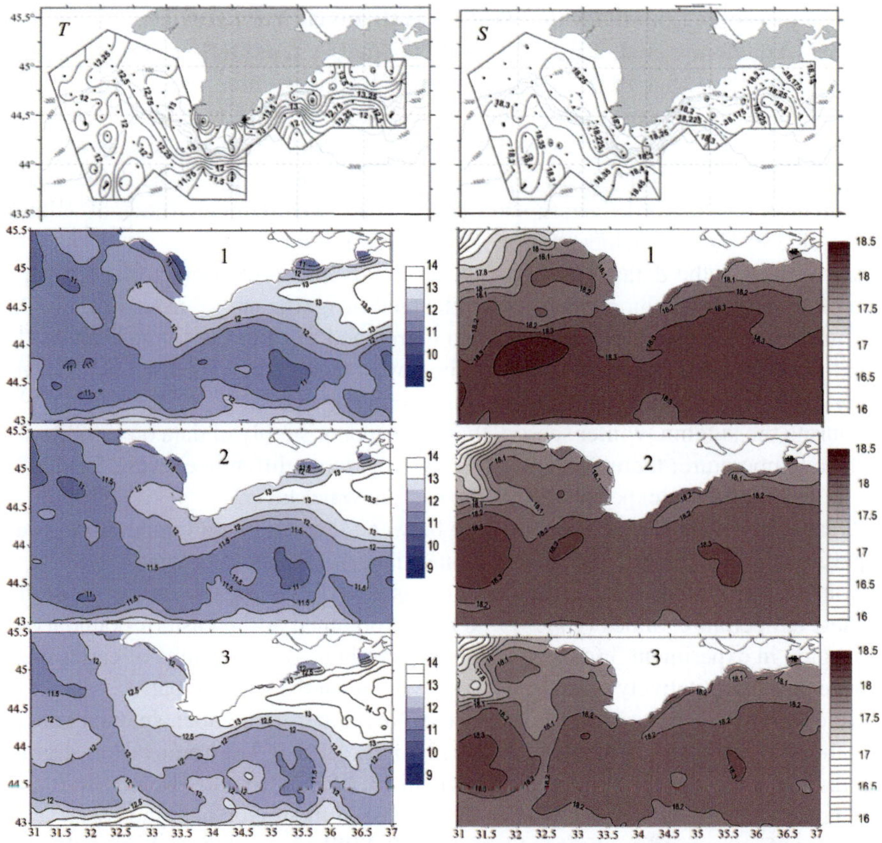

Fig. 3 Distribution of temperature and salinity for the surface layer according to data from cruise № 91 and model fields of temperature and salinity at the upper horizon for three experiments, averaged over the autumn–winter season (isotherms were drawn through 0.5 °C, isohalines through 0.1 PSU)

tendency for the temperature to increase to 25–26 °C was noted near the eastern coast of Crimea (including in the area of the anticyclone in the Yalta Bay) (Artamonov et al. 2017, 2018a). In all three experiments (Fig. 2) an area with warmer water (with a temperature of over 25 °C) formed near the eastern coast of Crimea, as recorded by observations in the summer season. "Tongues" of relatively cold waters appeared in the northwestern and eastern parts of the area (the temperature was 24.6–24.8 °C) in all experiments, but the results of experiment 2 were closer to observations.

Values of salinity varied in a wide range from 17.00 to 18.40 PSU (Artamonov et al. 2017, 2018a) according to measurements in July. The lowest values (up to 17 PSU) were noted in the west of the survey and were traces of desalination on the northwestern shelf of the Black Sea. Another area of relatively desalinated waters (up to 17.5 PSU) was observed in the northeastern part of the survey closer to the shores of Crimea. The origin of those desalinated waters was associated with the removal of waters from the Kerch Strait area to the northern periphery of the Rim Current (Artamonov et al. 2017, 2018a). In the deep-sea part of the survey, salinity increased to 18.3–18.4 PSU. According to the results of three model experiments (Fig. 2), areas of water with a salinity of less than 18 PSU were observed along the eastern coast, which was consistent with the measurement data. Zones of water with a salinity of less than 17.9 PSU were formed in the northwest, associated with desalination on the northwestern shelf of the Black Sea. Higher values of salinity (18.1–18.2 PSU) were achieved in the deep-water region.

According to the data of the 91st cruise in November–December, the temperature of upper quasi-homogeneous layer changed insignificantly—from 11.0 to 14.0 °C, while the area of the warmest waters stretched out like "a tongue" from the Kerch Strait along the coast of Crimea. The lowest temperature (below 11.5 °C) was observed in the south of the survey (Artamonov et al. 2018a, b).

During the autumn–winter season, there was a poor supply of data on satellite sea surface temperature; therefore, it was likely why some differences were formed in the fields in the three experiments for the surface water layer.

Quality correspondence was obtained between the results on temperature in all experiments (Fig. 3) and the observational data in the autumn–winter season. A region of warm waters, relatively adjacent, was observed along the eastern coast, the highest temperature values (14–14.5 °C) were obtained at the eastern boundary of the region in experiment 3 (closest to the observed ones). The results of experiment 3 were also quantitatively close to the observations in the western part.

According to the data of the 91st cruise in November–December, spatial changes in salinity were insignificant—from 18.15 to 18.4 PSU. The removal of slightly desalinated waters (with a salinity of less than 18.2 PSU) from the Kerch Strait area was noted (Artamonov et al. 2018a, b) in the east of the region near the coast.

Areas of water with a salinity of less than 18.2 PSU were observed along the east coast according to the results of three model experiments (Fig. 3). Zones of water with a salinity of less than 18 PSU were formed in the northwest, associated with desalination on the northwestern shelf of the Black Sea.

5 Vertical Temperature and Salinity Profiles for Different Types of State Equation

The thickness of upper quasi-homogeneous layer varied widely from 5 to 25 m in the summer season, according to observations (Artamonov et al. 2017, 2018a). The temperature difference was about 10–12 °C approximately between 5 and 30 m (in a jump layer). The maximum depth of the jump layer (25–30 m) was observed in the coastal area and the minimum (about 10 m) in the deep-water part.

Figure 4 showed the vertical profiles of temperature and salinity constructed for three experiments at typical stations where the maximum and minimum depth of the jump layer were noted.

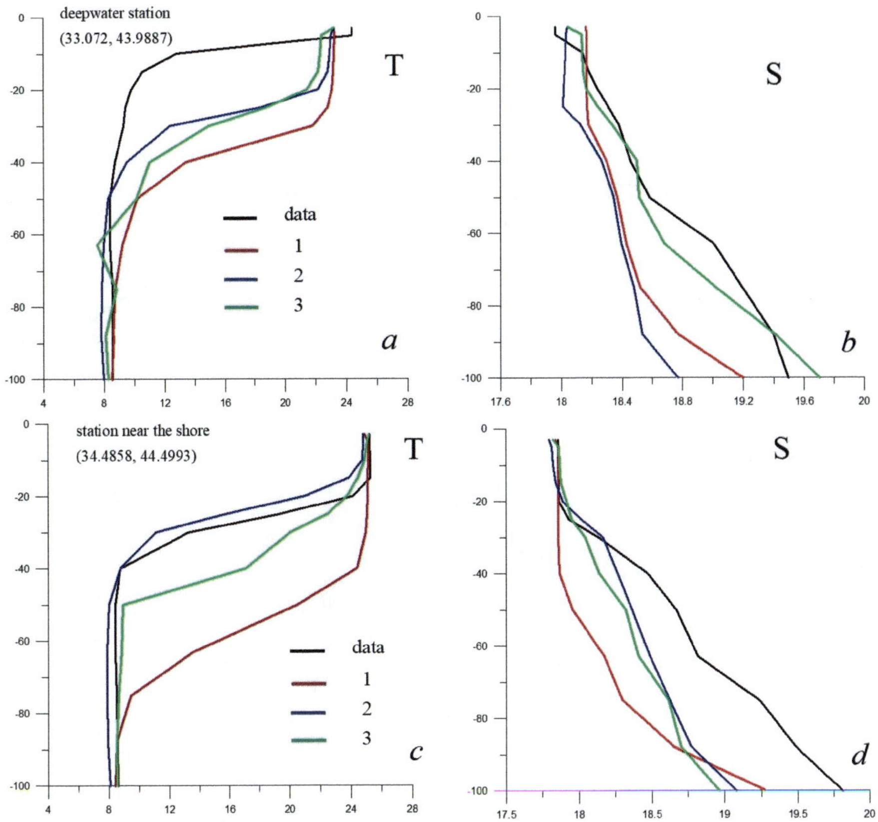

Fig. 4 Vertical temperature and salinity profiles for three experiments for the summer season and profiles, measured in 87 cruise

For the temperature in the deep-water station the results of all experiments contained large errors in temperature (Fig. 4a) in the 10–40 m layer (the depth of the jump layer was underestimated). For salinity (Fig. 4b) the results of experiment 3 were closer to the observational data.

The analysis of the profiles for the station, located near the coast, showed that for temperature (Fig. 4c) the results of experiment 2 were closest to the observational data. The results of experiments 1 and 3 underestimated the depth of the jump layer. For salinity (Fig. 4d) the upper quasi-homogeneous layer was reproduced close to the observational data for all experiments; below a depth of 40 m, the simulation results showed lower salinity values than the observations.

The thickness of upper quasi-homogeneous layer increased from 30–40 m in the deep-sea part of the survey to 60–65 m above the depth slope in the northwestern part of the survey in the autumn–winter period, according to observations. Figure 5 showed the vertical profiles of temperature and salinity obtained in three experiments at stations, where the maximum and minimum depth of the jump layer were noted.

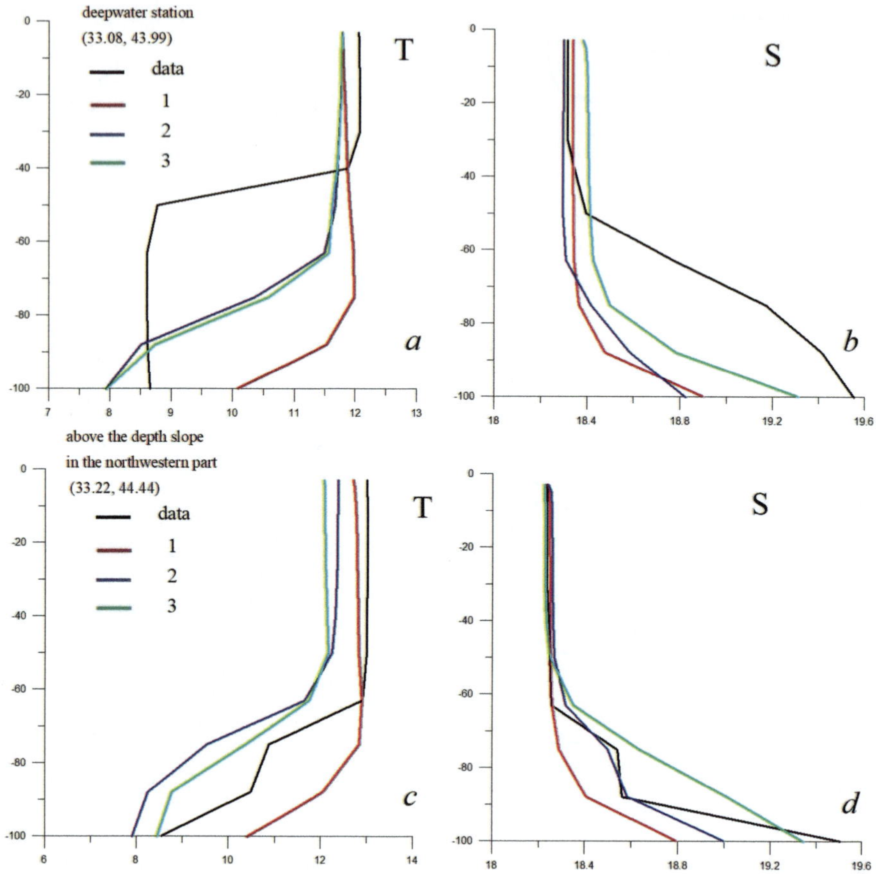

Fig. 5 Vertical profiles of temperature and salinity for three experiments for the autumn–winter season and profiles, measured in 91 cruise

The results of all experiments contained errors in temperature (Fig. 5a) and salinity (Fig. 5b) in the layer below 40 m (the depth of the jump layer was underestimated) in the deep-sea part of the survey. Analysis of the profiles for a typical station, located above the depth slope in the northwestern part (Fig. 5c, d), showed that the results of experiment 1 in the upper layer for temperature and salinity were closest to the observational data, but did not correctly reproduce the jump layer. The results of experiments 2 and 3 described the jump layer qualitatively more accurately.

6 Conclusion

Based on three numerical experiments, which differed in the form of the state equation, an assessment of the accuracy of the calculated values of T, S from the data of hydrological surveys in 2016 was given. The linear dependence of density on temperature and salinity was considered in experiment 1, the equation of state had the form of a second-degree polynomial in experiment 2, and the EOS-80 equation of state was considered in experiment 3.

From the analysis of the obtained standard deviations of temperature and salinity for layers 0–5 m, 5–30 m, 30–100 m depending on the type of equation of state and the season under consideration, the following was obtained. In the summer season, the results of experiment 1 contained the largest errors in temperature in the 5–30 m layer, in the 30–100 m layer the smallest deviations from the observational data for temperature were observed in experiment 3. In the autumn season, the results of experiment 1 contained the largest errors in temperature in the 30–100 m layer, the smallest deviations from the observational data for temperature for all layers were observed in experiment 2. In the autumn–winter season, the results of experiment 3 contained errors in temperature in the 30–100 m layer, the smallest errors in temperature in all layers were achieved in experiment 2. In all seasons, in the upper layer and the 5–30 m layer, the smallest deviations of salinity from the observational data were obtained in experiment 3, in the 30–100 m layer—in experiment 2.

When comparing the model fields of temperature and salinity in the upper water layer, averaged over the calculation period, with the distributions of thermohaline characteristics related to one moment in time (according to the data of the completed cruises), the following was obtained. In the summer season, the area of relatively cold salt waters in the northwest, noted by observations, was more accurately reproduced in experiment 2. In the autumn–winter season, the area of warm waters of slightly freshened waters along the eastern coast, noted by measurement data, was more accurately reconstructed in experiment 3.

For the summer season, for a typical station, located near the coast, the vertical profiles for temperature in experiment 2, for salinity in experiments 2 and 3 were closest to the observational data. For the autumn–winter season, for a typical station, located above the continental slope in the northwestern part, the vertical profiles for temperature and salinity in experiments 2 and 3 more accurately described the jump layer.

As expected, the worst estimates were obtained in the experiment 1 using a linear state equation. Comparison with observational data showed that the results of the experiments 2 and 3 were close to each other. According to the author, that conclusion was due to the fact that the density in the Black Sea was determined by salinity and taking into account terms with high degrees of temperature had small effect on the accuracy of density reproduction.

Acknowledgements The study was conducted within the framework of the state assignment on topic No. FNNN-2024-0001 (code "Oceanological processes").

References

Artamonov YV, Alekseev DV, Skripaleva EA et al (2017) Thermohaline structure of water near the Crimea coast and adjacent open water area of the Black Sea at summer 2016. Ecol Saf Coast Shelf Zones Sea (3):20–31 (in Russian)

Artamonov YuV, Skripaleva EA, Alekseev DV et al (2018a) Hydrological research in the northern part of the Black Sea in 2016 (87th, 89th and 91st cruises of R/V Professor Vodyanitsky). Phys Oceanogr 25(3):229–234. https://doi.org/10.22449/1573-160X-2018-3-229-234

Artamonov YV, Alekseev DV, Skripaleva EA et al (2018) Features of seasonal and synoptic variability of water structure in the zone of the rim current in the autumn-winter period of 2016. Ecol Saf Coast Shelf Zones Sea (1):32–43 (in Russian)

Bryan K, Cox MD (1972) An approximate equation of state for numerical models of ocean circulation. J Phys Oceanogr 2:510–514

Brydon D, Sun S, Bleck RA (1999) A new approximation of the equation of state for seawater, suitable for numerical ocean models. J Geophys Res Ocean 104:1537–1540

Demyshev SG (2012) A numerical model of online forecasting Black Sea currents. Izv Atmos Ocean Phys 48:120–132. https://doi.org/10.1134/S0001433812010021

Grekov AN, Grekov NA, Sychev EN (2019) New equations for calculating the density of sea water based on measurements of the speed of sound. Mechatron Autom Control 20(3):143–151. https://doi.org/10.33075/2220-5861-2017-1-12-18

Mamaev OI (1964) Simplified relationship between density, temperature and salinity of sea water. Izv AN SSSR Ser Geofiz 2:309–311 (in Russian)

Mamaev OI (1987) Thermohaline analysis of the waters of the World Ocean. Gidrometeoizdat Leningrad, p 296 (in Russian)

McDougall TJ, Jackett DR, Wright DG, Feistel R (2003) Accurate and computationally efficient algorithms for potential temperature and density of seawater. J Atmos Ocean Technol 20:730–741. https://doi.org/10.1175/1520-0426(2003)20%3c730:AACEAF%3e2.0.CO;2

Mellor GL, Yamada T (1982) Development of a turbulence closure model for geophysical fluid problems. Rev Geophys Space Phys 20(4):851–875

Roquet F, Madec G, Brodeau L, Nycander J (2015) Defining a simplified yet "realistic" equation of state for seawater. J Phys Oceanogr 45(10):2564–2579. https://doi.org/10.1175/JPO-D-15-0080.1

Wright DG (1997) An equation of state for use in ocean models: Eckart's formula revisited. J Atmos Oceanic Technol 14:735–740

GPSR Compliance

The European Union's (EU) General Product Safety Regulation (GPSR) is a set of rules that requires consumer products to be safe and our obligations to ensure this.

If you have any concerns about our products, you can contact us on ProductSafety@springernature.com

In case Publisher is established outside the EU, the EU authorized representative is:

Springer Nature Customer Service Center GmbH
Europaplatz 3
69115 Heidelberg, Germany

Batch number: 10704305

Printed by Printforce, the Netherlands